Lecture Notes in Physics

The Lecture Notes in Physics

The series Lecture Notes in Physics (LNP), founded in 1969, reports new developments in physics research and teaching – quickly and informally, but with a high quality and the explicit aim to summarize and communicate current knowledge in an accessible way. Books published in this series are conceived as bridging material between advanced graduate textbooks and the forefront of research and to serve three purposes:

- to be a compact and modern up-to-date source of reference on a well-defined topic

- to serve as an accessible introduction to the field to postgraduate students and nonspecialist researchers from related areas

- to be a source of advanced teaching material for specialized seminars, courses and schools

Both monographs and multi-author volumes will be considered for publication. Edited volumes should, however, consist of a very limited number of contributions only. Proceedings will not be considered for LNP.

Volumes published in LNP are disseminated both in print and in electronic formats, the electronic archive being available at springerlink.com. The series content is indexed, abstracted and referenced by many abstracting and information services, bibliographic networks, subscription agencies, library networks, and consortia.

Proposals should be sent to a member of the Editorial Board, or directly to the managing editor at Springer:

Christian Caron
Springer Heidelberg
Physics Editorial Department I
Tiergartenstrasse 17
69121 Heidelberg / Germany
christian.caron@springer.com

C. Massobrio
H. Bulou
C. Goyhenex (Eds.)

Atomic-Scale Modeling of Nanosystems and Nanostructured Materials

 Springer

Editors

Carlo Massobrio
Université Strasbourg
CNRS-UMR 7504
Inst. Physique et Chimie des
22 rue du Loess
67034 Strasbourg CX 2
France
carlo.massobrio@ipcms.u-strasbg.fr

Hervé Bulou
Université Strasbourg
CNRS-UMR 7504
Inst. Physique et Chimie des
22 rue du Loess
67034 Strasbourg CX 2
France
herve.bulou@ipcms.u-strasbg.fr

Christine Goyhenex
Université Strasbourg
CNRS-UMR 7504
Inst. Physique et Chimie des
22 rue du Loess
67034 Strasbourg CX 2
France
christine.goyhenex@ipcms.u-strasbg.fr

Massobrio, C. et al. (Eds.): *Atomic-Scale Modeling of Nanosystems and Nanostructured Materials*, Lect. Notes Phys. 795 (Springer, Berlin Heidelberg 2010),
DOI 10.1007/978-3-642-04650-6

ISSN 0075-8450   e-ISSN 1616-6361
ISBN 978-3-642-04649-0   e-ISBN 978-3-642-04650-6
DOI 10.1007/978-3-642-04650-6
Springer Heidelberg Dordrecht London New York

Library of Congress Control Number: 2009941648

Cover design: Integra Software Services Pvt. Ltd., Pondicherry

Printed on acid-free paper

Springer is part of Springer Science+Business Media (www.springer.com)

Preface

Understanding the structural organization of materials at the atomic scale is a long-standing challenge of condensed matter physics and chemistry. By reducing the size of synthesized systems down to the nanometer, or by constructing them as collection of nanoscale size constitutive units, researchers are faced with the task of going beyond models and interpretations based on bulk behavior. Among the wealth of new materials having in common a "nanoscale" fingerprint, one can encounter systems intrinsically extending to a few nanometers (clusters of various compositions), systems featuring at least one spatial dimension not repeated periodically in space and assemblies of nanoscale grains forming extended compounds. For all these cases, there is a compelling need of an atomic-scale information combining knowledge of the topology of the system and of its bonding behavior, based on the electronic structure and its interplay with the atomic configurations. Recent developments in computer architectures and progresses in available computational power have made possible the practical realization of a paradygma that appeared totally unrealistic at the outset of computer simulations in materials science. This consists in being able to parallel (at least in principle) any experimental effort by a simulation counterpart, this occurring at the scale most appropriate to complement and enrich the experiment. Focussing on the atomic scale, in which atoms (or better, ions) and electrons are represented by explicit degrees of freedom, the ambition of bridging the gap between the measurements carried out in a laboratory and the microscopic variables of statistical mechanics is more and more close to be fulfilled. Therefore, simulating materials and their behavior on a computer can be considered as of today a feasible task accessible to an increasingly large share of researchers.

The "Institut de Physique et de Chimie des Matériaux de Strasbourg" (IPCMS in what follows) has been one of the first on the European scientific scene to host and promote computer modeling in the context of a variety of research activities. All of them are focussed on material science as a combined discipline stemming from both chemistry and physics, the common denominator being the characterization of materials through the study of their structural, electronic, magnetic, and optical properties. Historically, IPCMS was created by the merging of five different departments, two of them linked to chemistry (organic and inorganic materials) and three to physics, involved in magnetism and metallic materials, surfaces and opto-electronics, respectively. Modeling began with a special interest in magnetism and electronic structure and rapidly expended to other areas such as dynamics and topology on surfaces, molecular dynamics on nanostructures and disordered systems, electron dynamics and behavior of liquid crystals. The period 2006–2009 has been characterized by the setup, in each department, of a team devoted to atomic-scale computer modeling of materials. Two main goals can be identified. First, a consistent number of experiments carried out within IPCMS are prone to be interpreted and/or complemented by calculations, due to the increased reliability of available schemes, (density functional theory as a main guideline, alternative descriptions of electronic structure, and chemical bonding being also available). Second, IPCMS scientists using the computer as the main working tool have proved able to propose and activate new research frontiers, by going well beyond a mere supportive action bound to back up experiments. A reveling example is provided in this book by the contribution dealing with biochemical processes, well suited to help establishing new bridges between biology and material science, unforeseen only a few years ago.

From the standpoint of methodology, computational science within IPCMS is able to offer a broad spectrum of techniques and applications. Looking for representative keywords, one would be tempted to select "atomic scale and density functional theory" on the side of methods and "nanostructures" on the side of applications. By atomic scale we intend a methodology rooted in the behavior of atoms and the underlying chemical bonding as coming out from modern and tractable theoretical tools, capable of describing with equal accuracy few atoms as well a few hundredths of them. This is case of density functional theory largely employed within IPCMS in the context of eletron dynamics and excited states, theory of magnetism, weak chemical interactions and first-principles molecular dynamics. Situations characterized by unaffordable system sizes are also considered. This is the case of surface topology and diffusion, for which sensible approximations of the interatomic interactions are put to good use in the search of diffusion mechanisms on surfaces. On the side of the applications, the focus on nanostructures is legitimated by experimental work on several fields related to nanoscience (femtosecond magnetism, spintronics, electron microscopy, widespread use of several surface probes like STM). Moreover, for each of the systems presented hereafter, specific atomic and/or molecular groups can be identified, irrespective of their being part of an isolated three-dimensional nanostructure, or, at the opposite end, of an interface resulting from distinct bulk-like arrangements having peculiar electronic and magnetic properties.

Taken altogether, the above considerations have strongly motivated the idea of producing a collection of research papers written by the computational scientists working at IPCMS. Also, we came to the conviction that "Atomic-scale modeling of nanosystems and nanostructured materials" could have been a good title since reflecting current ativity and future directions undertaken by the host institurion. This set of papers is designed to achieve optimal balance of methodological purposes, review parts, and recent applications. Each contribution contains enough methodological elements to be understood by a graduate student in chemistry or physics. Moreover, the presentation of the results is intended to be a valid instrument of scientific communication for both the specialists in each specific area and a first-time reader. Finally, being all the authors exposed in their everyday work to neighboring experimental activities, a special effort has been taken to make all chapters accessible to the non-specialist willing to interact and being exposed to simulations. We stress that the international reputation of all contributors is well established, with more than 1,100 papers published in internationally peer-reviewed journals and a consistent share of awarded invited talks, in the range of 40–50 every year. These data are unambiguous scientific facts demonstrating the quality and the impact on the international scene of computer modeling in material science within IPCMS. The contributions are presented in an order that reflect the size of the systems considered and the spatial range of the chemical and physical mechanisms under consideration. In this way, we start from the farthest apart to applied material science (electron dynamics and excitations) to end with the closest to bulk materials (magnetic compounds, disordered networks) featuring relevant phenomena on the nanoscale.

The promotors of the present initiative and the authors of the different chapters are indebted to the director of the IPCMS, Dr. Marc Drillon, for his encouragement and support during the course of the preparation of this book. Stimulating interactions also occurred with Prof. François Gautier, who was at the origin of the creation of IPCMS in the 1980s.

Strasbourg, July 2009   C. Massobrio
                                                                                     H. Bulou
                                                                              C. Goyhenex

Contents

**Effect of Spin–Orbit Coupling on the Magnetic Properties
of Materials: Theory** 227
M. Alouani, N. Baadji, S. Abdelouahed, O. Bengone, and H. Dreyssé

**Effect of Spin–Orbit Coupling on the Magnetic Properties of
Materials: Results** 309
M. Alouani, N. Baadji, S. Abdelouahed, O. Bengone, and H. Dreyssé

**Nanostructural Units in Disordered Network-Forming Materials and
the Origin of Intermediate Range Order** 343
C. Massobrio

Collective Electron Dynamics in Metallic and Semiconductor Nanostructures

G. Manfredi, P.-A. Hervieux, Y. Yin, and N. Crouseilles

Abstract We review different approaches to the modeling and numerical simulation of the nonlinear electron dynamics in metallic and semiconductor nanostructures. Depending on the required degree of sophistication, such models go from the full N-body dynamics (configuration interaction), to mean-field approaches such as the time-dependent Hartree equations, down to macroscopic models based on hydrodynamic equations. The time-dependent density functional theory and the local-density approximation – which have become immensely popular during the last two decades – can be understood as an upgrade of the Hartree approach allowing one to include, at least approximately, some effects that go beyond the mean-field. Alternative methods, based on Wigner's phase-space representation of quantum mechanics, are also described. Wigner's approach has the advantage of permitting a more straightforward comparison between semiclassical and fully quantum results. As an illustrative example, the many-electron dynamics in a semiconductor quantum well is studied numerically, using both a mean-field approach (Wigner–Poisson system) and a quantum hydrodynamical model. Finally, the above methods are extended to include the spin degrees of freedom of the electrons. The local-spin-density

G. Manfredi (✉)
Institut de Physique et Chimie des Matériaux de Strasbourg, 23 Rue du Loess, BP 43, F–67034 Strasbourg, France, giovanni.manfredi@ipcms.u-strasbg.fr

P.-A. Hervieux
Institut de Physique et Chimie des Matériaux de Strasbourg, 23 Rue du Loess, BP 43, F–67034 Strasbourg, France

Y. Yin
Institut de Physique et Chimie des Matériaux de Strasbourg, 23 Rue du Loess, BP 43, F–67034 Strasbourg, France

N. Crouseilles
Institut de Recherche en Mathématiques Avancées UdS, Strasbourg, France, crouseil@math.u-strasbg.fr

Manfredi, G. et al.: *Collective Electron Dynamics in Metallic and Semiconductor Nanostructures.* Lect. Notes Phys. **795**, 1–44 (2010)
DOI 10.1007/978-3-642-04650-6_1

approximation is used to investigate the linear electron response in metallic nanostructures. The modeling of nonlinear spin effects is sketched within the framework of Wigner's phase-space dynamics.

1 Introduction

Understanding the electron dynamics and transport in metallic and semiconductor nanostructures – such as metallic nanoparticles, thin films, quantum wells, and quantum dots – represents a considerable challenge for today's condensed matter physics, both fundamental and applied.

Experimentally, thanks to the recent development of ultrafast spectroscopy techniques, it is now possible to monitor the femtosecond dynamics of an electron gas confined in metallic nanostructures such as thin films [1–8], nanotubes [9], metal clusters [10, 11], and nanoparticles [6, 7, 12, 13]. Therefore, meaningful comparisons between experimental measurements and numerical simulations based on microscopic theories are becoming possible.

The dynamics of an electron gas confined in a metallic nanostructure is characterized by the presence of collective oscillations (surface plasmon) whose spectral properties depend on several conditions of temperature, density, and coupling to the environment. At lowest order, the linear response of the electron gas is simply given by the plasma frequency $\omega_p = (e^2 n / m \varepsilon_0)^{1/2}$ (up to a dimensionless geometrical factor) and does not depend on the temperature or the size of the nano-object. The plasma frequency represents the typical oscillation frequency for electrons immersed in a neutralizing background of positive ions, which is supposed to be motionless because of the large ion mass. The oscillations arise from the fact that, when some electrons are displaced (thus creating a net positive charge), the resulting Coulomb force tends to pull back the electrons toward the excess positive charge. Due to their inertia, the electrons will not simply replenish the positive region, but travel further away, thus re-creating an excess positive charge. This effect gives rise to coherent oscillations at the plasma frequency. Notice that, for a metallic nanostructure, the inverse plasma frequency is typically of the order of the femtosecond – this coherent regime can therefore be explored with the ultrafast spectroscopy techniques developed in the last two decades.

The coherence of such collective motions is progressively destroyed by Landau damping (i.e., by coupling to the internal degrees of freedom of the electron gas) and by electron–electron or electron–phonon collisions. The damping of the plasmon was observed experimentally in gold nanoparticles [14] and was studied theoretically in several works [15–17].

Although the linear response of the surface plasmon has been known for a long time, fully nonlinear studies have only been performed in the last decade and have revealed some interesting features. Our own contribution to this research area has mainly focused on the nonlinear electron dynamics in thin metal films, where the emergence of ballistic low-frequency oscillations has been pointed out [18–20].

On the other hand, the same type of collective electron motion is also observed in semiconductor nanostructures, such as quantum wells and quantum dots. Although the spatial and temporal scales differ by several order of magnitudes with respect to metallic nanostructures (due to the large difference in the electron density), the relevant dimensionless parameters take similar values in both cases [21]. For instance, the effective Wigner–Seitz radius is of order unity for both metallic and semiconductor nano-objects. Therefore, the electron dynamics can be investigated using similar models and both types of nano-objects are expected to share a number of similar dynamical properties.

In this review article, we will describe the collective electron dynamics in metallic and semiconductor nanostructures using different, but complementary, approaches. For small excitations (linear regime), the spectral properties can be investigated via quantum mean-field models of the TDLDA type (time-dependent local-density approximation), generalized to account for a finite electron temperature. In order to explore the nonlinear regime (strong excitations), we will adopt a phase-space approach that relies on the resolution of kinetic equations in the classical phase-space (Vlasov and Wigner equations). The phase-space approach provides a useful link between the classical and quantum dynamics and is well suited to model effects beyond the mean-field approximation (electron–electron and electron–phonon collisions). We will also develop a quantum hydrodynamic model based on velocity moments of the corresponding Wigner distribution function: this approach should lead to considerable gains in computing time in comparison with simulations based on conventional methods, such as density functional theory (DFT).

The above studies all refer to the *charge* dynamics in a semiconductor or metallic nanostructure, which has been intensively studied in the last three decades. In more recent years, there has been a surge of interest in the *spin* dynamics of the carriers, mainly for possible applications to the emerging field of quantum computing [22]. A promising approach to the development of a quantum computer relies on small semiconductor devices, such as quantum dots and quantum wells [23]. To implement basic qubit operations, most proposed schemes make use of the electron spin states, so that a thorough understanding of the spin dynamics is a necessary prerequisite. Nevertheless, in order to manipulate the electrons themselves, one must necessarily resort to electromagnetic fields, which in turn excite the Coulomb mean-field [24, 25]. The charge and spin dynamics are therefore closely intertwined and both must be taken into account for a realistic modeling of semiconductor-based qubit operations.

The ultrafast magnetization (spin) dynamics in ferromagnetic nanostructures has also attracted considerable experimental attention in the last decade. Pioneering experiments [26–28] on ferromagnetic thin films revealed that the magnetization experiences a rapid drop (on a femtosecond timescale) when the films are irradiated with an ultrafast laser pulse, after which it slowly regains its original value on a timescale close to that of the electron–phonon coupling. Despite many attempts [26–30], a clear theoretical explanation for these effects is still lacking. Here, we will illustrate how this problem can be addressed using some of the techniques developed for the electron dynamics, particularly quantum mean-field and phase-space methods, which will be generalized to include the spin degrees of freedom.

2 Models for the Electron Dynamics

Metallic and semiconductor nano-objects operate in very different regimes, as the electron density is several orders of magnitudes larger for the former. Consequently, the typical time, space, and energy scales can be very different, as illustrated in Table 1. However, if one takes into account the effective electron mass and dielectric constant, the relevant dimensionless parameters turn out to be rather similar [21]: for instance, from Table 1 we see immediately that the ratio of the screening length ($L_{screen} = v_F/\omega_p$, where v_F is the Fermi velocity) to the effective Bohr radius $a_B = 4\pi\varepsilon\hbar^2/me^2$ is of order unity. The same happens for the ratio of the plasmon energy $\hbar\omega_p$ to the Fermi energy E_F, so that the normalized Wigner–Seitz radius r_s is also of order unity for both cases.[1]

It is therefore not surprising that the electron dynamics of both types of nanostructures can be described by means of similar models. A bird's-eye view of the various relevant models is provided in Fig. 1. The diagram represents the various levels of modeling for the electron dynamics, both quantum (left column, dark gray) and classical (right column, light gray). The highest level of description is the N-body model, which involves the resolution of the N-particle Schrödinger equation in the quantum regime, or the N-particle Liouville equation for classical problems (the

Table 1 Typical time, space, and energy scales for metallic and semiconductor nanostructures

	Metal film	Quantum well
n_e	$10^{28}\,\text{m}^{-3}$	$10^{22}\,\text{m}^{-3}$
m	m_e	$m_* \simeq 0.07 m_e$
ε	ε_0	$\varepsilon \simeq 12\varepsilon_0$
L_{screen}	1 Å	100 Å
ω_p^{-1}	1 fs	1 ps
E_F	1 eV	1 meV
T_F	10^4 K	10 K
a_B	0.529 Å	100 Å
a_{latt}	5 Å	5 Å
r_s/a_B	5	3

[1] For a quantum well, all relevant lengths far exceed the semiconductor lattice spacing $a_{latt} \simeq$ 5 Å. This makes semiconductor systems a much better approximation to jellium (i.e., a continuum ionic density profile) than simple metals, for which the lattice spacing is comparable to the other electronic lengths.

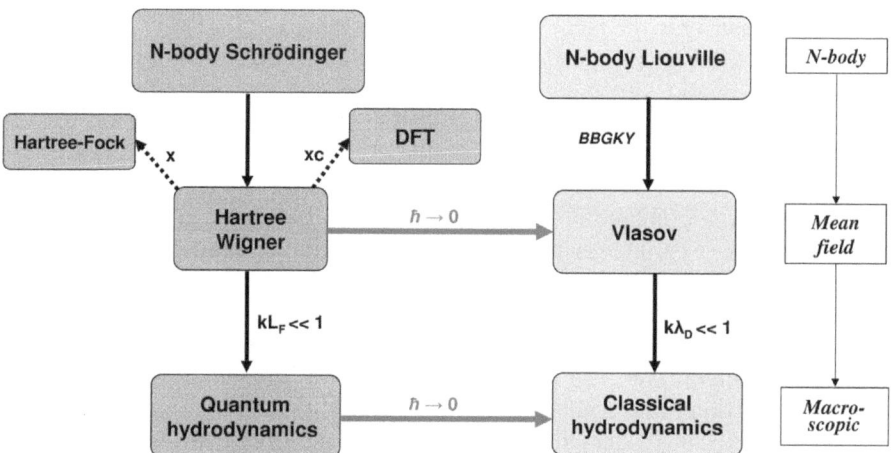

Fig. 1 Bird's-eye view of the models used to describe the electron dynamics. From top to bottom: N-body, mean-field, and macroscopic (hydrodynamic) theory. Left column (*dark gray*): quantum models; right column (*light gray*): classical models. Notation: x = exchange; xc = exchange and correlations; λ_D = Debye length (classical screening length); L_F = Thomas–Fermi screening length; k = typical wavevector; BBGKY = Bogoliubov, Born, Green, Kirkwood, Yvon hierarchy

latter is of course equivalent to Newton's equations of motion). This is a difficult task even classically, although molecular dynamics simulations that solve the exact N-body problem can nowadays attain a considerable level of sophistication. For Newton's equations with two-body interactions, the numerical complexity grows at most as N^2, and in some cases this can be reduced to a logarithmic dependence. Quantum mechanically, the N-body problem is virtually unmanageable, except for very small systems, because the size of the relevant Hilbert space grows exponentially with N. Nevertheless, exact simulations of the N-body Schrödinger equation can be performed using the so-called configuration interaction (CI) method. We have used this approach to study the exact electron dynamics in semiconductor quantum dots containing up to four electrons.

For larger systems, some rather drastic approximations need to be made if we want to end up with a mathematically and numerically tractable model. Most such reduced models are improvements on the so-called "mean-field approximation," which states that the motion of a single electron is determined by the positions and velocities of all other particles in the system. Such collective behavior is possible because of the long-range nature of electromagnetic forces. The mean-field approach can be viewed as a zeroth-order approximation to the N-body problem in which two-body (and higher order) correlations between the particles have been neglected. Classically, this procedure is known as the BBGKY hierarchy (from the names of Bogoliubov, Born, Green, Kirkwood, and Yvon) [31].

For classical systems of charged particles (plasmas), the mean-field dynamics is governed by the Vlasov equation, which describes the evolution of a one-particle probability density in the phase-space. The quantum analog of the Vlasov equation is provided by the time-dependent Hartree equations, which are actually one-body Schrödinger equations evolving in the mean-field potential. In both cases, the mean-field is obtained by solving Maxwell's equations, often reduced, in the electrostatic limit, to the sole Poisson's equation.

In this review, we concentrate on quantum-mechanical models. Several improvements have been proposed to the Hartree equations (which were derived in 1927, just 1 year after Schrödinger's seminal paper on the wave equation), most notably Fock's correction (1930). Indeed, the Hartree method does not respect the principle of antisymmetry of the wave function, although it does use the Pauli exclusion principle in its less stringent formulation, forbidding the presence of two electrons in the same quantum state. The Hartree–Fock equations respect the antisymmetry of the wave functions, thus leading to an extra interaction term between the electrons, termed the "exchange interaction."

A particularly successful extension of the mean-field approach is the density functional theory (DFT), which was developed by Hohenberg, Kohn, and Sham in the mid 1960s [32, 33]. Originally developed for the ground state at zero temperature, it has subsequently been extended to finite temperature and time-dependent problems. As its name suggests, DFT states that all the properties of a many-electron systems are determined by the electron spatial density, rather than by the wave functions. DFT allows one to introduce in the mean-field formalism effects that go beyond the strict mean-field approximation, particularly the exchange interaction described above. Indeed, DFT can deal with higher order correlations between the electrons, in principle exactly if the exact density functional were known. In practice, one has to make an educated guess for the appropriate correlation functional, which leads to various empirical approximations. Nevertheless, DFT has proven immensely useful for a wide range of electronic structure calculations.

The Hartree equations can be equivalently recast in a phase-space formalism by making use of the Wigner transformation, which was introduced by E. Wigner in 1932 [34]. The resulting Wigner function is a pseudo probability distribution, which can be used to compute expectation values just like its classical counterpart. Unfortunately, the Wigner function can take negative values, which precludes the possibility of interpreting it as a true probability density.

By taking velocity moments of the Wigner equation – and using some appropriate closure hypotheses – one can derive a set of quantum hydrodynamical (or fluid) equations that govern the evolution of macroscopic quantities such as the particle density, average velocity, pressure, and heat flux. Compared to the Wigner approach, the hydrodynamical one is obviously numerically advantageous, as it requires the resolution of a small number of equations in real (not phase) space. Generally speaking, hydrodynamical methods yield accurate results over distances that are larger than the typical electrostatic screening length, which is the Debye length $\lambda_D = (k_B T_e \varepsilon / e^2 n)^{1/2}$ for classical plasmas and the Thomas–Fermi screening length $L_F = v_F / \omega_p$ for degenerate electron gases (see Table 1).

In the following subsections, we shall present a brief overview of most of the quantum models introduced in Fig. 1.

2.1 Exact N-body Simulations: The Configuration Interaction (CI) Method

2.1.1 Method

In the Hartree–Fock model (HF), the many-body wave function is approximated by a single Slater determinant leading to a correlation between electrons having the same spin. However, electrons of different spin are not correlated in this approximation. This is why the difference between the exact value of the energy and the HF value is called the correlation energy. There are a number of quantum chemistry methods, which attempt to improve the description of the many-body wave function. The most important one is the so-called configuration interaction method (CI) [35] which is based on the minimization of the energy with respect to the expansion coefficients of a trial many-body wave function expressed as a linear combination of Slater determinants. With respect to the models based on density functional methods, the drawback of the CI method is its unfavorable scaling with the system size. Indeed, the dimension of a full CI calculation grows factorially with the number of electrons and basis functions.

From the above considerations, it is clear that CI calculations are restricted to confined systems with very few electrons (typically less than 10). In quantum chemistry, the "basis set" usually refers to the set of (nonorthogonal) one-particle functions used to build molecular orbitals. Concerning the computational methodology for confined electron systems (atoms, molecules, clusters, nanoparticles, quantum dots...) localized basis sets are the traditional choice and the most common type of basis functions is the Gaussian functions. It is worth noticing that, from the knowledge of the exact many-body wave-function, one can in principle (i) compute the temporal evolution of the system, including the dynamical correlations and (ii) obtain the true excited states of the system.

In the following, an application of the CI method in the field of semiconductor nanostructures and quantum dots is presented.

2.1.2 Application

Recent progress in semiconductors technology allows the realization of quantum systems composed of a small number of electrons (even a single electron!) confined in nanometer-scale potential wells. These systems, which provide highly tunable structures for trapping and manipulating individual electrons, are often named artificial atoms or quantum dots and are good candidates for the emerging technology of quantum computing. They have certain similarities with atoms in the sense that they have a discrete electronic structure that follows the well-known Hund's rule of atomic physics. However, in quantum dots the electrons are generally confined by harmonic or quasi-harmonic potentials, whereas atoms are characterized by Coulomb confinement potentials. The spectral properties of quantum dots are exotic with respect to the properties of atoms in the sense that most of the oscillator strength is concentrated almost exclusively on one dipolar transition. This property is a direct consequence of Kohn's theorem (KT) and does not depend on the number of electrons, the strength of the confinement, or the electron–electron interaction [36].

In a recent work [37], we have investigated quasi-two-dimensional Gaussian quantum dots containing up to four electrons within the framework of the CI method which allows in principle an exact treatment of the many-electron system. The Schrödinger equation for N-electrons confined by a potential V_{ext} is given by

$$H\Psi(1, ..., N) = E\Psi(1, ..., N),$$ (1)

where $(1, ..., N)$ represents the space $[\mathbf{r}_i = (x_i, y_i, z_i)]$ and spin coordinates of the electrons and

$$H = \sum_{i=1}^{N} -\frac{\hbar^2}{2m}\nabla_i^2 + \sum_{i>j}^{N} \frac{e^2}{4\pi\varepsilon|\mathbf{r}_i - \mathbf{r}_j|} + \sum_{i=1}^{N} V_{ext}(\mathbf{r}_i).$$ (2)

The confinement is modeled by an external one-particle anisotropic Gaussian potential given by

$$V_{ext}(\mathbf{r}_i) = -D\exp\left[-\gamma(x_i^2 + y_i^2)\right] + \frac{1}{2}m^2\omega_z^2 z_i^2.$$ (3)

It is worth noticing that for sufficiently large values of ω_z the electrons of the system are strongly compressed along the z direction. Therefore, in this situation, the system can be regarded as a quantum system confined by a two-dimensional Gaussian-type potential, i.e., as a quasi-two-dimensional Gaussian quantum dot. Since a Gaussian potential can be approximated close to its minimum by an harmonic potential, the potential of Eq. (3) is suitable for the modeling of anharmonic quantum dots. The anharmonicity of the confinement can be characterized by the depth of the Gaussian potential D and by the quantity $\omega = \sqrt{2D\gamma}/m$. Thus, when D is much larger than $\hbar\omega$ the Gaussian potential has many bound states and the potential curve follows closely the harmonic oscillator potential leading to a small anharmonicity of the system. On the other hand, when D is slightly larger than $\hbar\omega$ the Gaussian potential has only few bound states and, therefore, deviates strongly from the harmonic potential leading to a large anharmonicity. Also, a large (small) value of ω corresponds to a strong (weak) confinement with respect to the electron–electron interaction.

The wave function is approximated by a linear combination of Cartesian anisotropic Gaussian-Type Orbitals (c-aniGTO) [38, 39]. A c-aniGTO centered at (b_x, b_y, b_z) is defined as

$$\chi^{\mathbf{a},\zeta}(\mathbf{r}, \mathbf{b}) = x_{b_x}^{a_x} y_{b_y}^{a_y} z_{b_z}^{a_z} \exp(-\zeta_x x_{b_x}^2 - \zeta_y y_{b_y}^2 - \zeta_z z_{b_z}^2),$$ (4)

where $x_{b_x} = (x - b_x)$ etc... Following the quantum chemical convention the orbitals are classified as s-type and p-type, for $l = a_x + a_y + a_z = 0, 1, ...,$ respectively (this sum controls the value of the orbital angular momentum). The (b_x, b_y, b_z) parameters have been chosen to coincide with the center of the confining potential. This type of basis sets was found to be the most suitable one for expanding the eigenfunctions of an electron in an anisotropic harmonic oscillator potential. The calculations have been performed using the OpenMol Program.[2]

Energy spectra and oscillator strengths have been calculated for different strength of confinement ω and potential depth D. The effect of the electron–electron interaction on the distribution of oscillator strengths and the breakdown of the KT has been examined by focusing on the results with the same value of $D/\hbar\omega$, i.e., with the same anharmonicity.

A substantial red-shift has been observed for the dipole transitions corresponding to the excitation into the center-of-mass mode. The oscillator strengths, which are concentrated exclusively in the center-of-mass excitation in the harmonic limit, are distributed among the near-lying transitions as a result of the breakdown of the Kohn's theorem. The distribution of the oscillator strengths is limited to the transitions located in the lower energy region when ω is large (i.e., for strongly confined electrons) but it extends toward the higher energy region when ω becomes small (i.e., for weakly confined electrons).

The analysis of the CI wave functions shows that all states can be classified according to the polyad quantum number v_p [37]. The distribution of the oscillator strengths for large ω occurs among transitions involving excited states with the same value of v_p as the center-of-mass excited state, $v_{p,cm}$, while it occurs among transitions involving the excited states with $v_p = v_{p,cm}$ and $v_p = v_{p,cm+2}$ for small ω.

2.2 Time-Dependent Density Functional Theory (TDDFT) and the Local-Density Approximation (LDA)

Time-dependent density functional theory (TDDFT) extends the basic ideas of static density functional theory (DFT) to the more general situation of systems under the influence of time-dependent external fields. This dynamical approach relies on the electron density $n(r, t)$ rather than on the many-body wave function $\Psi(r_1, r_2, ..., r_N, t)$ of the system. In fact, the central theorem of the TDDFT is the Runge–Gross theorem [40–42] which tells us that all observables are uniquely determined by the density.

From the computational point of view, with respect to the resolution of the time-dependent Schrödinger equation (TDSE) of an N-electron system, the complexity is strongly reduced when using TDDFT. Indeed, the wave function depends on $3N+1$ variables $(r_1, r_2, ..., r_N, t)$ while the density depends only on 4 variables (r, t). This is one of the reasons why this method has become so popular. A practical

---

[2] see http://www.csc.fi/gopenmol

scheme for computing $n(r, t)$ is provided by the Kohn–Sham (KS) formulation of the TDDFT [32, 33]. In the latter, noninteracting electrons are moving in an effective local potential constructed in such a way that the KS density is the same as the one of the interacting electron system. The advantage of this formulation lies in its computational simplicity compared to other quantum chemical methods such as time-dependent Hartree–Fock or configuration interaction. The KS equations read as

$$i\hbar \frac{\partial}{\partial t}\phi_k(r, t) = \left(-\frac{\hbar^2}{2m}\nabla^2 + V_{\text{eff}}(r, t)\right)\phi_k(r, t), \tag{5}$$

with the KS density

$$n(r, t) = \sum_{k=1}^{\infty} f_k |\phi_k(r, t)|^2, \tag{6}$$

where f_k denotes the occupation numbers of the ground state, and

$$V_{\text{eff}}(r, t) = V_{\text{ext}}(r, t) + V_H(r, t) + V_{\text{xc}}(r, t). \tag{7}$$

In the above expression the first term is the external potential (ionic potential, laser field...), the second is the Hartree potential, which is a solution of the Poisson's equation, and the last term is the exchange–correlation potential.

The most popular choice for V_{xc} is the so-called adiabatic local-density approximation (ALDA) given by

$$V_{\text{xc}}(r, t) = \frac{d}{dn}[n\epsilon_{\text{xc}}(n)]_{n=n(r,t)}, \tag{8}$$

where $\epsilon_{\text{xc}}(n)$ is the exchange–correlation energy density for an homogeneous electron gas of density n. In this approach, the same functional used to calculate the properties of the ground state is employed in the dynamical simulations.

The validity of the local approximation has been discussed in many papers and textbooks [43]. This approximation works remarkably well for inhomogeneous electron systems. In contrast, the validity of the adiabatic approximation has been less thoroughly analyzed. Generally speaking, this approach is expected to hold for finite systems and for processes that evolve very slowly in time. The situation in bulk solids is more controversial since significant deficiencies in the description of absorption spectra have been noticed [44]. It was shown by Dobson [45] that ALDA fulfils the Kohn theorem when applied to a system of interacting electrons confined in an external parabolic potential. This theorem guarantees the existence of a collective state at the same frequency as the harmonic potential. It corresponds to a rigid oscillation of the many-body wave function around the center of the external potential.

Only a few attempts have been made to go beyond ALDA. To date, the most important ones are the work of Gross and Kohn [46] and that of Vignale and Kohn [47], the latter being the most promising in particular for studying electron relaxation phenomena [48]. Contrarily to ALDA, the approach of Gross and Kohn, which uses a frequency-dependent parametrization of the exchange–correlation kernel (see below), does not fulfil the Kohn theorem [52, 45]. This problem was further investigated by Vignale and Kohn [47], who proposed a new theory based on the so-called current density functional theory (CDFT). This model is described in detail in [49]. CDFT was originally derived by Vignale and Rasolt [50] to describe, within the framework of DFT, situations where strong magnetic fields and orbital currents cannot be ignored.

Few works have been devoted to the study of the nonlinear electron dynamics in finite metallic systems exposed to strong external fields. Indeed, the resolution of the time-dependent Kohn–Sham equations (5) is a very difficult task particularly for 3D systems. Some pioneering work on free simple metal clusters was performed by E. Suraud in Toulouse and P.-G. Reinhard in Erlangen [51]. More recently, Gervais et al. [52] have investigated the same problem in 3D geometry using a spherical basis expansion technique. This approach is restricted to small metal clusters. The interaction of strong femtosecond laser pulses with a C_{60} molecule (which possesses 240 delocalized electrons and can therefore be considered as a metallic nano-object [53]) has been investigated in [54] by employing a TDDFT approach. Still concerning the fullerene molecule, Cormier et al. [55] studied multiphoton absorption processes by solving numerically the associated time-dependent Schrödinger equation (TDSE) in the single active electron (SAE) approximation. This approximation consists in solving the equations (5) by using, instead of the time-dependent effective potential $V_{eff}(r, t)$ given in Eq. (7), the static effective potential of the ground state together with the time-dependent electric potential of the laser.

Let us now examine the linear regime, which has received much wider attention in the past.

Under the condition that the external field is weak, the simplest way to implement TDDFT is to work within the framework of the linear response theory. This approximation was first introduced by Zangwill and Soven [56] in the context of atomic physics for the study of photoionization in rare gases. Subsequently, this formalism has been successfully extended to the study of more and more complex electron systems: molecules [57], simple metal clusters [58], noble metal clusters [59], thin metal films [60], quantum dots [61], and condensed phase systems [44].

To date, in the field of nanoparticle physics, most applications of the time-dependent Kohn–Sham formalism have been performed at zero electron temperature. In order to interpret time-resolved pump-probe experiments carried out on noble metal nanoparticles, we have recently extended this approach to finite temperature. In the following we provide a brief overview of the model with the basic equations.

2.2.1 Ground State

The electron gas is assumed to be at thermal equilibrium with temperature T_e. In the Kohn–Sham formulation of the density functional theory at finite temperature within the grand-canonical ensemble [62–65], the ground-state electron density n of an N-electron system is written, in terms of single-particle orbitals ϕ_i and energies ε_i, as

$$n(\mathbf{r}) = \sum_{k=1}^{\infty} f_k \, n_k(\mathbf{r}) = \sum_{k=1}^{\infty} f_k \, |\phi_k(\mathbf{r})|^2, \tag{9}$$

where $f_k = \left[1 + \exp\left\{(\varepsilon_k - \mu)/k_B T_e\right\}\right]^{-1}$ are the Fermi occupation numbers and μ is the chemical potential. These orbitals and energies obey the Schrödinger equation

$$\left[-\frac{\hbar^2}{2m}\nabla^2 + V_{\text{eff}}(\mathbf{r})\right]\phi_i(\mathbf{r}) = \varepsilon_i \phi_i(\mathbf{r}) , \tag{10}$$

where $V_{\text{eff}}(\mathbf{r})$ is an effective single-particle potential given by

$$V_{\text{eff}}(\mathbf{r}) = V_{\text{ext}}(\mathbf{r}) + V_H(\mathbf{r}) + V_{\text{xc}}(\mathbf{r}) , \tag{11}$$

where $V_{\text{ext}}(\mathbf{r})$ is an external potential (e.g., due to the ionic background), $V_H(\mathbf{r})$ is the Hartree potential solution of Poisson's equation, and $V_{\text{xc}}(\mathbf{r})$ is the exchange–correlation potential defined by

$$V_{\text{xc}}(\mathbf{r}) = \frac{d}{dn}\left[n\omega_{\text{xc}}(n)\right]_{n=n(\mathbf{r})} , \tag{12}$$

where $\Omega_{\text{xc}}(n) \equiv \int n(\mathbf{r}) \, \omega_{\text{xc}}(n(\mathbf{r})) \, d\mathbf{r}$ is the exchange–correlation thermodynamic potential [66]. The temperature appears in the self-consistent procedure only through the occupation numbers and the exchange–correlation thermodynamic potential.

For low temperature (i.e., $T_e \ll T_F[n(\mathbf{r})]$ where $T_F[n(\mathbf{r})] = \frac{\hbar^2}{2mk_B}\left(3\pi^2 n(\mathbf{r})\right)^{2/3}$ is the local Fermi temperature), $\omega_{\text{xc}}(n)$ may be safely replaced by its value at $T_e = 0$, i.e., by $\epsilon_{\text{xc}}(n)$. The chemical potential is determined self-consistently by requiring the conservation of the total number of electrons from Eq. (9) [67, 68].

2.2.2 Excited States

In the usual first-order TDLDA at $T_e = 0$ in the frequency domain, the induced electron density $\delta n(\mathbf{r}; \omega)$ is related to $\delta V_{\text{ext}}(\mathbf{r}'; \omega)$, the Fourier transform (with respect to time) of the external time-dependent potential (generated, for instance, by the electric field of a laser beam), via the relation [56, 58, 69]

$$\delta n(r; \omega) = \int \chi(r, r'; \omega) \, \delta V_{\text{ext}}(r'; \omega) \, dr', \tag{13}$$

where $\chi(r, r'; \omega)$ is the retarded density correlation function or the dynamic response function. It is possible to rewrite the induced density as

$$\delta n(r; \omega) = \int \chi^0(r, r'; \omega) \, \delta V_{\text{eff}}(r'; \omega) \, dr', \tag{14}$$

with

$$\delta V_{\text{eff}}(r; \omega) = \delta V_{\text{ext}}(r; \omega) + \frac{e^2}{4\pi \varepsilon_0} \int \frac{\delta n(r'; \omega)}{|r - r'|} \, dr'$$
$$+ \int f_{\text{xc}}(r, r'; \omega) \, \delta n(r'; \omega) \, dr', \tag{15}$$

where the function $f_{\text{xc}}(r, r'; \omega)$ is the Fourier transform of the time-dependent kernel defined by $f_{\text{xc}}(r, t; r', t') \equiv \delta V_{\text{xc}}(r, t)/\delta n(r', t')$ and $\chi^0(r, r'; \omega)$ is the non-interacting retarded density correlation function. From Eqs. (13)–(15) we see that χ^0 and χ are related by an integral equation (Dyson-type equation)

$$\chi(r, r'; \omega) = \chi^0(r, r'; \omega) + \int \int \chi^0(r, r''; \omega)$$
$$\times K(r'', r'''; \omega) \, \chi(r''', r'; \omega) \, dr'' dr''', \tag{16}$$

with the residual interaction defined by

$$K(r, r'; \omega) = \frac{e^2}{4\pi \varepsilon_0 |r - r'|} + f_{\text{xc}}(r, r'; \omega). \tag{17}$$

In the *adiabatic* local-density approximation (ALDA) the exchange–correlation kernel is frequency-independent and local and reduces to [56, 69]

$$f_{\text{xc}}(r, r') = \frac{d}{dn} \, [V_{\text{xc}}(n)]_{n=n(r)} \, \delta(r - r') \, . \tag{18}$$

It should be mentioned that the functional, V_{xc} in the above equation is the same as the one used in the calculation of the ground state [see Eq. (12)]. For spin-saturated electronic systems, we have

$$\chi^0(\boldsymbol{r}, \boldsymbol{r}'; \omega) = 2 \sum_{jk} \left[f_j^0 - f_k^0 \right] \frac{\phi_j^*(\boldsymbol{r}) \phi_k(\boldsymbol{r}) \phi_k^*(\boldsymbol{r}') \phi_j(\boldsymbol{r}')}{\hbar\omega - (\varepsilon_k - \varepsilon_j) + i\eta}$$

$$= \sum_k^{occ} \phi_k^*(\boldsymbol{r}) \phi_k(\boldsymbol{r}') \, G_+(\boldsymbol{r}, \boldsymbol{r}'; \varepsilon_k + \hbar\omega) +$$

$$\sum_k^{occ} \phi_k(\boldsymbol{r}) \phi_k^*(\boldsymbol{r}') \, G_+^*(\boldsymbol{r}, \boldsymbol{r}'; \varepsilon_k - \hbar\omega), \tag{19}$$

where $\phi_k(\boldsymbol{r})$ and ε_k are the one-electron Kohn–Sham wave functions and energies, respectively. G_+ is the one-particle retarded Green's function and f_k^0 are the Fermi occupation numbers at $T_e = 0$ K (0 or 1). All the above quantities are obtained with the procedure described in the preceding subsection with $f_k = f_k^0$ in Eq. (9). In order to produce numerically tractable results, we have added a small imaginary part to the probe frequency, so that $\omega \to \omega + i\delta$ with $\eta = \hbar\delta$.

At finite electron temperature, the grand-canonical non-interacting retarded density correlation function reads [70]

$$\chi^0(\boldsymbol{r}, \boldsymbol{r}'; \omega; T_e) = \frac{1}{Z_G} \sum_{n,N} \exp\left\{ -\frac{1}{k_B T_e} [E_n(N) - N\mu] \right\}$$

$$\times \chi_{n,N}^0(\boldsymbol{r}, \boldsymbol{r}'; \omega; T_e), \tag{20}$$

where Z_G is the grand-canonical partition function

$$Z_G = \sum_{n,N} \exp\left\{ -\frac{1}{k_B T_e} [E_n(N) - N\mu] \right\} \tag{21}$$

with $E_n(N)$ the energy of the state $|nN\rangle$ having N electrons, μ the chemical potential and

$$\chi_{n,N}^0(\boldsymbol{r}, \boldsymbol{r}'; \omega; T_e) = \sum_m \frac{\langle nN |\hat{n}(\boldsymbol{r})| mN \rangle \, \langle mN |\hat{n}(\boldsymbol{r}')| nN \rangle}{\hbar\omega - (E_m(N) - E_n(N)) + i\eta}$$

$$- \frac{\langle nN |\hat{n}(\boldsymbol{r}')| mN \rangle \, \langle mN |\hat{n}(\boldsymbol{r})| nN \rangle}{\hbar\omega + (E_m(N) - E_n(N)) + i\eta}. \tag{22}$$

In the above expression $\hat{n}(\boldsymbol{r})$ is the particle density *operator* defined from the wave field operators by

$$\hat{n}(\boldsymbol{r}) = \hat{\psi}^+(\boldsymbol{r}) \hat{\psi}(\boldsymbol{r}), \tag{23}$$

with $\hat{\psi}^+(\boldsymbol{r}) = \sum_k \hat{a}_k^+ \phi_k^*(\boldsymbol{r})$ and $\hat{\psi}(\boldsymbol{r}) = \sum_k \hat{a}_k \phi_k(\boldsymbol{r})$. By using standard field theory techniques it is possible to show that

$$\chi^0(\boldsymbol{r}, \boldsymbol{r}'; \omega; T_e) = \sum_k f_k \, \phi_k^*(\boldsymbol{r})\phi_k(\boldsymbol{r}') \, G_+(\boldsymbol{r}, \boldsymbol{r}'; \varepsilon_k + \hbar\omega; T_e)$$

$$+ \sum_k f_k \, \phi_k(\boldsymbol{r})\phi_k^*(\boldsymbol{r}') \, G_+^*(\boldsymbol{r}, \boldsymbol{r}'; \varepsilon_k - \hbar\omega; T_e), \quad (24)$$

where $f_k = \left[1 + \exp\left\{(\varepsilon_k - \mu)/k_B T_e\right\}\right]^{-1}$. So far, we have assumed that the residual interaction (17) is temperature independent. This assumption is consistent with the use of $\omega_{\mathrm{xc}}(n) = \epsilon_{\mathrm{xc}}(n)$ in the calculation of the ground-state properties. Therefore, as for $T_e = 0$, the response function is solution of the Dyson equation (16) with χ^0 given by Eq. (24).

The above formalism can be employed to compute the photoabsorption by a metallic nanoparticle of size R. If the wavelength λ of the incoming light is such that $\lambda \gg R$ the dipolar approximation is valid. From the frequency-dependent dipole polarizability

$$\alpha(\omega; T_e) = \int \delta n(\boldsymbol{r}; \omega; T_e) \, \delta V_{ext}(\boldsymbol{r}; \omega) \, d\boldsymbol{r}, \quad (25)$$

one obtains the dipolar absorption cross-section [71]

$$\sigma(\omega; T_e) = \frac{\omega}{\varepsilon_0 c} \operatorname{Im}\left[\alpha(\omega; T_e)\right]. \quad (26)$$

As for the zero temperature case, the dipolar absorption cross-section fulfils the well-known Thomas–Reiche–Kuhn (TRK) sum rule

$$\int \sigma(\omega; T_e) \, d\omega = \frac{2\pi^2 N}{c}. \quad (27)$$

2.2.3 Application to Femtosecond Spectroscopy

Ultrafast spectroscopy using femtosecond laser pulses is a well-suited technique to study the electronic energy relaxation mechanisms in metallic nanoparticles (see [6, 12] and references therein). The experiments have been carried out with nanoparticles of noble metals containing several thousand atoms and embedded in a transparent matrix. By using a time-resolved pump-probe configuration it is possible to have access to the spectral and temporal dependence of the differential transmission $\frac{\Delta T}{T}(\tau, \omega)$, defined as the normalized difference between the probe pulse with and without the pump pulse. This quantity contains the information on the electron dynamics, which is measured as a function of the pump-probe time delay τ and of the laser frequency ω.

For pump-probe delays longer than a few hundred femtoseconds, the thermalization of the electrons is achieved, thus leading to an increase of the electron temperature of several hundred degrees. However, the electronic distribution is not in thermal equilibrium with the lattice, the thermal relaxation to the lattice being achieved in a few picoseconds via electron–phonon scattering. The energy exchange between the electrons and the lattice can be described by the two temperature model leading to a time-dependent electron temperature $T_e(t)$ [72, 73]

$$C_e \frac{\partial T_e}{\partial t} = -G(T_e - T_i) + P(t),$$

$$C_i \frac{\partial T_i}{\partial t} = G(T_e - T_i), \tag{28}$$

where $P(t)$ represents the laser source term, C_i (C_e) is the lattice (electron) heat capacity, and G is the electron–lattice coupling factor. In this simplified model, the two temperatures are assumed to be spatially uniform and therefore the heat propagation is neglected.

Provided that the relative changes of the dielectric function with respect to a non-perturbed system are weak (linear regime) and that they are only due to a modification of the electron temperature, one may identify the spectral dependence of the differential transmission measured for a given time delay as the difference of the linear absorption cross-sections evaluated at different electron temperatures. More precisely, the differential transmission is expressed as

$$\frac{\Delta T}{T}(\tau, \omega) = \frac{T[T_e(\tau), \omega] - T[T_e(0), \omega]}{T[T_e(0), \omega]} = -\Delta \tilde{\alpha}(\omega) \, l \tag{29}$$

$$= \frac{3}{2\pi R^2} \left[\sigma(\omega; T_e(0)) - \sigma(\omega; T_e(\tau)) \right], \tag{30}$$

where $l = 2R$ is the sample thickness (here, the diameter of the nanoparticle), $T[T_e(\tau), \omega]$ and $T[T_e(0), \omega]$ are the probe transmissions in the presence and absence of the pump, respectively, and $\Delta \tilde{\alpha}$ is the pump-induced absorption change. Obviously $T[T_e(0), \omega]$ corresponds to an absorption at room temperature $T_e(0) = 300$ K for the conditions where the pump-probe experiments have been performed. We have computed the optical spectrum of a closed-shell nanoparticle Ag_{2998} embedded in a transparent matrix (alumina $\varepsilon_m = 1.5$) for three values of the temperature. The diameter of the nanoparticle is 4.6 nm and the photon energy ranges from 2.2 eV to the interband threshold energy at 3.8 eV, i.e., in the spectral region associated to the surface plasmon of Ag nanoparticles. All these values correspond to typical experimental conditions performed in our group [6]. The results are presented in Fig. 2. The calculated oscillator strength is 90%. Indeed, due to the presence of the surface plasmon resonance, almost all the oscillator strength is concentrated in this energy range. A clear red-shift and broadening of the resonance as a function of the electron temperature is observed.

In the left panel of Fig. 3, the predictions of the normalized differential transmission [Eq. (30)] are presented as a function of the photon energy of the probe. The comparison is made for two electron temperatures $T_e = 600$ K and $T_e = 1200$ K. The asymmetric shape of $\Delta T/T$ around the resonance energy is related to a

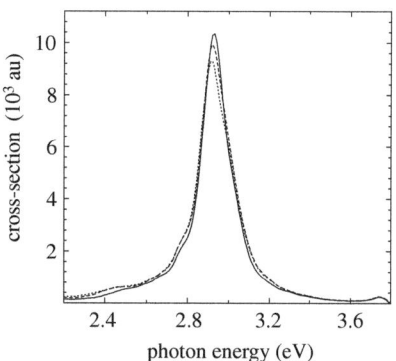

Fig. 2 TDLDA photoabsorption cross-section (in atomic units) of Ag_{2998} encapsulated in a transparent matrix ($\varepsilon_m = 1.5$) as a function of the photon energy. *Solid line*: $T_e = 0$ K; *dashed line*: $T_e = 300$ K; *dotted line*: $T_e = 1200$ K

combination of a red-shift and a broadening of the surface plasmon resonance. In the right panel of Fig. 3 the experimental spectrum of the normalized $\Delta T/T$ obtained for a pump-probe delay of $\tau = 2$ps is depicted. The pump pulse is set at 400 nm (second harmonic of a titanium sapphire laser amplified at 5 kHz) and the probe comes from a continuum generated in a sapphire crystal with the fundamental frequency of the amplified laser [6].

The asymmetric spectral shape of the differential transmission spectrum in Fig. 3, which is related to the shift and broadening of the plasmon, may have several origins. As pointed out in [6, 12, 74], the interband transition induces a modification of the real part of the dielectric function in this spectral region, the resonance being far enough from the interband threshold to induce significant changes of the corresponding imaginary part. As stressed in [12, 74], this is a strong indication that

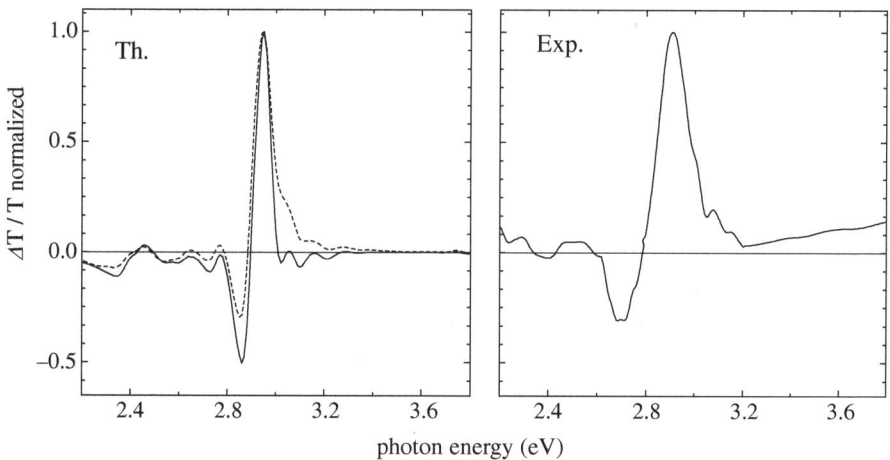

Fig. 3 *Left panel*: theoretical predictions of the normalized differential transmission for Ag_{2998} embedded in a transparent matrix as a function of the photon energy of the probe. *Solid line*: $T_e = 600$ K; *dotted line*: $T_e = 1200$ K. *Right panel*: Normalized experimental spectrum of $\Delta T/T$ of silver nanoparticles encapsulated in an alumina matrix for a pump-probe delay of 2 ps [6]

intraband processes also play an important role. Indeed, as clearly seen in Fig. 2, the conduction electrons contribution leads both to a shift and to a broadening. We can therefore conclude that one needs to consider both the interband and the intraband part on the same footing. Whereas this effect was previously taken into account in a phenomenological way via a shifted and broadened Lorentzian shape, here we have derived it directly from a quantum many-body approach based on the TDLDA at finite temperature.

2.3 Phase-Space Methods: From Hartree to Wigner and Vlasov

As we have seen in Sect. 2.1, the most fundamental model for the quantum N-body problem is the Schrödinger equation for the N-particle wave function $\Psi(r_1, r_2, \ldots, r_N, t)$. Unfortunately, the full Schrödinger equation cannot be solved exactly except for very small systems. A drastic, but useful and to some extent plausible, simplification can be achieved by neglecting two-body (and higher order) correlations. This amounts to assume that the N-body wave function can be factored into the product of N one-body functions:

$$\Psi(r_1, r_2, \ldots, r_N, t) = \psi_1(r_1, t)\,\psi_2(r_2, t)\ldots\psi_N(r_N, t). \tag{31}$$

For fermions, a weak form of the exclusion principle is satisfied if none of the wave functions on the right-hand side of Eq. (31) are identical.[3]

When the above assumption is made, the N-body Schrödinger equation reduces to a set of one-particle equations, coupled through Poisson's equation (*time-dependent Hartree model*):

$$i\hbar\frac{\partial\psi_\alpha}{\partial t} = -\frac{\hbar^2}{2m}\,\Delta\psi_\alpha - e\phi\psi_\alpha\,, \quad \alpha = 1\ldots N_{\text{orb}} \tag{32}$$

$$\Delta\phi = \frac{e}{\varepsilon}\left(\sum_{\alpha=1}^{N_{\text{orb}}} p_\alpha|\psi_\alpha|^2 - n_i(r)\right), \tag{33}$$

where $N_{\text{orb}} \geq N$ is the number of occupied orbitals, e and m are the absolute electron charge and mass, and ε is the dielectric constant; $n_i(r)$ is the ion density, which is supposed to be fixed and a continuous function of the position coordinate. This is

---

[3] A stronger version of the exclusion principle requires that $\Psi(r_1, r_2, \ldots, r_N, t)$ is antisymmetric, i.e., that it changes sign when two of its arguments are interchanged. This can be achieved by taking, instead of the single product of N wave functions as in Eq. (31), a linear combinations of all products obtained by permutations of the arguments, with weights ± 1 (Slater determinant) [75]. This is at the basis of Fock's generalization of the Hartree model.

known as the "jellium" hypothesis and is valid whenever the relevant length scales are significantly larger that the ionic lattice spacing $a_{latt} \sim 5\text{Å}$. As mentioned in Sect. 2, this is the case for semiconductor nanostructures, but not so for metals (see Table 1); nevertheless, the jellium models still yields reasonably accurate results for all but the smallest nano-objects.

The occupation probabilities p_α ($\sum_{\alpha=1}^{N_{orb}} p_\alpha = 1$) are defined to describe a Fermi–Dirac distribution at finite electron temperature, $p_\alpha = [1 + \exp(\beta(\varepsilon_\alpha - \mu))]^{-1}$, where $\beta = 1/k_B T_e$, μ is the chemical potential, and ε_α is the single-particle energy level. In practice, one first needs to obtain the ground-state equilibrium solution of Eqs. (32)–(33), which amounts to determining the N_{orb} occupation probabilities and the corresponding energy levels and wave functions. Subsequently, the equilibrium can be perturbed to study the electron dynamics. The numerical methods for the dynamics are quite standard, as the Eq. (32) are basically one-particle Schrödinger equations. We will not enter into the details of the numerical methods in this chapter: a list of relevant works on the Schrödinger equation can be found in [66].

We now show that the Hartree equations can be written in a completely equivalent form by making use of the Wigner transformation. The Wigner representation [34] is a useful tool to express quantum mechanics in a phase-space formalism (for reviews see [77–80]). The Wigner function is a function of the phase-space variables (x, v) and time, which, in terms of the single-particle wave functions, reads as

$$f(x, v, t) = \sum_{\alpha=1}^{N_{orb}} \frac{m}{2\pi\hbar} p_\alpha \int_{-\infty}^{+\infty} \psi_\alpha^* \left(x + \frac{\lambda}{2}, t\right) \psi_\alpha \left(x - \frac{\lambda}{2}, t\right) e^{imv\lambda/\hbar} \, d\lambda. \quad (34)$$

(we restrict our discussion to one-dimensional cases, but all results can easily be generalized to three dimensions). It must be stressed that the Wigner function, although it possesses many useful properties, is not a true probability density, as it can take negative values. However, it can be used to compute averages just like in classical statistical mechanics. For example, the expectation value of a generic quantity $A(x, v)$ is defined as

$$\langle A \rangle = \frac{\int \int f(x, v) A(x, v) dx dv}{\int \int f(x, v) dx dv} \quad (35)$$

and yields the correct quantum-mechanical value.[4] In addition, the Wigner function reproduces the correct quantum-mechanical marginal distributions, such as the spatial density:

---

[4] For variables whose corresponding quantum operators do not commute (such as $\hat{x}\hat{v}$), Eq. (35) must be supplemented by an ordering rule, known as Weyl's rule [80].

$$n(x, t) = \int_{-\infty}^{+\infty} f(x, v, t) \, dv = \sum_{\alpha=1}^{N_{orb}} p_\alpha \mid \psi_\alpha \mid^2 . \tag{36}$$

We also point out that, of course, not all functions of the phase-space variables are genuine Wigner functions, as they cannot necessarily be written in the form of Eq. (34). In general, although it is trivial to find the Wigner function given the wave functions that define the quantum mixture, the inverse operation is not generally feasible. Indeed, there are no simple rules to establish whether a given function of x and v is a genuine Wigner function. For a more detailed discussion on this issue, and some practical recipes to construct genuine Wigner functions, see [81].

The Wigner function obeys the following evolution equation:

$$\frac{\partial f}{\partial t} + v \frac{\partial f}{\partial x} +$$

$$\frac{em}{2i\pi\hbar^2} \int \int d\lambda \, dv' e^{im(v-v')\lambda/\hbar} \left[\phi \left(x + \frac{\lambda}{2} \right) - \phi \left(x - \frac{\lambda}{2} \right) \right] f(x, v', t) = 0 , \tag{37}$$

where $\phi(x, t)$ is the self-consistent electrostatic potential obtained self-consistently from Poisson's equation (33).

Developing the integral term in Eq. (37) up to order $O(\hbar^2)$ we obtain

$$\frac{\partial f}{\partial t} + v \frac{\partial f}{\partial x} + \frac{e}{m} \frac{\partial \phi}{\partial x} \frac{\partial f}{\partial v} = \frac{e\hbar^2}{24m^3} \frac{\partial^3 \phi}{\partial x^3} \frac{\partial^3 f}{\partial v^3} + O(\hbar^4). \tag{38}$$

In the limit $\hbar \to 0$ one recovers the classical Vlasov equation, well known from plasma physics (see Fig. 1). The Vlasov–Poisson system has been used to study the dynamics of electrons in metal clusters and thin metal films [51, 18–20]. It is appropriate for large excitation energies, for which the electrons' de Broglie wavelength is relatively small, thus reducing the importance of quantum effects in the electron dynamics. Nevertheless, for metallic nanostructures at room temperature, the equilibrium must be given by a Fermi–Dirac distribution, because the Fermi temperature is very high (see Table 1). For semiconductor nanostructures, $T_F \sim 10 - 50$K, so that a Maxwell–Boltzmann equilibrium is sometimes appropriate for moderate temperatures.

The Wigner equation must be coupled to the Poisson's equation for the electric potential:

$$\frac{\partial^2 \phi}{\partial x^2} = -\frac{e}{\varepsilon} [n_i(x) - n(x, t)] . \tag{39}$$

The resulting Wigner–Poisson (WP) system has been extensively used in the study of quantum transport [82–84]. Exact analytical results can be obtained by linearizing Eqs. (37) and (39) around a spatially homogeneous equilibrium given by $n_0 f_0(v)$ (Maxwell–Boltzmann or Fermi–Dirac distribution), where $n_0 = n_i = $ const. is the uniform equilibrium density. By expressing the fluctuating quantities as a sum of plane waves $\exp(ikx - i\omega t)$ with frequency ω and wave number k, the dispersion relation can be written in the form $\varepsilon(k, \omega) = 0$, where the "dielectric constant" ε reads, for the WP system,

$$\varepsilon_{WP}(\omega, k) = 1 + \frac{m\omega_p^2}{n_0 k} \int \frac{f_0(v + \hbar k/2m) - f_0(v - \hbar k/2m)}{\hbar k(\omega - kv)} \, dv, \qquad (40)$$

or equivalently

$$\varepsilon_{WP}(\omega, k) = 1 - \frac{\omega_p^2}{n_0} \int \frac{f_0(v)}{(\omega - kv)^2 - \hbar^2 k^4/4m^2} \, dv . \qquad (41)$$

This is just the Lindhard [85] dispersion relation, well known from solid-state physics. From Eq. (40), one can recover the Vlasov–Poisson dispersion relation by taking the classical limit $\hbar \to 0$

$$\varepsilon_{VP}(\omega, k) = 1 + \frac{\omega_p^2}{n_0 k} \int \frac{\partial f_0/\partial v}{\omega - kv} \, dv. \qquad (42)$$

The equivalence of the Hartree and Wigner–Poisson methods can be easily proven by comparing the linear results. For the Hartree equations (32), we linearize around a homogeneous equilibrium given by plane waves:

$$\psi_\alpha = \sqrt{n_0} \, \exp\left(i\frac{mu_{0\alpha}}{\hbar}x\right), \qquad (43)$$

each with occupation number p_α and energy $\epsilon_\alpha = mu_{0\alpha}^2/2$. The Hartree dielectric constant is found to be

$$\varepsilon_H(\omega, k) = 1 - \sum_{\alpha=1}^{N_{orb}} p_\alpha \frac{\omega_p^2}{(\omega - ku_{0\alpha})^2 - \hbar^2 k^4/4m^2}, \qquad (44)$$

which is a discrete form of the Wigner–Poisson dispersion relation (41).

2.3.1 Example — Ultrafast Electron Dynamics in Thin Metal Films

Several experiments have shown [2, 3] that electron transport in thin metal films occurs on a femtosecond timescale and involves ballistic electrons traveling at the Fermi velocity of the metal v_F. More recently, a regime of low-frequency nonlinear oscillations (corresponding to ballistic electrons bouncing back and forth on the film surfaces) was measured in transient reflection experiments on thin gold films [86].

These findings were corroborated by accurate numerical simulations based on the one-dimensional Vlasov–Poisson equations [18–20]. The electrons are initially prepared in a Fermi–Dirac equilibrium at finite (but small) temperature. They are subsequently excited by imposing a constant velocity shift $\delta v = 0.08v_F$ to the initial distribution, which is a rather strong excitation. This scenario is appropriate when no linear momentum is transferred parallel to the plane of the surface (i.e., $q_{\parallel} = 0$) and is relevant to the excitation of the film with optical pulses [87]. For $q_{\parallel} = 0$, only longitudinal modes (volume plasmon with $\omega = \omega_p$) can be excited.

As a reference case, we studied a sodium film with initial temperature $T_e = 0.008T_F \simeq 300$ K and thickness $L \simeq 120$ Å. The time evolution of the thermal E_{th} and center-of-mass E_{cm} energies was analyzed (Fig. 4). During an initial rapidly oscillating phase, E_{cm} is almost entirely converted into thermal energy (Landau damping). After saturation, a slowly oscillating regime appears, with period equal to $50\omega_p^{-1} \approx 5.3$fs, where $\omega_p = (e^2 n/m\varepsilon_0)^{1/2}$ is the plasmon frequency. This period is close to the time of flight of electrons traveling at the Fermi velocity and bouncing back and forth on the film surfaces (further details are provided in our previous work [18–20]).

The phase-space portrait of the electron distribution, shown in Fig. 5, clearly reveals that the perturbation starts at the film surfaces and then proceeds inward at the Fermi velocity of the metal. The structure formation at the Fermi surface, which has spread over the entire film for $\omega_p t > 150$, is responsible for the increase of the thermal energy (and thus the electron temperature) observed in Fig. 4. As no coupling to an external environment (e.g., phonons) is present, this excess temperature cannot be dissipated.

Quantum simulations of the electron dynamics using the Wigner–Poisson system were performed more recently: as expected, the Vlasov results were recovered in the large excitation regime $\delta v > 0.08v_F$. For smaller excitations, a different regime appears, in which the ballistic oscillations described above are no longer observed. Further work is in progress on this issue [88].

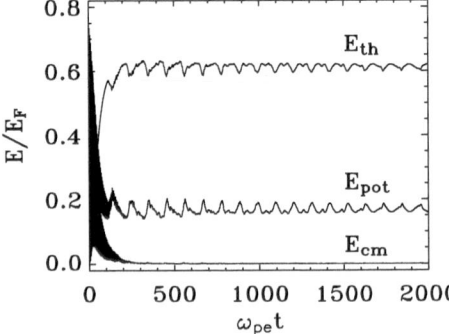

Fig. 4 Time evolution of the thermal, potential, and center-of-mass energies of the electron population in a thin sodium film

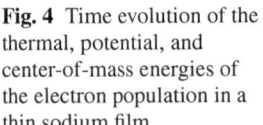

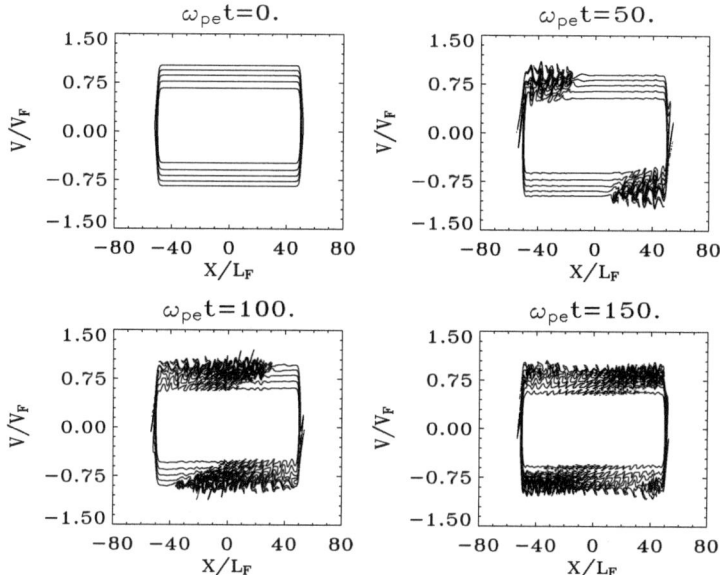

Fig. 5 Phase-space portrait of the electron distribution. Velocity is normalized to the Fermi velocity, and space to the Thomas–Fermi screening length $L_F = v_F/\omega_p$

2.3.2 Beyond the Mean-Field

The mean-field approach described above is appropriate to describe the electron dynamics on very short timescales ($<$100fs). On a longer timescale (0.1–1ps), the injected energy is redistributed among the electrons via electron–electron (e–e) collisions. Electron–phonon (e–ph) thermalization (i.e., coupling to the ionic lattice) is generally supposed to occur on even longer timescales. However, the results of [5, 89] on thin gold films have shown that nonequilibrium electrons start interacting with the lattice earlier than expected, so that a clear-cut separation between e–e and e–ph relaxation is not entirely pertinent.

The phase-space approach is particularly well suited to include corrections that go beyond the mean-field picture. This can be done with relative ease for semiclassical models (Vlasov), by using a Boltzmann-like e–e collision integral that respects Pauli's exclusion principle (Ühling–Uhlenbeck model) [90]:

$$\left(\frac{\partial f}{\partial t}\right)_{UU} = \int \frac{d^3\mathbf{p}_2 d\Omega}{(2\pi\hbar)^3}\,\sigma(\Omega)|v_{12}|(f_1 f_2 \overline{f}_3 \overline{f}_4 - f_3 f_4 \overline{f}_1 \overline{f}_2)\,, \qquad (45)$$

where v_{12} is the relative velocity of the colliding particles 1 and 2, $\sigma(\Omega)$ is the differential cross-section depending on the scattering angle Ω, and indices 3 and 4 label the outgoing momenta, $f_i = f(\mathbf{r}, \mathbf{p}_i, t)$ and $\overline{f}_i = 1 - f_i/2$. This collision term is similar to the well-known classical Boltzmann collision term but for Pauli blocking factors $\overline{f}_i \overline{f}_j$. As known from solid-state physics, this blocking factor plays a dramatic role for electronic systems [75]. At $T_e = 0K$, all collisions are Pauli blocked and the collisional mean-free path of the electrons becomes infinite. But if the system becomes excited, phase-space opens up and activates the collision term. The effect of the above e–e collision term on the semiclassical Vlasov dynamics in metal clusters was investigated numerically in [91].

It is conceptually harder to include collisions in fully quantum models. A significant constraint is that nonunitary corrections to the Wigner equation should be written in "Lindblad form" [92], which guarantees that the evolved Wigner function corresponds to a positive-definite density matrix.

The Ühling–Uhlenbeck collision term (45) is a complicated nonlinear integral, which is difficult to implement in a numerical code. It is therefore useful to construct some simplified collision terms that are more easily amenable to numerical treatment. In the following, we briefly illustrate two simple models of e–e and e–ph collisions that we have employed in our previous works.

2.3.2.1 Electron-Electron Collisions

To model e–e collisions, a relaxation term is added to the right-hand side of the Vlasov or Wigner equation:

$$\left(\frac{\partial f}{\partial t}\right)_{e-e} \equiv -\nu_{ee}(T_e)(f - f_\infty), \tag{46}$$

where ν_{ee} is the average e–e collision rate and $f_\infty(x, v)$ is a Fermi–Dirac distribution. The idea behind this model is that the electron distribution will eventually relax, on a timescale of the order ν_{ee}^{-1}, toward a Fermi–Dirac equilibrium f_∞ with total energy equal to that of the initial electron distribution $f(x, v, t = 0^+)$, including of course the initial excitation energy. For electrons near the Fermi surface, the e–e collision rate can be written as [93]

$$\nu_{ee}(T_e) = a(k_B T_e)^2, \tag{47}$$

where a is a (dimensional) proportionality constant. The latter has been estimated from numerical simulations of the electron dynamics in sodium clusters [91], yielding $a \simeq 0.4 \text{ fs}^{-1}\text{eV}^{-2}$, which is also compatible with the analytical prediction given by the random phase approximation [93]. The electron temperature is computed instantaneously during the simulation and plugged into the expression for the collision rate (47). It is important to underline that the above model for e–e collisions, though simple, is completely self-contained and requires no additional ad hoc parameters. The model has been applied to the electron dynamics in thin metal films. The slow ballistic oscillations of Fig. 4 are still observed, although they are damped on a timescale of the order of $500\omega_{pe}^{-1} \simeq 50\text{fs}$ (see Fig. 6).

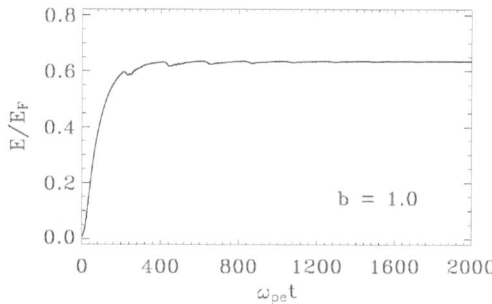

Fig. 6 Evolution of the thermal energy for a case with e–e collisions and $L = 100L_F \simeq 120\text{Å}$

2.3.2.2 Electron–Phonon Collisions

By coupling to the ionic lattice, the electrons progressively relax to a thermal distribution with a temperature equal to that of the lattice T_i. This relaxation time is generally termed τ_1 in the semiconductor literature. In addition, the lattice acts as an external environment for the electrons, leading to a loss of quantum coherence over a timescale τ_2 (decoherence time). The relaxation and decoherence times correspond, respectively, to the decay of diagonal and nondiagonal terms in the density matrix describing the electron population.

Such environment-induced decoherence can be modeled, in the Wigner representation, by an appropriate friction-diffusion term [94]:

$$\left(\frac{\partial f}{\partial t}\right)_{\text{e–ph}} = 2\gamma \frac{\partial(vf)}{\partial v} + D_v \frac{\partial^2 f}{\partial v^2} + D_x \frac{\partial^2 f}{\partial x^2} , \qquad (48)$$

where γ is the relaxation rate (inverse of the relaxation time τ_1), and D_v, D_x are diffusion coefficients in velocity and real space, respectively, which are related to the decoherence time τ_2 and depend on the lattice temperature T_i. The effect of the diffusive terms is to smooth out the fine structure of the Wigner function, thus suppressing interference phenomena, which are a typically quantum effect. Finally, we recall that, in order to preserve the positivity of the density matrix associated to the Wigner distribution function, the e–ph collision term (48) must be in Lindblad form [92]. This is automatically achieved [95] if the coefficients respect the inequality $D_v D_x \geq \gamma^2 \hbar^2 / 4m^2$.

2.4 Hydrodynamical Models: From Micro to Macro

Despite its considerable interest, the Wigner–Poisson (WP) formulation presents some intrinsic drawbacks : (i) it is a nonlocal, integro-differential system and (ii) its numerical treatment requires the meshing of the whole phase-space. Moreover, as is often the case with kinetic models, the Wigner–Poisson system gives more information than one is really interested in. For these reasons, it would be useful to obtain an accurate reduced model which, though not providing the same detailed information, is still able to reproduce the main features of the physical system under consideration.

In this section, we will derive an effective Schrödinger–Poisson (SP) system, which, in an appropriate limit, reproduces the results of the kinetic WP formulation [96]. In order to obtain the effective SP system, we will first derive a system of reduced hydrodynamic (or fluid) equations by taking moments of the WP system. It will be shown that the pressure term appearing in the fluid equations can be decomposed into a classical and a quantum part. With some reasonable hypotheses on the pressure term, the fluid system can be closed. For simplicity of notation, only one-dimensional problems will be considered, but the results can be easily extended to higher dimensions.

In order to derive a fluid model, we take moments of Eq. (37) by integrating over velocity space. Introducing the standard definitions of density, mean velocity, and pressure

$$n = \int f \, dv \,, \quad u = \frac{1}{n} \int f v \, dv \,, \quad P = m \left(\int f v^2 dv - n u^2 \right), \quad (49)$$

it is obtained

$$\frac{\partial n}{\partial t} + \frac{\partial (nu)}{\partial x} = 0 \,, \tag{50}$$

$$\frac{\partial u}{\partial t} + u \frac{\partial u}{\partial x} = \frac{e}{m} \frac{\partial \phi}{\partial x} - \frac{1}{mn} \frac{\partial P}{\partial x}. \tag{51}$$

We immediately notice that, surprisingly, Eqs. (50)–(51) do not differ from the ordinary evolution equations for a classical fluid. It can be shown, however, that quantum effects are actually hidden in the pressure term, which may be decomposed into a classical and a quantum part.

By using the definition of the Wigner function (34) and representing each state in terms of its amplitude $\sqrt{n_\alpha}$ and phase S_α

$$\psi_\alpha(x, t) = \sqrt{n_\alpha(x, t)} \exp \left(i S_\alpha(x, t)/\hbar \right), \tag{52}$$

we obtain that $P = P^C + P^Q$. The classical part of the pressure can be written as

$$P^C = mn \left[\sum_\alpha p_\alpha \frac{n_\alpha}{n} u_\alpha^2 - \left(\sum_\alpha p_\alpha \frac{n_\alpha}{n} u_\alpha \right)^2 \right] \equiv mn(\langle u_\alpha^2 \rangle - \langle u_\alpha \rangle^2), \quad (53)$$

where $mu_\alpha = \partial S_\alpha / \partial x$ [the u_α's should not be mistaken with the *global* mean velocity u defined in Eq. (49)]. This is the standard expression for the pressure as velocity dispersion, thus justifying the term "classical" pressure.

The quantum part of the pressure is written as

$$P^Q = \frac{\hbar^2}{2m} \sum_\alpha p_\alpha \left(\left(\frac{\partial \sqrt{n_\alpha}}{\partial x} \right)^2 - \sqrt{n_\alpha} \frac{\partial^2 \sqrt{n_\alpha}}{\partial x^2} \right). \tag{54}$$

It can be shown that, for distances larger that the Thomas–Fermi screening length L_F, one can replace n_α with n, the total density as defined in Eq. (49). In order to close the fluid system (50)–(51) one still has to express the classical pressure in terms of the density n. This is the standard procedure adopted in classical hydrodynamics: the relation $P^C(n)$ is the equation of state and depends on the particular conditions of the system, notably its temperature.

With these hypotheses, the Eq. (51) reduces to

$$\frac{\partial u}{\partial t} + u \frac{\partial u}{\partial x} = \frac{e}{m} \frac{\partial \phi}{\partial x} - \frac{1}{m} \frac{\partial W}{\partial x} + \frac{\hbar^2}{2m^2} \frac{\partial}{\partial x} \left(\frac{\partial^2(\sqrt{n})/\partial x^2}{\sqrt{n}} \right), \tag{55}$$

where we have defined the effective potential

$$W(n) = \int^n \frac{dn'}{n'} \frac{dP^C(n')}{dn'}. \tag{56}$$

Equations (50) and (55) constitute the quantum hydrodynamical approximation to the full Wigner (or Hartree) equation.

It is now possible to combine Eqs. (50) and (55) into an effective nonlinear Schrödinger equation. To this purpose, let us define the effective wave function

$$\Psi = \sqrt{n(x, t)} \exp(i S(x, t)/\hbar), \tag{57}$$

with $S(x, t)$ defined according to $mu(x, t) = \partial S(x, t)/\partial x$. We obtain that $\Psi(x, t)$ satisfies the equation

$$i\hbar \frac{\partial \Psi}{\partial t} = -\frac{\hbar^2}{2m} \frac{\partial^2 \Psi}{\partial x^2} - e\phi\Psi + W\Psi. \tag{58}$$

By linearizing Eqs. (50) and (55) around a homogeneous equilibrium, we obtain the following dispersion relation

$$\omega^2 = \omega_p^2 + v_0^2 k^2 + \frac{\hbar^2 k^4}{4m^2}, \tag{59}$$

where $mv_0^2 = (dP^C/dn)_{n=n_0}$. It can be proven that, by an appropriate choice of the equation of state $P^C(n)$, Eq. (59) reproduces correctly the leading terms of the Hartree or Wigner dispersion relation.

To summarize, we have shown that, under appropriate conditions, the Hartree or Wigner models can be reduced to a set of two hydrodynamical equations (50) and (55), or, equivalently, to a single nonlinear Schrödinger equation (58). The two hypotheses used for this reduction were that (i) all quantities vary on a length scale larger than L_F and (ii) the equation of state for the classical pressure is $P^C = P^C(n)$ (standard fluid closure).

2.4.1 Example — Thin Metal Films

We have studied the electron dynamics in a thin metal film using the above quantum hydrodynamical model [97]. A preliminary result is shown in Fig. 7, where we plot the evolution of the thermal and potential energies against time. In order to compare to the Vlasov simulations described in Sect. 2.3, the hydrodynamic equations are solved in the semiclassical limit, i.e., using a small value of the Planck constant normalized to E_F/ω_p (note, however, that here the initial excitation $\delta v = 0.22 v_F$ is larger compared to the case of Fig. 4, where $\delta v = 0.08 v_F$). The hydrodynamic results display some coherent oscillations at high frequency, which are a typical signature of quantum effects. Nevertheless, the initial increase of the thermal energy is clearly captured and the subsequent ballistic oscillations are still visible, particularly on the potential energy.

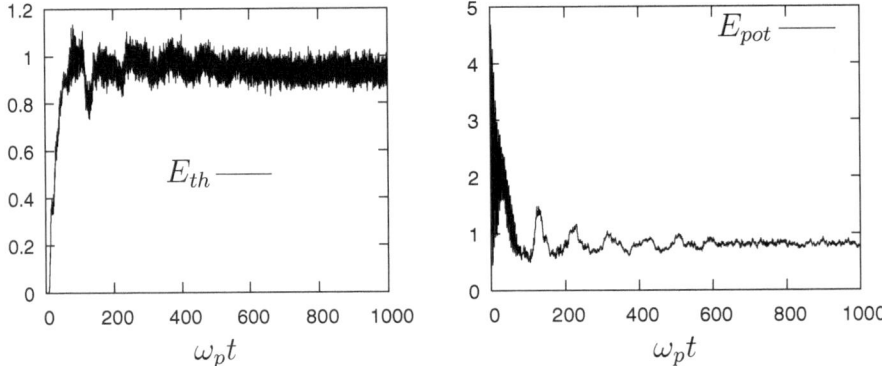

Fig. 7 Time evolution of the thermal and potential energies (normalized to E_F) of the electron population, obtained using a quantum hydrodynamics model

3 Spin Dynamics

The dynamics of magneto-optical processes in metallic nanostructures depends on the temporal and spatial characteristics that are being investigated. Short timescale ($t < 10^{-12}$ s) has only been explored recently. In 1996, the group of Jean-Yves Bigot in Strasbourg highlighted the existence of ultrafast demagnetization processes (within less than a hundred femtoseconds) induced by femtosecond laser pulses in ferromagnetic thin films [26–28]. These demagnetization processes are not yet fully understood.

From a theoretical point of view, very little is known on the time-dependent magneto-optical response of metallic nanostructures to an ultrafast optical pulse. The main difficulty is to provide an adequate description of the interplay between electronic and spin degrees of freedom in the metal. So far, only two theoretical models have been proposed to explain this effect [29, 30]. These works are based on two different mechanisms: in [29], the spin–orbit coupling is invoked, whereas in [30] phonon or impurity mediated spin-flip scattering is privileged. Unfortunately, the parameters employed in [29] are not realistic and the model developed in [30] is a phenomenological approach that does not allow quantitative predictions. From the above considerations it follows that there is a need for the development of efficient theoretical models able to explain in a quantitative manner the experimental findings.

A proper treatment of spin dynamics requires an extension of our model (TDLDA) to include spin degrees of freedom. In the following, the formalism of the time-dependent local-spin-density approximation (TDLSDA) in the linear regime (including also its extension to finite temperature) is presented. A second part will be devoted to the nonlinear dynamics.

3.1 Linear Response: Local-Spin-Density Approximation

The generalization of the linear TDLDA to spin-polarized electron systems has been performed by Rajagopal [98]. In the following we provide the basic equations of this approach including its extension to finite temperature.

Within the framework of DFT one can calculate the spin-density matrix $n_{\sigma\sigma'}(r)$ defined as

$$n_{\sigma\sigma'}(r) = \langle 0|\hat{\psi}_\sigma^+(r)\hat{\psi}_{\sigma'}(r)|0\rangle, \tag{60}$$

where $\hat{\psi}_\sigma^+(r)$ and $\hat{\psi}_\sigma(r)$ are the wave field operators corresponding to the creation and annihilation of an electron with spin σ at position r and $|0\rangle$ is the ground state of the system. When the system is subjected to a small local spin-dependent external potential $\delta V_{ext}^{\sigma\sigma'}(r;\omega)$ (this quantity describes the coupling of the charge and spin of the electrons to external electric and magnetic fields) the spin-density response function is defined through the equation:

$$\delta n_{\sigma\sigma'}(r;\omega) = \sum_{\sigma_1\sigma_2} \int \chi_{\sigma\sigma',\sigma_1\sigma_2}(r,r';\omega)\,\delta V_{ext}^{\sigma_1\sigma_2}(r';\omega)\,dr'. \tag{61}$$

For the sake of simplicity, we restrict ourself to the case of collinear magnetism, i.e., to the case of a uniform direction of magnetization. This restriction leads to a diagonal spin-density matrix ($n_{\sigma\sigma'} = n_\sigma\delta_{\sigma\sigma'}$) and simplified expressions. The spin-density response function defined in Eq. (61) reduces to

$$\delta n_\sigma(\boldsymbol{r}; \omega) = \sum_{\sigma'} \int \chi_{\sigma\sigma'}(\boldsymbol{r}, \boldsymbol{r}'; \omega) \, \delta V_{\text{ext}}^{\sigma'}(\boldsymbol{r}'; \omega) \, d\boldsymbol{r}', \qquad (62)$$

which can be rewritten as

$$\delta n_\sigma(\boldsymbol{r}; \omega) = \sum_{\sigma'} \int \chi_{\sigma\sigma'}^0(\boldsymbol{r}, \boldsymbol{r}'; \omega) \, \delta V_{\text{eff}}^{\sigma'}(\boldsymbol{r}'; \omega) \, d\boldsymbol{r}', \qquad (63)$$

with

$$\delta V_{\text{eff}}^\sigma(\boldsymbol{r}; \omega) = \delta V_{\text{ext}}^\sigma(\boldsymbol{r}; \omega)$$
$$+ \sum_{\sigma'} \int \left\{ \frac{e^2/4\pi\varepsilon_0}{|\boldsymbol{r} - \boldsymbol{r}'|} + f_{\text{xc}}^{\sigma\sigma'}(\boldsymbol{r}, \boldsymbol{r}'; \omega) \right\} \delta n_{\sigma'}(\boldsymbol{r}'; \omega) \, d\boldsymbol{r}'. \qquad (64)$$

In the above expression the function $f_{\text{xc}}^{\sigma\sigma'}(\boldsymbol{r}, \boldsymbol{r}'; \omega)$ is the Fourier transform of the time-dependent kernel defined by $f_{\text{xc}}^{\sigma\sigma'}(\boldsymbol{r}, t; \boldsymbol{r}', t') \equiv \delta V_{\text{xc}}^\sigma(\boldsymbol{r}, t)/\delta n_{\sigma'}(\boldsymbol{r}', t')$, and $\chi_{\sigma\sigma'}^0(\boldsymbol{r}, \boldsymbol{r}'; \omega)$ is the non-interacting retarded spin-density correlation function. For spin-polarized electron systems the exchange–correlation potential is defined as

$$V_{\text{xc}}^\sigma(\boldsymbol{r}) = \left[\frac{\partial}{\partial n_\sigma} \{n\omega_{\text{xc}}(n_+, n_-)\} \right]_{n_+=n_+(r); n_-=n_-(r)}, \qquad (65)$$

where $\Omega_{\text{xc}}[n_+, n_-] = \int n(r)\omega_{\text{xc}}(n_+(r), n_-(r)) \, d\boldsymbol{r}$ is the exchange-correlation thermodynamic potential and ω_{xc} the exchange–correlation thermodynamic potential per particle of the homogeneous electron gas calculated at the local density n and magnetization $m = n_+ - n_-$. By noting that

$$\frac{\partial}{\partial n_\sigma} \{n\omega_{\text{xc}}(n_+, n_-)\} = \frac{\partial}{\partial n} \{n\omega_{\text{xc}}(n, m)\} + \sigma \frac{\partial}{\partial m} \{n\omega_{\text{xc}}(n, m)\},$$

the expression (65) can be rewritten as [99]

$$V_{\text{xc}}^\sigma(\boldsymbol{r}) = \left[\frac{\partial}{\partial n} \{n\omega_{\text{xc}}(n, m)\} \right]_{n=n(r); m=m(r)} + \sigma \mu_B B_{\text{xc}}(\boldsymbol{r}), \qquad (66)$$

where $B_{\text{xc}}(\boldsymbol{r}) = \mu_B^{-1} \left[\frac{\partial}{\partial m} \{n\omega_{\text{xc}}(n, m)\} \right]_{n=n(r); m=m(r)}$ is the exchange–correlation magnetic field acting on spin, and $\mu_B = e\hbar/(2m)$ is the Bohr magneton. This is an *internal* magnetic field. The response functions χ^0 and χ are related by an integral equation (to be more precise, due to the spin degree of freedom, it is a matrix integral equation):

$$
\chi_{\sigma\sigma'}(r, r'; \omega) = \chi^0_{\sigma\sigma'}(r, r'; \omega) + \sum_{\sigma_1\sigma_2} \int\int \chi^0_{\sigma\sigma_1}(r, r''; \omega)
$$

$$
\times\ K^{\sigma_1\sigma_2}(r'', r'''; \omega)\, \chi_{\sigma_2\sigma'}(r''', r'; \omega)\, dr''dr''', \tag{67}
$$

with the residual interaction defined by

$$
K^{\sigma_1\sigma_2}(r, r'; \omega) = \frac{e^2}{4\pi\varepsilon_0|r - r'|}\delta_{\sigma_1\sigma_2} + f_{\mathrm{xc}}^{\sigma_1\sigma_2}(r, r'; \omega). \tag{68}
$$

As for TDLDA, in the *adiabatic* local-density approximation (ALDA) the exchange–correlation kernel is frequency-independent and local and reduces to

$$
f_{\mathrm{xc}}^{\sigma\sigma'}(r, r') = \left[\frac{\partial^2[n\omega_{\mathrm{xc}}(n, m)]}{\partial n_\sigma \partial n_{\sigma'}}\right]_{n=n(r);m=m(r)} \delta\left(r - r'\right). \tag{69}
$$

It should be mentioned that the functional ω_{xc} in the above expression should be the same as the one used in the calculation of the ground state [see Eq. (65)]. By using the same field-theory techniques employed previously for TDLDA (see Sect. 2.2), one can show that the free response function reads

$$
\chi^0_{\sigma\sigma'}(r, r'; \omega; T_e) = \delta_{\sigma\sigma'} \sum_k f_k^\sigma\, \phi_k^{\sigma*}(r)\phi_k^\sigma(r')\, G_+^\sigma(r, r'; \varepsilon_k^\sigma + \hbar\omega; T_e)
$$

$$
+ \sum_k f_k^\sigma\, \phi_k^\sigma(r)\phi_k^{\sigma*}(r')\, G_+^{\sigma*}(r, r'; \varepsilon_k^\sigma - \hbar\omega; T_e), \tag{70}
$$

where $\phi_k^\sigma(r)$ and ε_k^σ are the one-electron Kohn–Sham wave functions and energies, respectively. G_+^σ is the one-particle retarded Green's function for the spins σ and $f_k^\sigma = \left[1 + \exp\left\{(\varepsilon_k^\sigma - \mu)/k_B T_e\right\}\right]^{-1}$. Similarly to TDLDA, we have assumed that the residual interaction (68) is temperature independent. Thus, it is consistent with the use of $\omega_{\mathrm{xc}}(n, m) = \epsilon_{\mathrm{xc}}(n, m)$ in the calculation of the ground-state properties.

From the above formalism one can compute the dipolar absorption cross-section

$$
\sigma(\omega; T_e) = \frac{\omega}{\varepsilon_0 c}\, \mathrm{Im}\left[\alpha(\omega; T_e)\right], \tag{71}
$$

where α is the frequency-dependent dipole *electric* polarizability defined as

$$
\alpha(\omega; T_e) = \int \left[\delta n_+(r; \omega; T_e) + \delta n_-(r; \omega; T_e)\right] \delta V_{\mathrm{ext}}(r; \omega)\, dr. \tag{72}
$$

By analogy, one defines a quantity which is constructed from the local magnetization (instead of the local density)

$$
\sigma_m(\omega; T_e) = \frac{\omega}{\varepsilon_0 c}\, \mathrm{Im}\left[\alpha_m(\omega; T_e)\right], \tag{73}
$$

where α_m is the frequency-dependent dipole *magnetic* polarizability defined as

$$\alpha_m (\omega; T_e) = \int \left[\delta n_+(\boldsymbol{r}; \omega; T_e) - \delta n_-(\boldsymbol{r}; \omega; T_e) \right] \delta V_{\text{ext}}(\boldsymbol{r}; \omega) \, d\boldsymbol{r} \ . \tag{74}$$

On can show that σ_m fulfils the following sum rule

$$\int \sigma_m (\omega; T_e) \, d\omega = \frac{2\pi^2 M(T_e)}{c} \ , \tag{75}$$

where $M = N^+ - N^-$ is the total magnetization of the system (N^+ being the number of spins up and N^- the number of spins down). It is worth mentioning that M is generally temperature dependent [100].

3.2 Nonlinear Response: Phase-Space Methods

In order to investigate the nonlinear regime of the charge and spin dynamics, a phase-space approach is particularly interesting. In this paragraph, we will construct a Wigner equation that includes spin effects in the local-density approximation and show that its classical limit takes the form of a Vlasov equation.

The starting point for the derivation is the time-dependent Kohn-Sham (KS) equations described in Sect. 3.1. In terms of the Pauli 2-spinors

$$\Psi_i(\mathbf{r}, \mathbf{t}) = \begin{pmatrix} \Psi_i^\uparrow(\mathbf{r}, \mathbf{t}) \\ \Psi_i^\downarrow(\mathbf{r}, \mathbf{t}) \end{pmatrix} ,$$

the KS equations can be written as

$$i\hbar \frac{\partial \Psi_i}{\partial t} = \left[\left(-\frac{\hbar^2}{2m} \nabla^2 + V(\mathbf{r}, t) \right) \mathbf{I} + \mu_B \boldsymbol{\sigma} \cdot \mathbf{B}(\mathbf{r}, t) \right] \Psi_i(\mathbf{r}, t) \tag{76}$$

where $V(\mathbf{r}, t) = V_{\text{ext}}(\mathbf{r}, t) + V_H(\mathbf{r}, t) + V_{\text{xc}}^0(\mathbf{r}, t)$, μ_B is Bohr's magneton, $\boldsymbol{\sigma} = (\sigma_x, \sigma_y, \sigma_z)$ are the 2×2 Pauli matrices, and $\mathbf{I}$ is the identity matrix. Here, V_{ext} is an external potential (e.g., ionic jellium, external electric field, ...), V_H is the Hartree potential that obeys Poisson's equation, and V_{xc}^0 is the scalar part of the exchange–correlation potential. The magnetic field $\mathbf{B} = \mathbf{B}_{\text{ext}} + \mathbf{B}_{\text{xc}}$ is composed of an external part and an "internal" part that stems from the exchange and correlation energy [see Eq. (66)]. In the so-called collinear approximation, the latter reduces to $\mathbf{B}_{\text{xc}} = B_{\text{xc}}\hat{z}$.

Equation of Motion for the Density Matrix

By defining the density matrix

$$\rho^{\eta\eta'}(\mathbf{r}, \mathbf{r}') = \sum_i \Psi_i^{\eta}(\mathbf{r})\Psi_i^{\eta'*}(\mathbf{r}'), \tag{77}$$

where $\eta = \uparrow, \downarrow$, the KS equations (76) can be written in the following compact form (Von Neumann equation):

$$i\hbar \frac{\partial \rho}{\partial t} = [H, \rho], \tag{78}$$

where

$$\rho = \begin{pmatrix} \rho^{\uparrow\uparrow} & \rho^{\uparrow\downarrow} \\ \rho^{\downarrow\uparrow} & \rho^{\downarrow\downarrow} \end{pmatrix} ; \quad H = \begin{pmatrix} h^{\uparrow\uparrow} & h^{\uparrow\downarrow} \\ h^{\downarrow\uparrow} & h^{\downarrow\downarrow} \end{pmatrix}. \tag{79}$$

The only nondiagonal terms in the Hamiltonian come from the external or internal magnetic field $\mathbf{B}$.

We now introduce the following basis transformation for the Hamiltonian:

$$H = h_0 \mathbf{I} + \mathbf{h} \cdot \boldsymbol{\sigma}, \tag{80}$$

where $\mathbf{h} = (h_x, h_y, h_z)$ and

$$h_0 = \frac{h^{\uparrow\uparrow} + h^{\downarrow\downarrow}}{2} , \quad h_x = \frac{h^{\uparrow\downarrow} + h^{\downarrow\uparrow}}{2} \tag{81}$$

$$h_z = \frac{h^{\uparrow\uparrow} - h^{\downarrow\downarrow}}{2} , \quad h_y = \frac{h^{\downarrow\uparrow} - h^{\uparrow\downarrow}}{2i}. \tag{82}$$

For the Hamiltonian of Eq. (76), we have

$$h_0(r) = -\frac{\hbar^2}{2m}\nabla^2 + V(r, t) \tag{83}$$

$$h_\alpha(r) = \mu_B B_\alpha(r, t), \quad \alpha = x, y, z. \tag{84}$$

The same transformation (with identical notation) is also applied to the density matrix. With these definitions, the equations of motion for ρ_0 and ρ_α read as

$$i\hbar\partial_t\rho_0 = [h_0, \rho_0] + \sum_{\alpha=x,y,z} [h_\alpha, \rho_\alpha] \tag{85}$$

$$i\hbar\partial_t\rho_\alpha = [h_0, \rho_\alpha] + [h_\alpha, \rho_0]. \tag{86}$$

3.2.1 "Spin" Wigner and Vlasov Equations

By making use of the Wigner transformation

$$f_0(\mathbf{r}, \mathbf{v}, t) = \frac{m}{2\pi\hbar} \int d\lambda \rho_0 \left(\mathbf{r} - \frac{\lambda}{2}, \mathbf{r} + \frac{\lambda}{2}\right) e^{im\mathbf{v}\lambda/\hbar}, \tag{87}$$

$$f_\alpha(\mathbf{r}, \mathbf{v}, t) = \frac{m}{2\pi\hbar} \int d\lambda \rho_\alpha \left(\mathbf{r} - \frac{\lambda}{2}, \mathbf{r} + \frac{\lambda}{2}\right) e^{im\mathbf{v}\lambda/\hbar}, \tag{88}$$

one can easily obtain the equations of motion for the Wigner functions:

$$\frac{\partial}{\partial t} f_0 + \mathbf{v} \frac{\partial}{\partial \mathbf{r}} f_0 -$$

$$\frac{m}{2i\pi\hbar^2} \int d\lambda \int d\mathbf{v}' e^{im(\mathbf{v}-\mathbf{v}')\lambda/\hbar} \left[V\left(\mathbf{r} + \frac{\lambda}{2}\right) - V\left(\mathbf{r} - \frac{\lambda}{2}\right)\right] f_0(\mathbf{r}, \mathbf{v}', t) -$$

$$\sum_\alpha \frac{m\mu_B}{2i\pi\hbar^2} \int d\lambda \int d\mathbf{v}' e^{im(\mathbf{v}-\mathbf{v}')\lambda/\hbar} \left[B_\alpha\left(\mathbf{r} + \frac{\lambda}{2}\right) - B_\alpha\left(\mathbf{r} - \frac{\lambda}{2}\right)\right] f_\alpha(\mathbf{r}, \mathbf{v}', t) = 0,$$

$$\frac{\partial}{\partial t} f_\alpha + \mathbf{v} \frac{\partial}{\partial \mathbf{r}} f_\alpha -$$

$$\frac{m}{2i\pi\hbar^2} \int d\lambda \int d\mathbf{v}' e^{im(\mathbf{v}-\mathbf{v}')\lambda/\hbar} \left[V\left(\mathbf{r} + \frac{\lambda}{2}\right) - V\left(\mathbf{r} - \frac{\lambda}{2}\right)\right] f_\alpha(\mathbf{r}, \mathbf{v}', t) -$$

$$\frac{m\mu_B}{2i\pi\hbar^2} \int d\lambda \int d\mathbf{v}' e^{im(\mathbf{v}-\mathbf{v}')\lambda/\hbar} \left[B_\alpha\left(\mathbf{r} + \frac{\lambda}{2}\right) - B_\alpha\left(\mathbf{r} - \frac{\lambda}{2}\right)\right] f_0(\mathbf{r}, \mathbf{v}', t) = 0.$$

The corresponding Vlasov equations are obtained in the classical limit $\hbar \to 0$:

$$\frac{\partial}{\partial t} f_0 + \mathbf{v} \frac{\partial}{\partial \mathbf{r}} f_0 - \frac{1}{m} \frac{\partial V}{\partial \mathbf{r}} \frac{\partial f_0}{\partial \mathbf{v}} - \frac{\mu_B}{m} \sum_\alpha \frac{\partial B_\alpha}{\partial \mathbf{r}} \frac{\partial f_\alpha}{\partial \mathbf{v}} = 0, \tag{89}$$

$$\frac{\partial}{\partial t} f_\alpha + \mathbf{v} \frac{\partial}{\partial \mathbf{r}} f_\alpha - \frac{1}{m} \frac{\partial V}{\partial \mathbf{r}} \frac{\partial f_\alpha}{\partial \mathbf{v}} - \frac{\mu_B}{m} \frac{\partial B_\alpha}{\partial \mathbf{r}} \frac{\partial f_0}{\partial \mathbf{v}} = 0, \tag{90}$$

with $\alpha = x, y, z$.

Within the collinear approximation, the equations for $\alpha = x, y$ vanish. In this case, it is more convenient revert to the original representation and use

$$f_\uparrow = f_0 + f_z,$$
$$f_\downarrow = f_0 - f_z.$$

The corresponding Vlasov equations then become

$$\frac{\partial}{\partial t} f_\uparrow + \mathbf{v} \frac{\partial}{\partial \mathbf{r}} f_\uparrow - \frac{1}{m} \left(\frac{\partial V}{\partial \mathbf{r}} + \mu_B \frac{\partial B_z}{\partial \mathbf{r}} \right) \frac{\partial f_\uparrow}{\partial \mathbf{v}} = 0, \tag{91}$$

$$\frac{\partial}{\partial t} f_\downarrow + \mathbf{v} \frac{\partial}{\partial \mathbf{r}} f_\downarrow - \frac{1}{m} \left(\frac{\partial V}{\partial \mathbf{r}} - \mu_B \frac{\partial B_z}{\partial \mathbf{r}} \right) \frac{\partial f_\downarrow}{\partial \mathbf{v}} = 0. \tag{92}$$

The above Wigner and Vlasov equations can be used to study the nonlinear spin dynamics in a ferromagnetic nanoparticle or thin film, using numerical techniques similar to those employed for the electron dynamics. In their present form, these equations preserve the total spin and thus cannot be used to describe the loss of magnetization observed in experiments [26–28]. A proper generalization, along the lines of the e–e and e–ph collision operators detailed in Sect. 2.3, would be necessary to account for these effects.

4 Numerical Example: The Nonlinear Many-Electron Dynamics in an Anharmonic Quantum Well

In order to illustrate qualitatively the practical implementation of the models described in the previous sections, we concentrate on a specific – and relatively simple – example. We consider an electron population confined in a one-dimensional anharmonic well defined by the potential

$$V_{\mathrm{conf}}(x) = \frac{1}{2} \omega_0^2 m_* x^2 + \frac{1}{2} K x^4, \tag{93}$$

where m_* is the effective electron mass. The frequency ω_0 can be related to a fictitious homogeneous positive charge of density n_0 via the relation $\omega_0^2 = e^2 n_0 / m_* \varepsilon$. The total potential seen by the electrons is the sum of the confining potential V_{conf} and the Hartree potential, which obeys Poisson's equation

$$\frac{\partial^2 V_H}{\partial x^2} = \frac{e^2}{\varepsilon} \int_{-\infty}^{\infty} f \, dv, \tag{94}$$

where e is the absolute electron charge and ε is the effective dielectric constant. As initial condition, we take a Maxwell–Boltzmann distribution with Gaussian density profile

$$f_0(x, v) = \frac{\bar{n}_e}{\sqrt{2\pi k_B T_e / m_*}} \exp\left(-\frac{m_* v^2 + m_* \omega_0^2 x^2}{2 k_B T_e} \right), \tag{95}$$

with temperature T_e and peak density $\bar{n}_e$.

The electron dynamics is mainly determined by two dimensionless parameters: (i) the "filling fraction" $\eta = \bar{n}_e/n_0 = \omega_p^2/\omega_0^2$, which is a measure of self-consistent effects (in the limit case $\eta = 0$, corresponding to very dilute electron densities, the Hartree potential is negligible); and (ii) the normalized Planck constant $H = \hbar\omega_0/k_B T_e$, which determines the importance of quantum effects. Notice that a small value of H corresponds to a large electron temperature.

We use typical parameters for semiconductor quantum wells [101, 102]: effective electron mass and dielectric constant $m_* = 0.067m_e$ and $\varepsilon = 13\varepsilon_0$; volume density $n_0 = 10^{16}\,\mathrm{cm}^{-3}$, oscillator energy $\hbar\omega_0 = 3.98\,\mathrm{meV}$, and oscillator length $L_{\mathrm{ho}} = \sqrt{\hbar/m_*\omega_0} \simeq 17\mathrm{nm}$. For $\eta = 1$, this yields a maximum surface density for the electrons $n_s = 4.64 \times 10^{10}\,\mathrm{cm}^{-2}$ and a maximum Fermi temperature $T_F = 29.3\,\mathrm{K}$. A low electron temperature $T_e \simeq 46\,\mathrm{K}$ then yields $H \simeq 1$, whereas at room temperature $T_e \simeq 300\,\mathrm{K}$ one has $H \simeq 0.15$.

The electron dynamics is excited by shifting the electron density of a finite distance $\delta x = L_{\mathrm{ho}}$. We will primarily be interested in the relaxation of the electric dipole, defined as the center of mass of the electron population: $d(t) = \int\int fx\,dx\,dv/\int\int f\,dx\,dv$ and of the average kinetic energy $E_{\mathrm{kin}} = \frac{1}{2}\int\int f\, m_*v^2\,dx\,dv/\int\int f\,dx\,dv$.

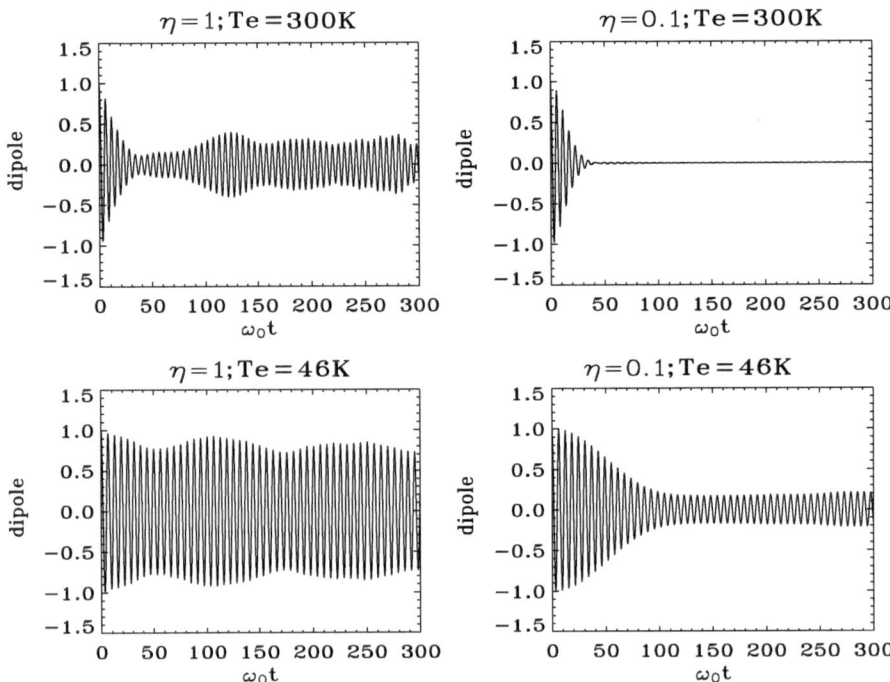

Fig. 8 Evolution of the electric dipole (in units of $L_{\mathrm{ho}} = 17$ nm) obtained from the Wigner–Poisson model, for several values of η and the electron temperature. Time is normalized to the oscillator frequency

First, we present results obtained from the numerical resolution of the Wigner equation (37), coupled to Poisson's equation (94). The results were obtained with a numerical code that combines the split-operator method with fast Fourier transforms in the velocity coordinate [103]. We explore the electron dynamics for different values of the two relevant dimensionless parameters, H and η. The anharmonicity parameter appearing in the confining potential (93) is fixed to $K = 0.1$ (in units where $\hbar = m_* = \omega_0 = 1$). If the confinement were purely harmonic (i.e., $K = 0$), the dipole would simply oscillate at the frequency ω_0 irrespective of the value of the filling fraction. This result goes under the name of Kohn's theorem [36], and we have checked that it holds for our numerical simulations. When the confinement is not harmonic, the dipole should decay because of phase mixing effects.

The numerical results are shown in Fig. 8 (dipole) and Fig. 9 (kinetic energy). The fast oscillations correspond to the center of mass of the electron gas oscillating in the anharmonic well. For low electron densities and large temperatures ($\eta = 0.1$, $T_e = 300$ K), the dipole relaxes to the bottom of the well, $d \simeq 0$, whereas the kinetic energy relaxes to a constant asymptotic value. This is a semiclassical regime where the energy spectrum is almost continuous: the observed relaxation is due to phase mixing effects.

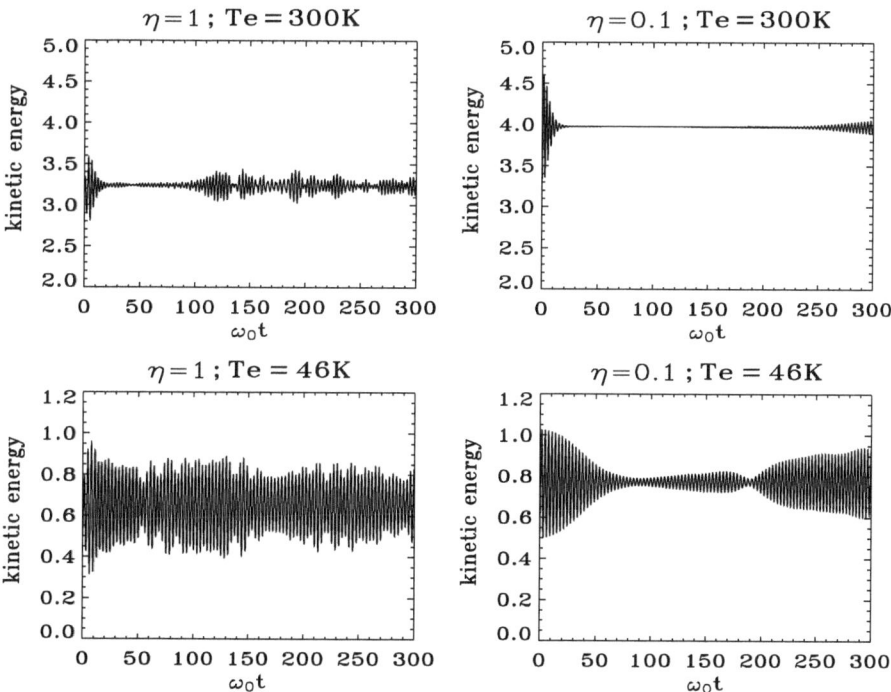

Fig. 9 Evolution of the kinetic energy (normalized to $\hbar\omega_0 = 3.98$ meV) obtained from the Wigner–Poisson model, for several values of η and the electron temperature. Time is normalized to the oscillator frequency

Decreasing the temperature ($T_e = 46$ K) while keeping the density low ($\eta = 0.1$) produces a revival that occurs after the kinetic energy has initially relaxed. This is a typically quantum effect resulting from the discrete nature of the energy spectrum. The revival is clearly visible on the kinetic energy, but not so much on the dipole. When the electron density is large ($\eta = 1$), self-consistent electron–electron interactions (Hartree potential) prevent the dipole and the kinetic energy from relaxing completely, even at large temperatures.

Next, we have added a dissipative term to the Wigner equation, in order to model electron–phonon (e–ph) collisions. This model has been discussed in Sect. 2.3. The relaxation rate is chosen to be $\gamma = 0.001\omega_0$, yielding a realistic relaxation time $\tau_1 = \gamma^{-1} \simeq 165$ps. The velocity–space diffusion coefficient is $D_v = \gamma v_{th}$, where the thermal velocity is $v_{th} = \sqrt{k_B T_e / m_*}$. The relaxation time τ_2 depends on the velocity scale: for instance, a velocity scale Δv is damped on a timescale $\tau_2 = \tau_1 \Delta v / v_{th}$. Therefore, for velocity scales smaller than the thermal velocity, the decoherence time is always smaller than the relaxation time, in accordance with experimental findings.

We simulated the low-temperature scenario ($T_e = 46$ K) in the presence of e–ph collisions and observed that the revival occurring in the kinetic energy for $\eta = 0.1$ is now suppressed (see Fig. 10). For large densities, however, the coherence of the

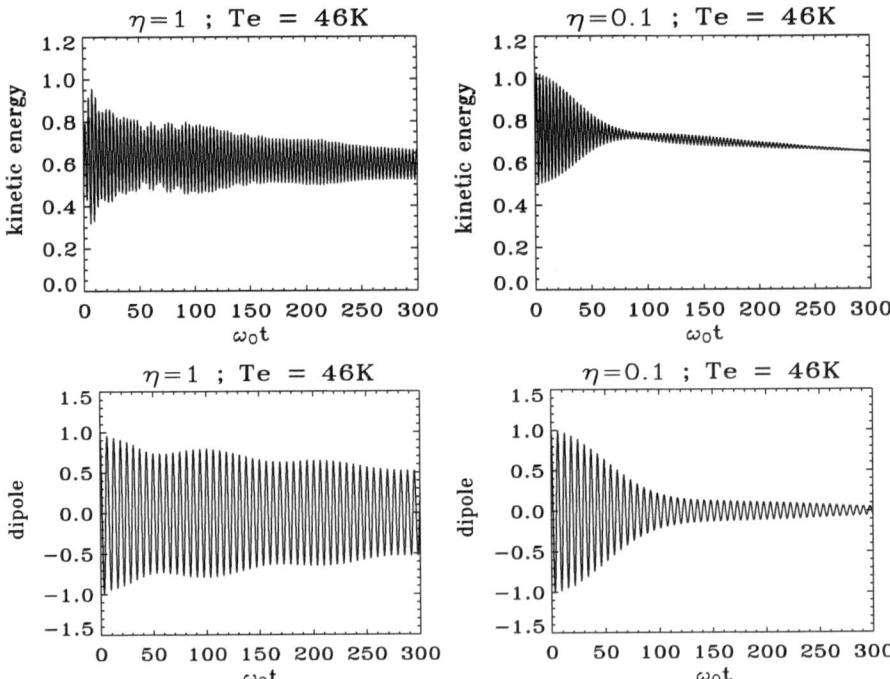

Fig. 10 Evolution of the kinetic energy (*top panels*) and electric dipole (*bottom panels*), from the Wigner–Poisson model including e–ph collisions. Same normalizations as in Figs. 8 and 9

electron motion is not lost, and the relaxation of the dipole and the kinetic energy is only marginally faster compared to the collisionless regime.

Finally, we want to consider the zero temperature case. For doing this, we resort to the hydrodynamical model described in Sect. 2.4. The relevant dimensionless parameters now are η and r_{s0}, the normalized Wigner–Seitz radius computed with the background density n_0. For $n_0 = 10^{16}\,\text{cm}^{-3}$, one has $r_{s0} = 2.8$. In Fig. 11 we plot the evolution of the electric dipole for different values of the filling fraction. Now, even for low electron densities, the dipole oscillates indefinitely without any appreciable decay. For larger electron densities, the motion is even more regular. It appears, therefore, that the dynamics becomes more and more regular as the electron temperature decreases, i.e., when quantum effect become more important. As mentioned above, this is essentially due to phase mixing effect, which become increasingly important in the semiclassical regime, where the energy levels are almost continuous.

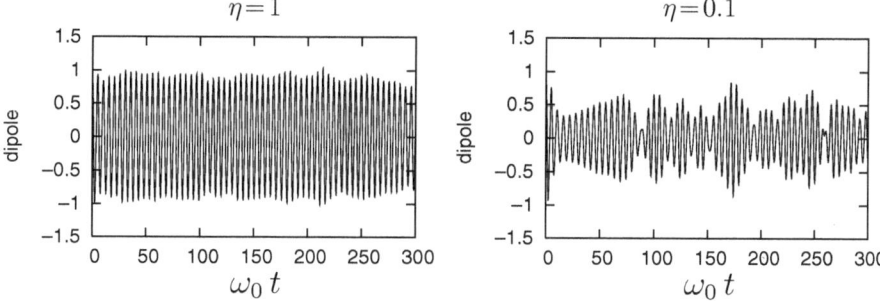

Fig. 11 Evolution of the electric dipole for $\eta = 1$ (*left frame*) and $\eta = 0.1$ (*right frame*), obtained from the quantum hydrodynamic model at $T_e = 0$

5 Conclusions and Perspectives

In this review chapter, we have presented some of the most common theoretical models used to describe the charge and spin dynamics in metallic and semiconductor nanostructures. Three levels of description have been identified (see Fig. 1): (i) the full quantum N-body problem, which can only be addressed for small systems by using, for instance, the Configuration Interaction (CI) method; (ii) mean-field models (Hartree and Wigner) and their generalizations to include exchange and correlations (Hartree–Fock, density functional theory); and (iii) quantum hydrodynamical models, which describe the electron dynamics via a small number of macroscopic variables, such as the density and the average velocity.

Each of these quantum-mechanical approaches has its classical counterpart: classical N-body models have been developed for molecular dynamics computations, as well as for gravitational N-body problems; classical mean-field models are ubiquitous in plasma physics (Vlasov–Maxwell equations) and in the study of self-gravitating objects such as star clusters, galaxies, or even the entire universe; classi-cal hydrodynamics hardly needs mentioning, as it is in itself an extremely wide field of research.

For each approach, we have stressed the difference between the linear and the nonlinear response. The former is valid for weak excitations and presupposes that the response is directly proportional to the excitation. Linear response theory is generally represented in the frequency domain. In contrast, nonlinear effects kick in for large excitations and are best described in the time domain (this is because the time–frequency Fourier transform is a linear operation, thus not adapted to describe nonlinear relations). Although a vast literature on the linear electronic response is available and dates back from the works of Drude in the early twentieth century, nonlinear effects have only been investigated in the last two decades, mainly with computer simulations.

The mean-field level of description is perhaps the most widely used, as it incor-porates, at least to lowest order, some of the features of the N-body dynamics, but still avoids the formidable complexity of the full problem. A particularly chal-lenging open problem is the inclusion of dynamical correlations within mean-field models. Dynamical correlations differ from the correlations that are included in time-dependent density functional theory (TDDFT), inasmuch as they cannot be described by a slowly varying density functional, as is done in ALDA (adi-abatic local-density approximation). Whereas adiabatic correlations are described within an essentially Hamiltonian formulation and thus cannot model irreversible effects, dynamical correlations are responsible for the relaxation of the electron gas toward thermodynamical equilibrium. Some recent results have been obtained using a generalization of TDDFT that relies on the electron *current* as well as the electron density [48]. The phase-space approach, via the Wigner formulation, also appears promising to model effects beyond the mean-field, as we have illustrated in Sect. 2.3.

Another important issue, which was not mentioned earlier in this review, is the inclusion of relativistic corrections in the above models for the electron dynam-ics. Spin–orbit coupling (which is an effect appearing at second order in v/c) is sometimes taken into account in a semi-phenomenological way within the Pauli equation. However, other terms occurring at the same order are often neglected without further justification. A consistent derivation of relativistic effects to a certain order in v/c can of course be carried out, starting from the Dirac equation, for the case of a single particle in an external electromagnetic field [104, 105]. For a many-body system, this issue is much trickier and is the object of current investigations.

Nanostructures are by definition finite-size objects. Due to the presence of bound-aries and interfaces, the electron dynamics can thus display novel and unexpected features compared to bulk matter. For example, as the elastic and inelastic scattering length (~ 10–50 nm for bulk metals) are much longer than the size of the system, an electron – or a group of electrons – can travel coherently through the length of the system, thus leading to ballistic transport between the surfaces. The theoretical tools to study finite-size nano-objects are also relatively recent and have been developed alongside the experimental breakthroughs that made these objects widely available.

If the electron dynamics in nanosized objects has received considerable attention for the last 30 years, the *spin* dynamics is a much younger field of research, both experimentally and theoretically. Nevertheless, the already existing applications to memory storage and processing, and the still speculative, but highly enthralling, developments in quantum computing, have stimulated a large number of works in this direction. In Sect. 3 we have illustrated how the models for the electron dynamics can be extended to include the spin degrees of freedom, both in the linear and in the nonlinear regimes. An outstanding question concerns the demagnetization processes observed in ferromagnetic thin films irradiated with femtosecond laser pulses, for which a clear theoretical explanation is still lacking.

The field of optical control of spins in semiconductor nanostructures is also a very active research area. It is nowadays possible to fabricate and optically probe individual semiconductor quantum dots doped with one or more magnetic impurities [106, 107]. One of the major interest of this type of structure is the possibility to control magnetism via optical processes acting on the charge carriers. Thus, ferromagnetism becomes optically manipulable on an ultrafast timescale. This is particularly interesting for the elaboration of future fast-access magnetic storage devices. We are currently working on quasi one- and two-dimensional nonparabolic quantum dots containing up to four electrons and doped with a finite number of localized magnetic impurities. Within the framework of the CI method and the Anderson model, we aim at investigating the influence of the impurities on the energy spectra and oscillator strengths with special emphasis on the breakdown of the Kohn theorem.

Finally, another procedure that has attracted particular attention over the last decade is the low-density doping of semiconductor nanostructures with magnetic impurities such as manganese ions. The resulting materials (named DMS, for diluted magnetic semiconductors) can display Curie temperatures as high as 80 K [108] and possibly larger [109]. The spin of the Mn ions is coupled to the spin degrees of freedom of the electrons and holes, whose dynamics can be optically excited. DMS thus offer the possibility of using laser pulses to control the magnetization dynamics of semiconductor nanostructures.

Given the wealth of fundamental issues and practical applications, the interplay of charge and spin effects in nanosized objects is bound to remain a major area of research in the coming years.

References

1. G. L. Eesley, Phys. Rev. Lett. **51**, 2140 (1983).
2. S. D. Brorson, J. G. Fujimoto, and E. P. Ippen, Phys. Rev. Lett. **59**, 1962 (1987).
3. C. Suárez, W. E. Bron, and T. Juhasz, Phys. Rev. Lett. **75**, 4536 (1995).
4. R. H. M. Groeneveld, R. Sprik, and A. Lagendijk, Phys. Rev. B **51**, 11433 (1995).
5. C.-K. Sun, F. Vallée, L. H. Acioli, E. P. Ippen, and J. G. Fujimoto, Phys. Rev. B **50**, 15337 (1994).
6. J.-Y. Bigot, V. Halté, J.-C. Merle, and A. Daunois, Chem. Phys. **251**, 181 (2000).
7. M. Bauer and M. Aeschlimann, J. Electr. Spectr. **124**, 225 (2002).

8. W. Rudolph, P. Dorn, X. Liu, N. Vretenar, and R. Stock, Appl. Surf. Sci. **208–209**, 327 (2003).
9. J-S. Lauret, C. Voisin, G. Cassabois, C. Delalande, Ph. Roussignol, O. Jost, and L. Capes, Phys. Rev. Lett. **90**, 057404 (2003).
10. R. Schlipper, R. Kusche, B. v. Issendorff, H. Haberland, Appl. Phys. A **72**, 255–259 (2001).
11. E. E. B. Campbell, K. Hansen, K. Hoffmann, G. Korn, M. Tchaplyguine, M. Wittmann, and I. V. Hertel, Phys. Rev. Lett. **84**, 2128 (2000).
12. C. Voisin, D. Christofilos, N. Del Fatti, F. Vallée, B. Prével, E. Cottancin, J. Lermé, M. Pellarin, and M. Broyer, Phys. Rev. Lett. **85**, 2200 (2000).
13. M. Nisoli, S. Stagira, S. De Silvestri, A. Stella, P. Tognini, P. Cheyssac, and R. Kofman, Phys. Rev. Lett. **78**, 3575 (1997).
14. B. Lamprecht, J. R. Krenn, A. Leitner, and F. R. Aussenegg, Phys. Rev. Lett. **83**, 4421 (1999).
15. U. Kreibig and M. Vollmer, *Optical Properties of Metal Clusters* (Springer, New York, 1995).
16. R. A. Molina, D. Weinmann, and R. A. Jalabert, Phys. Rev. B **65**, 155427 (2002).
17. D. F. Zaretsky, Ph. A. Korneev, S. V. Popruzhenko, and W. Becker, J. Phys. B **37**, 4817 (2004).
18. G. Manfredi and P.-A. Hervieux, Phys. Rev. B **70**, 201402(R) (2004).
19. G. Manfredi and P.-A. Hervieux, Phys. Rev. B 72, 155421 (2005).
20. G. Manfredi and P.-A. Hervieux, Optics Lett. **30**, 3090 (2005).
21. J. F. Dobson, Phys. Rev. B **46**, 10163 (1992).
22. D. Loss, D. P. DiVincenzo, Phys. Rev. A **57**, 120 (1998).
23. P. Zoller et al., Eur. J. Phys. **36**, 203 (2005).
24. J. Gorman, D. G. Hasko, and D. A. Williams, Phys. Rev. Lett. **95**, 090502 (2005).
25. J. R. Petta, A. C. Johnson, C. M. Marcus, M. P. Hanson, and A. C. Gossard, Phys. Rev. Lett. **93**, 186802 (2004).
26. E. Beaurepaire, J.-C. Merle, A. Daunois, J.-Y. Bigot, Phys. Rev. Lett. **76**, 4250 (1996).
27. J.-Y. Bigot, L. Guidoni, E. Beaurepaire, and P. N. Saeta, Phys. Rev. Lett. **93**, 077401 (2004).
28. L. Guidoni, E. Beaurepaire, and J.-Y. Bigot, Phys. Rev. Lett. **89**, 017401 (2002).
29. G. Zhang and W. Hübner, Phys. Rev. Lett. **85**, 3025 (2000).
30. B. Koopmans, J. J. M. Ruigrok et al., Phys. Rev. Lett. **95**, 267207 (2005).
31. R. Balescu, *Equilibrium and Nonequilibrium Statistical Mechanics* (Wiley, New York, 1975).
32. P. Hohenberg and W. Kohn, Phys. Rev. B **136**, B864 (1964).
33. W. Kohn and L. J. Sham, Phys. Rev. **140**, A1133 (1965).
34. E. P. Wigner, Phys. Rev. **40**, 749 (1932).
35. C. D. Sherrill and H. F. Schaefer, Adv. Quantum Chem. **34**, 143 (1999).
36. W. Kohn, Phys. Rev. **123**, 1242 (1961).
37. T. Sako, P.-A. Hervieux, and G. H. F. Diercksen, Phys. Rev. B **74**, 045329 (2006).
38. T. Sako and G. H. F. Diercksen, J. Phys. B: At. Mol. Opt. Phys. **36**, 1433 (2003).
39. T. Sako and G. H. F. Diercksen, J. Phys. B: At. Mol. Opt. Phys. **36**, 1681 (2003).
40. E. Runge and E. K. U. Gross, Phys. Rev. Lett. **52**, 997 (1984).
41. M. E. Casida, *Recent Developments and Applications of Modern Density Functional Theory*, vol. 4, ed. J. M. Seminario (Elsevier, Amsterdam, 1996).
42. E. K. U. Gross, J. F. Dobson, and M. Petersilka, *Topics in Current Chemistry* (Springer, Berlin, 1996), pp. 81–172.
43. S. Lundqvist and N. H. March, *Theory of the Inhomogeneous Electron Gas* (Plenum Press, New York, 1983).
44. G. Onida, L. Reining, and A. Rubio, Rev. Mod. Phys. **74**, 601 (2002).
45. J. F. Dobson, Phys. Rev. Lett. **73**, 2244 (1994).
46. E. K. U. Gross and W. Kohn, Adv. Quantum Chem. **21**, 255 (1990).
47. G. Vignale and W. Kohn, Phys. Rev. Lett. **77**, 2037 (1996).
48. R. D'Agosta and G. Vignale, Phys. Rev. Lett. **96**, 016405 (2006).
49. G. Vignale and W. Kohn, *Electronic Density Functional Theory*, ed. J. Dobson, M. P. Das, and G. Vignale (Plenum Press, New York, 1997).
50. G. Vignale and M. Rasolt, Phys. Rev. Lett. **59**, 2360 (1987).

51. F. Calvayrac, P.-G. Reinhard, E. Suraud, and C. Ullrich, Phys. Rep. **337**, 493 (2000).
52. B. Gervais, E. Giglio, A. Ipatov, J. Douady, Comput. Mater. Sci. **35**, 359 (2006).
53. L. F. Ruiz, P.-A. Hervieux, J. Hanssen, M. F. Politis, and F. Martin, Int. J. Quantum Chem. **86**, 106 (2002).
54. D. Bauer, F. Ceccherini, A. Macchi, and F. Cornolti, Phys. Rev. A **64**, 063203 (2001).
55. E. Cormier, P.-A. Hervieux, R. Wiehle, B. Witzel, and H. Helm, Eur. Phys. J. D **26**, 83 (2003).
56. A. Zangwill and P. Soven, Phys. Rev. A **21**, 1561 (1980).
57. Z. Levine and P. Soven, Phys. Rev. Lett. **50**, 2074 (1983).
58. W. Eckardt, Phys. Rev. B **31**, 6360 (1985).
59. J. Lermé et al., Eur. Phys. J. D **4**, 95 (1998).
60. A. G. Eguiluz, Phys. Rev. Lett. **51**, 1907 (1983).
61. E. Lipparini and Ll. Serra, Phys. Rev. B **57**, R6830 (1998).
62. N. D. Mermin, Phys. Rev. **137**, A1441 (1965).
63. W. Kohn and L. J. Sham, Phys. Rev. **140**, A1133 (1965).
64. U. Gupta and A. K. Rajagopal, Phys. Rep. **87**, 259 (1982).
65. W. Yang, Phys. Rev. A **38**, 5504 (1988).
66. L. D. Landau and E. M. Lifchitz, *Statistical Physics* (Pergamon Press, Oxford, 1969).
67. P.-A. Hervieux, A. Benabbas, V. Halté, and J.-Y. Bigot, Eur. Phys. J. D **24**, 185 (2003).
68. P.-A. Hervieux and J.-Y. Bigot, Phys. Rev. Lett. **92**, 197402 (2004).
69. M. Petersilka et al., Phys. Rev. Lett. **76**, 1212 (1996).
70. W. Yang, Phys. Rev. A **38**, 5512 (1988).
71. G. D. Mahan and K. R. Subbaswamy, *Local Density Theory of Polarizability* (Plenum Press, New York, 1990).
72. J. G. Fujimoto et al., Phys. Rev. Lett. **53**, 1837 (1984).
73. L. Jiang and H.-L. Tsai, J. Heat Transfer **127**, 1167 (2005).
74. F. Vallée, J. Phys. Chem. B **105**, 2264 (2001).
75. N. W. Ashcroft and N. D. Mermin, *Solid State Physics* (Saunders College Publishing, Orlando, 1976).
76. T. N. Truong et al., J. Chem. Phys. **96**, 2077 (1992).
77. J. E. Moyal, Proc. Cambridge Philos. Soc. **45**, 99 (1949).
78. V. I. Tatarskii, Sov. Phys. Usp. **26**, 311 (1983).
79. V. I. Tatarskii, Usp. Fis. Nauk. **139**, 587 (1983).
80. M. Hillery, R. F. O'Connell, M. O. Scully, and E. P. Wigner, Phys. Rep. **106**, 121 (1984).
81. G. Manfredi and M. R. Feix, Phys. Rev. E **53**, 6460 (1996).
82. N. C. Kluksdahl, A. M. Kriman, D. K. Ferry, and C. Ringhofer, Phys. Rev. B **39**, 7720 (1989).
83. P. A. Markowich, C. A. Ringhofer, and C. Schmeiser, *Semiconductor Equations* (Springer, Vienna, 1990).
84. J. E. Drummond, *Plasma Physics* (McGraw- Hill, New York, 1961).
85. J. Lindhard, K. Dan. Vidensk. Selsk. Mat. Fys. Medd. **28**, 1 (1954).
86. X. Liu, R. Stock, and W. Rudolph, Phys. Rev. B **72**, 195431 (2005).
87. M. Anderegg, Phys. Rev. Lett. **27**, 1575 (1971).
88. R. Jasiak, G. Manfredi, and P.-A. Hervieux, New J. Phys. **11**, 063042 (2009).
89. W. S. Fann, R. Storz, H. W. K. Tom, and J. Bokor, Phys. Rev. B **46**, 13592 (1992).
90. E. A. Uehling et al., Phys. Rev. **43**, 552 (1933).
91. A. Domps, P.-G. Reinhard, and E. Suraud Phys. Rev. Lett. **81**, 5524 (1998).
92. G. Lindblad, Commun. Math. Phys. **48**, 119 (1976).
93. D. Pines and P. Nozières, *The Theory of Quantum Liquids* (W. A. Benjamin, New York, 1966).
94. W. H. Zurek, Rev. Mod. Phys. **75**, 715 (2003).
95. A. Isar, A. Sandulescu, H. Scutaru, E. Stefanescu, and W. Scheid, Int. J. Mod. Phys. E **3**, 635 (1994).
96. G. Manfredi and F. Haas, Phys. Rev. B **64**, 075316 (2001).
97. N. Crouseilles, P.-A. Hervieux, and G. Manfredi, Phys. Rev. B **78**, 155412 (2008).

 98. A. K. Rajagopal, Phys. Rev. B **17**, 2980 (1978).
 99. M. I. Katsnelson and A. I. Lichtenstein, J. Phys.: Condens. Matter **16**, 7439 (2004).
100. E. Maurat and P.-A. Hervieux, New Journal of Physics **11**, 103031 (2009).
101. H. O. Wijewardane and C. A. Ullrich, Appl. Phys. Lett. **84**, 3984 (2004).
102. G. Manfredi and P.-A. Hervieux, Appl. Phys. Lett. **91**, 061108 (2007).
103. N. Suh, M. R. Feix, and P. Bertrand, J. Comput. Phys. **94**, 403 (1991).
104. L. D. Landau and E. M. Lifschitz, *Quantum Electrodynamics* (Pergamon Press, Oxford, 1983).
105. P. Strange, *Relativistic Quantum Mechanics* (Cambridge University Press, Cambridge, 1998).
106. L. Besombes et al., Phys. Rev. Lett **93**, 207403 (2004).
107. Y. Leger et al., Phys. Rev. Lett. **97**, 107401 (2006).
108. J. Wang et al., Phys. Rev. Lett. **98**, 217401 (2007).
109. T. Dietl et al., Science **287**, 1019 (2000).

commensurable systems. A complete study of C_{60} dimers is also presented with ature perspective for the study of C_{60} molecular crystals. We will conclude with 1 overview of this work, discussing interaction and transport at metal–organics terfaces from the point of view of applications in the field of molecular electronics.

Introduction

Noncovalent interactions, such as hydrogen bonding or van der Waals (also called ispersion [1, 2]) interactions, become more and more important in modern research. These interactions are of special relevance not only in physics and chemistry but also in biological science. One can find numerous examples in the study of carbon nanostructures like graphene, carbon nanotubes (CNT), or fullerenes (C_{60}) [3–6], π-conjugated molecules and physisorption processes on metallic surfaces [7–9], are gases dimers [10, 11], water molecules dynamics [12, 13], colloidal chemistry, nteractions between biological membranes as protein folding, helicoidal structure of DNA through hydrogen bonding [14], molecular recognition, etc.

Nevertheless, a first-principle determination of such interactions remains an important challenge, especially regarding the case of extended systems. Indeed, these interactions are weak with respect to the covalent interaction (the corresponding energy ranges from some meV to hundreds of meV) and long range (up to some nanometers in some cases). This challenge is related to the complexity of the dispersion interactions, but also to the need to describe accurately the weak "chemical" interaction (see below) between the interacting subsystems. This interaction is related to the overlaps of the electronic densities and becomes complicated to handle when these overlaps are too small. The van der Waals interaction, associated with virtual electronic excitations, is a pure quantum-mechanical effect that can be viewed as an interaction between instantaneous fluctuating dipoles which leads to a long-range correlation energy. In that manner, hydrogen bonding, which is of high importance in biological systems, is also often denominated as van der Waals interaction, due to its dipolar origin. It has to be precised here that this dipolar interaction does not take into account permanent dipole interactions, as it is often the case, and which leads to strong confusion with respect to the physical nature of the van der Waals interaction. Moreover, in most cases, the van der Waals interaction is approximated in the dipolar limit, but one has to bear in mind that a quadrupolar term can also contribute to this energy and can be even dominant in some cases.

Consequently, due to the quantum nature of van der Waals interaction, it is justified and even necessary to develop a first-principle method, in order to characterize precisely the systems where this interaction is important. Therefore, one important problem for such a method is the ability to treat at the same time dispersion and covalent interactions. Indeed, the usual ab initio methods like standard density functional theory (DFT) are able to describe pretty well strong covalent bonding, but usually fail in describing weak interactions accurately. Various techniques have been proposed to overcome this problem and can be classified into three categories, which are semiempirical models, quantum chemistry, and DFT-based models. In

Weak Chemical Interaction and van der Forces: A Combined Density Functiona and Intermolecular Perturbation Theor Application to Graphite and Graphitic S

Y.J. Dappe, J. Ortega, and F. Flores

Abstract In this contribution we address the theoretical understan chemical interactions and of the van der Waals forces, in conjunctio developments in this area and selected applications to nanostructure section, we highlight the importance of these interactions, in physics and also in biology, and we recall early treatments of these issues, as der Waals and London. After a brief review of the existing methods interactions, we present a model based on DFT (for each van der Waa independent system) and an intermolecular perturbation theory that us orbitals basis set. We will first detail a weak overlap expansion (LC perturbation treatment to determine the weak chemical interaction. T show how to implement the van der Waals interaction in the DFT solut dipolar approximation in a perturbation theory. We apply this model t system for weak interactions, i.e., the interaction between two planes In the framework of a minimal basis set that describes each indepen and the weak chemical repulsion, we show that it is necessary to take atomic dipole transitions involving high excited states like $3d$ orbitals describe the van der Waals interaction. We demonstrate how the delic between chemical repulsion and van der Waals attractive interaction giv librium geometry and the binding energy of the system. Moreover, as a of this work, we obtain the adsorption energy of a carbon nanotube on gr adsorption energy of a C_{60} molecule on a carbon nanotube, and the ener molecule encapsulated in a carbon nanotube. This gives us the opportunit

Y.J. Dappe (✉)
Institut de Physique et Chimie des Matériaux de Strasbourg, 23 Rue du Loess, BP Strasbourg Cedex 2, France, yannick.dappe@ipcms.u-strasbg.fr

J. Ortega
Departamento de Física Teórica de la Materia Condensada, Universidad Autónoma Campus de Cantoblanco, 28049 Madrid, Spain, jose.ortega@uam.es

F. Flores
Departamento de Física Teórica de la Materia Condensada, Universidad Autónoma Campus de Cantoblanco, 28049 Madrid, Spain, fernando.flores@uam.es

Dappe, Y.J. et al.: *Weak Chemical Interaction and van der Waals Forces: A Combine Functional and Intermolecular Perturbation Theory – Application to Graphite and Systems.* Lect. Notes Phys. **795**, 45–79 (2010)
DOI 10.1007/978-3-642-04650-6_2   © Springer-Verlag Berlin Heidel

Sect. 1, we will give an overview of the most important existing techniques up to now to describe van der Waals and weak chemical interaction. Special emphasis will be given to DFT-based models due to the high relevancy of DFT in electronic structure calculations. In Sect. 2, we discuss a DFT-based model combined with intermolecular perturbation theory to calculate van der Waals interaction and its application to a reference system such as graphite. This formalism is built in a localized orbital DFT frame, and we will present here our LCAO-S^2 specific model for calculating weak chemical interactions, as well as our approach to calculate the van der Waals forces. The obtained results underline the need to take into account dipolar transitions with high-excited states. In Sect. 4, we propose an extension of the previous model, with application to graphitic systems like graphene, carbon nanotubes (CNT), or fullerenes (C_{60}). We focus especially on lateral interactions between CNT and adsorption of C_{60}, as well as interaction in C_{60}-dimers or molecular crystals. In all these results, we analyze the power law of the van der Waals interaction and try to deduce some useful parameters for classical molecular dynamics. Then, we will conclude with an overview of this work, discussing interaction and transport at metal–organics interfaces from the point of view of applications to the field of molecular electronics.

2 Theory and Existing Models

2.1 General

Due to the complexity of the van der Waals interaction and, as we have underlined before, its quantum nature and its nonlocal characteristics, it represents a real challenge to calculate such interaction. Therefore, a lot of different models have been proposed, with no real satisfactory solution until now, and the reader is referred to review articles from Grimme or Dobson, for example, for more general information about it [15–17]. Our purpose here is to make a brief account of the state of the art of the existing methods with their advantages and difficulties, before explicating our DFT-based model in the next section. Before such description, one has to understand first why standard electronic structure calculation methods do not give correct results. Consider, for example, the Hartree–Fock approximation [18]. In this case, one can determine the Slater determinant of an N-electron system with the minimal energy. Nevertheless, when one tries to treat van der Waals interactions in such a frame, this method is not valid any more, because being a mean-field theory, it does not yield the properties associated with the long-range electronic fluctuations.

2.2 The Lennard-Jones Potential

A first standard way to overcome this problem, which has been extensively and is still used, is the description of both weak chemical interaction and van der Waals interaction by semiempirical methods, like the definition of a general pair interaction potential. The Lennard-Jones potential [19–22] is probably the most commonly used:

$$u(r) = 4\varepsilon \left[-\left(\frac{\sigma}{r}\right)^6 + \left(\frac{\sigma}{r}\right)^{12} \right],$$

(1)

where the parameters ε and σ are chosen empirically to fit the studied material. The r^{-6} term represents the van der Waals interaction, this power being typical from dipole–dipole interactions as we will discuss it later. The r^{-12} term is more arbitrary and is chosen to describe the repulsion at short range, whose origin is mainly the Pauli exclusion principle. This potential describes the interaction energy between two atoms at a distance r. To obtain the total binding energy of the system, one has to sum all the pair energies of the two subsystems, taking into account the inter-atomic distance, which means the geometry of the system. This empirical method has been successful in providing a unified and consistent description of properties which depend on weak chemical interactions and van der Waals dipolar interactions. But although this potential is widely used and has given numerous interesting results, it presents a strong disadvantage: the parameters have to be adjusted for each considered material, and it cannot take into account structural modification of this material, at the atomic level, for example. For example, the deformation of a molecule during an adsorption process, which obviously will affect the interatomic potential, cannot be described correctly through this potential. Moreover, the validity of the fitted parameters is often questionable, because the arbitrarity of the r^{-12} term has often to be compensated by the dipolar term, leading to a bad estimation of the r^{-6} term, the pure van der Waals energy [23, 4].

2.3 Quantum Chemistry Methods

Another way of treating weak interactions is the one proposed in quantum chemistry. The idea is to determine the two interacting subsystems accurately from first-principle methods and then to treat the weak interaction in the frame of perturbation theory [24–27]. As the van der Waals interaction is much weaker than the covalent interaction, this approach is totally justified. Nevertheless, it is important to pay attention to the way this method is applied. In standard perturbation theory, the electronic wavefunction of the total system is not antisymmetric, which can generate some problems. To overcome this difficulty, a lot of different perturbation methods have been developed. A first group is called symmetrized perturbation theory [28–32] where the zero-order wavefunction of the total system is the antisymmetrized product of the wavefunctions of each subsystem. The goal is therefore to handle correctly the non-orthogonality of the basis orbitals in each subsystem. The second group is known as symmetry adapted perturbation theory (SAPT) [33–35], where the zero-order wavefunction of the total system is a simple product of the wavefunctions of each isolated subsystem and the antisymmetry is taken into account at each order of the perturbation expansion. This method is very useful and has provided many interesting results on molecular systems. Nevertheless, it presents the disadvantage, as most of the quantum chemistry methods for weak interaction determination, to be computationally much time and memory consuming, which limits its application to small molecular systems. Similarly, we have to underline the existence of second-order Møller–Plesset theory (MP2) [36–44], which is also a quantum chemistry perturbation theory, but where the zero-order Hamiltonian is the original one, minus the Fock Hamiltonian. This method is very accurate and works quite well for small molecular systems. However, when one wants to deal with extended systems or wants to go further in the perturbation development, the computation time increases in a drastic way and the expressions become really complex to handle. For the sake of completeness, we can also talk about single and double coupled clusters methods, with perturbative triple corrections [CCSD(T)] which go even further in perturbation order [45].

2.4 The DFT-Based Methods

Density functional theory is probably one of the most widely used techniques to describe electronic systems [46, 47]. In principle, according to the Hohenberg–Kohn theorem, with the appropriate functional, it is possible to calculate the ground state energy of every electronic system. Consequently, one should be able with the correct functional to evaluate also weak interactions like van der Waals. Of course, all the problem lies in the determination of the appropriate functional, and especially in the determination of the exchange–correlation potential. Many approximations have been developed to evaluate the exchange–correlation energy and consequently the related potential. The most famous ones are based on a local approximation, as the local density approximation (LDA) or the generalized gradient approximation (GGA) which also takes into account the gradient of the local electronic density [48]. As it was already discussed before, weak chemical interactions as well as van der Waals interactions are weak and long-range interactions that in the frame of DFT cannot be reproduced. In particular, long-range interactions are described by an exponential decay in LDA, far from the r^{-6} power of van der Waals. Even the GGA, as a correction to the LDA in order to obtain a "less local" approximation, fails to describe that power law. On the chemical point of view, the main difficulty is due to the large distance appearing in these weak interacting systems, which results in very small electronic density overlaps. In LDA, for example, this small density is averaged in the whole space, like the homogeneous electron gas, which is obviously not a good representation of the realistic electronic density [49–52]. On the physical point of view, as the van der Waals interaction can be seen as a field interaction between virtual dipoles, these local approximations are not able to describe this process. Moreover, these dipole–dipole interactions can also be seen as exchange of virtual photons between the two systems, involving transitions with highly virtual states, whose description remains out of the range of standard DFT [15, 53].

Despite all these difficulties, there exist many attempts within the DFT frame to describe dispersion interactions. These attempts range from semiempirical extension of DFT until fully first-principle calculation involving the determination of a new functional. In the semiempirical extension, a pairwise atom–atom van der Waals interaction term is added to the DFT calculation. Due to the dipolar nature of van der Waals interaction, the term is of the form $E_{vdW} = -f_d(R)C_6/R^6$, where R is the distance between the given pair of atoms and $f_d(R)$ is a damping function that goes to zero for short distances. C_6 is a coefficient which is adjusted normally to experimental results, as in the case of the Lennard-Jones potential and which depends on the nature of the atom. This is especially the case of the DFT-D approach from Grimme et al. [54, 55] and other approaches [56–59]. The main disadvantage of these methods has been discussed previously in the semiempirical methods section. This model does not take into account appropriately the repulsive part coming from the overlap of the electronic densities, because it is assumed to be correctly treated in the frame of DFT: this introduces inaccuracies for very small overlaps.

The fully first-principle calculation is focused on the determination of a new functional, like the work from Lundqvist et al. [60, 61, 3] able to recover dispersion interactions like van der Waals [16, 62–65]. As it was discussed before, weak interactions are not easy to handle in the frame of DFT, which converts this search of a functional in a very difficult task. Moreover, the obtained functionals present also the disadvantage of being really expensive with respect to computational time and resources. In midway between these two approaches, one can find hybrid methods, involving DFT and perturbation theory, or DFT and short/long-range separation of the interaction. In the first case, the two interacting systems are treated separately with DFT, and the weak interaction is added as a perturbation, with all the difficulties related to the wavefunction symmetry discussed previously. In the second case, the use of the error function allows to decompose the Coulombic interaction into the following way:

$$\frac{1}{r_{12}} = \frac{1 - \text{erf}(\alpha r_{12})}{r_{12}} + \frac{\text{erf}(\alpha r_{12})}{r_{12}}, \tag{2}$$

where r_{12} is the distance between the two interacting systems and α a fitting parameter. The idea of such decomposition is to express the exchange-correlation functional as a sum of two functionals, one of short range (like LDA, for example) and the other of long range, to recover van der Waals interaction, for example [66–70]. The obtained exchange–correlation energy can be written in the following way:

$$E^{xc}[\rho(\mathbf{r})] = E^{xc}_{LDA}[\rho(\mathbf{r})] + E^{xc}_{vdW}[\rho(\mathbf{r})]. \tag{3}$$

This way of doing things can bring intricacies for intermediate distances, where it is difficult to say if it is a covalent zone (where LDA can describe correctly the interaction) or if it is a weak interaction range, where dispersion interactions are dominant.

In the following section we will present our DFT-based model developed to describe the interaction between two planes of graphene; this will be generalized in the fourth section to all kinds of graphitic materials like carbon nanotubes or C_{60} molecules.

3 DFT and Intermolecular Perturbation Theory: LCAO-S^2 + vdW, Application to Graphene

3.1 General Frame: The LCAO-OO Formalism

In this section, we present an approach that combines DFT in a localized orbital formalism with intermolecular perturbation theory to take into account weak chemical interactions as well as van der Waals interaction. Our general theoretical framework is the linear combination of atomic orbitals-orbital occupancy (LCAO-OO) method [71–73], which allows us to connect local orbital DFT with intermolecular perturbation theory in second quantization formalism. The use of the second quantization formalism, among other advantages, prevents from the symmetry problems encountered in usual perturbation theory as it was mentioned previously. Our approach has been previously used to analyze van der Waals forces in rare gas dimers [74] and the interaction of two graphene layers [53]; in this section we outline the main ideas of our approach, as applied to this last case (details can be found in [53]).

Let us start with the general LCAO-OO Hamiltonian

$$\hat{H} = \sum_{v,\sigma}(\epsilon_v + V_{vv,\sigma}^{ps})\hat{n}_{v,\sigma} + \sum_{\mu \neq v,\sigma}(t_{\mu v,\sigma} + V_{\mu v,\sigma}^{ps})\hat{c}_{\mu\sigma}^{\dagger}\hat{c}_{v\sigma} +$$

$$+ \frac{1}{2}\sum_{v\omega\sigma\mu\lambda\sigma'}O_{\omega\lambda}^{v\mu}\hat{c}_{v\sigma}^{\dagger}\hat{c}_{\mu\sigma'}^{\dagger}\hat{c}_{\lambda\sigma'}\hat{c}_{\omega\sigma}, \tag{4}$$

where the creation and annihilation operators $\hat{c}^+$, $\hat{c}$ as well as the occupation number operator $\hat{n} = \hat{c}^+\hat{c}$ are defined in a Löwdin orthonormal basis set $\{\phi_\mu\}$. This orthonormal basis set is defined from an original basis set of optimized atomic-like orbitals $\{\psi_v\}$ by the so-called Löwdin orthogonalization procedure

$$\phi_\mu = \sum_v (S^{-1/2})_{\mu v}\psi_v, \tag{5}$$

where $S_{\mu v} = < \psi_\mu \mid \psi_v >$ is the overlap matrix.

In Eq. (4) $\varepsilon_{i\mu} + V_{i\mu,i\mu}^{PS}$ and $t_{i\mu,jv} + V_{i\mu,jv}^{PS}$ define the one-electron terms (with the pseudopotential (PS) contributions included) and

$$O_{\omega\lambda}^{v\mu} = \int \phi_v(\bar{r})\phi_\omega(\bar{r})\frac{1}{\mid \bar{r} - \bar{r}' \mid}\phi_\mu(\bar{r}')\phi_\lambda(\bar{r}')d\bar{r}d\bar{r}' = (v\omega|\mu\lambda) \tag{6}$$

are the electron–electron terms. As we underlined before, the use of second quantization ensures that antisymmetry is properly included in the calculation of the interlayer interaction. In the LCAO-OO formalism, Hamiltonian (4) is rewritten as

$$\hat{H} = \hat{H}_0 + \delta\hat{H}, \tag{7}$$

$$\hat{H}_0 = \sum_{\nu\sigma}(\epsilon_\nu + V^{ps}_{\nu\nu,\sigma})\hat{n}_{\nu\sigma} + \sum_{\nu\neq\mu,\sigma}\hat{T}_{\nu\mu,\sigma}\hat{c}^\dagger_{\nu\sigma}\hat{c}_{\mu\sigma} + \sum_\nu U_\nu\hat{n}_{\nu\uparrow}\hat{n}_{\nu\downarrow} +$$
$$+\frac{1}{2}\sum_{\nu,\mu\neq\nu,\sigma}\left[J_{\nu\mu}\hat{n}_{\nu\sigma}\hat{n}_{\mu\bar{\sigma}} + (J_{\nu\mu} - J^x_{\nu\mu})\hat{n}_{\nu\sigma}\hat{n}_{\mu\sigma}\right], \tag{8}$$
$$\hat{T}_{\nu\mu,\sigma} = [t_{\nu\mu} + V^{ps}_{\nu\mu,\sigma} + \sum_{\lambda,\sigma'}h_{\lambda,\nu\mu}\hat{n}_{\lambda\sigma'} - \sum_\lambda h^x_{\lambda,\nu\mu}\hat{n}_{\lambda\sigma}].$$

In $\hat{H}_0$ the many-body terms are written explicitly showing the contributions depending on one, two, and three different orbitals. In particular, $U_\nu = (\nu\nu \mid \nu\nu)$, $J_{\nu\mu} = (\nu\nu \mid \mu\mu)$, $J^x_{\nu\mu} = (\nu\mu \mid \nu\mu)$, $h_{\lambda,\nu\mu} = (\lambda\lambda \mid \nu\mu)$, $h^x_{\lambda,\nu\mu} = (\lambda\nu \mid \lambda\mu)$, see Eq. (6). A deeper interpretation of the energy associated with each term can be found in [71–73]. The vdW interaction $\hat{H}_{vdW}$ is included in $\delta\hat{H}$. Regarding our system of interest, the graphene–graphene interaction, $\hat{H}_0$, takes into account the covalent interaction inside each graphene plane and the weak chemical interaction between graphene layers.

3.2 DFT Solution for Each Subsystem: The Fireball Code

Our treatment is based on a DFT solution for each isolated subsystem (in the present case each plane of graphene) to which we add intermolecular perturbation theory as discussed later. This DFT solution, obtained in the frame of the orbital occupancy method, is based on an alternative approach to DFT, in which instead of the traditional electronic density $\rho(r)$, we use the orbital occupancy $n_{\mu\sigma}$ as the central quantity [75]:

$$\rho(r) \Longrightarrow \{n_{\mu\sigma}\}.$$

In usual DFT, the Hohenberg–Kohn theorem tells that the total energy of the fundamental state of an electronic system is a functional of the electronic density. In our formalism, this total energy is now a function of the orbital occupation number, $E = E\left[\{n_{\mu\sigma}\}\right]$. We can then rewrite Kohn–Sham-like equations to solve the new effective one-electron problem [71, 75]. Similarly to standard density-based DFT, all the difficulty lies in the determination of the exchange–correlation energy, or potential, which is here a function of the orbital occupancies, $E^{XC}[\{n_{\nu\sigma}\}]$. In this chapter we outline how van der Waals and weak chemical interaction can be incorporated in DFT using the LCAO-OO formalism. A more detailed description of the LCAO-OO method can be found in [71–73].

In this work, for simplicity reasons, we describe each isolated subsystem using the DFT code Fireball [76–80], which can be viewed as an efficient simplified version of the more general LCAO-OO formalism. In similarity to the LCAO-OO method, self-consistency is achieved in Fireball in terms of the occupation numbers $n_{\mu\sigma}$, using a self-consistent version of the Harris functional [77] instead of the traditional Kohn–Sham functional based on the electronic density. To define these occupation numbers, we use an optimized atomic-like orbital basis set. In [81] an optimized minimal basis set for carbon was obtained, considering various carbon phases as well as several hydrocarbon molecules; this basis set was optimized for the covalent interactions in those systems (i.e., the basis set optimization did not take into account weak interactions). In particular for the sp^3 basis set of the carbon, the optimized numerical atomic-like orbitals ψ are

$$\psi(\mathbf{r}) = A\left[c\psi_0(\mathbf{r}) + (1 - c)\psi_1(\mathbf{r})\right], \tag{9}$$

(A is a normalizing constant) where $\psi_0(\mathbf{r})$ is the standard Fireball orbital for a neutral atom [76] and $\psi_1(\mathbf{r})$ corresponds to a doubly excited (2+) atom. In both calculations we have used a cutoff radius of $R_c = 4.5$ a.u. for the s and p orbitals. The parameter c is chosen to optimize the total energy for a single graphene layer; at the same time, the optimized orbitals yield significantly improved structural parameters: the C–C distance inside the graphene layer is 1.43 Å, close to the experimental distance 1.42 Å; for the sake of comparison, the non-optimized basis set, i.e., $c = 1$ in Eq. (9), gives a value of 1.48 Å for the in-plane C–C distance. The optimized orbitals are shown in Fig. 1.

A comparison of the Fireball and LCAO-OO approaches has been made recently, using the same optimized basis set, and we have found that both yield similar results

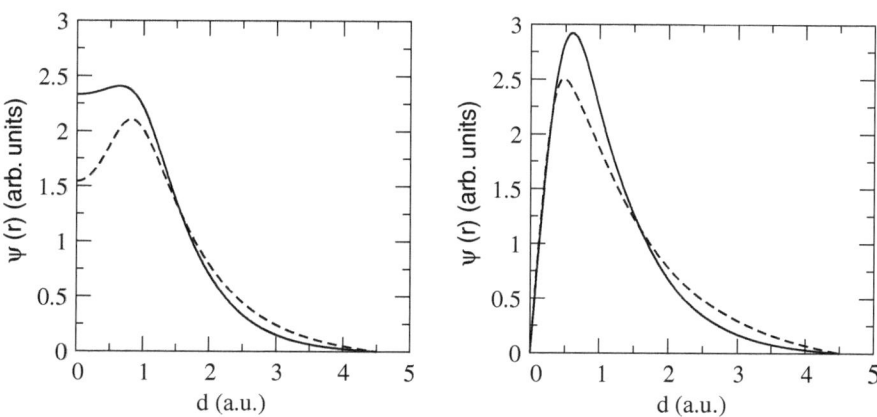

Fig. 1 Optimized (*solid line*) and standard (*dashed line*) Fireball atomic-like orbitals (radial component) for carbon, see Eq. (9): (*left*) s orbital , (*right*) p orbital

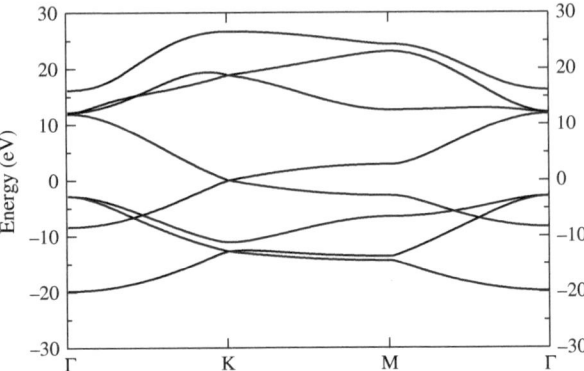

Fig. 2 Band structure of an isolated plane of graphene calculated with Fireball. The origin of energy corresponds to the Fermi level of the system

[73]. Finally, we mention that in the Fireball calculations pseudo-potentials are used [82, 83], and the LDA exchange–correlation energy is calculated using the multicenter weighted exchange–correlation density approximation (McWEDA) [79, 80].

In this frame, the eigenstates are therefore defined by

$$\varphi_n(\boldsymbol{k}) = \sum_i c_{ni}(\boldsymbol{k})\phi_i^0 = \sum_i a_{ni}(\boldsymbol{k})\psi_i \tag{10}$$

and eigenvalues $\varepsilon_n(\boldsymbol{k})$, as well as the orbital occupation numbers $\{n_{i,\sigma}\}$, for the effective DFT problem for each independent layer. In Eq. (10) $\boldsymbol{k}$ is the momentum parallel to the graphene planes, n the band index, and ϕ_i^0 the orthonormal basis orbitals within each layer, i.e., obtained using Eq. (5) for each isolated layer. A representation of the band structure of a graphene plane obtained with Fireball is shown in Fig. 2.

3.3 Weak Chemical Interaction: The LCAO-S^2 Method

As mentioned before, the weak interaction between two graphene layers is mainly due to two contributions: an attractive van der Waals interaction and a repulsive "weak chemical interaction," which is often neglected or assimilated to van der Waals interaction, and which can be viewed as a residue of the strong covalent interaction which occurs at smaller distances. This repulsion arises mainly from orthogonalization effects between the molecular wavefunctions of each subsystem, which means that it is directly related to the overlap (S) between these wavefunctions. As these overlaps are really small in this case (for example, in the case of two graphene planes, at the equilibrium distance, all the overlaps between atomic-like orbitals in different layers are less than 0.01), we will use an overlap (S^2) expansion to obtain the corresponding energy [84–87] as illustrated in Fig. 3. This expansion is based on a development in S of the $S^{-1/2}$ term appearing in the Löwdin orthogonalization presented in Eq. (5).

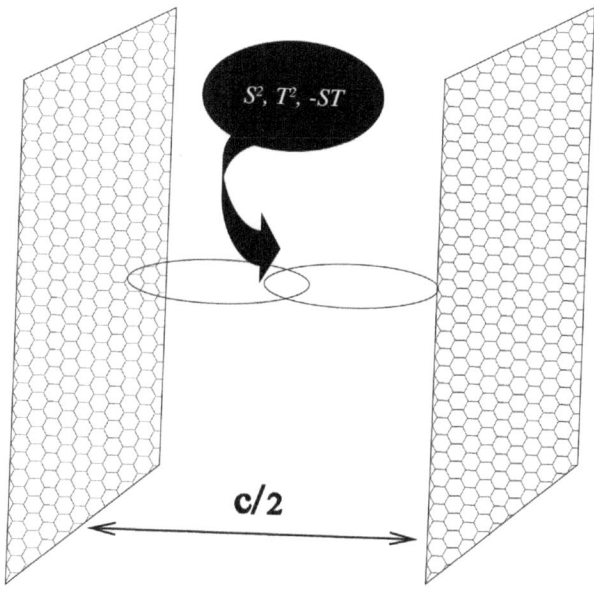

Fig. 3 Representation of the weak chemical interaction in the LCAO-S^2 model between two graphene planes

Now we will analyze this "chemical" intermolecular interaction starting from our Hamiltonian $\hat{H}_0$ defined in Eq. (8); this Hamiltonian includes the contributions of each subsystems, here the two planes of graphene, and an intermolecular contribution that contains both one-electron and many-body contributions [the van der Waals interaction is included in the term $\delta\hat{H}$ of Eq. (8)]. For the systems considered in this chapter, the many-body terms yield an almost negligible contribution, as we will see in the results section; a detailed explanation of how to calculate this contribution can be found in [53]. In this chapter we focus now on the important one-electron term which yield the main contribution to the "chemical" interaction.

Let us consider the eigenstates $\{\varphi_n(\boldsymbol{k})\}$ (first layer) and $\{\varphi_m(\boldsymbol{k})\}$ (second layer) obtained from the DFT calculation for each independent graphene layer, see Eq. (10). These eigenstates are already orthogonal to the rest of eigenstates for the same layer, but there exists an overlap $S_{nm}(\boldsymbol{k})$ between eigenstates in different layers. Due to orthogonalization requirements, this overlap induces a shift of the occupied eigenenergies of each independent plane of graphene, leading to a repulsion energy between the planes:

$$\delta^S\varepsilon_n(\boldsymbol{k}) = -\sum_m \frac{1}{2}\left[S_{nm}(\boldsymbol{k})T_{mn}(\boldsymbol{k}) + T_{nm}(\boldsymbol{k})S_{mn}(\boldsymbol{k})\right]$$
$$+\frac{1}{4}\sum_m |S_{nm}(\boldsymbol{k})|^2(\varepsilon_n(\boldsymbol{k}) - \varepsilon_m(\boldsymbol{k})),$$

$$\delta^S \varepsilon_m(\mathbf{k}) = -\sum_n \frac{1}{2} \left[S_{mn}(\mathbf{k}) T_{nm}(\mathbf{k}) + T_{mn}(\mathbf{k}) S_{nm}(\mathbf{k}) \right]$$

$$+ \frac{1}{4} \sum_n |S_{nm}(\mathbf{k})|^2 (\varepsilon_m(\mathbf{k}) - \varepsilon_n(\mathbf{k})), \tag{11}$$

where $S_{nm}(\mathbf{k})$ and $T_{mn}(\mathbf{k})$ are the overlap and hopping integrals, respectively, between eigenvectors m and n in different layers, for a given $\mathbf{k}$. The *orthogonal* hopping terms $T_{mn}(\mathbf{k})$ are obtained from the *non-orthogonal* hopping terms T_{mn}^0 between eigenstates n and m of the isolated graphene layers:

$$T_{mn}(\mathbf{k}) = T_{mn}^0(\mathbf{k}) - \frac{1}{2} S_{mn}(\mathbf{k}) \left[\varepsilon_m(\mathbf{k}) + \varepsilon_n(\mathbf{k}) \right], \tag{12}$$

where the second term in the right is the correction due to the small overlap between eigenstates in different layers, $S_{nm}(\mathbf{k})$. The hoppings T_{mn}^0 are directly calculated using the local-orbital code Fireball [76–80] that we use for the DFT calculation of each layer.

The effect of the hopping $T_{mn}(\mathbf{k})$ matrix elements can be calculated in a standard intermolecular second-order perturbation theory:

$$\delta^T \varepsilon_n(\mathbf{k}) = \sum_m \frac{|T_{mn}(\mathbf{k})|^2}{\varepsilon_n(\mathbf{k}) - \varepsilon_m(\mathbf{k})},$$

$$\delta^T \varepsilon_m(\mathbf{k}) = \sum_n \frac{|T_{mn}(\mathbf{k})|^2}{\varepsilon_m(\mathbf{k}) - \varepsilon_n(\mathbf{k})}. \tag{13}$$

Thus, we obtain the following "one-electron" contribution to the interaction energy:

$$E_{\text{one-electron}} = 2 \sum_{n=\text{occ.}} (\delta^S \varepsilon_n + \delta^T \varepsilon_n) + 2 \sum_{m=\text{occ.}} (\delta^S \varepsilon_m + \delta^T \varepsilon_m), \tag{14}$$

where a factor of 2 has been included to take into account the spin degeneracy and only filled states are considered. The different contributions to the chemical interaction between two graphene layers will be shown and discussed in the result section.

3.4 van der Waals Interaction

We will now discuss how to incorporate van der Waals interaction in our formalism. Before describing the method we use to evaluate this interaction, we will try to precise what is exactly this interaction. If we refer to the London vision of dispersion or dipolar interactions like van der Waals [1, 2, 88], we can consider three categories of interactions: the first one is the interaction between permanent dipoles, like what we have in the water molecule, for example. The second case is the interaction between a permanent dipole and an electronic density. In such a case, the permanent dipole polarizes the electronic cloud which gives rise to an induced dipole; this induced dipole interacts itself with its originating permanent dipole. These two cases are not of interest for the situations we consider like graphene–graphene interaction (since there is no permanent dipole in the graphite), and more generally for the study of pure van der Waals interactions which do not involve permanent dipoles. The third category is the interaction between two induced dipoles, and this is the case we will consider in that work.

As pointed out above, in the LCAO-OO method the van der Waals interaction is included in $\delta\hat{H}$:

$$\delta\hat{H} = \frac{1}{2} \sum_{\nu\omega\sigma\mu\lambda\sigma'} O_{\omega\lambda}^{\nu\mu} \hat{c}_{\nu\sigma}^{\dagger} \hat{c}_{\mu\sigma'}^{\dagger} \hat{c}_{\lambda\sigma'} \hat{c}_{\omega\sigma}, \tag{15}$$

where μ, ν, ω, and λ refer to four different orbitals. This term is of course really difficult to handle in a general way and includes, in particular, the van der Waals contribution, which in our approach corresponds to the following term:

$$\hat{H}^{\text{vdW}} = \sum_{i,j,\alpha,\beta,\sigma_1,\sigma_2} J_{i,j;\alpha,\beta}^{\text{vdW}} \hat{c}_{i,\sigma_1}^{+} \hat{c}_{j,\sigma_1} \hat{c}_{\alpha,\sigma_2}^{+} \hat{c}_{\beta,\sigma_2}, \tag{16}$$

with $J_{i,j;\alpha,\beta}^{\text{vdW}} = (ij \mid \alpha\beta)$, see Eq. (6), where i, j orbitals ($i \neq j$) belong to the first graphene layer and α, β ($\alpha \neq \beta$) to the second graphene layer. In our work we have used an atom–atom approximation, keeping in Eq. (16) only the terms with i, j orbitals in the same atom, and α, β in the same atom of the other layer, and have neglected all the other interlayer interactions from $\delta\hat{H}$. Also orbitals $\{\psi\}$ have been used for the calculation of $J_{i,j;\alpha,\beta}^{\text{vdW}}$, instead of orbitals $\{\phi\}$ for simplicity.

This first approximation is discussed by Dobson et al. [15] and seems not to be valid in the metallic case. In the case of graphene–graphene interaction as well as for metallic CNTs, the question is more polemic at large distance due to the zero gap in the K-point and the weak metallic character at graphitic equilibrium distance. All this problem is related to the screening of the van der Waals interaction between the subsystems. In the metallic case, the energies associated to the dipolar transitions are really small, and therefore the screening is important. In the graphene case as we will see later, the energies associated to the virtual fluctuations which lead to van der Waals interaction are very important (up to 50 eV for the $3d$ band). Consequently, we obtain a high-frequency screening which can be neglected here, the dielectric function $\varepsilon(\mathbf{r}, \omega)$ going to unity. In that manner, we can say that there is no important collective effect like plasmon frequency shift [89] which validates our atom–atom approximation. Of course, this approach is not valid in the case of two interacting metals with high screening.

The next step is the calculation of the four-center Coulombic integral $J_{i,j;\alpha,\beta}^{\text{vdW}}$. In the present case it can be easily calculated using a dipolar approximation. This approximation is totally justified, due to the standard equilibrium distances around 3 Å, which leads, as already discussed, to really small overlaps. In this classical dipole–dipole approximation, the resulting $J_{i,j;\alpha,\beta}^{\text{vdW}}$ integral is calculated as

$$J_{i,j;\alpha,\beta}^{\text{vdW}} = \frac{1}{R^3} \left(<i|x|j><\alpha|x'|\beta> + <i|y|j><\alpha|y'|\beta> \right.$$
$$\left. -2 <i|z|j><\alpha|z'|\beta> \right) \tag{17}$$

(R is the distance between the two atoms assumed in this expression along the z-axis), which depends on the different dipolar matrix elements in each atom. Of course, this approximation would not be valid anymore if we would like to evaluate van der Waals interaction at really short distances, but this is not the case for this work. Moreover, as it was discussed above, this $J_{i,j;\alpha,\beta}^{\text{vdW}}$ represents the bare van der Waals interaction, without screening, as approximated in our model.

The van der Waals energy between the two subsystems is then calculated using second-order perturbation theory. The van der Waals energy is weak with respect to the covalent energy, which justifies the use of this approximation. In that frame, we can now easily find the following van der Waals interaction energy, in terms of the different eigenstates, $k_n = \varphi_n(k)$ and $k_m = \varphi_m(k)$, of the two layers [see Eq. (10)]:

$$E^{\text{vdW}} = 4 \sum \frac{\left| W(k_{n_1}, k_{n_2}, k_{m_1}, k_{m_2}) \right|^2}{(\varepsilon(k_{n_1}) - \varepsilon(k_{n_2}) + \varepsilon(k_{m_1}) - \varepsilon(k_{m_2}))}, \tag{18}$$

where the sum runs through occupied eigenstates k_{n_1}, k_{m_1} and empty eigenstates k_{n_2}, k_{m_2}, the factor 4 includes the spin degeneracy of both layers, and

$$W(k_{n_1}, k_{n_2}, k_{m_1}, k_{m_2}) = \sum_{i,j;\alpha,\beta} c_i(k_{n1}) c_j^*(k_{n2}) c_\alpha(k_{m1}) c_\beta^*(k_{m2}) J_{i,j;\alpha,\beta}^{\text{vdW}} \tag{19}$$

with $i \neq j$ on the same atom of the first layer and $\alpha \neq \beta$ on the same atom of the second layer.

Momentum conservation imposes the following condition:

$$k_{n_1} + k_{m_1} = k_{n_2} + k_{m_2}.$$

As $J_{i,j;\alpha,\beta}^{\text{vdW}}$ only includes terms having (i, j) or (α, β) in the same atom, Eq. (18) can be approximated by

$$E^{\text{vdW}} = 4 \sum_{i,j,\alpha,\beta} (J_{i,j,\alpha,\beta}^{\text{vdW}})^2 \int \frac{\rho_i(\varepsilon_1)\rho_j(\varepsilon_2)\rho_\alpha(\varepsilon_3)\rho_\beta(\varepsilon_4)}{(\varepsilon_1 - \varepsilon_2 + \varepsilon_3 - \varepsilon_4)} d\varepsilon_1 d\varepsilon_2 d\varepsilon_3 d\varepsilon_4, \tag{20}$$

where $\rho(\varepsilon)$ represents the local density of states per spin on each orbital; the integrals in ε_1, ε_3 run through the occupied states and the integrals in ε_2, ε_4 along the empty states. This expression can be further simplified to express the result in terms of the occupation numbers of each state:

$$E^{\text{vdW}} = 4 \sum_{i,j,\alpha,\beta} (J_{i,j,\alpha,\beta}^{\text{vdW}})^2 \frac{n_i(1 - n_j)n_\alpha(1 - n_\beta)}{(\overline{e}_i - \overline{e}_j + \overline{e}_\alpha - \overline{e}_\beta)}. \tag{21}$$

In this expression, n_i are the orbital occupation numbers (per spin):

$$n_i = \int_{\text{occupied}} \rho_i(\varepsilon)d\varepsilon \tag{22}$$

and

$$\overline{e}_i = \int_{\text{occupied}} \varepsilon\rho_i(\varepsilon)d\varepsilon \bigg/ \int_{\text{occupied}} \rho_i(\varepsilon)d\varepsilon \tag{23}$$

$$\overline{e}_j = \int_{\text{empty}} \varepsilon\rho_j(\varepsilon)d\varepsilon \bigg/ \int_{\text{empty}} \rho_j(\varepsilon)d\varepsilon \tag{24}$$

are average occupied and empty levels. Using these expressions and the DFT band structures of each plane obtained previously with Fireball, we can evaluate the van der Waals energy between the planes. By combining this energy with the weak chemical repulsion obtained in the LCAO-S^2 approach, we can determine the binding energy of two graphene planes and compare it accurately with experimental results and other theoretical determinations. This is the goal of the next section.

3.5 Results and Discussion for the Graphene–Graphene Interaction

In this section we present the results obtained with the theory described above as applied to the interaction of two graphene layers in the graphite parallel configuration. One has to bear in mind that we study here the so-called AB stacking configuration for the two graphene planes, which is the most favorable configuration energetically. In these calculations we have used an optimized sp^3 basis set for carbon, as discussed above, and each graphene layer is calculated using the local-orbital DFT–LDA code Fireball [76–80].

First of all we present the results of weak chemical energy calculations. This energy is represented in Fig. 4. We can see the different contributions discussed above, versus the interplane distance, i.e., the one-electron part and the many-body part (exchange and Hartree; notice that the exchange term also arises from the LCAO-S^2 expansion; more details can be found in [53]). This figure shows that the Hartree and exchange contributions are marginally attractive, but much smaller than the repulsion arising from the "one-electron" terms: the interaction energy without vdW is repulsive for all distances. The inset in Fig. 4 shows the contributions

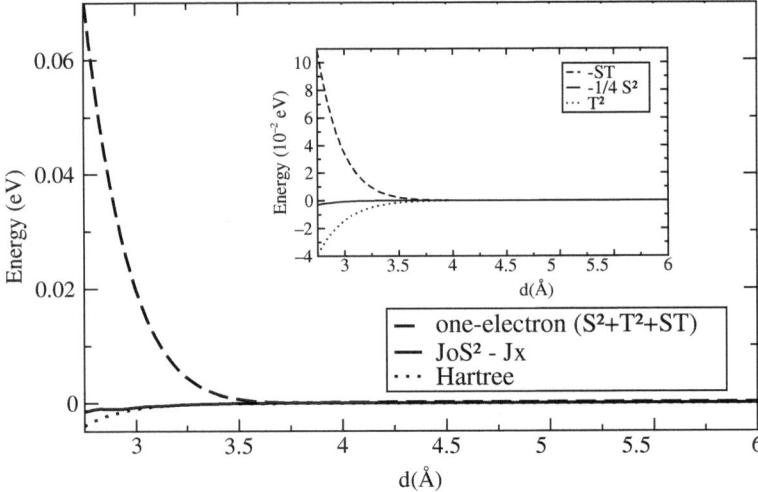

Fig. 4 One-electron, Hartree and exchange integral $(J_o S^2 - J_x)$ contributions (see text) to the interaction energy (per atom in the unit cell) for two graphene layers as a function of the graphene–graphene distance. *Inset*: decomposition of the one-electron term (see text)

from the different one-electron terms: the repulsion is due to the orthogonalization effects associated with the $-ST$ contribution [first term on the right in Eq. (11)] that dominates over the attractive term due to the hoppings [Eq. (13)] and the almost negligible contribution from the S^2 term [second term on the right in Eq. (11)].

As a general remark, we can say that this calculation is really close to a GGA calculation, which would be slightly better than LDA in this case, as underlined by Rydberg et al. [3], with high precision since the overlaps remain very small, which means near the equilibrium distance and over. Of course, in a standard Lennard-Jones representation, this term would be represented by a r^{-12} power, fitted empirically. In our first-principles approach the decay is obviously better, and we can see that the decay has rather an exponential form, which is due to the dependence of the overlaps with the distance. By using this LCAO-S^2 expansion, we also avoid numerical problems due to average of the electronic density in the whole space when it is really small compared to the density in the plane. Moreover, we have to say that a standard LDA calculation with the Fireball code could not reproduce this result. In this case we find a minimum of attractive energy around some meV. We also stress that what is called usually van der Waals determination in DFT–LDA is in fact similar to our calculation of the weak chemical energy. Although theoretically DFT should include all kinds of electronic interactions, the van der Waals energy is not included in the LDA formalism and should be the object of a specific calculation as what we present here.

We will now study the results for the van der Waals contribution in the graphene–graphene interaction. As we can see from Eq. (21), all what we need to determine the van der Waals energy lies in the DFT resolution of each graphene plane (energy eigenvalues and charges occupations) plus the dipole matrix elements. Within our minimal basis set, the vdW interaction in Eqs. (20) and (21) involves only the dipole–dipole interactions related to the term $< s|z|p_z >$ (or $< s|x|p_x >$, etc.). For our optimized basis (see Fig. 1) we obtain the value

$$< s|z|p_z > = 0.44 \text{ Å}. \tag{25}$$

The corresponding vdW energy is represented in Fig. 5, together with the chemical interaction, represented as well in the total energy (sum of vdW and chemical). This figure clearly shows that the total interaction energy is a balance between the repulsion from the "chemical" interaction energy (sum of the different terms in Fig. 4) and the vdW attraction. As shown in this figure, the minimal basis calculation yields only a weak attraction between graphene layers, with an interlayer energy of ~ 7 meV per atom in the unit cell. This value is clearly insufficient when compared to the experimental binding energy for graphite (see below). Let us remind here that the present calculation has been achieved in a minimal sp^3 basis set for carbon. This minimal basis set has been chosen as we told previously because it reproduces quite well various characteristics of carbon phases and gives of course adequate characteristics for one plane of graphene. That means we consider here only $2s \rightarrow 2p$ transitions, which is obviously not enough, when looking at the obtained result. Thus, we have analyzed the improvement of the calculation of the vdW energy, taking into account further atomic dipoles transitions in Eq. (21) following an idea developed previously for rare gas interaction [74]. First, we have analyzed the dipole–dipole contributions associated with a double basis set $sp^3s^*p^{*3}$. The excited states s^* and p^* are obtained within the subspace used to optimize the minimal basis set, see Eq. (9), as those orthogonal to the corresponding s or p orbital. The following dipole terms are obtained for these orbitals:

$$< s|z|p_z^* > = -0.14 \text{ Å}, \tag{26}$$
$$< s^*|z|p_z > = -0.22 \text{ Å}. \tag{27}$$

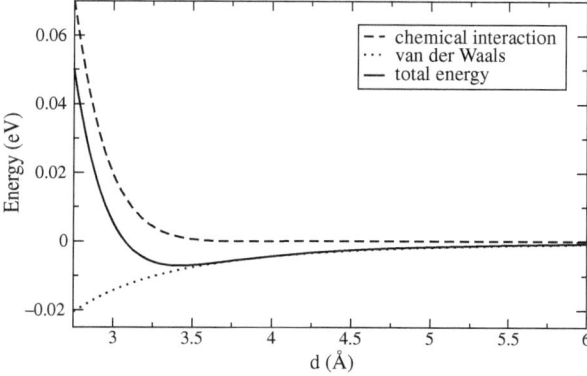

Fig. 5 Chemical (sum of different contributions shown in Fig. 4), van der Waals and total interaction energies (per atom in the unit cell) for two graphene layers for the minimal basis set calculation

The inclusion of the transitions related to dipoles of this type yields an increase for the vdW energy by a factor of 1.6 (60% increase) as compared with the minimal basis vdW energy. As shown below, important dipole–dipole vdW transitions are still missing. In order to determine this missing dipoles, and as excited states are not really well described in DFT (moreover a complete calculation with extended basis set would increase dramatically the computation time), we can use quantum mechanics sum rules. For example, for the s orbital

$$<s|z^2|s> = \sum_n <s|z|n><n|z|s> \tag{28}$$

we can analyze the saturation of the associated transitions. The direct calculation yields a value of 0.23 Å^2 for $<s|z^2|s>$ that can be compared with

$$<s|z^2|s> = <s|z|p_z><p_z|z|s> + <s|z|p_z^*><p_z^*|z|s> +... \tag{29}$$

The dipole values obtained above in Eqs. (25)–(27) suggest that the dipole transitions involving the s orbital are already well represented by the $<s|z|p_z>$- and $<s|z|p_z^*>$-like terms (the first two terms shown in Eq. (29) yield already 0.21 Å^2). This is really consistent, since there are no transition between s and d orbitals, which would be the next missing dipole.

Let us analyze in a similar way the transitions involving the p orbitals. The corresponding sum rule is

$$<p_z|z^2|p_z> = <p_z|z|s><s|z|p_z> + <p_z|z|s^*><s^*|z|p_z> +... \tag{30}$$

The direct calculation yields $<p_z|z^2|p_z> = 0.42$ Å^2, while the $<p_z|z|s>$ and $<p_z|z|s^*>$ terms only add up to 0.24 Å^2, 57% of the total value; this result suggests that important dipole transitions are still missing in the calculation of the vdW energy. The obvious candidates are the dipolar transitions involving d orbitals.

In order to estimate the contribution of these transitions to the vdW energy we need to obtain a value for the $<p_z|z|d_{z^2}>$ term (all other dipole terms involving p and d orbitals can be easily obtained from this one). We can estimate this dipole as follows. First, assuming that this term provides complete saturation in the sum rule, Eq. (30), gives an upper limit of 0.42 Å for this term. Second, we can use the saturation of the sum rule for the s orbital, Eq. (29), as a guide to obtain a more approximate value for $<p_z|z|d_{z^2}>$. For example, we may assume that the contribution from d states to the sum-rule Eq. (30) is split between $<p_z|z|d_{z^2}>$ and $<p_z|z|d_{z^2}^*>$ in a similar way as the one from p orbitals in the sum rule Eq. (29) is split between the terms $<s|z|p_z>$ and $<s|z|p_z^*>$; following these arguments we obtain a value of $<p_z|z|d_{z^2}> \simeq 0.32 - 0.35$ Å. The inclusion of the $<p_z|z|d_{z^2}>$ term in Eq. (30) using these values yields 82–87% saturation for the $<p_z|z^2|p_z>$ sum rule.

We still need to discuss how to calculate the energies, $\bar{e}_{s*}$, $\bar{e}_{p*}$, and $\bar{e}_d$, for the s^*, p^*, and d orbitals, see Eq. (21). Regarding $\bar{e}_{s*}$, and $\bar{e}_{p*}$, we have recalculated the graphene electronic band structure using the extended basis s, p, s^*, and p^*; this calculation yields new empty bands that allows us to obtain

$$\overline{e}_{s*} = \int_{empty} \varepsilon \rho_{s*}(\varepsilon) d\varepsilon / \int_{empty} \rho_{s*}(\varepsilon) d\varepsilon, \tag{31}$$

$$\overline{e}_{p*} = \int_{empty} \varepsilon \rho_{p*}(\varepsilon) d\varepsilon / \int_{empty} \rho_{p*}(\varepsilon) d\varepsilon. \tag{32}$$

Regarding the d orbitals, which belong to the same shell as the s^* and p^* orbitals, we have assumed $\overline{e}_d$ to be a little larger (5 eV) than $\overline{e}_{p*}$: these small changes do not modify practically the van der Waals energy ($\overline{e}_{p*}$ is around 50 eV above E_F, i.e., the (s^*, p^*, d) shell is resonating with the continuous spectrum).

With these values, we can now calculate the contribution of the many different dipole–dipole $p \rightarrow d$ transitions to the total vdW energy. Surprisingly, these transitions represent $\sim 55\%$ of the total vdW energy, while the minimal basis contribution (transitions between s and p orbitals) is only $\sim 28\%$ of the total vdW energy. The remaining vdW energy is associated with $s \rightarrow p^*$ and $p \rightarrow s^*$ transitions.

Figure 6 shows the total vdW energy that is obtained when the dipole–dipole $s \rightarrow p^*$, $p \rightarrow s^*$, and $p \rightarrow d$ transitions discussed above are also included in the calculation, using Eq. (21). We represent as well the total energy obtained by adding this vdW energy to the chemical energy calculated using the minimal basis (see Fig. 5). We obtain an interlayer graphene–graphene equilibrium distance of 3.1–3.2 Å (depending on the value of $<p_z|z|d_{z^2}>$) and an interlayer energy of 30–36 meV per atom in the unit cell, i.e., a binding energy of 60–72 meV. These results correspond to an interlayer binding energy of 120–145 meV per unit cell; each unit cell contains four atoms, two in each layer. These values can be compared with the experimental evidence for graphite. Neglecting interlayer interactions beyond

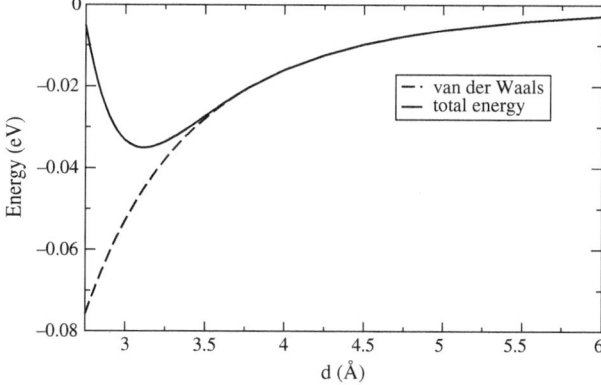

Fig. 6 van der Waals and total interaction energy (per atom in the unit cell) when the dipole–dipole $s \rightarrow p^*$, $p \rightarrow s^*$, and $p \rightarrow d$ transitions are included in the calculation of the vdW energy, as discussed in the text. In this figure we have used the value $< p_z|z|d_{z^2} >= 0.34$ Å

the first neighboring layers in graphite, our results for the graphene–graphene case suggest an interlayer binding energy for graphite of the order of 60–72 meV/atom, a value that can be compared with the different experimental estimates: 43 meV/atom [19] (see also [58]), 35 ± 10 meV/atom [90], and 52 ± 5 meV/atom [91]; notice that the two last experimental numbers are indirect measurements of the interlayer binding energy for graphite, one through the analysis of collapsed multiwall carbon nanotubes [90], the other through the study of the interaction of polyaromatic hydrocarbons with graphite using thermal desorption spectroscopy [91]. Regarding the interlayer distance, our results for the graphene–graphene case suggest an interlayer distance for graphite of 3.1–3.2 Å, to be compared with the experimental value of 3.34 Å. It is also worth comparing our results to those obtained using the vdW-DF [61] technique, where a fully non-local correlation functional $E_c^{nl}[\rho]$ completes the GGA-DFT calculation. Recent calculations for graphite using this technique yield [92] 45.5 meV/atom and 3.6 Å (24 meV/atom and 3.76 Å using the previous version of the functional [3]). Finally, we mention that our calculation for the vdW interaction presents a $-1/d^4$ behavior as a function of the interlayer distance, corresponding to parallel 2D insulators. Here we mention again a comment made by Dobson et al. [93, 16], regarding the power law of van der Waals interaction. For graphene–graphene interaction, they find a power law of $-1/d^3$ at long distance, for two π-conjugated layers. This treatment is based on a description of interacting plasmons between the two planes, yielding a shift of the plasmon frequency in each plane. All the question here is to describe van der Waals interaction as a collective effect, by using a plasmon description, or a sum of discrete effects, like the use of pairwise approximation. In our work, we do not include the long-distance behavior, $-1/d^3$, associated with surface plasmons as explained previously. Our vision is to treat the two graphene planes like two molecular systems in interaction. This can be done thanks to the equilibrium distance, around 3 Å, which is still quite small with respect to the plane extension and because we neglect screening effect due to the high-energy virtual dipolar transitions. Therefore, we believe that we can treat it in a local perspective, claiming that collective effects are fully operative for d larger than 4–5 Å. In the work of Dobson et al., the term $-1/d^4$ corresponding to short distances is neglected as they are mainly interested in behavior at large distances.

To summarize this part about graphene–graphene interaction, we would like to underline some points. First, the weak interaction between π-conjugated systems has two components: the well-known attractive van der Waals part and a weak chemical repulsion due to the small overlap between the electronic densities. The balance between these two contributions gives the binding energy of the system. Second, our original LCAO-S^2 approach is really accurate to describe the repulsive weak chemical energy. Third, we have developed a fast and simple description of the van der Waals interaction, based on a pairwise atomic interaction, the dipolar approximation, and a perturbation treatment. Fourth, the very important and original result is that more than half of the van der Waals energy is due to virtual transitions, i.e., transitions with high-excited states like here the d band of carbon. And finally, this result is in really good agreement with experimental or other theoretical determinations. In the following part, we will use this theory validated by the graphene–graphene result to study weak interactions between other graphitic systems like carbon nanotubes or fullerenes.

4 Graphitic Systems: The Case of Carbon Nanotubes and Fullerenes

In this part we will present some very recent results about van der Waals interaction in more general graphitic systems, for example, carbon nanotubes (CNT), fullerenes (C_{60}), and their adsorption on graphene [101]. We use the same formalism as previously developed for the interaction between graphene planes, i.e., the LCAO-S^2 approach, combined to our van der Waals perturbation treatment. In all cases, each isolated subsystem is totally determined within our DFT formalism, and using the Fireball code for numerical applications. In the following, we will detail weak lateral interaction between single-wall CNTs, for various radii, the case of C_{60}-dimers, adsorption of C_{60} on graphene or CNT, and finally, double-wall CNTs and encapsulation of C_{60} in CNT. This work brings new results in this field and represents as well a first step for such studies as many aspects still need to be improved or deeply studied.

4.1 Weak Lateral Interaction Between CNT: Influence of the Radius

In this section we study the lateral interaction between two CNTs for various radius. The atomic configuration for two CNTs 4×4 is represented in Fig. 7. We then represent in Fig. 8 the evolution of the lateral binding energy between two CNTs, for different radius as CNT 4×4, 6×6, 8×8, 10×10, and 12×12. We represent also the equilibrium position of two graphene planes, which should be reached by interaction of CNTs with infinite radius.

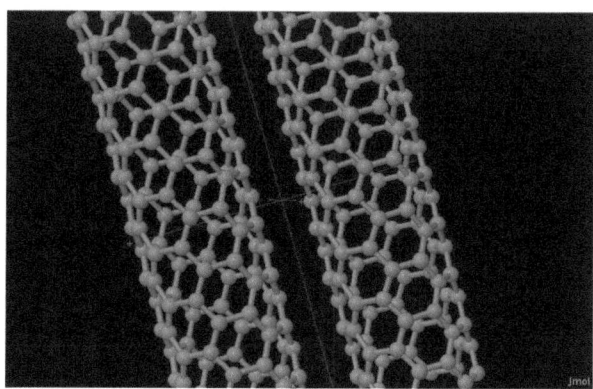

Fig. 7 Atomic configuration of two CNT 4×4 in parallel, for weak lateral interaction study

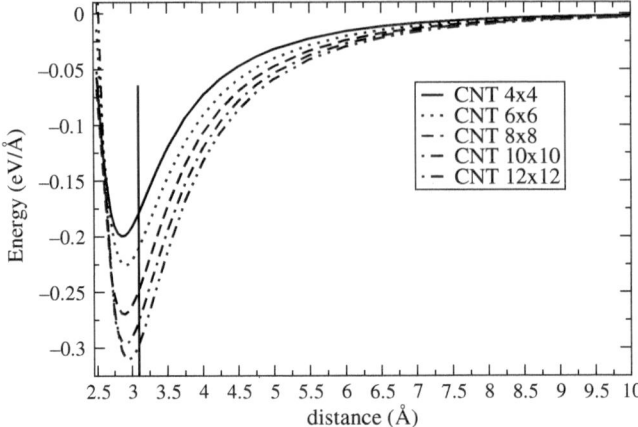

Fig. 8 Evolution of the lateral binding energy of CNT as a function of the wall to wall distance, and for various CNT radii. The *vertical line* represents the equilibrium distance between two graphene planes

The various interaction energies per unit length are shown in Table 1. We can observe the increase of the binding energy per unit length with the radius of the CNT which is related obviously with the increase in the number of atoms in interaction. Therefore, if we consider an effective surface as defining the interaction between two CNTs, we can assume geometrically, for quite large CNT, that this effective surface is proportional to the square root of the radius. In Fig. 9, we represent the same lateral energies as before, divided by the square root of the CNT radius. We can observe that the normalized energy curves are almost similar, deviations occurring only in the case of small CNT (4 × 4, for example), where the radius is not big enough to validate the geometrical approximation. Another interesting point to consider is the mutual orientation of the CNTs. It is not our goal here to make a complete study of this problem, but just to point out the main idea. We did the calculation of the lateral energy between 12 × 12 CNTs, in an AA stacking and in an AB stacking, in analogy with the possible graphene–graphene configurations; we can see on Fig. 10 that this gives rise to a slight difference in binding energy and minimum (about 16 meV/Å for the binding energy and 0.05 Å for the equilibrium

Table 1 Evolution of the equilibrium positions and the energy minimum of CNT lateral interaction, as a function of the CNT dimensions

CNT dimensions	Equilibrium position (Å)	Energy (meV/Å)
4 × 4	2.85	199
6 × 6	2.9	226
8 × 8	2.9	269
10 × 10	2.9	296
12 × 12	2.95	310

Fig. 9 Evolution of the
lateral binding energy,
normalized with the square
root of the CNT radius, as a
function of the wall to wall
distance, and for various
CNT radii

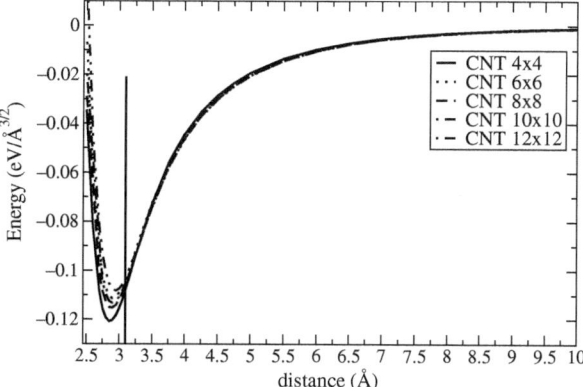

position), as the AA configuration here is more repulsive, similarly to the graphene
case.

Finally, we also observe that the power law of the van der Waals interaction in the
energy interaction tail is $1/d^4$ between two CNTs, d being the wall to wall distance,
as is the case for graphene–graphene interaction.

4.2 Binding Energy of C_{60} Dimers

In this part, we are interested in the interaction between two fullerenes (C_{60}). This
is an interesting problem which has been already addressed by different methods
[95, 23, 6]. The binding energy of such system is still unclear, as we can find theoretical determinations ranging from 80 to 554 meV per dimer. Moreover, it is not well
established if this binding energy is van der Waals like or if it is slightly covalent. In
this work, following the formalism we have developed for graphene–graphene interaction, we consider that these two interactions coexist as in the previous systems,

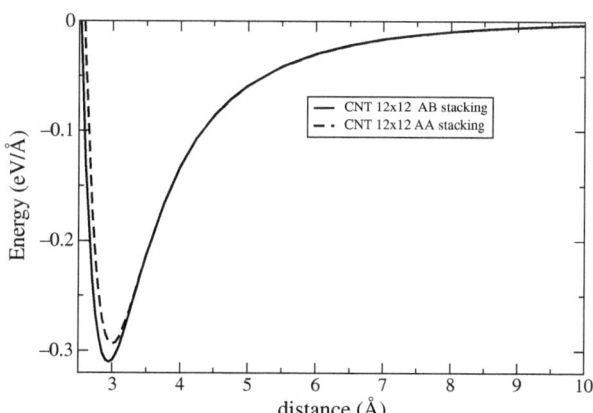

Fig. 10 Comparison of AB
and AA stacking for the
lateral interaction energy
between two CNT 12 × 12

Fig. 11 Atomic configuration
of the studied C_{60} dimer

and the balance between both is determined within our DFT + intermolecular perturbation theory. In Fig. 11 we represent the configuration of the studied system.

We did not focus really on the mutual orientation of the two C_{60}, as our goal is
to get a first idea of the binding energy of such system. Nevertheless, in a future
work, we will look more precisely at this detailed structure. The point here is to see
how our approach compares with previous theoretical determinations. Our result is
represented in Fig. 12, as the total binding energy versus the wall to wall distance.
We present as well the detail of the two contributions, which are the "chemical"
repulsion and the van der Waals energy.

We find a minimum energy of 440 meV for the whole dimer, at an equilibrium
position of 3 Å. This result constitutes an intermediate result between previous
DFT–LDA determination which underestimates the binding energy because of the
lack of van der Waals interaction, and the Lennard-Jones potential determination,
which overestimates the van der Waals part to compensate an incorrect determination of the repulsive part. We also observe a power law in $1/d^4$ for the van der Waals
interaction in this system.

Fig. 12 C_{60} dimer binding
energy within our model.
"Chemical" repulsion and
van der Waals contribution
are also represented

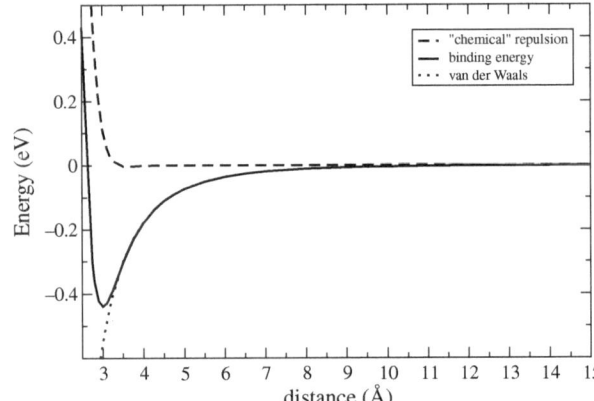

It is also interesting to present the interaction between C_{60} in a C_{60}-molecular crystal, in the simple cubic crystalline form. This structure is represented in Fig. 13.

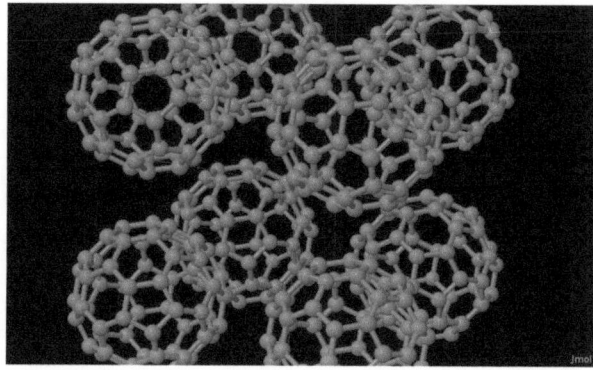

Fig. 13 Atomic configuration of the studied C_{60}-molecular crystal

We find a minimum binding energy of 1.74 eV per C_{60} molecule, for an equilibrium lattice parameter of 9.7 Å, which is in quite good agreement with experimental determination. This result has to be taken with caution, as well-known rotational effects have not been considered here. Moreover, a deeper study has to be achieved, since the crystalline lattice is known to be simple cubic at low temperature and face centered cubic at ambient temperature, from neutron and X-ray diffusive diffusion determination [96]. These two forms are related by an orientational ordering transition [97, 98], which still has to be explored within our approach. This problem will be the subject of a future work where van der Waals forces will be introduced within a molecular dynamics calculation.

4.3 Adsorption of C_{60} on Graphene and CNT

In this part, we present the adsorption of a C_{60} molecule on graphene and on a CNT 10×10. This work is also a first step in the study of weakly adsorbed molecules on metallic surfaces as well as the study of organics-doped CNTs. The C_{60} adsorption on graphene has already given us elements to estimate the equilibrium position of $C_{60}/Au(111)$ in order to determine interface dipole and charge transfer [99].

The adsorption configuration of C_{60} on graphene is represented in Fig. 14, and the adsorption energy curve as well as the "chemical" repulsion and the van der Waals energy, calculated in our approach are represented in Fig. 15.

In this case, we find an equilibrium position of 2.9 Å, a bit less than for graphene–graphene interaction, and a minimum energy of about 1 eV per C_{60} molecule. This result is due to a lower repulsion between C_{60} and graphene, because of the curvature of the molecule, while the van der Waals interaction, which is long range, remains the same. We find here a power law of $1/d^3$ at short distances, which goes to $1/d^4$ for distances bigger than the C_{60} dimensions.

In Fig. 16, we represent the configuration of C_{60} adsorbed on CNT 10×10. This CNT has been chosen for its relative similar size to C_{60}, a bigger CNT would give a result close to the one obtain for C_{60} on graphene. In Fig. 17, we represent the

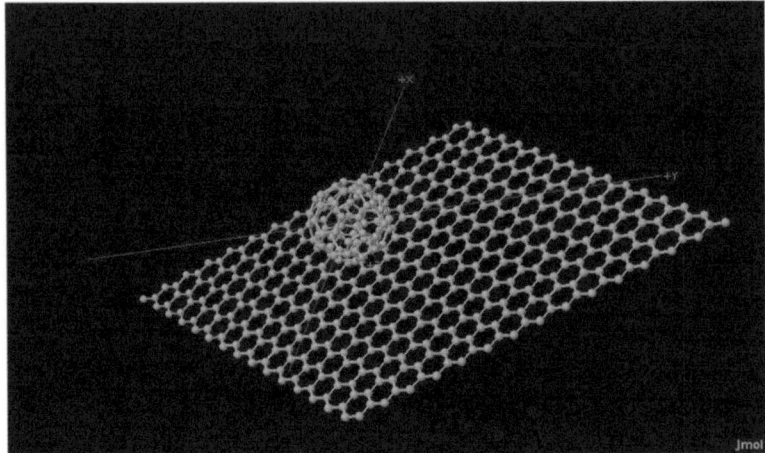

Fig. 14 Atomic configuration of the C_{60} adsorption on graphene

Fig. 15 Binding energy, "chemical" repulsion, and van der Waals energy of C_{60} adsorbed on graphene, calculated with our approach

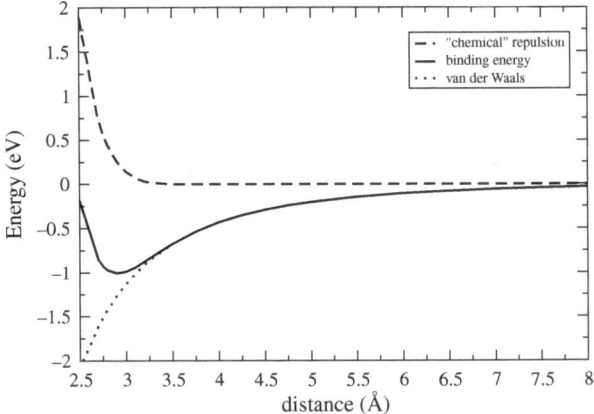

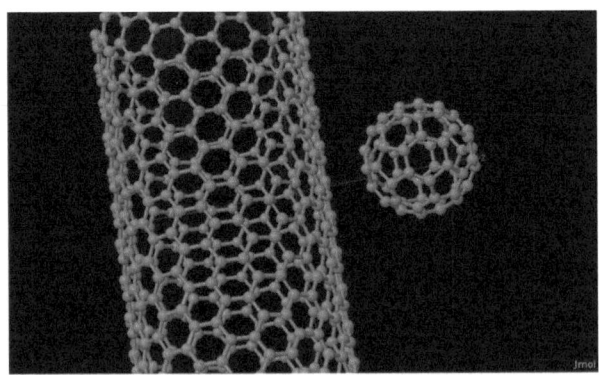

Fig. 16 Atomic configuration of the C_{60} adsorption on CNT 10×10

Fig. 17 Binding energy, "chemical" repulsion, and van der Waals energy of C_{60} adsorbed on CNT 10×10, calculated with our approach

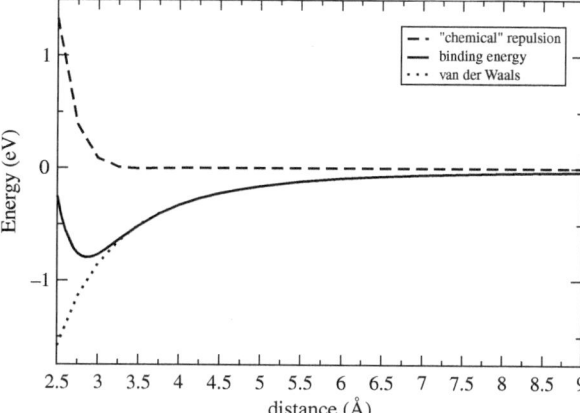

evolution of the binding energy per C_{60} molecule, with respect to the wall to wall distance with the CNT.

Here we find an equilibrium position at 2.85 Å and a minimum of energy of 796 meV per C_{60} molecule. The power law is found to be $1/d^{\frac{7}{2}}$ for $d \ll R$ (R being the C_{60} radius), which is close to the result encountered for adsorption on graphene.

In these two situations, more work has still to be done, for example, about mutual orientation between the two systems or about doped CNTs with organic molecules [100]. This will be explored in the future also.

4.4 Encapsulation of C_{60} in CNT and Double-Wall CNT

The problem of encapsulation of molecules and especially C_{60} is very important nowadays as these systems present interesting charge transfer properties [101, 102]. For example, it has been observed that a C_{60} encapsulated in a CNT, also called peapod [103, 104], presents an excess of electronic charge, resulting in a negative net charge. These properties are very interesting for molecular electronics, and the study of various molecules inserted in CNT is a hot topic. However, if the binding energy can be estimated experimentally, there is still an intense theoretical controversy to determine it with great accuracy, as the nature of the bond remains difficult to understand. Many attempts have been done to determine this interaction, mostly with Lennard-Jones calculations [105, 106] which do not really bring a physical comprehension of this interaction. We can also find a very recent paper where calculations have been achieved in a pure DFT formalism [107], without any inclusion of van der Waals interaction, which seems surprising.

We present here the result of LCAO-S^2 + van der Waals calculation for a C_{60} molecule inserted in a CNT 10×10. The geometry of the configuration is represented in Fig. 18.

Fig. 18 Atomic configuration
of a C_{60} molecule
encapsulated in a CNT
10×10

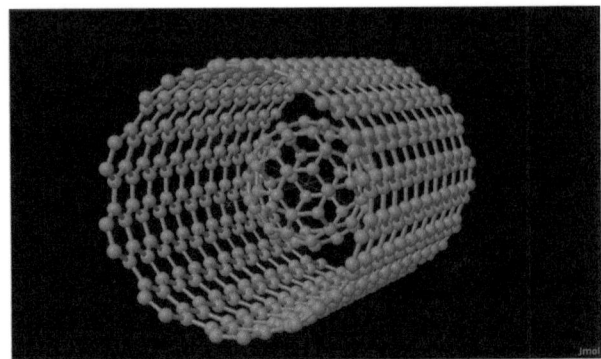

We have then represented the evolution of the binding energy of this C_{60} molecule
as a function of the distance between the center of the molecule with the axis of the
CNT. The result is shown in Fig. 19.

From this result, we observe first that the minimum energy of C_{60} is not centered
in the CNT, but is situated at about 0.2 Å from the center for the CNT 10×10
and 1.9 Å for the CNT 12×12. This is due to the balance between the repulsive
weak chemical interaction and the attractive van der Waals force. The curves here
are quite flat, but if we increase the diameter of the tube we can clearly see these
radial minima (here there are only two positions, as we represent the evolution along
a diameter). Regarding the minimum of energy, we observe total binding energies
for the molecule of 4.05 eV (CNT 10×10) and 2.28 eV (CNT 12×12), which is in
good agreement with previous calculations from Girifalco et al. [106] but reveals a
stronger cohesion energy than the ones calculated in DFT [107]. From these results,
we stress that the interaction of C_{60} with CNT is driven by weak interactions, among
which one can find van der Waals, which cannot be reproduced correctly in the
frame of standard DFT. The different results with the two sizes of CNT are due
to the number of effective interacting atoms of the CNT with the molecule; this

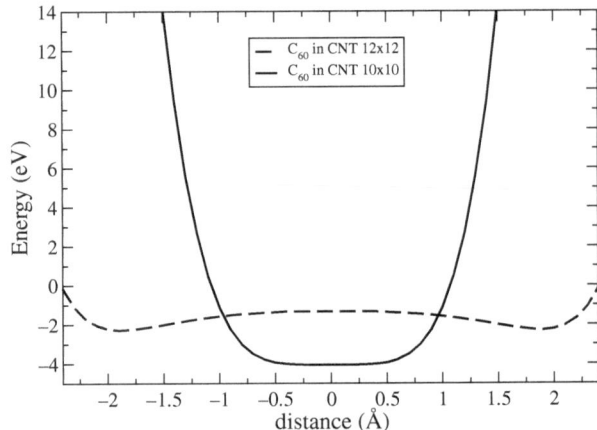

Fig. 19 Binding energy of
C_{60} encapsulated in a CNT
10×10 and in a CNT 12×12
in function of the distance
between the center of the
molecule and the CNT axis

Fig. 20 Atomic configuration of a CNT 4 × 4 inserted in a CNT 10 × 10

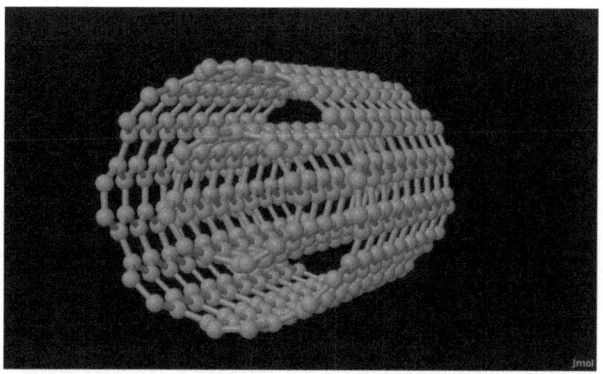

number is smaller for bigger CNT, and we tend to the situation of C_{60} adsorbed on graphene. Another interesting point is to calculate the variation of energy for a translation of the C_{60} molecule along the axis of the CNT. We observe that there are only very small variations (about some meV) of the binding energy of C_{60} in the CNT, due to the corrugation, which means that the translation is practically costless energetically. This kind of result has already been observed in bioorganic molecules, with covalent binding energy [104], but it has not been demonstrated theoretically in the case of peapods until now.

We have proceeded to the same study with CNT 4 × 4 inserted in CNT 10 × 10. This system is comparable to the C_{60} considered before, as the diameters of both systems are similar. Moreover, this study opens the way to a more general study of multiwall CNT, which has still to be done, since there is no clear interpretation of the binding energy in that case either. The atomic configuration is represented in Fig. 20.

The energy of such system, per unit length, is represented in the Fig. 21, as a function of the interaxis distance.

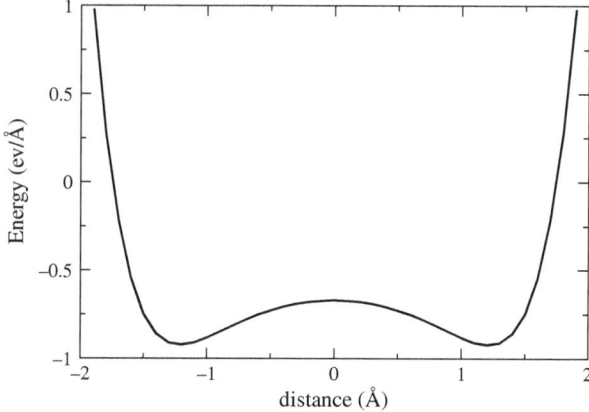

Fig. 21 Binding energy of CNT 4 × 4 inserted in a CNT 10 × 10 versus the interaxis distance

Here also we can observe this radial minimum (here two positions are represented, since we show the evolution along a diameter) previously seen with the C_{60} molecule, which is situated at 1.3 Å from the central axis. The binding energy is 0.92 eV/Å, which is fully comparable with the one obtained with C_{60} if we take about 6 Å for the length of the CNT 4 × 4.

These problems of encapsulations present many interesting applications, as underlined before, and these two model systems constitute a first step for future study of complex molecules inserted in CNT.

4.5 C_6 analysis of the van der Waals Interaction in Graphitic Materials

Before concluding this part of the review on graphitic materials, we would like to make a summary of the results encountered here. With such objective, we have represented in Table 2 the values of the C_6 coefficient (the van der Waals interaction between two atoms has a $1/r^6$ dependence, as we determined it from dipolar interaction) obtained for all our calculations, for a pair of two carbon atoms. These coefficients represent an average value obtained over all pairs of atoms.

Table 2 Evolution of the average C_6 coefficients obtained for the different systems considered in this work

System	C_6 coefficient (eV/Å^6)
graphene–graphene	−13.8
CNT–CNT (all diameters)	−14.7
C_{60}-C_{60}	−15.1
C_{60}-graphene	−14.9
C_{60}-CNT	−14.9

The main observation we can make is that the variation of our C_6 value is really small (about 1 eV/Å^6), which confirms what we have already developed previously: in our approach, the repulsive weak chemical energy is determined with high accuracy, and the van der Waals contribution is determined in an independent way, from DFT data. It means that it is not necessary to compensate one interaction (or one potential) with the other to obtain a good equilibrium position and a correct binding energy. This is a disadvantage of some previous work [23], where one can observe variations of the C_6 coefficient of about 5 eV/Å^6.

We will not present C_{12} variations for the weak chemical repulsion, as it obviously does not correspond to the reality. Historically this $1/r^{12}$ variation is supposed to represent only the Pauli repulsion, which is observed at distances much shorter than the ones considered here. It is indeed more realistic to study an exponential decay of this part, as it follows the overlaps evolution, but it can be either approximated from a GGA calculation with reasonable accuracy.

5 Summary

In the present work, we have tried to present some important concepts about weak chemical interactions and van der Waals forces and a new fast and efficient method to determine them from DFT plus perturbation theory. The main important idea about weak interactions, which are often attributed only to van der Waals forces, is that we have a balance between two kinds of interactions. The first one, generally repulsive, is what we call the weak chemical interaction, which is due to the weak electronic overlap between the two interacting systems. The second one, attractive at large distance, is the van der Waals interaction. The balance of these two interactions gives a weak potential well, at an equilibrium position larger than the usual covalent interaction. Another precision has to be made about the term "weak interaction". Its sense is weak with respect to the covalent interaction, but as we have seen for the C_{60} molecule inserted in a CNT, we can reach some eV for the whole interacting molecule, which is not so weak indeed.

It is well known that DFT, and especially the LDA approximation, fails to describe these weak interactions. LDA results giving reasonable values for such interactions can only be a compensation of errors in the calculation. On the other hand, Lennard-Jones-like calculation, which has given most of the known results in these weak interacting systems, is based on fitted parameters, which depend on the material and its phase. In this method, we can often see that the C_6 parameter associated to the van der Waals interaction has to compensate the random value of the C_{12} parameter associated to the repulsive part in order to get agreement with the experiment.

We have presented a fast and efficient method, based on DFT calculations in a localized-orbitals basis set, using an intermolecular perturbation theory. The numerical part is achieved by the use of the Fireball ab initio tight-binding molecular dynamics code. The weak chemical interaction is correctly determined from DFT as a shift of the molecular levels of each subsystems. The van der Waals part is obtained as the interaction between fluctuating atomic dipoles, using DFT, perturbation theory, and quantum mechanics sum rules. This approach has been tested and validated on a reference system, i.e., the interaction between two graphene planes. An important result is the major contribution of dipolar transitions with high-excited states, the $3d$ states for the carbon atom, which represents more than a half of the total van der Waals energy.

We have then extended the application of this method to more general graphitic systems, like the lateral interaction between CNTs, C_{60}-dimers and molecular crystals, adsorption of C_{60} on CNT and graphene, and finally insertion of C_{60} or CNT in CNT. As we said before, this work is preliminar to a more complete determination in such systems. We just picked up some interesting cases to show the good agreement between our approach and previous determinations, as well as the robustness and the potential of this method. Of course, there is an important zoology in these systems which still has to be explored and will be the object of further works.

This method opens also many lines for future perspectives. On a methodological point of view, two goals have to be reached, which are first the generalization of the method to every kind of chemical specie (to go beyond carbon atom interaction), and second, the derivation of the forces from this energy determination to develop a molecular dynamics code including van der Waals forces. As we underlined before, there are many potential applications to these developments. In graphitic systems, as CNTs, for example, there is still a huge unknown world to discover with the inclusion of molecules in the CNT, for chemical applications, electronics, or even medicine with some CNT-based molecular treatment.

Another very important line which will be developed more and more in the next years is the field of molecular electronics. Indeed, it is well known that the conductance between a metal and a molecule reaches a maximum at distances between 2.5 and 4–5 Å, which is typically the distance where our method is more efficient. This is due to the barrier felt by the electrons when crossing the interface. In case of hybridization at shorter distance, the electrons are mainly reflected, while when the interaction becomes weaker, and the molecule far from the surface, the electrons are transmitted. This is the case of π-conjugated molecules, like metallophthalocyanines, for example, whose interaction with metallic surface study is in big progress recently [109–111]. Here the knowledge of the equilibrium shape of the molecule is of fundamental importance to determine the transport properties. Moreover, the influence of van der Waals interaction on the DOS at the Fermi level still remains a big scientific challenge and represents a key for the comprehension of these transport properties. We expect that the new method presented here, as well as its future developments, can bring some answers to all these problems.

Acknowledgments This work is supported by Spanish MEyC under contract MAT2007-60966 and NAN2004-09183-C1007 and the Juan de la Cierva Programme (MCyT). We acknowledge fruitful discussions with M. A. Basanta, Ch. Joachim, and M. Alouani.

References

1. F. London, Z. Phys. Chem. Abt. B **11**, 222 (1930).
2. F. London, Z. Phys. **63**, 245 (1930).
3. H. Rydberg, M. Dion, N. Jacobson, E. Schröder, P. Hyldgaard, S. I. Simak, D. C. Langreth, and B. I. Lundqvist, Phys. Rev. Lett. **91**, 126402 (2003).
4. A. N. Kolmogorov and V. H. Crespi, Phys. Rev. B **71**, 235415 (2005).
5. L. Henrard, E. Hernández, P. Bernier, and A. Rubio, Phys. Rev. B **60**, R8521 (1999).
6. F. Tournus, J.-C. Charlier, and P. Mélinon, J. Chem. Phys. **122**, 094315 (2005).
7. F. Tournus and J.-C. Charlier, Phys. Rev. B **71**, 165421 (2005).
8. Q. Wu and W. Yang, J. Chem. Phys. **116**, 515 (2002).
9. J. S. Arellano, L. M. Molina, A. Rubio, M. J. López, and J. A. Alonso, J. Chem. Phys. **117**, 2281 (2002).
10. M. J. Allen and D. J. Tozer, J. Chem. Phys. **117**, 11113 (2002).
11. K. T. Tang and J. P. Toennies, J. Chem. Phys. **118**, 4976 (2003).
12. R. N. Barnett and U. Landman, Phys. Rev. B **48**, 2081 (1993).
13. J. Ortega, J. P. Lewis, and O. F. Sankey, Phys. Rev. B **50**, 10516 (1994).
14. N. Kurita, H. Inoue, and H. Sekino, Chem. Phys. Lett. **370**, 161 (2003).
15. S. Grimme, J. Antony, T. Schwabe, and Ch. Mück-Lichtenfeld, Org. Biomol. Chem. **5**, 741 (2007).

16. J. F. Dobson, J. Wang, B. P. Dinte, K. McLennan, and H. M. Le, Int. J. Quantum Chem. **101**, 579 (2005).
17. M. A. Basanta, PhD Thesis, Autonomus University of Madrid (2005).
18. A. L. Fetter and J. D. Walecka, *Quantum Theory of Many Particle-Systems* (McGraw-Hill, Inc., New York, 1971).
19. L. A. Girifalco and R. A. Lad, J. Phys. Chem. **25**, 693 (1956).
20. V. I. Zubov, N. P. Tretiakov, J. N. Teixeira Rabelo, and J. F. Sanchez Ortiz, Phys. Lett. A **194**, 223 (1994).
21. R. S. Ruoff and A. P. Hickman, J. Phys. Chem. **97**, 2494 (1993).
22. J. Song and R. L. Cappelletti, Phys. Rev. B **50**, 14678 (1994).
23. L. A. Girifalco, M. Hodak, and R. S. Lee, Phys. Rev. B **62**, 13104 (2000).
24. A. J. Stone, *The Theory of Intermolecular Forces* (Oxford University Press, Oxford, 2000 reprint).
25. A. Szabo and N. S. Ostlund, *Modern Quantum Chemistry* (Dover Publications, New York, 1989), p. 446.
26. C. Cohen-Tannoudji, B. Diu, and F. Laloë, *Quantum Mechanics*, vols. I and II, (John Wiley and Sons, New York, 1977), p. 1525.
27. N. W. Ashcroft and N. D. Mermin, *Solid State Physics* (Saunders College Publishing,Philadelphia, 1976), p. 826.
28. M. V. Basilevsky and M. M. Berenfeld, Int. J. Quantum Chem. **6**, 23 (1972).
29. V. Kvasnicka, V. Laurinc, and I. Hubac, Phys. Rev. A **10**, 2016 (1974).
30. I. C. Hayes and A. J. Stone, Mol. Phys. **53**, 83 (1984).
31. P. R. Surjan, C. Pérez del Valle, and L. Lain, Int. J. Quantum Chem. **64**, 43 (1997).
32. V. Lukes, V. Laurinc, and S. Biskupic, Int. J. Quantum Chem. **75**, 81 (1999).
33. B. Jeziorski and W. Kolos, Int. J. Quantum Chem. **12**, Suppl. 1, 91 (1977).
34. S. Rybak, B. Jeziorski, and K. Szalewicz, J. Chem. Phys. **95**, 6576 (1991).
35. K. Patkowski, B. Jeziorski, and K. Szalewicz, J. Chem. Phys. **120**, 6849 (2004).
36. C. Møller and M. S. Plesset, Phys. Rev. **46**, 618 (1934).
37. D. Cremer, *Møller-Plesset Perturbation Theory in Encyclopedia of Computational Chemistry*, vol. 3, ed. P. von Rague-Schleyer (John Wiley, New York, 1998), p. 1706.
38. E. Engel, A. Höck, and R. M. Dreizler, Phys. Rev. A **61**, 032502 (2000).
39. A. J. Misquitta and K. Szalewicz, Chem. Phys. Lett. **357**, 301 (2002).
40. A. J. Misquitta, B. Jeziorski, and K. Szalewicz, Phys. Rev. Lett. **91**, 033201 (2003).
41. A. Hesselmann and G. Jansen, Chem. Phys. Lett. **367**, 778 (2003).
42. I. C. Gerber and J. G. Ángyán, Chem. Phys. Lett. **416**, 370 (2005).
43. J. G. Ángyán, I. C. Gerber, A. Savin, and J. Toulouse, Phys. Rev. A **72**, 012510 (2005).
44. A. J. Misquitta, R. Podeszwa, B. Jeziorski, and K. Szalewicz, J. Chem. Phys. **123**, 214103 (2005).
45. J. A. Pople, M. Head-Gordon, and K. Raghavachari, J. Chem. Phys. **87**, 5968 (1987).
46. P. Hohenberg and W. Kohn, Phys. Rev. **136**, B 864 (1964).
47. W. Kohn and L. J. Sham, Phys. Rev. **140**, A 1133 (1965).
48. M. C. Payne, M. P. Teter, D. C. Allan, T. A. Arias, and J. D. Joannopoulos, Rev. Mod. Phys. **64**, 1045 (1992).
49. T. Mikaye, F. Aryasetiawan, T. Kotani, M. van Schilfgaarde, M. Usuda, and K. Terakura, Phys. Rev. B **66**, 245103 (2002).
50. J. F. Dobson and J. W. Wang, Phys. Rev. B **69**, 235104 (2004).
51. D. M. Ceperley and B. J. Alder, Phys. Rev. Lett. **45**, 566 (1980).
52. J. P. Lewis, K. R. Glaesmann, G. A. Voth, J. Fritsch, A. A. Demkov, J. Ortega, and O. F. Sankey, Phys. Rev. B **64**, 195103 (2001).
53. Y. J. Dappe, M. A. Basanta, J. Ortega, and F. Flores, Phys. Rev. B **74**, 205434 (2006).
54. S. Grimme, J. Comput. Chem. **25**, 1463 (2004).
55. S. grimme, J. Comput. Chem. **27**, 1787 (2006).
56. X. Wu, M. C. Vargas, S. Nayak, V. Lotrich, and G. Scoles, J. Chem. Phys. **115**, 8748 (2001).

57. M. Elstner, P. Hobza, Th. Frauenheim, S. Suhai, and E. Kaxiras, J. Chem. Phys. **114**, 5149 (2001).
58. M. Hasegawa and K. Nishidate, Phys. Rev. B **70**, 205431 (2004).
59. U. Zimmerli, M. Parrinello, and P. Koumoutsakos, J. Chem. Phys. **120**, 2693 (2004).
60. D. C. Langreth, M. Dion, H. Rydberg, E. Schröder, P. Hyldgaard, and B. I. Lundqvist, Int. J. Quantum Chem. **101**, 599 (2005).
61. M. Dion, H. Rydberg, E. Schröder, D. C. Langreth, and B. I. Lundqvist, Phys. Rev. Lett. **92**, 246401 (2004).
62. W. Kohn, Y. Meir, and D. E. Makarov, Phys. Rev. Lett. **80**, 4153 (1998).
63. J. F. Dobson and J. Wang, Phys. Rev. Lett. **82**, 2123 (1999).
64. P. García-González and R. W. Godby Phys. Rev. Lett. **88**, 056406 (2002).
65. J. A. Alonso and A. Mañanes, Theor. Chem. Acc. **117**, 467 (2007).
66. J. Toulouse, F. Colonna, and A. Savin, Phys. Rev. A **70**, 0622505 (2004).
67. P. Gori-Giorgi and A. Savin. Phys. Rev. A **71**, 032513 (2005).
68. M. Kamiya, T. Tsuneda, and K. Hirao, J. Chem. Phys. **117**, 6010 (2002).
69. T. Leininger, H. Stoll, H. J. Werner, A. Savin, Chem. Phys. Lett. **275**, 151 (1997).
70. Y. Andersson, D. C. Langreth, and B. I. Lundqvist, Phys. Rev. Lett. **76**, 102 (1996).
71. F. J. García-Vidal, J. Merino, R. Pérez, R. Rincón, J. Ortega, and F. Flores, Phys. Rev. B **50**, 10537 (1994).
72. P. Pou, R. Pérez, F. Flores, A. Levy Yeyati, A. Martin-Rodero, J. M. Blanco, F. J. García-Vidal, and J. Ortega, Phys. Rev. B **62**, 4309 (2000).
73. Y. J. Dappe, R. Oszwaldowski, P. Pou, J. Ortega, R. Pérez, and F. Flores, Phys. Rev. B **73**, 235124 (2006).
74. M. A. Basanta, Y. J. Dappe, J. Ortega, and F. Flores, Europhys. Lett. **70**, 355 (2005).
75. K. Schönhammer, O. Gunnarsson, and R. M. Noack, Phys. Rev. B **52**, 2504 (1995).
76. O. F. Sankey and D. J. Niklewski, Phys. Rev. B **40**, 3979 (1989).
77. A. A. Demkov, J. Ortega, O. F. Sankey, and M. P. Grumbach, Phys. Rev. B **52**, 1618 (1995).
78. J. P. Lewis, K. R. Glaesemann, G. A. Voth, J. Fritsch, A. A. Demkov, J. Ortega, and O. F. Sankey, Phys. Rev. B **64**, 195103 (2001).
79. P. Jelinek, H. Wang, J. P. Lewis, O. F. Sankey, and J. Ortega, Phys. Rev. B **71**, 235101 (2005).
80. O. F. Sankey and D. J. Niklewski, Phys. Rev. B **40**, 3979 (1989).
81. M. A. Basanta, Y. J. Dappe, P. Jelinek, and J. Ortega, Comput. Mater. Sci. **39**, 759 (2007).
82. N. Troullier and J. L. Martin, Solid States Commun. 74, 613 (1990).
83. N. Troullier and J. L. Martin, Phys. Rev. B **43**, 1993 (1991).
84. E. C. Goldberg, A. Martín-Rodero, R. Monreal, and F. Flores, Phys. Rev. B **39**, 5684 (1989).
85. F. J. García-Vidal, A. Martín-Rodero, F. Flores, J. Ortega, and R. Pérez, Phys. Rev. B **44**, 11412 (1991).
86. J. Ortega, J. P. Lewis, and O. F. Sankey, Phys. Rev. B **50**, 10516 (1994).
87. J. Ortega, J. P. Lewis, and O. F. Sankey, J. Chem. Phys. **106**, 3696 (1997).
88. J. N. Israelashvili, *Intermolecular and Surface Forces*, 2nd ed. (Academic, New York, 1992).
89. F. García-Moliner and F. Flores, *Introduction to the Theory of Solid Surfaces* (Cambridge University Press, Cambridge, 1979).
90. L. X. Benedict, N. G. Chopra, M. L. Cohen, A. Zettl, S. G. Louie, and V. H. Crespi, Chem. Phys. Lett. **286**, 490 (1998).
91. R. Zacharia, H. Ulbricht, and T. Hertel, Phys. Rev. B **69**, 155406 (2004).
92. S. D. Chakarova-Kack, E. Schröder, B. I. Lundqvist, and D. C. Langreth, Phys. Rev. Lett. **96**, 146107 (2006).
93. J. F. Dobson, A. White, and A. Rubio, Phys. Rev. Lett. **96**, 073201 (2006).
94. Y. J. Dappe, J. Ortega, and F. Flores, Phys. Rev. B **79**, 165409 (2009).
95. G. C. La Rocca, Europhys. Lett. **25**, 5 (1994).
96. P. Launois, *Research Habilitation* (Paris-Sud University, Orsay, 1999).
97. P. A. Heiney, J. E. Fischer, A. R. McGhie, W. J. Romanow, A. M. Denenstein, J. P. McCauley, Jr., A. B. Smith, III, and D. E. Cox, Phys. Rev. Lett. **66**, 2911 (1991).
98. W. I. F. David, R. M. Ibberson, J. C. Matthewman, K. Prassides, T. J. S. Dennis, J. P. Hare, H. W. Kroto, R. Taylor, and D. R. M. Walton, Nature **353**, 147 (1991).

99. E. Abad, J. Ortega, Y. J. Dappe, and F. Flores, Appl. Phys. A **95**, 119 (2009).
100. T. Pankewitz and W. Klopper, J. Phys. Chem. C **111**, 18917 (2007).
101. J. Lu, S. Nagase, S. Zhang, and L. Peng, Phys. Rev. B **68**, 121402 (2003).
102. M. Yudasaka, K. Ajima, K. Suenaga, T. Ichihashi, A. Hashimoto, and S. Iijima, Chem. Phys. Lett. **380**, 42 (2003).
103. A. Gloter, K. Suenaga, H. Kataura, R. Fujii, T. Kodama, H. Nishikawa, I. Ikemoto, K. Kikuchi, S. Suzuki, Y. Achiba, and S. Iijima, Chem. Phys. Lett. **390**, 462 (2004).
104. M. M. Calbi, S. M. Gatica, and M. W. Cole, Phys. Rev. B **67**, 205417 (2003).
105. H. Ulbricht, G. Moos, and T. Hertel, Phys. Rev. Lett. **90**, 095501 (2003).
106. L. A. Girifalco and M. Hodak, Phys. Rev. B **65**, 125404 (2002).
107. S. Okada, Phys. Rev. B **77**, 235419 (2008).
108. B. Toudic, P. Garcia, Ch. Odin, Ph. Rabiller, C. Ecolivet, E. Collet, Ph. Bourges, G. J. McIntyre, M. D. Hollingsworth, and T. Breczewski, Science **319**, 69 (2008).
109. S. Boukari, A. Ghaddar, Y. Henry, J. Arabski, V. Da Costa, M. Bowen, J. Le Moigne, and E. Beaurepaire, Phys. Rev. B **76**, 033302 (2007).
110. S. Kera, M. Casu, K. Bauschpies, D. Batchelor, T. Schmidt, and E. Umbach, Surf. Sci. **600**, 1077 (2006).
111. A. R. Rocha, V. Garcia-Suarez, S. W. Bailey, C. J. Lambert, J. Ferrer, and S. Sanvito, Nat. Mater. **4**, 335 (2005).

Reactive Simulations for Biochemical Processes

M. Boero

Abstract After a brief review of the hybrid QM/MM molecular dynamics scheme and its coupling to the metadynamics method, I will show how such a combination of computational tools can be used to study chemical reactions of general biological interest. Specifically, by using such a reactive hybrid paradigm, where the QM driver is a Car–Parrinello Lagrangian dynamics, we have inspected the ATP hydrolysis reaction in the anti-freezing protein known as heat shock cognate protein (Hsc70) and the unconventional propagation of protons across peptide groups in the H-path of the bovine cytochrome c oxidase. While the former represents a fundamental reaction operated by all living beings in a wealth of processes and functions, the second one is involved in cell respiration. For both systems accurate X-ray data are available, yet the actual reaction mechanism escapes experimental probes. The simulations presented here provide the complementary information missing in experiments, offer a direct insight into the reaction mechanisms at a molecular level, and allow to understand which pathways nature can follow to realize these processes fundamental to living organisms.

1 Introduction

Modeling biochemical reactions represents one of the recent challenges in molecular simulations. The attention that this specific field of atomic scale calculations has gained, especially in the recent years, stems from the fact that biomolecules are becoming an important target in biochemistry and molecular engineering. Although still at a pioneering stage, proteins and nucleic acids are at the crossroad of biology, chemistry, and nanotechnology because of the wealth of functions exerted, their self-assembly properties, and the possibility of being synthesized in a laboratory in specifically designed ways. However, experimental probes are often insufficient to

M. Boero (✉)

Institut de Physique et Chimie des Matériaux de Strasbourg,
23 rue du Loess, F-67037 Strasbourg, France; Center for Computational Sciences,
University of Tsukuba, 1-1-1 Tennodai, Tsukuba, Ibaraki 305-8577, Japan; CREST–
Japan Science and Technology Agency, Honcho 4-1-8, Kawaguchi, Saitama, Japan
boero@comas.frsc.tsukuba.ac.jp; mauro.boero@ipcms.u-strasbg.fr

Boero, M.: *Reactive Simulations for Biochemical Processes*. Lect. Notes Phys. **795**, 81–98 (2010)
DOI 10.1007/978-3-642-04650-6_3   © Springer-Verlag Berlin Heidelberg 2010

recover a detailed atomic scale information suitable to address all the issues raised by these rather complicated systems.

One possibility to recover the missing information and to perform atomic scale virtual experiments is represented by molecular simulations. In particular, the introduction of first-principles molecular dynamics by R. Car and M. Parrinello more than 20 years ago [1] made it possible to simulate electronic structure changes of complex system at finite temperature. Such an approach has been applied to a variety of systems [2] from inorganic solid state chemistry to aqueous reactions and biochemical systems. Nonetheless, two main bottlenecks have to be faced when large biomolecules are involved.

On the one hand, the size of a typical biochemical system, namely a protein or a nucleic acid, often exceeds 10,000 atoms. In addition to that, they are generally immersed in several thousands of solvating water molecules that easily make the size of the system not affordable at a full quantum level. On the other hand, classical model potentials can afford systems of these sizes, but at the price of neglecting the electronic structure. Since breaking and formation of chemical bonds are always occurring when chemical reactions are involved, taking into account the electronic structure and the modifications it undergoes as a consequence of these processes is of fundamental importance.

A second bottleneck is represented by the timescale on which biochemical reactions occur. These times can range from microseconds or milliseconds to seconds and even longer times when large activation barriers have to be overcome. Such times are far beyond the reach of any quantum simulation, generally limited to few tens of picoseconds, and also classical simulations, which are of the order of a few hundreds of nanoseconds.

The scope of the present work is to give an overview on these two issues. After a brief summary of a widely used hybrid QM/MM molecular dynamics scheme, able to overcome the size problem at least in a class of systems where the active site region does not extend to thousands of atoms, a short description of the metadynamics method will be given. The latter is useful to simulate rare events and to solve, for the class of problems discussed in the reminder of the chapter, the timescale problem.

Emphasis will be given on the coupling and joint use of QM/MM and metadynamics, showing how such a combination of computational tools can be used to study chemical reactions of general biological interest. Specifically, by using such a reactive hybrid paradigm, where the QM driver is a Car–Parrinello dynamics, we have inspected the ATP hydrolysis reaction in the anti-freezing protein known as heat shock cognate protein (Hsc70) [3–5], the unconventional propagation of protons across peptide groups in the H-path of the bovine cytochrome c oxidase [6–8] and the charge hopping process along DNA [9, 10].

These three selected themes are of fundamental importance in many respects. The first one represents the basic reaction operated by all living beings in a wealth of processes and functions [11]. The second one is at the basis of cell respiration and one of the forefront research topics in nanobiology and biochemistry [12]. The third one is crucial in charge localization and transport both for the oxidative damage of DNA in living organisms [13] and because of potential applications of synthetic DNA strands in nanoelectronics [14]. For these systems accurate structural X-ray data are available, yet the actual reaction mechanism escapes experimental probes. The simulations presented here provide the complementary information missing in experiments, offer a direct insight into the reaction mechanisms at a molecular level, and allow to understand which pathways nature can follow to realize these processes fundamental to living organisms.

2 Bridging Length and Timescales

2.1 Hybrid QM/MM Approaches

The original idea of dividing a large system into two (or more) subsystems, each computed at a different level of accuracy, dates back to 1978 and was due to Momany [15]. Namely, a small portion of the system, where the important chemical reactions and electronic structure modifications occur, is represented quantum mechanically (QM) while the rest of the system, important from a structural point of view, but chemically inert, is described by classical molecular mechanics (MM) [16, 17], which, for the applications described here, is represented by the well-assessed and benchmarked Amber force field [18, 19].

In this type of approaches, the delicate part is represented by the interface between the two subsystems. By using a DFT-based first-principles approach, as in Car–Parrinello molecular dynamics, to represent the QM subsystem, a compromise between accuracy and computational workload can be achieved. Yet a fully Hamiltonian coupling can be obtained by extending the Car–Parrinello Hamiltonian H^{CP} as

$$H^{QM/MM} = H^{CP}[\rho, \{\mathbf{R}_I\}] + H^{MM}[\{\mathbf{r}_J\}] + H^{int}[\rho, \{\mathbf{r}_J\}], \qquad (1)$$

where ρ is the electron density provided by the DFT-based QM driver, $\mathbf{R}_I$ ($I = 1, ..., QM$) the Cartesian positions of the QM atoms, and $\mathbf{r}_J$ ($I = 1, ..., MM$) the Cartesian positions of the rest of the (MM) system described by a force field. The coupling functional

$$H^{int}[\rho, \{\mathbf{r}_J\}] = \sum_{J=1}^{MM} q_J \int \frac{\rho(\mathbf{x})}{|\mathbf{x} - \mathbf{r}_J|} d^3x \qquad (2)$$

represents one of the major computational costs, since the sum runs over all the MM classical atoms and the integration has to be performed numerically in the whole space. Such a partitioning of the whole system is schematically shown in Fig. 1. To reduce the workload due to this particle–mesh interaction one can observe that large $|\mathbf{x} - \mathbf{r}_J|$ distance leads to negligible contributions both because the denominator in

Fig. 1 Hybrid QM/MM
partitioning scheme. The QM
subsystem is generally
embedded inside a larger MM
system and the two worlds
interact with each other via
the coupling Hamiltonian
H^{int}. If electrons are
represented on a grid inside
the QM box, as in the case of
plane waves, H^{int} is a
particle–mesh interaction
hamiltonian

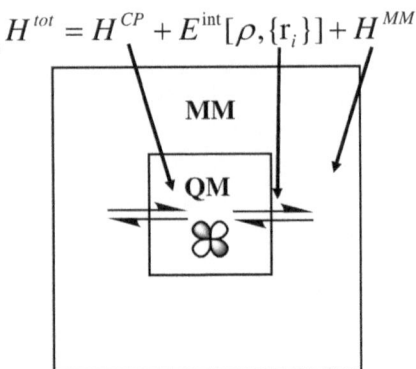

$$H^{tot} = H^{CP} + E^{\text{int}}[\rho, \{\mathbf{r}_i\}] + H^{MM}$$

Eq. (2) becomes very large and because classical q_J charges very far from the QM subsystem are screened by all the interposed atoms, molecules, and residues.

It is then possible to divide the calculations in different domains, one close to the QM subsystem, $r < r_1 \sim 10 - 15$ Å , where the coupling functional is computed as in Eq. (2) but for a reduced subset of atoms MM$'$ < MM, one at intermediate distances $r_1 < r < r_2$ where a rescaled electrostatic potential (RESP) scheme is adopted [20] and whose extension depends on the system under study, and a longer range domain, $r > r_2$, where a standard multipolar expansion is adopted. The RESP term replaces the particle–mesh integral with a classic-like expression

$$H_{\text{RESP}}^{\text{int}} = \sum_{J=1}^{\text{MM}''} q_J \sum_{I=1}^{\text{QM}} \frac{q_I^{\text{RESP}}(\rho)}{|\mathbf{R}_I - \mathbf{r}_J|}, \tag{3}$$

and MM$''$ < MM is the subset of classical atoms inside this intermediate region $r_1 < r < r_2$. The quantity indicated as q_I^{RESP} is constructed by fitting the point-like Coulomb potential to the true electrostatic potential (ESP)

$$V_J = \int \frac{\rho(\mathbf{x})}{|\mathbf{x} - \mathbf{r}_J|} d^3 x \tag{4}$$

smoothed at short distances to avoid spurious overpolarization effects. In addition to that, the charge is restrained to values close to the so-called Hirshfeld charge [21] defined as

$$q_I^{\text{Hirshfeld}} = \int \rho(\mathbf{x}) \frac{\rho^{at}(|\mathbf{x} - \mathbf{R}_I|)}{\sum_{I'} \rho^{at}(|\mathbf{x} - \mathbf{R}_{I'}|)} d^3 x - Z_I, \tag{5}$$

where ρ is the charge density provided by the DFT used for the QM subsystem, ρ^{at} the atomic valence charge density, and

$$Z_I = \int \rho^{at}(|\mathbf{x} - \mathbf{R}_I|)d^3x \qquad (6)$$

is the bare valence of the Ith atom. In practice, we minimize on the fly, during the dynamics, the quantity

$$\chi = \sum_{J=1}^{MM''} \left(\sum_{I=1}^{QM} \frac{q_I^{RESP}}{|\mathbf{R}_I - \mathbf{r}_J|} - V_J \right)^2 - w_q \sum_{I=1}^{QM} \left(q_I^{RESP} - q_I^{Hirshfeld} \right)^2, \qquad (7)$$

where w_q is a weight factor generally ranging from 0.1 to 0.25. As can be noticed, this expression can be easily rewritten as

$$\chi = \sum_{J=1}^{MM''} \left(\sum_{I=1}^{QM} A_I^J q_I^{RESP} - T^J \right)^2, \qquad (8)$$

where the matrix elements A_I^J are equal to $1/|\mathbf{r}_I - \mathbf{r}_J|$ for $J \in MM''$ and equal to $w_q \delta_{IJ}$ for $J \in QM$. Analogously, the vector T^J is equal to V_J for $J \in MM''$ and to $w_q q_J^{Hirshfeld}$ for $J \in QM$. Hence, the minimization of χ becomes a trivial least square procedure and, as such, it reduces to a simple matrix inversion, doable "on the fly," that does not add significant computational cost to the simulation.

2.2 Metadynamics and Free Energy Sampling

Simulating chemical reactions and activated processes is computationally challenging because of the long timescales involved, due to relatively high free energy barriers separating reactants and products. The various methods proposed over the years to overcome this difficulty can be classified into two groups: (i) path sampling/optimization methods [22–25], assuming known initial and final states and (ii) biasing potential approaches [26–29] not relying on a priori knowledge of the final states but requiring one or more well-defined reaction coordinates. This latter case is the one more often used in biochemical simulations, since X-ray data can generally provide the coordinates of the initial structure representing the reactant side.

The first step is the identification of suitable order parameters, or collective variables $s_\alpha(\mathbf{q})$, with $\alpha = 1, ..., p$, which are generally analytic functions of a subset $\mathbf{q} = \{\mathbf{R}_I, \psi_i, \mathbf{r}_J, ...\}$ of QM and/or MM atomic coordinates, electronic wavefunctions ψ_i, molecular dipoles, and so on. This step depends clearly on the system and type of reaction. As a thumb rule, all the slowly varying degrees of motion and in particular those that involve breaking and formation of chemical bonds (hence QM variables) must be accounted for in the selected set of collective variables. Specific cases will be discussed in the next section.

These s_α can then be included as additional dynamical variables completing the Hamiltonian of the system as

$$H^{\text{Tot}} = H^{\text{QM/MM}} + \frac{1}{2} \sum_\alpha M_\alpha \dot{s}_\alpha^2(t) + \frac{1}{2} \sum_\alpha k_\alpha \left[s_\alpha(t) - s_\alpha(\mathbf{q}) \right]^2 - V(s_\alpha, t), \quad (9)$$

where the harmonic potential restrains the oscillations around the analytical expression of the collective variables and a time-dependent penalty potential $V(s_\alpha, t)$ has the scope of pushing the system to escape the initial local minimum and preventing this same system to visit again a region of the phase-space already explored [28, 29].

In practice, after performing an ordinary molecular dynamics run and exploring the local phase-space, a Gaussian penalty function is added to the global potential and subsequent additions of these functions, accumulated in $V(s_\alpha, t)$, allow the system to explore the free energy landscape spanned by the variables s_α as schematically illustrated in Fig. 2. The history-dependent potential has the explicit form

$$V(s_\alpha, t) = \int_0^t |\dot{\mathbf{s}}(t')| \delta \left(\frac{\dot{\mathbf{s}}(t')}{|\dot{\mathbf{s}}(t')|} [\mathbf{s}(t') - \mathbf{s}] \right) \cdot A(t') \exp \left(-\frac{(\mathbf{s}(t') - \mathbf{s})^2}{2(\Delta s)^2} \right) dt', \quad (10)$$

where $\mathbf{s} = (s_1, ..., s_\alpha, ...)$, the Dirac delta ensures the continuity of the trajectory and velocities and the amplitude $A(t')$ has the dimensions of an energy. In a dynamical simulation, a new Gaussian contribution is added to the potential $V(s_\alpha, t)$ every given time interval, amounting to few hundreds of ordinary molecular dynamics steps. A careful tuning of the two input parameters $A(t')$ and Δs is required, since these two quantities determine how accurate is the free energy landscape exploration and how fine is the sampling of the reaction coordinates, respectively. Since

Fig. 2 Metadynamics scheme. The local minimum is filled with subsequent small Gaussian functions until saturation. The system then escapes the local minimum and goes exploring another part of the free energy landscape

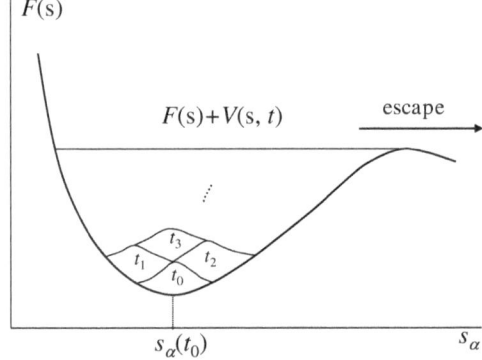

$s_\alpha(t)$ are dynamical variables entering the extended Lagrangian, they have their own equations of motion

$$M_\alpha \ddot{s}_\alpha(t) = k_\alpha \left[s_\alpha(t) - s_\alpha(\mathbf{q}) \right] + \frac{\partial V(s_\alpha, t)}{\partial s_\alpha(t)} \quad (11)$$

that can be solved numerically by finite difference within the Verlet algorithm, i.e.,

$$s_\alpha(t + \Delta t) = 2s_\alpha(t) - s_\alpha(t - \Delta t) + \frac{1}{2M_\alpha} \left\{ k_\alpha \left[s_\alpha(t) - s_\alpha(\mathbf{q}) \right] + \frac{\partial V}{\partial s_\alpha(t)} \right\}. \quad (12)$$

In this way, since $s_\alpha(t)$ represent also the center of the small Gaussian contributions to the penalty potential $V(s_\alpha, t)$, the center of each Gaussian and its dynamical evolution is self-selected by the code, driven by the associated Euler–Lagrange equations of motion and does not rely on any user's choice.

The free energy hypersurface $F(s_\alpha)$ results as

$$F(s_\alpha) = -\lim_{t \to \infty} V(s_\alpha, t) + \text{const.}, \quad (13)$$

where the limit has to be intended in the sense that the metadynamics is continued and Gaussians accumulated in the history-dependent potential until the selected portion of the phase-space, spanned by $\{s_\alpha\}$, is completely explored. It can be demonstrated that the addition of these Gaussian functions leads to a proper flat saturation of the free energy (hyper)surface, provided that they are sufficiently small [30–33].

3 Reactive Biochemical Systems

The first application of a combined QM/MM and metadynamics approach to a system of general biochemical interest targeted the adenosine triphosphate (ATP) hydrolysis in a particular protein called heat shock cognate-70 (Hsc70) [34]. Its curious name from the original discovery in cells exposed to high temperature. As a response to the external increment of temperature, this protein is produced and used to attenuate the external shock and to recover possible thermal damages to the cell [4, 5]. Furthermore, Hsc70 has a wealth of important biological roles: protects cells from stress, maintains cellular homeostasis, influences metabolism and muscular adaptation, prevents ischemia, and acts as an anti-freezing agent [35, 36]. This specific protein is structurally formed by three main blocks, an N-terminal ATP synthase (ATPase) domain, a peptide-binding site, and a C-terminus. The chemically active part is the ATP domain where three phosphate groups PO_4 are located and chemically bound to each other in a chain as illustrated in Fig. 3.

The system used in hybrid QM/MM–metadynamics simulations is the active mutant of Hsc70 named T13G as provided by the RCSB Protein Data Bank [37]. This corresponds to a set of Cartesian coordinates for the bare protein, obtained by accurate X-ray crystallography. Yet, apart from few water molecules trapped in the

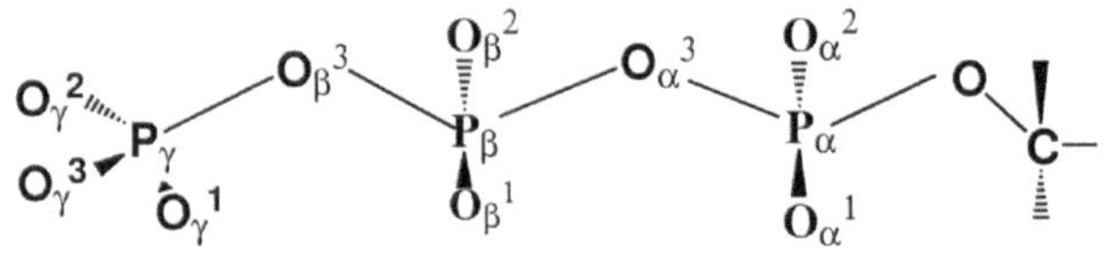

Fig. 3 The three-phosphate chain of the ATP domain, common to any ATP system, with the standard labeling of the P and O atoms

pocket where the ATP domain is located, the natural environment in which proteins and nucleic acids are present in vivo is not provided by these experiments. For this reason, after downloading these set of coordinates, all the surrounding solvating water has to be added and the system carefully equilibrated at fisiological (room) temperature. For this reason we first solvated and equilibrated the protein at 300 K within a classical molecular dynamics framework using a standard Amber force field [19]. After reaching the thermal equilibrium, the run was continued within the QM/MM approach for about 5 ps to allow also for the electronic degrees of freedom to equilibrate.

The Hsc70 T13G mutant is shown in Fig. 3. Since the target of the study was the ATP hydrolysis reaction, $ATP+H_2O \rightarrow ADP+P_i$, the QM subsystem had to include the three phosphorous atoms named P_α, P_β, and P_γ and their chemically bonded O atoms, the metal cations K^+ and Mg^{2+}, and their hydration water molecules. Furthermore, the sp^3 carbon atom chemically bound to P_α was included in the QM calculation since it is easier to saturate its empty valence, across which the QM/MM boundary passes, with a monovalent link atom [38, 39]. This amounts to a total QM subsystem of 35 atoms. The reaction path sampling, performed via metadynamics, made use of two collective variables; the first one is the distance $s_1 = O_\beta^3 - H_{water}$ between the O atom bridging P_β and P_γ (O_β^3) and a proton of a peculiar water molecule identified as the catalytic one and indicated as pH_2O in Fig. 3. The second one was the distance $s_2 = P_\gamma - O_{water}$ between the terminal phosphorous atom and the oxygen atom belonging to the catalytic H_2O molecule which is expected to dissociate and to participate actively to the reaction.

The simulations have been able to show that this specific water molecule, initially belonging to the hydration shell of the metal cation Mg^{2+}, present in the vicinity of the β-phosphate, is indeed the trigger of the reaction. This justifies its name, found in the literature and proposed few years ago on the basis of the X-ray data analysis: *putative catalytic water* [40].

Upon dissociation, this water molecule donates one of its protons to the P-bridging O_β^3 atom and this process starts the whole hydrolysis reaction. The free energy landscape is shown in Fig. 4, where the snapshots (a), (b), and (c) indicate the initial ATP configuration, the transition state, and the final product, respectively. Two important issues were unraveled by these simulations: (i) the pathway followed by both the proton H^+ and the hydroxyl anion OH^- after the dissociation of the putative catalytic water and (ii) the cooperative role of the metal cations K^+ and Mg^{2+} in completing the reaction and promoting the release of the inorganic phosphate

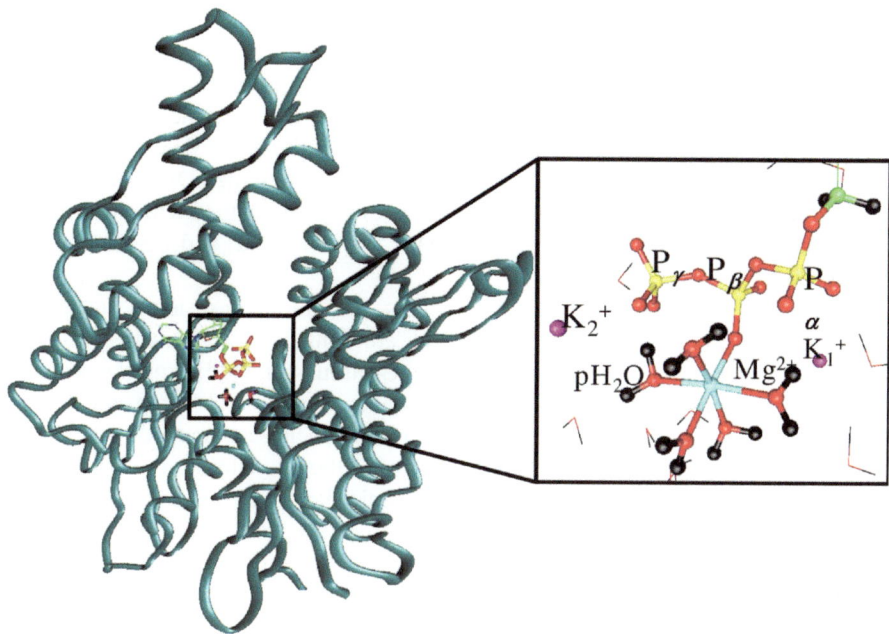

Fig. 4 The Hsc70 protein with the ATP center (QM subsystem) shown with stick and balls in the *inset*. The ribbon is the ideal line connecting all the sp^3 C atoms along the protein. The main atoms involved in the reaction are labeled according to the standard notation

P_i via an exchange of the OH^- hydroxyl anion between their respective hydration shells. This is deeply different from the proton wire mechanisms [41] evidenced in other proteins, such as actin [42], and lowers significantly the free energy barrier, that for this specific reaction turns out to be $\Delta F^* \sim 3 - 4$ kcal/mol, with an overall energy difference $\Delta F = 6.9$ kcal/mol between ATP and ADP, in good agreement with the experimental outcome (7.1 kcal/mol). This is summarized by the free energy landscape and the snapshots reported in Fig. 4. It is worthy of note that the use of a different unique reaction coordinate, namely the coordination number of the terminal phosphorous atom P_γ with any possible water molecule around, gave an analogous reaction mechanism, triggered by the dissociation of exactly the putative catalytic H_2O molecule, with numerically identical activation and overall free energies [34]. Such a small activation barrier means a very high reaction rate. This is essential, for instance, to release the energy provided by the ATP-to-ADP conversion and to prevent the freezing of the cell and the consequent crystallization of the physiological water into ice that could break the cell membranes and kill the cell.

A similar computational approach was used to study the proton transfer process through transmembrane proteins, a fundamental step in cell respiration. One of the most representative and better characterized protein is cytochrome c oxidase (CcO), known to act as a one-directional proton pump. Recent researches focused on the identification, mainly based on analyses of X-ray crystallographic structure, of possible proton transfer pathways in this peculiar protein present both in mammals and in aerobic bacteria [6–8]. Its specific action is to pump protons across the inner mitochondrial membrane or bacterial cytoplasmatic membrane and this process is coupled with O_2 reduction. On the basis of the X-ray data, two pathways driven by two distinct hydrogen bond (H-bond) networks, called D-path and K-path (Fig. 5), have been proposed [43, 44]. These pathways are named with the letter that identifies their respective main residues, D91 (aspartic acid) and K319 (lysine) in bovine CcO.

These networks are formed by side chains of amino acid residues and crystallographically ordered H_2O molecules. Both are considered to be active pathways along which protons can be transferred via a Grotthuss-like [45] proton wire mechanism [41]. In particular, both the K- and D-paths involve an H^+ propagation along a fully connected H-bond network until the proton reaches a heme redox center (see Fig. 5). Then the proton continues its trip inside the cytochrome c oxidase and is eventually pumped outside the inner cytoplasmic membrane. These two pathways are realized by a series of Eigen–Zundel transitions [46, 47], typical of proton propagation in H-bond networks, in which the proton is transferred sequentially from a donor to an acceptor.

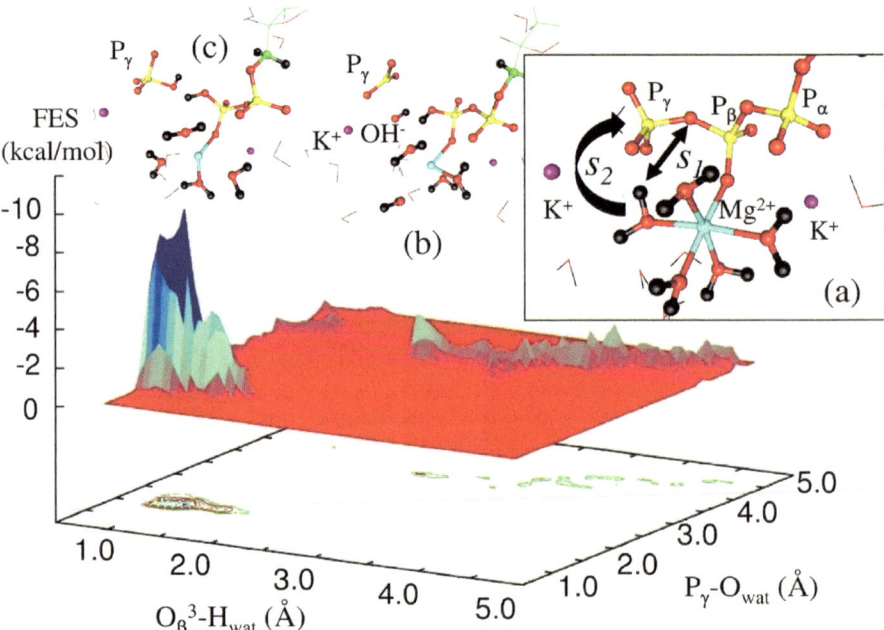

Fig. 5 ATP hydrolysis in Hsc70. The initial system (**a**) reaches a transition state (**b**) via deprotonation of a specific water molecule. This results in the bond cleavage $P_\beta–O_\beta^3–P_\gamma \rightarrow P_\beta–O_\beta^3 \, P_\gamma$ with subsequent ADP formation and release of the inorganic phosphate (**c**). The two collective variables used in the metadynamics are the distances $s_1 = O_\beta^3 – H_{water}$ and $s_2 = P_\gamma – O_{water}$ shown by the *arrows*

Extensive studies have been published on the subject [48, 49]; without any claim of completeness, let us recall that whenever a single proton is released in water, its positive charge experiences an electrostatic attraction from surrounding negative charges. Now, in water, negative charges are represented by lone pair electrons located on the O atoms of H_2O molecules. The proton can then approach a water monomer and form an extra O—H bonds becoming an hydronium OH_3^+. Such a system is characterized by a net charge $Q = +1$ and structurally appears as a triangular pyramidal structure in which the O atom is the vertex and the three bonded protons form the base and are hydrogen bonded to the O atoms of surrounding water molecules as sketched in the left panel of Fig. 6. Such a configuration is termed Eigen complex.

Since three O—H bonds, for a single O atom, do not represent a steady state, one of the three protons of OH_3^+ departs on a picosecond timescale and the hydronium reverts to a water molecule. Here a first difficulty in the proton transfer process arises: the proton that departs from OH_3^+ is not necessarily the one that arrives. As a consequence, the motion of a proton in a H-bond network is not the motion of a single particle in a medium. In jumping from one molecule to the next one, the proton experiences two almost equivalent attractive forces from the lone pair electrons of the two H-bonded H_2O monomers. As a result, for a certain time, again of the order of the picosecond, the proton jumps between the two molecules with a rapid switch between an O—H σ-bond and a hydrogen bond in a typical Grotthuss mechanism [45]. The proton sits on average between two H_2O molecules in a Zundel complex (right panel of Fig. 6) and can be seen as a shared proton rather than an

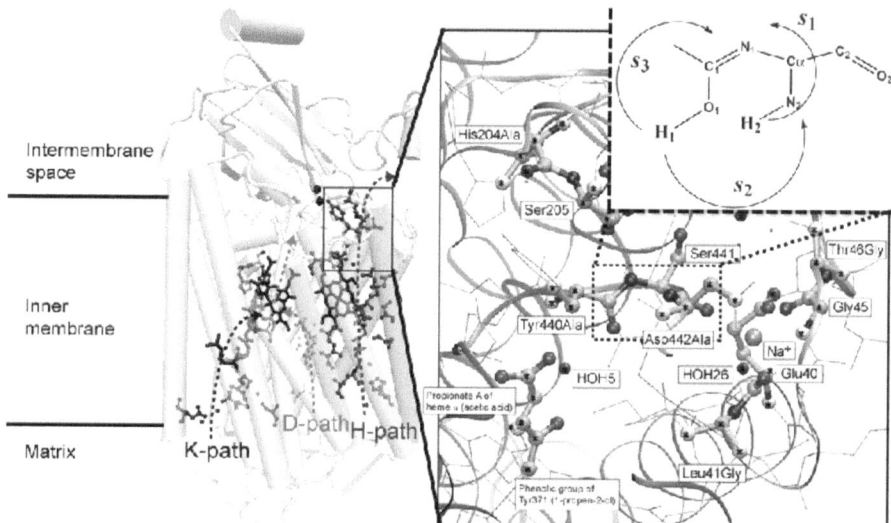

Fig. 6 Schematic view of cytochrome c oxidase and the three possible proton propagation paths. The K-path (*on the left*) and the D-path (*at the center*) drive the proton to a heme redox center. Along the H-path (*on the right*), instead, the proton is directly expelled outside the inner membrane. The *insets* show the QM system and the specific collective variable used

atom chemically bound to a given O site. Given this scenario, the propagation of a proton in a H-bond network can only be described as a series of continuous switches between an Eigen and a Zundel complex. In this respect, the traditional concept of *transition state* is not applicable [50].

In the case of CcO the H-bonds of the migrating hydronium OH_3^+ are formed not only with water molecules, present along the path and crystallographically identified, but also with O or N atoms belonging to side chains and residues facing the proton pathway.

As far as the H-path is concerned, its novelty stems from the presence of a peptide group in the pathway that interrupts the regular H-bond network and could, in principle, hinder the proton propagation. Contrary to the two paths described above, although the H-path still consists mainly of an H-bond network, a peptide group located between two residues, Tyr440 and Ser441, and shown in the inset of Fig. 5, is present on the pathway and breaks the H-bond network into two parts [51]. The traveling proton finds then an obstacle that cannot be overcome by an ordinary Eigen–Zundel mechanism. In fact, the crossing of a chemical bond, instead of the propagation along the weaker H-bonds, by a proton represents a non-trivial process that made for a long time questionable the viability of the H-path.

Recent simulations within the metadynamics approach [52, 53] were performed targeting this particular step in the H-path in an attempt at investigating whether or not such an anomalous proton transfer could occur. Specifically, two possible alternative pathways were investigated. The first one involves a double-proton transfer, one from the nitrogen site N_1 to N_2 and one from the oxygen site O_1 to N_2. Such a simulation involved the simultaneous use of two collective variables s_1 and s_2 indicated in the inset of Fig. 5: s_1 is the distance H_2 to N_1, while s_2 is the distance H_1 to N_2. These distances account for the proton transfer from N_2 to N_1 (s_1), necessary to pump a proton above the peptide group, and for a second proton transfer from O_1 to N_2, necessary to restore the protonation state of the N_2 site and to make O_1 ready to accept the next proton coming (via Grotthuss mechanism) from the inner membrane of the system.

In a second simulation, a direct proton transfer from O_1 to N_1 was simulated, using as a unique collective variable s_3, corresponding to the distance of the proton H_1, initially bound to the acceptor site O_1, to the site N_1 to be protonated. Such a setup corresponds to a direct jump of the incoming proton across the peptide bond, without including any other atom but the two directly involved in the H^+ propagation process.

The results have shown that, provided that the double-proton transfer pathway occurs, protons can indeed cross this blocking peptide group without inducing any permanent conformational change to the system, which keeps its experimental structure at the end of the reaction. This makes the simulated mechanism compatible with all the known X-ray data. This kind of reaction is known as tautomerization and has been shown to occur in several systems characterized by an analogous chemical structure, such as, for instance, in polyglycine enol-to-keto transformations [54]. Indeed, this is the actual rate limiting step in the peptide group crossing process and involves the overcoming of a relatively small free energy barrier (13 kcal/mol).

Conversely, to realize the direct proton transfer from O_1 to N_1, important geometrical modifications, inducing large bond stresses, are required. As a result a very large free energy barrier, amounting to more than 60 kcal/mol, characterizes this pathway, making it very unlikely. Moreover, large structural modifications are required for this pathway, thus making it incompatible with the experimental crystallographic structure.

These simulations have represented the first atomic-level inspection and the first support to the H-pathway in bovine CcO. As such, they have been able to complement the experiments and to offer a direct insight into a crucial part of the proton propagation mechanism not directly accessible to experimental probes.

Another interesting case in which proton transfer processes are crucial, yet peculiar, is represented by charge propagation along a double-stranded DNA system. Charge transfer processes in native and synthetic DNA have been at the center of intense studies. Besides its basic biological importance [13, 55], potential nanotechnological applications are being pioneered [56, 57]. In fact, DNA is a stable polymer with a strong self-assembly one-dimensional character that seems well suited for the engineering of molecular junctions and nanoarrays [58, 59].

Recent experiments [9, 10, 60] have provided a novel insight into charge transfer processes in DNA; the general picture that emerges is that the charge displacement, especially holes h^+, from a Guanine–Cytosine (G:C) base pair to another one along the strand is accompanied by a deprotonation of the G-base of the base pair where the charge is localized after having been displaced. In this respect, here the proton transfer is not the main actor, but rather the helper, a sort of *gate* able to open or to close the door to the passage of electronic charge along the DNA double helix.

In an attempt at recovering the microscopic picture of this mechanism, hybrid QM/MM simulations on a fully hydrated DNA system, coupled to the metadynamics approach, were performed [61, 62]. An equilibrated and fully hydrated double-stranded B-DNA amounting to 20,265 atoms, 5,902 solvating water molecules, and 238 atoms in the QM subsystem was adopted. The QM part consists of five base pairs corresponding to the segment $5'-\text{GTGGG}-3'$, namely a G:C base pair separated by other three G:C pairs by one intercalated A:T (adenine–thymine), similar to the sequence used in the cited experiments.

The collective variable selected to simulate this process was the coordination number of the N atom of a C-base, labeled as N_1 in Fig. 7 with nearby H-bonded protons. If the proposed mechanism is correct, then protonating this N site should lead to a charge localization on the G-base H-bonded to this C upon deprotonation of this G.

Indeed, the mechanism reproduced by the simulation and shown in Fig. 7 provided a direct evidence for the proposed hole transfer mechanism. Namely, the charge is transferred from a G:C base pair to another one in a coherent single step, involving a double-proton exchange process. Starting from a regular G:C configuration, shown in the left panel in the scheme of Fig. 7, a temporary deprotonation of C occurs indicated in the middle panel as a shift of the upper black ball from G to C; this step, although being a transient stage, is crucial since it confers an acid character to the G-base that becomes ready to release protons. In the absence of

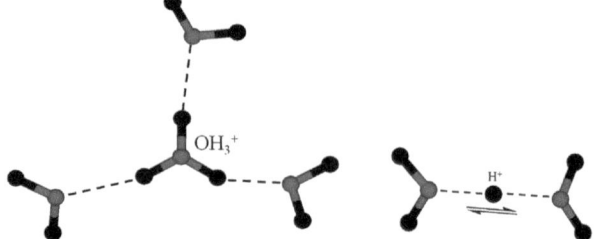

Fig. 7 Eigen complex $[H_3O \cdot (H_2O)_3]^+$ *(left panel)*, with the OH_3^+ at the center, surrounded by its three hydrogen bonded H_2O molecules and Zundel complex $[H \cdot (H_2O)_2]^+$ *(right panel)* in which the extra proton is shared between two water molecules. H_2O molecules are represented as v-shaped stick and balls (H atoms are *black* and O atoms *dark-gray*), *dashed* lines indicate H-bonds, and the shared proton in the Zundel complex is labeled as H^+ and shown as a *black ball*

this auxiliary proton transfer, a G-base is pretty stable and no proton transfer would occur, hence the gate to the charge transfer would not open. On the other hand, in the conditions induced by this temporary proton transfer, the N_1 site of C accepts a proton from G that, in turn, becomes deprotonated. Then the temporarily shifted (from C to G) H^+ is donated back to C, leaving behind a deprotonated G_{-H} base.

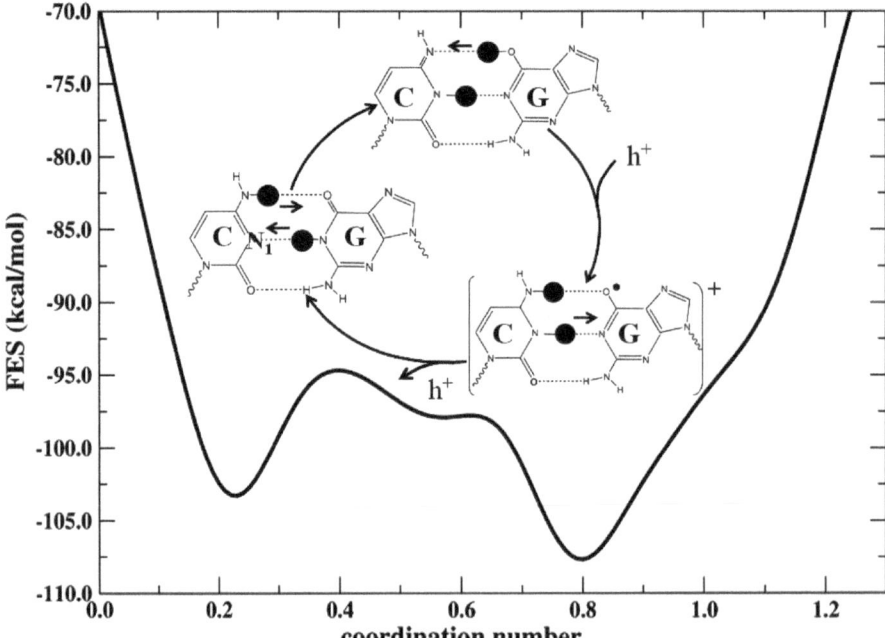

Fig. 8 Free energy profile for the double-proton exchange mechanism for a hole (h^+) transfer in a G:C base pair of DNA obtained by metadynamics, assuming as a collective variable the coordination number with H of the N atom of the C-base labeled as N_1 and facing its H-bonded G-base N site. The two protons involved in the process are shown as *black balls*. *Dashed lines* indicate H-bonds

When this deprotonation occurs, the charge displaces along the DNA and localizes on this $G_{-H}:C^+$ site as shown in Fig. 8.

The free energy barrier for this process was estimated to be about 7 kcal/mol, as can be easily seen from the reported curve, with the final minimum lower in energy by 2–3 kcal/mol. Upon hole extraction and re-protonation of this site, the relative stability of the two minima is reversed and the charge can migrate elsewhere, as indicated by the schematic loop in the figure.

These results have found recently an indirect confirmation in the analysis of the oxidation of a guanine nucleobase to its radical cation in DNA oligomers [63]. These experiments were able to show that oxidation processes are responsible for an increase in the acidity of the N amino proton of G. This increase of acidity, in turn, is able to lead to a spontaneous proton transfer from the G-base to the N_1 site of the paired C-base.

Hence, the displacement of protons from one base to the nearby one along the H-bonds between the two coupled bases plays the role of a sort of gate, opening the door to the flowing charge and promoting its transfer along the double-stranded DNA. Once the gate is open, the electronic charge can either flow along the phosphate backbone or make a direct hopping to $G_{-H}:C^+$. These results have been regarded as a first support to the mechanism proposed on experimental grounds and, at the same time, as a complement to the experimental results obtained both from H/D isotopic substitution and from electron paramagnetic resonance [9, 10, 60].

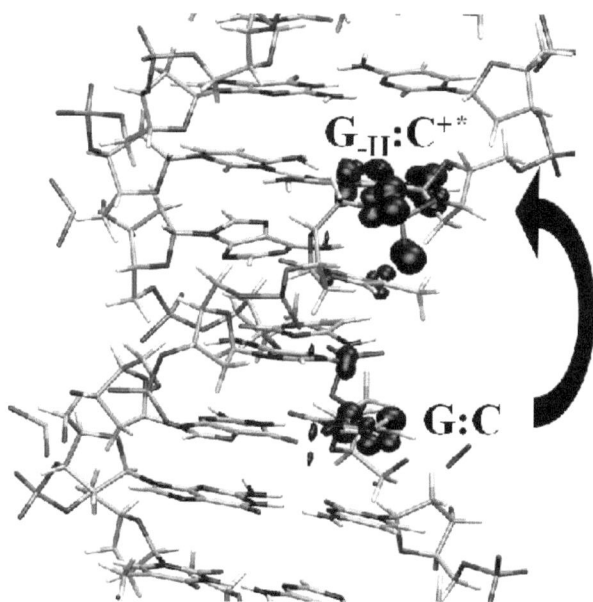

Fig. 9 Migration of an injected hole charge from the initial G:C base pair to the proton-shift site $G_{-H}:C^+$ along the double-stranded DNA system. The *thick black arrow* indicates the direction of the charge migration, represented by the black isosurface

4 Conclusions

The overview presented here, covering just a limited number of examples of computational approaches suitable to study chemical and catalytic reactions in complex biological systems, is far from being complete and exhaustive of such a rapidly developing field. Within the limitations of the present contribution, I just aimed at offering to the reader a hint about what is possible nowadays, with the present computer facilities and methodologies, to theoretically investigate biomolecular systems. The examples reported refer to some of the more recent approaches and algorithms that can be used in the simulation of biochemical reactions of practical importance and how they can be used to either complement the experimental information or to perform virtual experiments aimed at making predictions and to support the experimental investigation.

Specifically, the selected examples, taken from recent applications, are shown to offer the reader a direct view of the kind of information that can be obtained by applying these methods to specific problems of general biological and nanotechnological interest. Many more examples are available in the rich literature, part of which is reported in the references included, and that can provide a wider scenario and a more complete panorama to researchers interested in this field.

In a sense, the few examples discussed here, as well as the ones reported so far in the literature, show also the limitations that still exist in dealing with complex biosystems and how simulations must be carefully focused in order to achieve some useful outcome. Indeed, the combined use of hybrid approaches and free-energy sampling methods allows to inspect in great detail non-trivial biochemical reactions, taking into account the whole system (and its solvent) without relying on smaller models extracted from the whole system. In fact, smaller models are often insufficient to reproduce all the features of the real system. Yet, the study of chemical reactions is limited to the small portion of the system tractable by quantum mechanical approaches and this, in turn, has to cope with the many-body approximations involved in these first-principle/ab initio methods.

References

1. R. Car and M. Parrinello, Phys. Rev. Lett. **55**, 2471 (1985).
2. For a comprehensive review, see the monographic issue ChemPhysChem **6**, 1671–1952 (2005).
3. Y. Liu, L. Gampert, K. Nething, and J. M. Steinacker, Front. Biosci. **11**, 2802 (2006).
4. T. G. Chapell, W. J. Welch, D. M. Schlossman, K. B. Palter, M. J. Schlesinger, and J. E. Rothman, Cell **45**, 3 (1986).
5. K. M. Flaherty, C. DeLuca-Flaherty, and D. B. McKay, Nature **346**, 623 (1990).
6. C. Ostermeir, A. Harrenga, U. Ermler, and H. Michel, Proc. Natl. Acad. Sci. U.S.A. **94**, 10547 (1997).
7. M. Svensson-Ek, J. Abramson, G. Larsson, S. Törnroth, P. Brzezinski, and S. Iwata, J. Mol. Biol. **321**, 29 (2002).
8. T. Tsukihara, K. Shimokata, Y. Katayama, H. Shimada, K. Muramoto, H. Aoyama, M. Mochizuki, K. Shinazawa-Itoh, E. Yamashita, M. Yao, Y. Ishimura, and S. Yoshikawa, Proc. Natl. Acad. Sci. U.S.A. **100**, 15304 (2003).

9. B. Giese, J. Amaudrut, A. Köhler, M. Spormann, and S. Wessely, Nature **412**, 318 (2001).
10. V. Shafirovich, A. Dourandin, and N. E. Geacintov, J. Phys. Chem. B **105**, 8431 (2001).
11. P. D. Boyer, *Nobel Lectures, Chemistry 1996–2000*, ed. I. Grenthe (World Scientific Publishing, Singapore, 2003), p. 120.
12. T. E. Decoursey, Physiol. Rev. **83**, 476 (2003).
13. T. Douki, J. L. Ravanat, D. Angelov, J. R. Wagner, and J. Cadet, Top. Curr. Chem. **236**, 1 (2004).
14. M. di Ventra and M. Zwolak, *Encyclopedia of Nanoscience and Nanotechnology*, ed. H. Singh-Nalwa (American Scientific Publishers, New York, 2004).
15. F. J. Momany, J. Phys. Chem. **82**, 592 (1978).
16. H. M. Senn and W. Thiel, Angew. Chem. Int. Ed. **48**, 1198 (2009).
17. M. Boero, J. M. Park, Y. Hagiwara, and M. Tateno, J. Phys.: Condens. Matter **19**, 365217 (2007).
18. C. I. Bayly, P. Cieplak, W. D. Cornell, and P. A. Kollman, J. Phys. Chem. **97**, 10269 (1993).
19. W. D. Cornell, P. Cieplak, C. I. Bayly, I. R. Gould, K. R. Merz, D. M. Ferguson, D. C. Spellmeyer, T. Fox, J. W. Caldwell, and P. A. Kollman, J. Am. Chem. Soc. **117**, 5179 (1995).
20. A. Laio, J. Vande Vondele, and U. Röthlisberger, J. Phys. Chem. B **106**, 7300 (2002).
21. F. L. Hirshfeld, Theor. Chim. Acta **44**, 129 (1977).
22. C. Dellago, P. G. Bolhuis, and D. Chandler, J. Chem. Phys. **108**, 1964 (1998).
23. G. Henkelman, B. P. Uberuaga, and H. Jonsson, J. Chem. Phys. **113**, 9901 (2000).
24. D. Passerone and M. Parrinello, Phys. Rev. Lett. **87**, 8032 (2001).
25. D. Branduardi, F. L. Gervasio, and M. Parrinello, J. Chem. Phys. **126**, 054103 (2007).
26. H. Gubmüller, Phys. Rev. E **52**, 2893 (1995).
27. M. Sprik and G. Ciccotti, J. Chem. Phys. **109**, 7737 (1998).
28. A. Laio and M. Parrinello, Proc. Natl. Acad. Sci. U.S.A. **99**, 12562 (2002).
29. M. Iannuzzi, A. Laio, and M. Parrinello, Phys. Rev. Lett. **90**, 238302 (2003).
30. A. Laio, A. Rodriguez-Fortea, F. L. Gervasio, M. Ceccarelli, and M. Parrinello, J. Phys. Chem. B **109**, 6714 (2005).
31. M. Boero, M. Tateno, K. Terakura, and A. Oshiyama, J. Chem. Theory Comput. **1**, 925 (2005).
32. A. Laio and F. L. Gervasio, Rep. Prog. Phys. **71**, 126601 (2008).
33. A. Barducci, G. Bussi, and M. Parrinello, Phys. Rev. Lett. **100**, 020603 (2008).
34. M. Boero, T. Ikeda, E. Ito, and K. Terakura, J. Am. Chem. Soc. **128**, 16798 (2006).
35. D. E. Rancourt, V. K. Walker, and P. D. Davies, Mol. Cell Biol. **7**, 2188 (1987).
36. M. I. Qureshi, S. Qadir, and L. Zolla, J. Plant Physiol. **164**, 1239 (2007).
37. X-ray crystallographic coordinates from RCSB Protein Data Bank (http://www.rcsb.org/pdb/home/home.do), accession code "2bup".
38. U. C. Singh and P. A. Kollman, J. Comput. Chem. **7**, 718 (1986).
39. M. J. Field, P. A. Bash, and M. Karplus, J. Comput. Chem. **11**, 700 (1990).
40. M. C. Sousa and D. MacKay, Biochemistry **7**, 15392 (1998).
41. R. Pomès and B. Roux, J. Phys. Chem. **100**, 2159 (1996).
42. J. Akola and R. O. Jones, J. Phys. Chem. B **110**, 8121 (2006).
43. R. B. Gennis, Biochim. Biophys. Acta **1365**, 241 (1998).
44. P. Brzezinski and G. Larsson, Biochim. Biophys. Acta 1605, 1 (2003).
45. C. J. T. von Grotthuss, Ann. Chim. **LVII**, 54 (1806).
46. M. Eigen, Angew. Chem. Int. Ed. **3**, 1 (1964).
47. G. Zundel and H. Metzger, Z. Physik. Chem. **58**, 225 (1968).
48. D. Marx, M. E. Tuckerman, J. Hutter, and M. Parrinello, Nature **397**, 601 (1999).
49. M. Boero, T. Ikeshoji, and K. Terakura, ChemPhysChem **6**, 1775 (2005).
50. D. Marx, ChemPhysChem **7**, 1848 (2006).
51. S. Yoshikawa, Adv. Protein Chem. **60**, 341 (2002).
52. K. Kamiya, M. Boero, M. Tateno, K. Shiraishi, and A. Oshiyama, J. Am. Chem. Soc. **129**, 9663 (2007).
53. K. Kamiya, M. Boero, M. Tateno, K. Shiraishi, and A. Oshiyama, J. Phys.: Condens. Matter **19**, 3652209 (2007).
54. K. Kamiya, M. Boero, K. Shiraishi, and A. Oshiyama, J. Phys. Chem. B **110**, 4443 (2006).

55. D. T. Odom and J. K. Barton, Biochemistry **40**, 8727 (2001).
56. N. Forbes, Comput. Sci. Eng. **2**, 83 (2000).
57. C. Dekker and M. A. Ratner, Phys. World **14**, 29 (2001).
58. J. Chen and N. C. Seeman, Nature **360**, 631 (1991).
59. Y. Zhang and N. C. Seeman, J. Am. Chem. Soc. **116**, 1661 (1994).
60. B. Giese and W. Wesseley, Chem. Commun. **20**, 2108 (2001).
61. F. L. Gervasio, M. Boero, and M. Parrinello, Angew. Chem. Int. Ed. **45**, 5606 (2006).
62. M. Boero, F. L. Gervasio, and M. Parrinello, Mol. Simul. **33**, 57 (2007).
63. A. K. Ghosh and G. B. Schuster, J. Am. Chem. Soc. **128**, 4172 (2006).

Molecular Dynamics Simulations
of Liquid-Crystalline Dendritic Architectures

**C. Bourgogne, I. Bury, L. Gehringer, A. Zelcer, F. Cukiernik, E. Terazzi,
B. Donnio, and D. Guillon**

Abstract We report here a few examples of the self-organization behaviour of
some novel materials based on liquid-crystalline dendritic architectures. The original design of the molecules imposes the use of all-atomic methods to model
correctly every intra- and intermolecular effects. The selected materials are octo-

C. Bourgogne (✉)
Institut de Physique et Chimie des Matériaux de Strasbourg UMR 7504 (CNRS-UDS) 23 rue du
Loess, BP 43, 67034 Strasbourg Cedex, France

I. Bury
Institut de Physique et Chimie des Matériaux de Strasbourg UMR 7504 (CNRS-UDS) 23 rue du
Loess, BP 43, 67034 Strasbourg Cedex, France

L. Gehringer
Institut de Physique et Chimie des Matériaux de Strasbourg UMR 7504 (CNRS-UDS) 23 rue du
Loess, BP 43, 67034 Strasbourg Cedex, France

A. Zelcer
INQUIMAE, Departamento de Química Inorgánica, Analítica y Química Física Facultad de Cien-
cias Exactas y Naturales, Universidad de Buenos Aires Pab. II, Ciudad Universitaria, C1428EHA
Buenos Aires, Argentina

F. Cukiernik
INQUIMAE, Departamento de Química Inorgánica, Analítica y Química Física Facultad de Cien-
cias Exactas y Naturales, Universidad de Buenos Aires Pab. II, Ciudad Universitaria, C1428EHA
Buenos Aires, Argentina

E. Terazzi
Institut de Physique et Chimie des Matériaux de Strasbourg UMR 7504 (CNRS-UDS) 23 rue du
Loess, BP 43, 67034 Strasbourg Cedex, France,
emmanuel.terazzi@ipcms.u-strasbg.fr

B. Donnio
Institut de Physique et Chimie des Matériaux de Strasbourg UMR 7504 (CNRS-UDS) 23 rue du
Loess, BP 43, 67034 Strasbourg Cedex, France,
bertrand.donnio@ipcms.u-strasbg.fr

D. Guillon
Institut de Physique et Chimie des Matériaux de Strasbourg UMR 7504 (CNRS-UDS) 23 rue du
Loess, BP 43, 67034 Strasbourg Cedex, France,
daniel.guillon@ipcms.u-strasbg.fr

Bourgogne, C. et al.: *Molecular Dynamics Simulations of Liquid-Crystalline Dendritic
Architectures*. Lect. Notes Phys. **795**, 99–122 (2010)
DOI 10.1007/978-3-642-04650-6_4   © Springer-Verlag Berlin Heidelberg 2010

pus dendrimers with block anisotropic side-arms, segmented amphiphilic block codendrimers, multicore and star-shaped oligomers, and multi-functionalized manganese clusters. The molecular organization in lamellar or columnar phases occurs due to soft/rigid parts self-recognition, hydrogen-bonding networks or from the molecular shape intrinsically.

1 Introduction

The arborescent dendritic structure is one of the most pervasive, prolific, and influential natural pattern that can be observed on earth, at all dimension length scales (from nanometre to kilometre), at once in the inert, the virus, and the living worlds [1–4]. Such a natural hyperbranched architecture has reached an unrivalled level of perfection and provides maximum interfaces for efficient contacts and interactions, as well as for optimum information collection, transport, and distribution. The elaboration and the synthesis of such aesthetically challenging architectures have been driven by the need to mimic the macroscopic natural branching networks and to convey their functions at the molecular level. Dendrimers and dendrons (or elementary dendritic units) may be considered as polymers with geometrically restricted structures [3, 4].

From another point of view, molecular engineering of liquid crystals (LC) is an important issue for controlling the self-assembling ability and the self-organizing processes of single moieties into periodically ordered meso- and nanostructures [5]. Moreover, ordered supramolecular assemblies can considerably enhance the functions of the single molecule [6, 7]. Dendrimers, dendrons, dendronized and hyperbranched polymers have proved particularly versatile candidates as novel and original scaffoldings for the elaboration of new LC functional materials and research in this area has experienced an outstanding development during the last years, overlapping polymer chemistry and supramolecular chemistry [8–14]. In particular, the dendritic structure appeared as an interesting framework where mesomorphism can be modulated by very subtle modifications of the intrinsic dendritic connectivity. LC dendrimers are now representative of an important class of mesogens where new types of mesophases and original morphologies may be discovered [8–14].

Along this line, understanding the mesomorphic behavior of molecules with novel architectures or complex chemical systems has always been a challenge in spite of the development of accurate analysis techniques such as small angle X-ray diffraction (XRD) or dilatometry. Theoretical simulations of liquid crystals took benefit from the considerable increase of the speed of computers in the recent years and can be now used as a supplementary tool to help experimentalists discriminating between several possible structures. There are many different techniques that can be successfully used for the simulation of liquid crystals [15–17] that differ from each other by the range of length and timescales of the model: from fully atomistic molecular model to mesoscopic physical systems, and from femtoseconds to microseconds timescales. Among molecular-based methods [18] we will focus only on atomistic methods in the present work.

Most of the computer simulations of liquid crystals are focused on the modelling of well-known materials such as nematic cyanobiphenyls and the calculation of their theoretical properties, i.e., phase transition temperature or nematic order parameter. The common techniques used to perform the thermodynamic equilibration of the systems are molecular dynamics (MD) or Monte Carlo (MC) methods, and many published studies are based on coarse-grained or mesoscale models. Such simulations often use a physical representation of the whole molecule as objects with a given shape (cylinder, rod, disc, sphere, etc., or combinations of them) interacting with a Gay–Berne potential [19], or as particles driven by the dissipative particles dynamics (DPD) technique [20]. They are unbeatable to modelize a large number of molecules during a long time and thus to predict the formation of original mesophases [21]. Nevertheless the efficiency of these methods drops if they are used with molecules with shape or electrostatic properties different from those they were specifically designed for, although they can fit many kinds of molecules [22], and even dendrimers [23].

Due to the very long simulation times – tens of nanosecond – and the number of molecules in the model – hundreds – that are necessary to achieve a good level of accuracy in the calculation of bulk properties of the material (even for simple rod-shaped molecules), atomistic-level models are very consuming in computational resources and therefore quite uncommon for liquid crystal modelling. Fully atomistic models (in which every atom of the molecule is explicitly described) although have been proved successful for the calculation of some physical properties [24–26] with a lot of development in tuning the force fields [27] to fit the transition temperatures [28, 29] or/and with using a hybrid approach where some groups of atoms are represented by a single site (united atom) [30, 31]. All this development is necessary to reduce the consumption of computational resources and increase the time step of the MD simulations, and therefore permits much longer calculations. Nevertheless, fitting the force fields also takes a long time and must be done on well-known materials, so it may not be safe to use these methods on novel molecules.

If the aim of the study is not to study a phase transition in extenso or a property that has to be averaged from a huge number of conformations, we can save a large amount of calculation time by studying only a local arrangement of the system during a reasonable time window and to construct the starting conformation of the model in a clever way so that we can reach the thermodynamic stability even with all-atom simulations.

The materials discussed in this chapter have been studied using the Discovery molecular mechanics software from Accelrys. For a single molecule, atomic charges are first calculated from a semi-empirical quantum mechanics AM1 calculation or set by the typing rules of the software if no subsequent problem is dreaded about this. Then models were built as periodic boundary condition (PBC) cells in pseudo-2D or 3D symmetries by assembling by hand a given number of molecules, according to X-ray diffraction experimental information (periodicities and molecular areas, Fig. 1). *For lamellar phases*, mono or bilayers are built in a cell with x and y

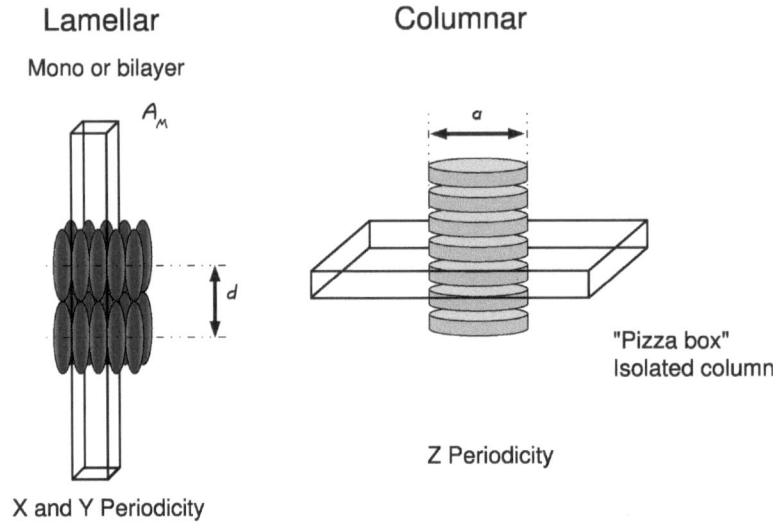

Fig. 1 Schematic representation of the PBC used for the lamellar and columnar phases

dimensions set according to the experimental molecular area, and z dimension set arbitrarily large so that no interactions can occur in that direction (this simulates a 2D periodicity along the xy plane). *Concerning the columnar phases*, assemblies of molecules are constrained in the z-axis in a flat cell with large x and y dimensions (kind of pizza box) that simulates an individual column with a periodicity in the z-direction. The energy relaxation and the MD thermodynamic equilibrium of the systems were performed in the NPT ensemble using some general-purpose and well-validated force fields available in the software (cvff, pcff, or esff). The resulting structures are then compared to experimental data to validate the self-association assumptions used to build the model.

In this chapter, we will report on a few examples of self-organization behaviour of some novel materials based on liquid-crystalline dendritic architectures. The original design of the molecules imposes the use of all-atomic methods to model correctly every intra- and intermolecular effects. The selected materials are octopus dendrimers with block anisotropic side-arms, segmented amphiphilic block codendrimers, multicore and star-shaped oligomers, and multi-functionalized manganese clusters. The molecular organization in lamellar or columnar phases which has been observed occurs due to soft/rigid parts self-recognition, hydrogen-bonding networks, or from the molecular shape intrinsically.

2 Octopus Dendrimers

The insertion of rigid and linear segments within dendritic scaffolds leads to the class of the so-called main-chain systems. Regarding the structure of these dendrimers, the junctions are no single and spherical atoms (C, N, P, Si, etc) but consist of anisotropic molecular moieties instead [3, 4]. These units are linked together

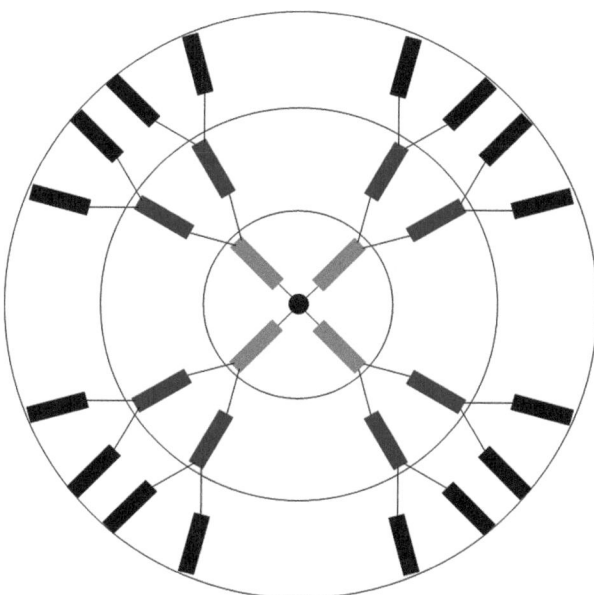

Fig. 2 Schematic representation of a main-chain dendrimer. The *rectangles* represent the rigid and linear segments (anisotropic mesogenic moieties) and the *thin lines* the flexible aliphatic spacers

through long and flexible alkyl spacers, generating therefore the dendritic matrix (Fig. 2).

As far as self-organization into mesophases is concerned, the anisotropic groups present at every level of the dendritic hierarchy are responsible for some loss of conformational freedom with respect to their homologues with more flexible and deformable core, such dendrimers being forced to adopt more constrained molecular conformations. Consequently, the anisometric branches do not radiate isotropically, but, on the contrary, favour preferentially an anisotropic order at an early stage of the organizing process by a gain in the enthalpy of the system, in order to produce the most stable structure. More precisely, for the octopus dendrimers, these conformations result from the coupling of segmented and symmetrical dendritic branches containing mesogenic moieties at each generational level (junction) onto a small tetra-podand core.

The dendrimers bearing one terminal aliphatic chain per peripheral unit (Fig. 3, **1**, X: $=/\equiv$ and Y: $\equiv$, $R^1 = R^3 = H$, and $R^2 = OC_{12}H_{25}$) exhibit smectic phases; in contrast, none of the precursory branches was mesomorphic. The mesomorphic behaviour was explained by the elongated (*prolate*) molecular conformation adopted by the dendrimers. Indeed, the lamellar periodicities were rather large (10–12 nm), consistent with a fully stretched molecular conformation, and with the mesogenic groups homogeneously distributed on both sides of the tetravalent core [32]. The formation of the smectic phases resulted therefore from the parallel disposition of these giant rod-like supermolecules into layers. In this case, the structure of the smectic phases is quite unique and consists of a highly segregated, sublevelled, molecular

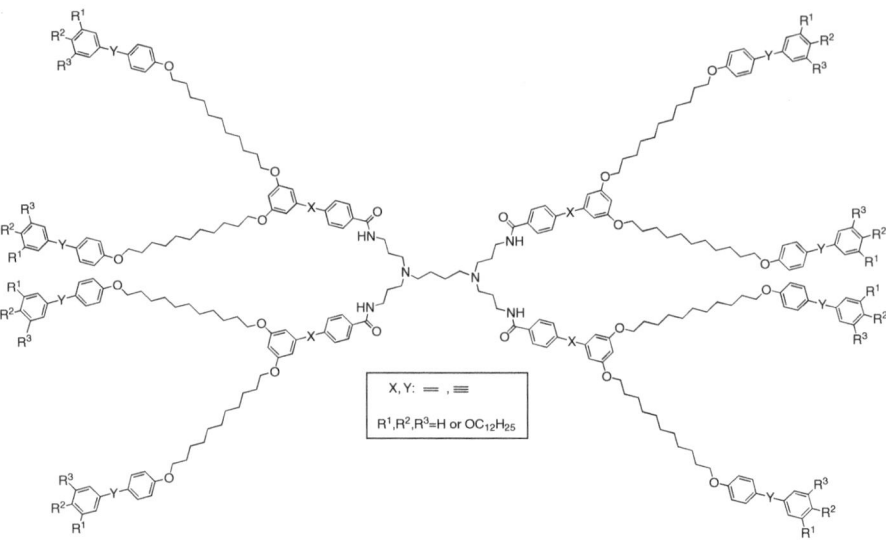

Fig. 3 General chemical structure of the octopus dendrimers, **1**

organization made of an internal sublayer containing tilted rigid segments (segments of the first generation, randomly tilted to compensate the molecular area), flanked by outer slabs inside which the mesogenic groups are arranged perpendicular to the layer (Fig. 4); these various sublayers are separated by the aliphatic continuum.

Molecular modelling supports this view of strongly segregated multilayer structures, with interfaces between the various molecular parts. Indeed, as already mentioned, one can realistically assume an elongated conformation for the dendrimer, where the rigid parts are colinear to the layer normal. To allow the formation of lamellar mesophases, the overall molecular structure ought to adopt such a parallel conformation, with the elementary mesogenic units arranged in a pseudo-parallel fashion, with necessarily half of the mesogenic units extending up and the other half extending down the molecular centre [33–35]. As such, the molecular model of the smectic layer consists of a tetragonal periodic cell in which the cross-section was set to match roughly the molecular area of four mesogens in a smectic phase (10×10 Å^2). The third parameter of the cell was fixed at 150 Å, much longer than the length of the fully extended dendrimer in order to simulate a single layer and to allow the molecules to expand or shrink freely. The molecular dynamics simulation was then performed on this model to evaluate the dendrimer conformation within the smectic layer. The result of these calculations is represented by the molecular snapshot in Fig. 4 and the estimated molecular length (ca. 90–100 Å) was found to be in very good agreement with the periodicities measured by XRD (ca. 100 Å).

Thus, the morphology of the smectic phases generated by such multiblock molecules is rather unusual in that it possesses a two-level molecular organization each being dependent of the other one. It consists of an internal sublayer made of tilted rigid segments with no correlation of the tilt and outer slabs inside which the

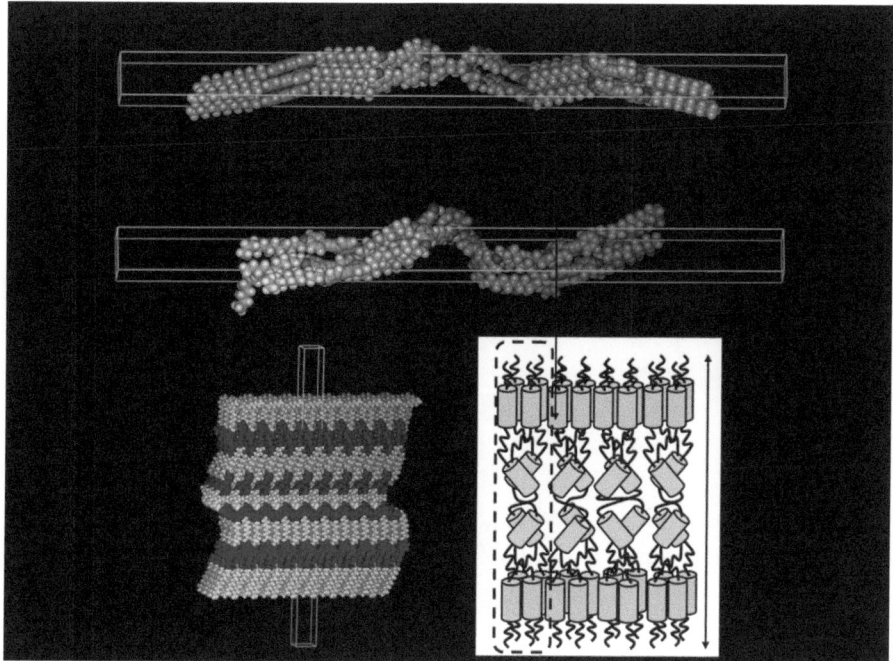

Fig. 4 Modelling and schematic representation of the multilayered smectic phase structure

mesogenic groups are arranged perpendicular to the layer. Obviously, all the "hard–soft" interfaces are not so well defined due to thermal fluctuations in mesophases.

In contrast, the increase of the chain/end-group ratio prevents such a parallel disposition of the pro-mesogenic groups, and the octopus dendrimers bearing two or three aliphatic chains per outer mesogenic unit (Fig. 3, **1**: X, Y: =/≡, and $R^1 = R^3 = OC_{12}H_{25}$ and $R^2 = H$, or $R^1 = R^2 = OC_{12}H_{25}$ and $R^3 = H$, or $R^1 = R^2 = R^3 = OC_{12}H_{25}$) formed systematically an hexagonal columnar mesophase (Col_h phase) [32, 36]. The formation of these columnar superstructures is a consequence of the mismatch between the surface areas of the aromatic cores and the cross-sections of the aliphatic chains, resulting in the curvature of all the interfaces, as in polycatenar liquid crystals [37–39] and dendromesogens [40]. Consequently, the octopuses adopt a wedge-like conformation (or cone-like), authorized by the great flexibility of the zeroth generation polypropyleneimine (G0-PPI core), allowing the pro-mesogenic groups to be radially distributed with uniform interfaces at the different generational levels. With such a folded or fan conformation, the dendrimers self-assemble into (supra)molecular discs or columns which further self-organize into a hexagonal net (Fig. 5). Depending on the number of terminal chains per terminal mesogenic units (2 or 3), three- or two-folded dendrimers, respectively, can perfectly pave the hexagonal lattice, consistent with the hexagonal parameters (9–10 nm). Considering the diblock, alternated chemical nature of these dendrimers, a leek or onion morphology for the columns is most likely (intracolumnar segregation) (Fig. 6).

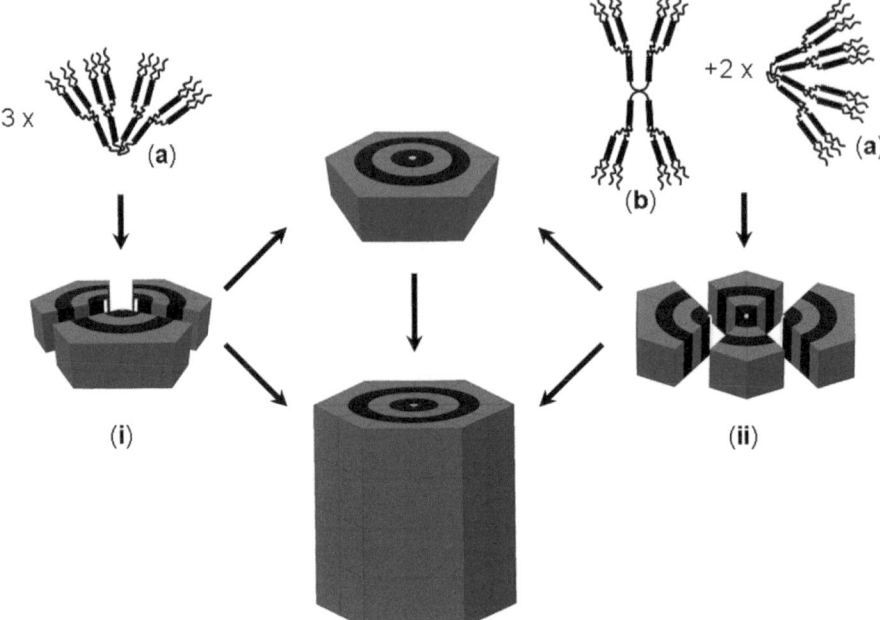

Fig. 5 Schematic representation of two possible molecular conformations (**a, b**) of the octopus dendrimers **1** and their self-assembling processes into columns (**i, ii**)

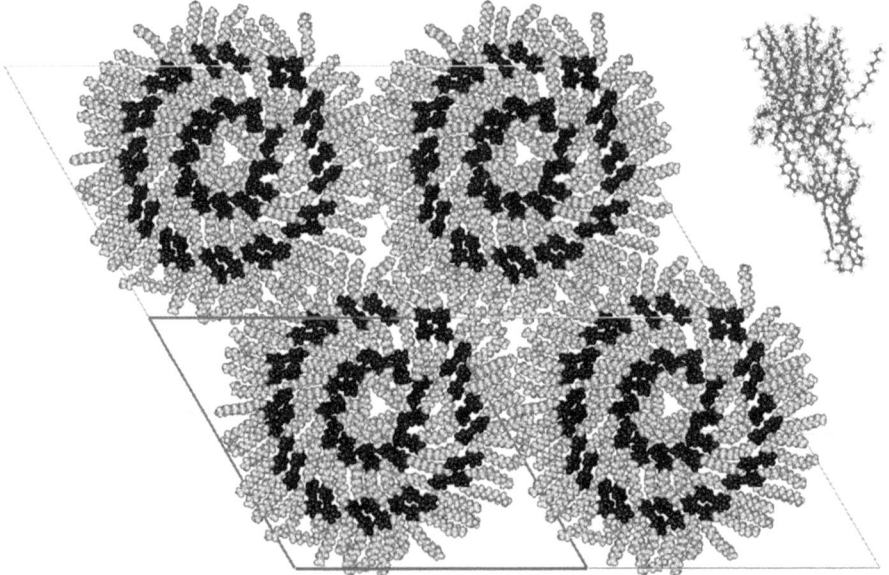

Fig. 6 Snapshot of the molecular conformation and self-organization of octopus dendrimers (**1**) in the columnar obtained by molecular dynamics calculation. The parts in *grey* and *black* represent the aliphatic and aromatic parts, respectively. *Inset*: wedge-like conformation of the dendrimer

This proposed model of a multi-segregated internal structure of the columns was supported by molecular dynamics: for one dendrimer, a periodic molecular model was built from the experimental X-ray data, that is, a hexagonal lattice with a 97 Å parameter and a thickness of 4.5 Å, paved with three molecules in a flattened wedge conformation. The result of the calculation (Fig. 6) evidenced a good filling of the available volume, acknowledged by a calculated density of 0.95, and good segregation at the molecular level by means of interlocked concentric crowns of anisotropic units belonging to the same generation (each crown, separated by "inert" aliphatic films, being further stabilized by intermolecular interactions). An enhancement of the micro-segregation over the entire simulation experiment time was also observed, contributing to the stabilization of the onion structure. Furthermore, the compensation of the molecular areas at the various interfaces at every level of the arborescence, which implies the tilt of the internal and external rigid segments with respect to the radial directions, was also shown in the modelization (Fig. 6).

3 "Janus-Like" Diblock Codendrimers

The second example describes the self-organization of some amphiphilic diblock codendrimers (supramolecular dendromesogens) where the two compartments of the dendrimers have different chemical nature and affinity. The concept of amphiphilicity in synthetic dendrimers is quite recent and has been successfully used for example to obtain unimolecular micelles for drug encapsulation and transport [41], to mimic the aggregation of globular proteins [42], or in therapeutic applications for their recognition ability towards protein receptors or biomedical materials [43, 44]. Such amphiphilic dendrimers have also a potential use as catalysts of organic reactions in aqueous solutions and phase-transfer agents [45, 46].

In this section, we consider the self-assembling behaviour of new amphiphilic codendrimers of the Janus type [47] which can be considered as building blocks for self-ordered mesostructures [48] into Col_h mesophases. Various structural parameters were selectively modified in order to establish relationships between the molecular structure (control of the hydrophilic/hydrophobic balance and hydrogen-bonding ability) and the self-organization properties into liquid-crystalline phases. These parameters include the generation numbers of both the hydrophobic (polybenzyl ether block) and hydrophilic (polyol block) parts which have been changed independently and the terminal chain substitution (Fig. 7).

Dendrimers having the second dendritic generation hydrophilic lobe(six hydroxyl groups,compounds **2** and **3**) were mesomorphic, exhibiting a columnar hexagonal Col_h [49]. The Col_h phase was deduced from the analysis of the X-ray diffractograms.

For these dendritic hexol compounds, the great size of their polar parts with respect to their apolar parts, and particularly true for the G1 compounds (**2**), could in principle favour the formation of normal phases (i.e. with a positive interfacial curvature), and therefore, self-organization into a columnar phase resembling the H_1 phase observed in lyotropic systems [50] could not be immediately excluded.

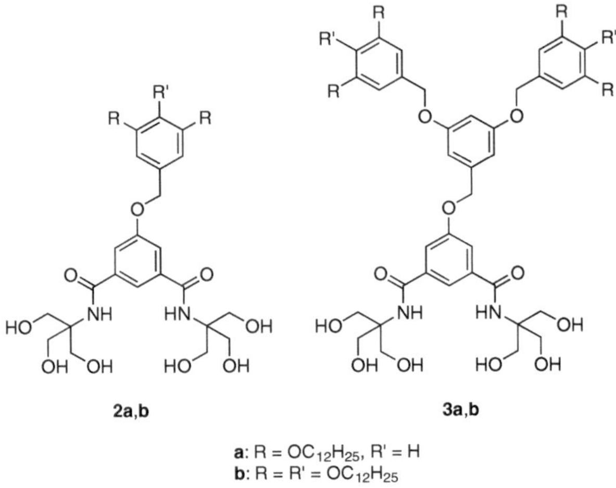

Fig. 7 Molecular structures of diblock amphiphilic dendrimers **2** and **3**

Recall that in such a normal phase, the polar heads of the molecules are directed outwards the core to form a hydrophilic outer-wall interface, whereas the internal core is filled by the lipophilic fragments. In the case of the lyotropic systems, such an arrangement is even more facilitated since it occurs in the presence of a polar solvent like water (binary system). In these conditions, the solvent contributes to the swelling of the polar part, to the dilution of the H-bonding network, to the smoothing of the interface by counterbalancing the geometric constraints, and to the fluidity of the system by filling the empty cavities. However, this normal arrangement seems very unfavourable in anhydrous materials and up to now only two examples of normal phases (Col$_h$ and cubic phases) have been reported in thermotropic systems [51, 52]. Indeed, if such an organization into normal columns was true here, the honeycomb structure, which would be extended in the third direction in the present case, generated by the H-bonding network would be too rigid to be compatible with a fluid mesophase, contrasting with the high fluidity of the mesophases as observed by polarized light optical microscopy (POM). Consequently, the classical inverted model has thus been considered for the description of the molecular arrangement of these amphiphiles.

The geometrical structural requirement necessary to induce an inverted columnar phase (negative curvature of the interface, i.e. when the apolar chains irradiate out of the columnar spines) is a priori satisfied only for the G2 derivatives (**3**). Indeed, four or six side-chains cover the periphery of the dendron and form the wide end, whereas the narrow end is constituted of polar hydroxy groups. When the chains are molten, the mismatch at the polar/apolar interface is further enhanced, favouring the microsegregation of these tapered molecules into cylindrical structures. Thus, it is first assumed that all dendrimers self-assemble in such a way that the polar apices segregate to form the polar columnar core of the cylindrical micelle, further stabilized by a dense and interlocked H-bonding network. The wall of the inner columnar core (hydroxyl–dendritic interface) is surrounded by the hydrophobic branches, with the aliphatic chains radiating laterally. In order to pave the hexagonal lattice efficiently, the molecules, radially oriented, are packed side by side to form a thick disc slice as in a pizza.

To validate this assumption and to propose a model for the self-organization within the Col_h phase, it was thus necessary to perform molecular dynamics. Using a geometrical approach, the hexagonal cell 4.5 Å-thick contains 5.35 molecules **2a** (D = 45.7 Å, V_{mol} = 1 520 Å^3 at T = 130°C). Considering the general hypothesis of a mesophase based on supramolecular columns with a polar interior core, there is no effective possibility to arrange 5.35 of these mesogens in such a cell which respects both a perfect paving of the hexagonal 2D lattice (**2a** possessing only two aliphatic chains) and a good agreement between molecular and mesostructure dimensions. Indeed, the bulky polar fragment occupies a large volume not easily compensated by the reduced number of aliphatic chains per molecule. A solution to this problem was given by computer modelling. The optimization of the geometry by molecular dynamics at 130°C, temperature of the X-ray experiment, led to a hexagonal lattice (with D = 45.7 Å fixed before calculation) with a thickness of 13.5 Å and containing 16 molecules in this cell (these quantities correspond to the same ratio of 5.35 molecules per stratum 4.5 Å-thick, and thus to the same packing density). Comparing the molecular dimensions and the lattice size, the best compromise is found when the 16 molecules are placed into two strata of 6.75 Å-thick, each stratum containing 8 molecules disposed in such a way that they form a cylinder and occupy the available space homogeneously (Fig. 8). In this simulation, the relative disposition of the molecules is not random, the latter adapting their shape in order for the polar segments to be localized in the central part of the discs, i.e. in the interior of the column (Fig. 8) to allow strong H-bonding interactions. The density ratio calculated at 130°C estimated from MD and XRD was found to be very

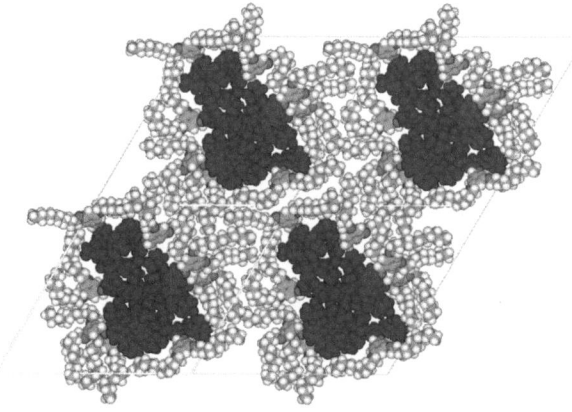

Fig. 8 Snapshot showing the organization of **2a** in the Col_h phase (in *dark polar* central core). Only one layer (6.75 Å) is represented, the apparent empty zones, being actually filled by neighboured layers

close to unity (0.995), supporting this arrangement. The relative size of the region occupied by the polar parts is also rather large with respect to that filled by the aliphatic and aromatic parts, therefore suggesting here that the dominant driving force for the mesophase formation is clearly the hydrogen interactions between the hydroxy groups, necessary to stabilize the columnar polar spine, the chains forming the infinite continuum.

Similarly, we also resorted to molecular dynamics to understand the packing of **3a** in the Col_h phase. In this case, the cross-section of the polar part is comparable to that of the aliphatic part, and thus a lamellar structure was expected instead of the Col_h phase. It appeared that for a lattice parameter (D = 49.8 Å, V_{mol} = 2 505 Å^3 at T = 100°C), a slice 4.5 Å-thick would contain four molecules, which is not satisfying with an efficient paving of the lattice (discrepancy between molecular dimensions and columnar cross-section). However, the result of the molecular dynamics calculation suggested that it was preferable to consider a stratum with a thickness of 9.0 Å containing eight molecules to obtain good filling of the available volume, with the relative disposition of the hydroxy groups which, through efficient H-bonding interactions, are able to ensure the stability of the polar column, as shown in Fig. 9. The density ratio estimated from MD and XRD for this packing at 100°C (1.036) shows the good agreement with the model.

As for **2a**, the main driving force for the Col_h formation of **3a** is the importance of the H-bonding interactions which stabilized the overall architecture, as proven by the modelization. For both **2a** and **3a**, a good agreement is observed between the results obtained by XRD and those issued from MD. Despite the a priori incompatible molecular conformation, either an inverted triangular shape (**2a**) or a quasi-cylindrical shape (**3a**, quasi-equivalence of the two lobes, with a near zero-interfacial curvature), a good segregation in columns is nevertheless being achieved between the polar hydroxy parts and the aliphatic parts: the columns are stabilized by H-bonding interactions (**2a**, **3a**) and van der Waals interactions

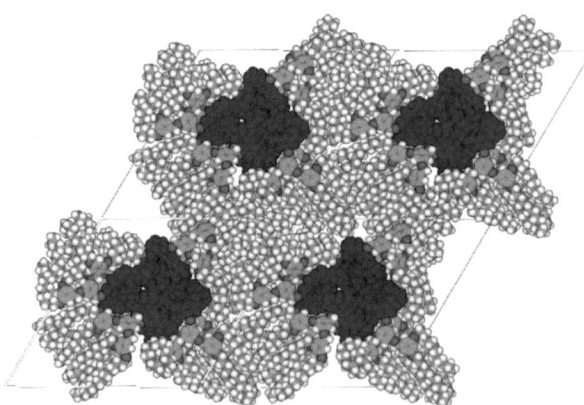

Fig. 9 Snapshot showing the packing of **3a** after MD in the hexagonal 2D lattice of the Col_h phase

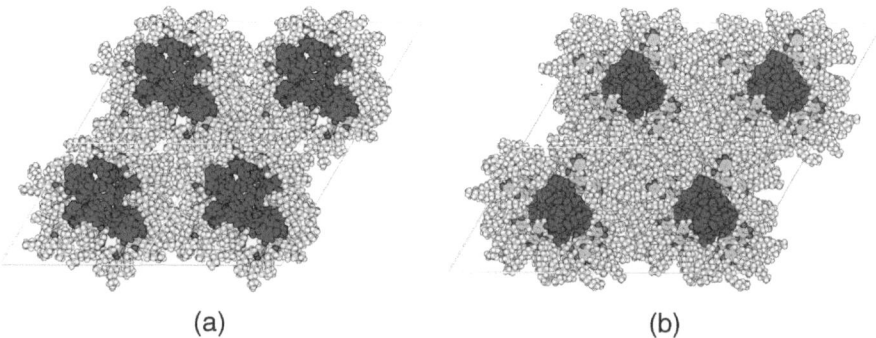

(a)   (b)

Fig. 10 Snapshots of packing of **2b** (**a**) and **3b** (**b**) after MD in the hexagonal 2D lattice of the Col_h phase

(**3a**), the latter then self-organizing into a hexagonal honeycomb packing, i.e. Col_h phase.

Conducting this analysis with **2b** (D = 50.35 Å, V_{mol} = 1860 Å^3 at T = 140°C) and **3b** (D = 54.15, V_{mol} = 3150 Å^3 at T = 100°C) led to similar idealized representations (Fig. 10). Ten molecules are needed over a thickness of 8.47 Å in the case of **2b**, while for **3b**, eight molecules are necessary to fill a cylinder 9.76 Å thick. MD shows the good agreement between molecular and mesophase dimensions, with an almost perfect paving of the 2D lattices. As can be seen by the MD snapshots (Figs. 8–10), the filling of the available space is more efficient with increasing terminal aliphatic chains (**a**→**b**) and with increasing generation (**2**→**3**), as expected.

Of course, let us recall that all these molecular models are static and local representations only of a more dynamic and macroscopic reality, but are significantly descriptive and useful to explain the supramolecular organization of all these dendrimers within the hexagonal columnar phases. This is why for instance the polar cores do not appear circular but rather distorted in shape instead. Averaging these local arrangements over larger assemblies with random (not correlated) main orientation leads effectively in the timescale of the experiment to columns with circular polar cross-sections, compatible with the symmetry of the mesophase.

4 Self-Organized Hybrid Siloxane-Triphenylene Star-Shaped Heptamer

The third example concerns flat, π-conjugated discotic molecules [53, 54], already used in phase compensation films of liquid-crystalline devices [55]. They offer a unique possibility as potential 1D charge-carrier systems [56] due to their ability to self-assemble into long-range 1D intermolecular columnar stackings and then into columnar organizations, preferentially the hexagonal columnar phase (Col_h). Electronic interactions as well as electrons and excitons migrations are indeed strongly favoured within the columns since the stacking periodicity along the column is much shorter than the intercolumnar distance. As such, discotic liquid crystals are seen as promising organic semiconductors for applications in the domain of molecular electronics, optoelectronics, photoconductivity, photovoltaic, and electro-luminescent devices [57–61]. Moreover, the liquid-crystalline state also offers many alternative advantages to organic monocrystals in that they can be more easily macroscopically aligned (monodomains) and processed, and structural defects can be self-healed because of the molecular fluctuations. However, the limited efficiency of the charge-carrier mobility (10^{-3} to 0.1 $cm^2V^{-1}s^{-1}$ compared to that found in graphite of 3 $cm^2V^{-1}s^{-1}$) [62], an essential parameter to estimate for an optimal design, is in part due to the lack of long-range order of the intra-columnar stacking in the liquid-crystalline mesophase (topological defects, thermal fluctuations, and molecular diffusion). The improvement of these properties requires perfectly stable monodomains of the materials, ideally operational at ambient temperature. Various methods for increasing the extent of ordering, facilitating the processing, and improving the performances of the charge mobility have been employed. Among these, the freeze-in of the columnar order into stable, room temperature glasses appear to be an attractive strategy since the anisotropic properties and macroscopically aligned monodomains can be easily vitrified and the ordering preserved. For such prospects, triphenylene-containing liquid crystals (oligomers, elastomers, networks) have been the most extensively studied discotic materials [53, 54, 57, 59–61, 63, 64] and have been reported to address these points.

More recently, hybrid molecular systems that combine a siloxane part with an organic disc-like group are now being considered. In general, the attachment of a flexible siloxane part to a mesogenic structure, via an alkyl spacer, maintains the liquid-crystalline property but considerably reduces the transition temperatures with respect to the aliphatic analogues. Moreover, the bulkiness of these groups disfavours crystallization and such hybrid materials show a strong tendency to freeze-in the mesophase on cooling due to a strong supercooling effect. Consequently, the mesophase temperature range becomes more accessible than their siloxane-free analogues.

In this context, we have been interested in the design and the synthesis of new hybrid discotic oligomeric materials and in the investigation of their thermal behaviour and self-organization. In order to obtain stabilized columnar mesophases where lateral slippage of molecules from one column to the adjacent one is strongly prevented, a large oligomeric molecule (star-like heptamer, Fig. 11) has been prepared and studied [65]. As deduced from XRD experiments, this compound showed a columnar hexagonal phase between 38 and 111°C. The hexagonal lattice parameter a in the liquid-crystalline state is 22.2 Å and the presence of a signal at ca. 3.6 Å proves that there is a long-range intramolecular stacking order along the columns. This heptamer shows no signal at 6.4 Å (signature of interactions between siloxane parts) in the XRD patterns, suggesting that there is in fact no contact between siloxane chains of different molecules. In addition, a signal at twice the intercolumnar

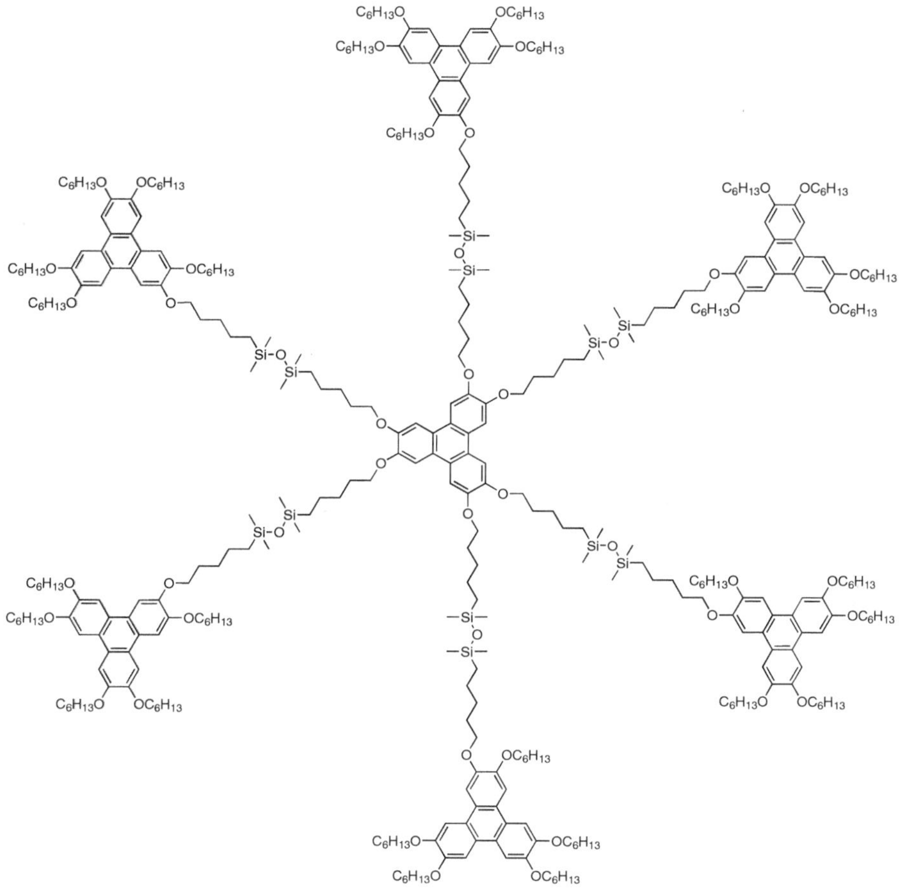

Fig. 11 Chemical structure of the discotic heptamer, **4**

distance shows that there is some superstructure besides the hexagonal columnar arrangement.

When considering the possible packing modes (models **I**, **II**, and **III** in Fig. 12a), in which none, one or two pairs of triphenylene units are interdigitated between each pair of heptamers, the only packing consistent with X-ray experiments is **II** (each heptamer shares one peripheral triphenylene with six adjacent neighbours), where the distance between planes containing the cores is twice that of the regular hexagonal lattice, in agreement with the superstructure. Within this packing mode, the siloxane fragments are segregated into precise locations, leading to columns with a different environment consecutive to the specific repartition of siloxane. It should be noted that such a packing prevents the contact between siloxane chains, as shown in Fig. 12b. In model **I** (each heptamer shares two peripheral triphenylene with three adjacent neighbours) the interplanar distance is 1.5 times the intercolumnar

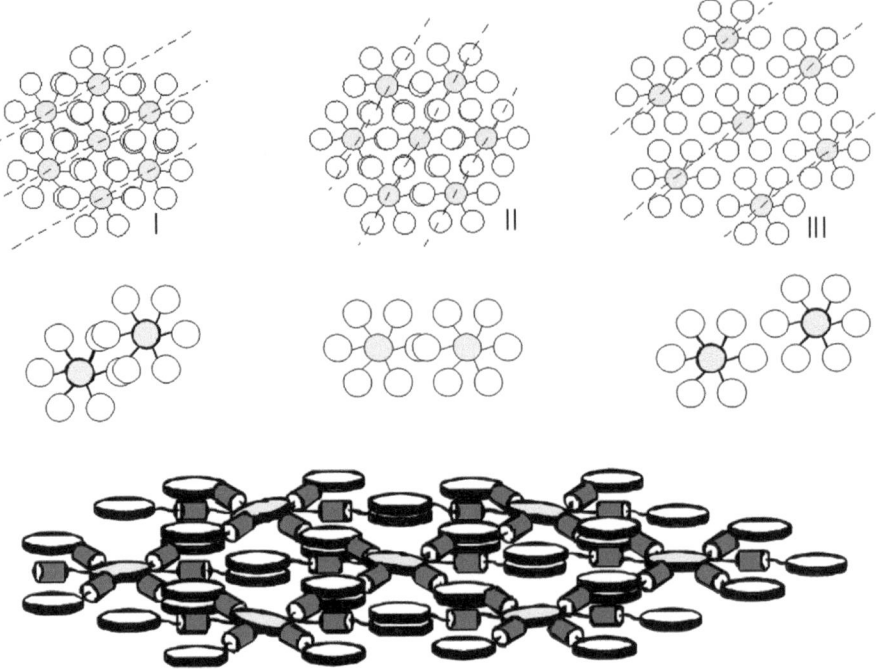

Fig. 12 Possible packing modes of the heptamer and packing model of the heptamer in the Col$_h$ phase. *Grey circles* correspond to the central triphenylene bearing six peripheral siloxane fragments, and the *white circles* correspond to the peripheral triphenylene moieties

distance, while in model **III** (the "ideal packing") the hexagonal pattern is scaled up by a factor of $\sqrt{7}$.

The driving force for such a high degree of order (superstructure) is very often considered as resulting from microsegregation due to strong attractive interactions, usually between aromatic parts. In the case of the present heptamer, the steric interactions between spacers give rise to the described superstructure, instead of the "ideal stacking" one. Such a view of the molecular packing was confirmed by molecular dynamics simulation (Fig. 13). In agreement with the X-ray experimental data, the siloxane fragments are totally "miscible" with the aliphatic fragments, leading to indiscernible triphenylene columns.

This molecular model of a single heptamer-based interlocked columns or "fibres" made from seven columns was built according to Fig. 14. The model resulted in a quadratic cell of dimension 110×110 Å (much larger than the expected diameter of the fibre so that interactions between adjacent fibres can be avoided in the MD simulations) and 25.2 Å in thickness (7×3.6 Å), corresponding to the height of seven triphenylene cores stacked on top of each other. The model is first minimized in energy in order to relax some local intermolecular interactions, and a first MD simulation is performed as a 50 ps isotherm at 80°C. The data for the distances

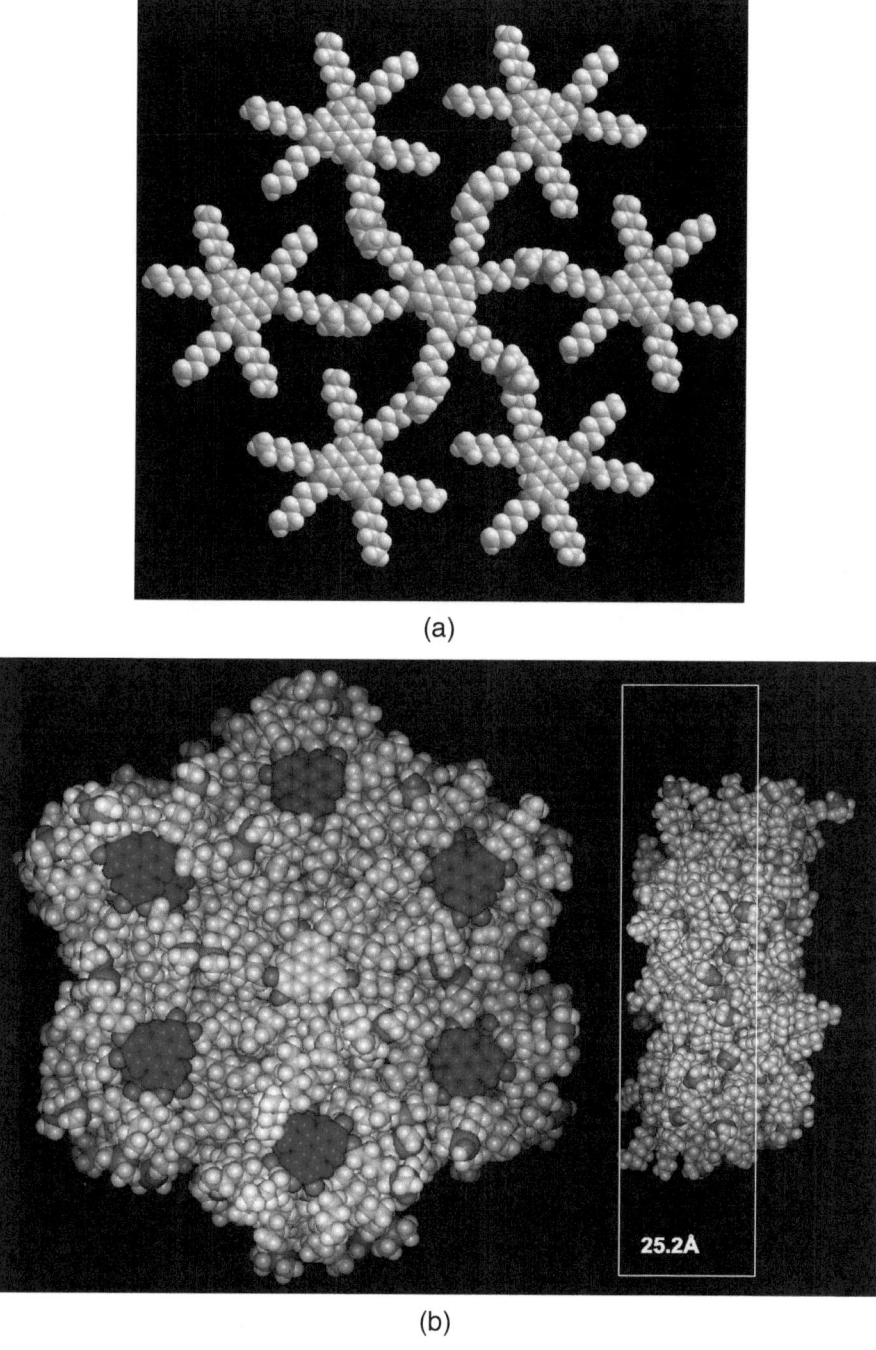

(a)

(b)

Fig. 13 (a) Idealized CPK model of the heptamer, (b) snapshot of the heptamer **4** in the Col_h phase obtained by MD simulation

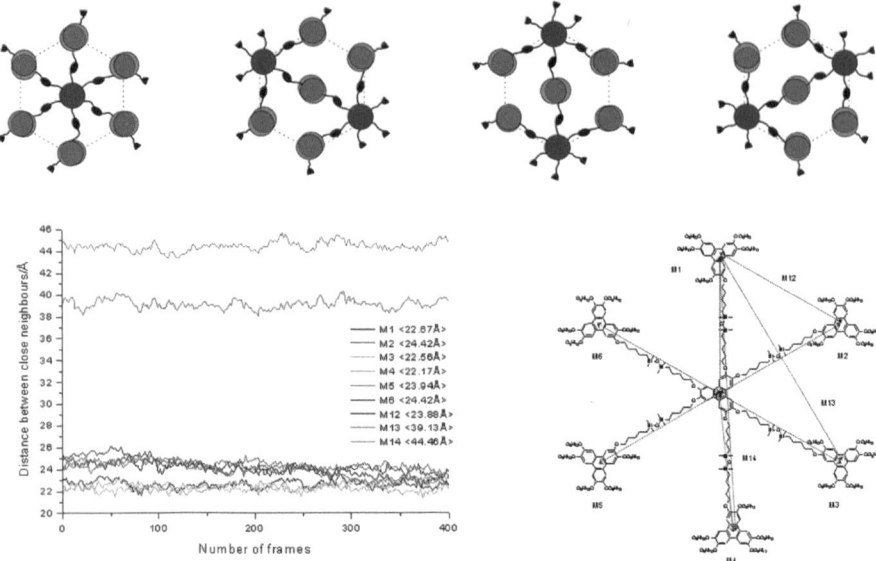

Fig. 14 Scheme of the four successive strata used to build the fibre-like molecular model. Each stratum contains a full molecule of the heptamer and/or some fraction of it (sectioned at the middle of the siloxane group) in order to simulate the participation of a neighbouring molecule in the stacking of a given rod while sharing one of its side triphenylene moieties. Each triphenylene stack in the resulting cell contains six side groups (drawn in clear) and one central part (drawn in dark). Evolution of the intercolumnar distances as a function of time (as given in the molecular scheme) and average distances during the 200 ps simulation

calculations given in the graph of Fig. 14 were then collected from a further 200 ps isotherm simulation. It can be clearly seen that the average intercolumnar distances converge towards the measured lattice parameter ($a = 22.2$ Å).

5 Single Molecule Magnet

Single molecule magnets (SMMs) are materials able to retain a magnetization at the molecular level below a certain temperature known as blocking temperature [66]. These compounds currently elicit a sustained research activity: being quantum objects they are envisioned as the future qubits of a quantum computer while their magnetic properties make them the ultimate storage bits of a molecular magnetic memory. Schematically, these hybrid molecules are made from metal ions bond together by various organic ligands and numerous types of compounds have been reported to show SMM behaviour: $4f$-coordination compounds [67, 68], polymetallic cages [69], and oxometallaclusters [70–76]. In spite of its poor stability against water and temperature, the so-called Mn_{12} is probably the most studied SMM for at least two reasons: its synthesis is cheap and easy and it long held the record of the highest blocking temperature [72].

On the other hand, the use of functional molecules in macroscopic devices (bottom-up technology) requires that some degree of low-dimensionality self-organization (1D or 2D) is imparted to the molecules. It is well known that self-organization can only take place in a system that keeps some fluidity during the process so that positioning errors can be corrected automatically. Liquid crystals are precisely molecular assemblies where order coexists with fluidity [5–7].

Using molecular engineering, we were able to endow the Mn_{12} molecule with liquid-crystalline properties while preserving the peculiar magnetic properties of the original core, a side effect of this added functionality being a much improved thermal stability [77]. The formation of positionally ordered mesophases is a first step on the route to organizing SMMs in view of their eventually being part of a functional nanodevice. In order to counterbalance the a priori unfavourable molecular shape and bulkiness of the Mn_{12} cluster core and the important geometrical constraints, we applied the classical strategy used to obtain thermotropic mesophases, consisting in the covalent grafting of a mesomorphic promoter onto the inorganic cluster via a flexible aliphatic spacer [78–82], to improve interfaces and areas compatibilities between both moieties, and to enhance microsegregation. Thus, a mesomorphic, dodecanuclear manganese complex $[Mn_{12}O_{12}(O_2CR)_{16}(H_2O)_4]$ was obtained by the replacement of the 16 LC-inert acetate groups (R=Me) holding the cluster together by anisotropic gallate-derived moieties (**5**, Fig. 15) [77]. As a result, a 1D (smectic) organized lamellar mesophase was indeed induced.

The corresponding functionalized cluster exhibited a homogeneous and fluid birefringent optical texture from 40°C upwards. X-ray diffraction scans show two equidistant sharp reflections in the 1:2 ratio (42.3 and 21.0 Å, 00l reflections with l = 1, 2), indicative of a lamellar order. A broad and diffuse scattering, corresponding to the short-range order of the molten chains, was detected at 4.5 Å (h_{ch}). Another diffraction signal, at ca. 28.5±0.5 Å (d), with a different line shape and not commensurate with the peaks coming from the lamellar structure, having therefore a different origin, was also present. This peak was assigned to some intralayer order associated to a specific arrangement of the diffracting centres, i.e. the magnetic

$$[Mn_{12}O_{12}(CH_3CO_2)_{16}(H_2O)_4] + 16HL \quad \rightleftharpoons \quad [Mn_{12}O_{12}(L)_{16}(H_2O)_4] + 16CH_3CO_2H$$

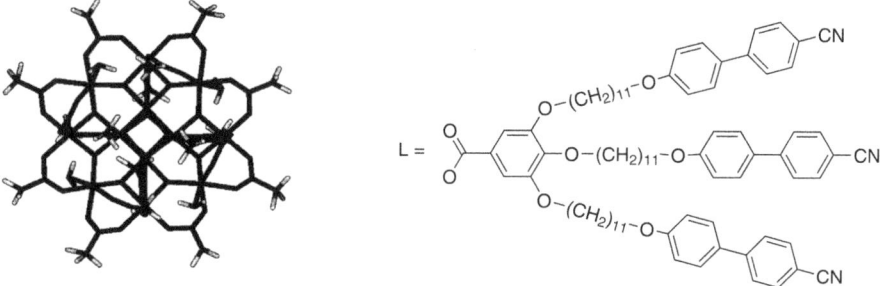

Fig. 15 Chemical structure of the gallate derivative substituent and of the Mn_{12} cluster forming **5**

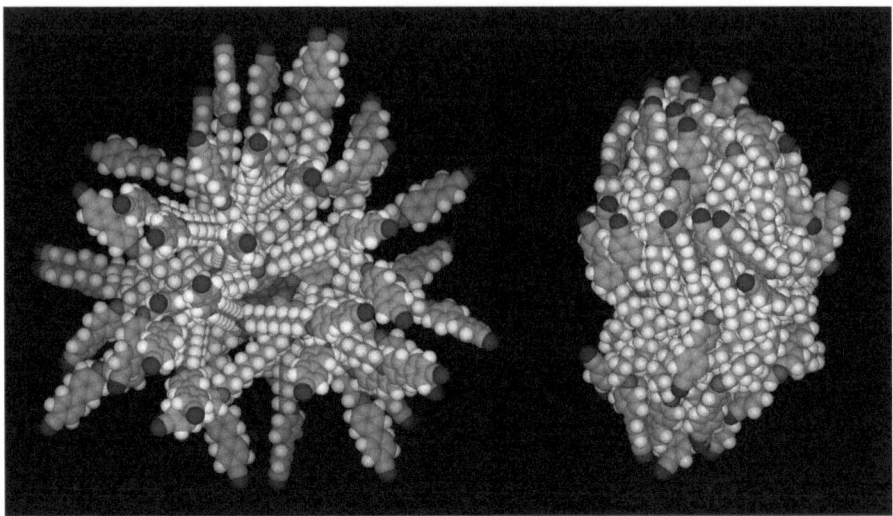

Fig. 16 Molecular dynamics simulation in the substituted Mn_{12} cluster

cores. The features of the peak (intensity as compared to d_{002} and FWHM) imply that this order is short-ranged and that the position of the diffracting centres is not correlated from layer to layer. Such an intralayer organization would be sterically induced by the bulky cyanobiphenyl groups. The intensity of the 001 reflection is weak compared to that of 002, implying a modulation of the electronic density within the lamellar periodicity. This suggests an organization in which the peripheral mesogenic groups are equally distributed on either side of the metallic cluster in a compact manner (Fig. 16, cylindrical molecular conformation), resembling the microsegregated smectic structures formed by mesogenic end-capped dendrimers [8–14].

This supramolecular organization has been validated by molecular dynamics simulations. A molecular model of a smectic bilayer was built, based on a cell of 200 Å height (larger than the measured value so as not to artificially constrain the thickness) and an area S $\sim$ 810 Å^2, containing two superimposed molecules. As none of the available force fields for the molecular dynamics studies were able to simulate correctly the ring shape of the Mn_{12} cluster, each manganese atom of a given cluster was restricted to its relative position to the others, according to the X-ray diffraction structure of a reference compound. All non-Mn atoms of the cluster are allowed to move freely, as well as the two molecules in the cell. An average interlayer periodicity of 41.9 Å was obtained, in excellent agreement with the measured value (d = 42.3 Å). Still using molecular modelling, we attempted to precise the nature of the aforementioned in-plane order. Positional order in molecular 2D assemblies is most often hexagonal or square-like. Considering a hexagonal packing of the Mn_{12} cores, the average distance between neighbouring magnetic centres would then be 30.5 Å (S = $\sqrt{3}/2 \times a_{Hex}^2$), and the intralayer order 26.5 Å, smaller than the measured d value. However, if a square-like pattern is considered,

Fig. 17 Views of the lamellar packing of the cluster [$Mn_{12}L_{16}$]. Side view of the lamellar packing of the cyanobiphenyl groups and clusters in the smectic layer (*top*). Top view of the in-layer lateral arrangement of the clusters forming a square-like lattice (*bottom*)

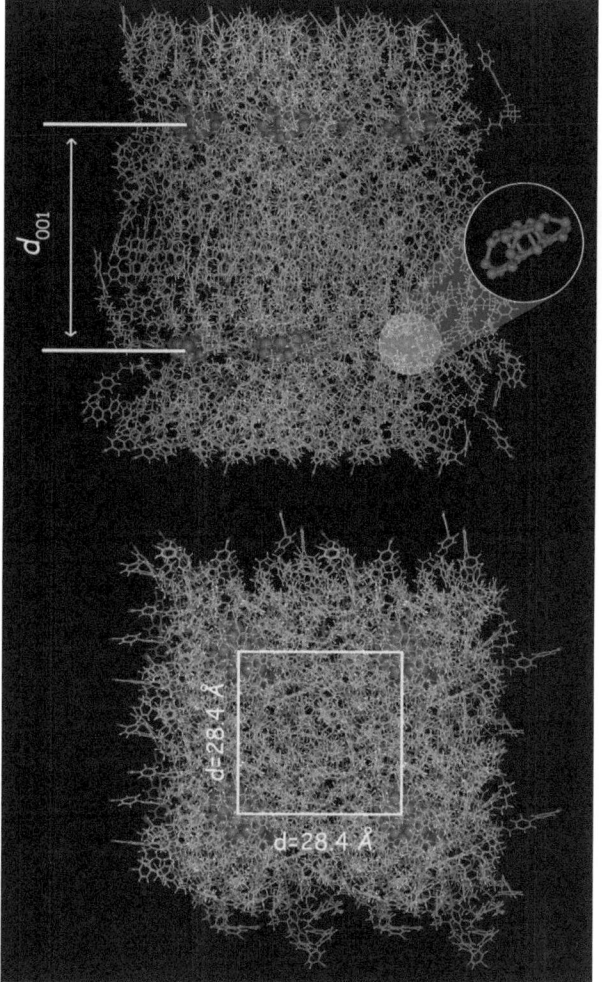

an average core-to-core distance of 28.4 Å ($\sqrt{S}$) is found, in better accordance with the experimental data (Fig. 17). The intrinsic fourfold molecular cluster symmetry could favour the square-like packing. In the fluid phase, both patterns are likely to coexist, with the square lattice prevailing over the hexagonal one.

Thus, this smectic arrangement can be described as follows: the two incompatible segments (mesogens and aliphatic spacers) form alternating layers, with tilt and interdigitation of the mesogens between successive periods, and the Mn_{12} cores are located in the aliphatic sublayers and arranged into a square-like planar array. This supramolecular organization resembles that of the so-called filled mesh phases observed with some facial amphiphiles where ionic clusters (formed by the laterally attached alkali metal carboxylates) are organized in a hexagonal array within the smectic phase [83]: the carboxylates are organized in the layers of the calamitics with or without correlation between the layers (filled mesh phases with random, 3D rhombohedral and channelled-layer sub-organizations). Here, the Mn_{12} clusters are squarely packed in the alkyl sublayers, with no long-range correlation between the layers, and can also accordingly be described as a filled random mesh smectic phase.

References

1. D. Astruc, C. R. Acad. Sci. Ser. II Paris **322**, 757 (1996).
2. U. Boas and M. H. Heegaard, Chem. Soc. Rev. **33**, 43 (2004).
3. G. R. Newkome, C. N. Moorefield, and F. Vögtle, *Dendrimers and Dendrons: Concepts, Synthesis and Applications* (Wiley & Sons, Weinheim, 2001).
4. J. M. J. Fréchet and D. A. Tomalia (eds.), *Dendrimers and Other Dendritic Polymers, Wiley Series in Polymer Science* (Wiley& Sons, Chichester, 2001).
5. D. Demus, J. W. Goodby, G. W. Gray, H. -W. Spiess, and V. Vill (eds.) *Handbook of Liquid Crystals* (Wiley-VCH, Weinheim, 1998).
6. T. Kato, N. Mizoshita, and K. Kishimoto, Angew. Chem. Int. Ed. **45**, 38 (2006).
7. J. W. Goodby, I. M. Saez, S. J. Cowling, V. Görtz, M. Draper, A. W. Hall, S. Sia, G. Cosquer, S.-E. Lee, and E. P. Raynes, Angew. Chem. Int. Ed. **47**, 2754 (2008).
8. S. A. Ponomarenko, N. I. Boiko, and V. P. Shibaev, Polym. Sci. C **43**, 1 (2001).
9. D. Guillon and R. Descheneaux, Curr. Opin. Solid State Mater. Sci. **6**, 515 (2002).
10. M. Marcos, A. Omenat, and J.-L. Serrano, C. R. Chim. **6**, 947 (2003).
11. J. Barberá, B. Donnio, L. Gehringer, D. Guillon, M. Marcos, A Omenat, and J.-L. Serrano, J. Mater. Chem. **15**, 4093 (2005).
12. B. Donnio and D. Guillon, Adv. Polym. Sci. **201**, 45 (2006).
13. B. Donnio, S. Buathong, I. Bury, and D. Guillon, Chem. Soc. Rev. **36**, 1495 (2007).
14. M. Marcos, R. Martin-Rapun, A. Omenat, and J.-L. Serrano, Chem. Soc. Rev. **36**, 1889 (2007).
15. M. R. Wilson, J. M. Ilnytskyi, L. M. Stimson, and Z. E. Hughes, Computer simulations of liquid crystal polymers and dendrimers. *Computer Simulations of Liquid Crystals and Polymers*, eds. P. Pasini, C. Zannoni, and S. Žumer (Kluwer, The Netherlands, 2004), pp. 57–78.
16. C. M. Care and D. J. Cleaver, Rep. Prog. Phys. **68**, 2665 (2005).
17. M. R. Wilson, Int. Rev. Phys. Chem. **24**, 421 (2005).
18. M. R. Wilson, Chem. Soc. Rev. **36**, 1881 (2007).
19. J. G. Gay and B. J. Berne, J. Chem. Phys. **74**, 3316 (1981).
20. R. D. Groot and P. B. Warren, J. Chem. Phys. **107**, 4423 (1997).
21. B. Glettner, F. Liu, X. Zeng, M. Prehm, U. Baumeister, M. Walker, M. A. Bates, P. Boesecke, G. Ungar, and C. Tschierske, Angew. Chem. Int. Ed. **47**, 9063 (2008).
22. C. Zannoni, AIP Conf. Proc. **963**, 520 (2007).
23. M. R. Wilson, J. M. Ilnytskyi, and L. M. Stimson, J. Chem. Phys. **119**, 3509 (2003).
24. D. L. Cheung, S. J. Clark, and M. R. Wilson, Chem. Phys. Lett. **356**, 140 (2002).
25. D. L. Cheung, S. J. Clark, and M. R. Wilson, J. Chem. Phys. **121**, 9131 (2004).
26. J. Pelaez and M. R. Wilson, Phys. Rev. Lett. **97**, 267801 (2006).
27. D. L. Cheung, S. J. Clark, and M. R. Wilson, Phys. Rev. E **65**, 051709 (2002).
28. R. Berardi, L. Muccioli, and C. Zannoni, ChemPhysChem **5**, 104 (2004).
29. I. Cacelli, L. De Gaetani, G. Prampolini, and A. Tani, J. Phys. Chem. B **111**, 2130 (2007).
30. A. J. McDonald and S. Hanna, J. Chem. Phys. **124**, 164906 (2006).
31. G. Tiberio, L. Muccioli, R. Berardi, and C. Zannoni, ChemPhysChem **10**, 125 (2009).
32. L. Gehringer, C. Bourgogne, D. Guillon, and B. Donnio, J. Am. Chem. Soc. **126**, 3856 (2004).
33. J. Barberá, M. Marcos, and J. L. Serrano, Chem. Eur. J. **5**, 1834 (1999).
34. M. Marcos, R. Giménez, J. L. Serrano, B. Donnio, B. Heinrich, and D. Guillon, Chem. Eur. J. **7**, 1006 (2001).

35. B. Donnio, J. Barberá, R. Giménez, D. Guillon, M. Marcos, and J. L. Serrano, Macromoelcules **35**, 370 (2002).
36. L. Gehringer, D. Guillon, and B. Donnio, Macromolecules **36**, 5593 (2003).
37. J. Malthête, H. T. Nguyen, and C. Destrade, Liq. Cryst. **13**, 171 (1993).
38. H. T. Nguyen, C. Destrade, J. Malthête, Adv. Mater. **9**, 375 (1997).
39. D. Fazio, C. Mongin, B. Donnio, Y. Galerne, D. Guillon, and D. W. Bruce, J. Mater. Chem. **11**, 2852 (2001).
40. V. Percec, C. M. Mitchell, W. D. Cho, S. Uchida, M. Glodde, G. Ungar, X. Zeng, Y. Liu, V. S. K. Balagurusamy, and P. A. Heiney, J. Am. Chem. Soc. **126**, 6078 (2004) and references therein.
41. C. J. Hawker, K. L. Wooley, and J. M. J. Fréchet, J. Chem. Soc. Perkin Trans. **1** 1287 (1993).
42. D. J. Pesak and J. S. Moore, Tetrahedron **53**, 15331 (1997).
43. E. R. Gillies and J. M. J. Fréchet, J. Am. Chem. Soc. **124**, 14137 (2002).
44. K. Aoi, K. Itoh, and M. Okada, Macromolecules **30**, 8072 (1997).
45. Y. Pan and W. T. Ford, Macromolecules **32**, 5468 (1999).
46. Y. Pan and W. T. Ford, Macromolecules **33**, 3731 (2000).
47. I. M. Saez and J. W. Goodby, J. Mater. Chem. **15**, 26 (2005).
48. J. Ropponen, S. Nummelin, and K. Rissanen, Org. Lett. **6**, 2495 (2004).
49. I. Bury, B. Heinrich, C. Bourgogne, D. Guillon, and B. Donnio, Chem. Eur. J. **12**, 8396 (2006).
50. J. M. Seddon, Biochim. Biophys. Acta **1031**, 1 (1990).
51. K. Borisch, C. Tschierske, P. Göring, and S. Diele, Chem. Commun. **2711** (1998).
52. S. Fischer, H. Fischer, S. Diele, G. Pelzl, K. Jankowski, R. R. Schmidt, and V. Vill, Liq. Cryst. **17**, 855 (1994).
53. A. N. Cammidge, R. J. Bushby, *Handbook of Liquid Crystals*, vol. 2B, Chap. VII, (Wiley-VCH, Weinheim, 1998), pp. 693–748.
54. S. Kumar, Chem. Soc. Rev. **35**, 83 (2006).
55. K. Kawata, Chem. Rec. **2**, 59 (2002).
56. A. M. van de Craats and J. M. Warman, Adv. Mater. **13**, 130 (2001).
57. N. Boden, B. Movaghar, *Handbook of Liquid Crystals*, vol. 2B, Chap. IX (Wiley-VCH, Weinheim, 1998), pp. 781–798.
58. M. D. Watson, A. Fechtenkötter, and K. Müllen, Chem. Rev. **101**, 1267 (2001).
59. R. J. Bushby and O. R. Lozman, Curr. Opin. Sol. State Mater. Sci. **6**, 569 (2002).
60. S. Laschat, A. Baro, N. Steinke, F. Giesselmann, C. Hügele, G. Scalia, R. Judele, E. Kapatsina, S. Sauer, A. Schreivogel, and M. Tosoni, Angew. Chem. Int. Ed. **46**, 4832 (2007).
61. S. Sergeyev, W. Pisula, and Y. H. Geerts, Chem. Soc. Rev. **36**, 1902 (2007).
62. A. M. van de Craats, J. M. Warman, K. Müllen, Y. Geerts, and J. D. Brand, Adv. Mater. **10**, 36 (1998).
63. R. J. Bushby and O. R. Lozman, Curr. Opin. Colloid. Interface Sci. **7**, 343 (2002).
64. S. Kumar, Liq. Cryst. **32**, 1089 (2005).
65. A. Zelcer, B. Donnio, C. Bourgogne, F. D. Cukiernik, and D. Guillon, Chem. Mater. **19**, 1992 (2007).
66. R. Winpenny (ed.), *Single-Molecule Magnets and Related Phenomena, Structure and Bonding*, vol. 122 (Springer-Verlag, Berlin, 2006).
67. S. Osa, T. Kido, N. Matsumoto, N. Re, A. Pochaba, and J. Mrozinski, J. Am. Chem. Soc. **126**, 420 (2004).
68. C. M. Zaleski, E. C. Depperman, J. W. Kampf, M. L. Kirk, and V. L. Pecoraro, Angew. Chem. Int. Ed. **43**, 3912 (2004).
69. R. E. P. Winpenny, J. Chem. Soc., Dalton Trans. 1 (2002).
70. R. Sessoli, Mol. Cryst. Liq. Cryst. **274**, 145 (1995).
71. D. Gatteschi, R. Sessoli, and A. Cornia, Chem. Commun. 725 (2000).
72. D. Gatteschi and R. Sessoli, Angew. Chem. Int. Ed. **42**, 268 (2003), and references therein.
73. O. Roubeau and R. Clérac, Eur. J. Inorg. Chem. 4325 (2008).

74. G. Aromí, S. M. J. Aubin, M. A. Bolcar, G. Christou, H. J. Eppley, K. Folting, D. N. Hendrickson, J. C. Huffman, R. C. Squire, H.-L. Tsai, S. Wang, and M. W. Wemple, Polyhedron **17**, 3005 (1998).
75. E. K. Brechin, Chem. Commun. 5141 (2005).
76. C. J. Milios, S. Piligkos, and E. K. Brechin, Dalton Trans. 1809 (2008).
77. E. Terazzi, C. Bourgogne, R. Welter, J.-L. Gallani, D. Guillon, G. Rogez, and B. Donnio, Angew. Chem. Int. Ed. **47**, 490 (2008).
78. N. Tirelli, F. Cardullo, T. Habicher, U. W. Suter, and F. Diederich, J. Chem. Soc., Perkin Trans. **2**, 193 (2000).
79. R. Deschenaux, B. Donnio, and D. Guillon, New J. Chem. **31**, 1064 (2007).
80. T. Cardinaels, K. Driesen, T. N. Parac-Vogt, B. Heinrich, C. Bourgogne, D. Guillon, B. Donnio, and K. Binnemans, Chem. Mater. **17**, 6589 (2005).
81. I. Aprahamian, T. Yasuda, T. Ikeda, S. Saha, W. R. Dichtel, K. Isoda, T. Kato, and J. F. Stoddart, Angew. Chem. Int. Ed. **46**, 4675 (2007).
82. E. D. Baranoff, J. Voignier, T. Y., V. Heitz, J.-P. Sauvage, and T. Kato, Angew. Chem. Int. Ed. **46**, 4680 (2007).
83. B. Chen, X. B. Zeng, U. Baumeister, S. Diele, G. Ungar, and C. Tschierske, Angew. Chem. Int. Ed. **43**, 4621 (2004).

Surface Diffusion on Inhomogeneous Surfaces

H. Bulou, C. Goyhenex, and C. Massobrio

Abstract We address the issue of topology and diffusion on inhomogeneous surfaces by employing realistic interatomic potentials and a set of atomic-scale simulations tools, such us structural optimization and molecular dynamics. We focus on heterogeneous combinations of transition metals substrate/adsorbate systems, motivated by experimental evidence pointing to non-trivial diffusion processes on short and extended spatial scales. The applications described in this chapter have in common the existence of strong non-homogeneous structural effects at the substrate level, resulting in surface reconstruction and preferential sites for adsorption and diffusion. Specifically, we analyze the migration processes on the Pt(111) and Au(111) substrate by referring mostly to the behavior of Co atoms, for which the account of both hopping and site exchanges was found to be crucial. The theoretical framework underlying our results is fully elucidated and validated through a detailed description of the interatomic potential construction, the molecular dynamics method, and the strategy for an effective search of the diffusion paths. This scheme allows to capture the diffusion mechanisms on both the short and the extended length scales.

1 Introduction

This chapter aims at reviewing the current theoretical understanding of single and collective diffusion on inhomogeneous surfaces. Our motivation is twofold. Real

H. Bulou (✉)
Institut de Physique et Chimie des Matériaux de Strasbourg, 23 rue du Loess, BP 43, F-67034 Strasbourg Cedex 2, France, herve.bulou@ipcms.u-strasbg.fr

C. Goyhenex
Institut de Physique et Chimie des Matériaux de Strasbourg, 23 rue du Loess, BP 43, F-67034 Strasbourg Cedex 2, France, christine.goyhenex@ipcms.u-strasbg.fr

C. Massobrio
Institut de Physique et Chimie des Matériaux de Strasbourg, 23 rue du Loess, BP 43, F-67034 Strasbourg Cedex 2, France, carlo.massobrio@ipcms.u-strasbg.fr

Bulou, H. et al.: *Surface Diffusion on Inhomogeneous Surfaces*. Lect. Notes Phys. **795**, 123–159 (2010)
DOI 10.1007/978-3-642-04650-6_5

surfaces used in experiments are very different from ideal, defect-free surfaces due to the presence of defects such as dislocations, holes, or steps, acting as preferential nucleation centers. Moreover, inhomogeneous surfaces (from the standpoint of structure and/or chemistry) are increasingly used as substrates on which the matter can be organized at the nanoscale by atomic deposition.

Developing a reliable model able to account equally well for structure and dynamics on inhomogeneous surfaces requires an approach founded on an affordable and accurate description of the interatomic interactions. To this purpose, we adopt the *n*-body interatomic potentials detailed in Sect. 2 by considering the classes of materials and the physical properties that such potentials can model realistically.

Section 3 is devoted to our simulation methodology. We show that the concerted use of classical molecular dynamics (to follow in real time the migration steps) and energy minimization procedures (to obtain potential energies along diffusion pathways) leads to a comprehensive description of the atomic-scale processes involving supported atoms and clusters. We shall also focus on the predictive power of such methods in the framework of diffusion on inhomogeneous surfaces.

In Sect. 4, we show how inhomogeneous surfaces can be described in connection with the pristine homogeneous regular crystallographic planes. We begin with the observation that atomic rearrangements at pure surfaces, induced by a periodicity breaking, lead to reconstructed surfaces with particular regular patterns. Reconstructions may be also obtained by depositing a material A on another type of material B, A and B featuring a strong lattice misfit driving "heteroepitaxial-induced" reconstructions. Vicinal surfaces will be also considered as particular inhomogeneous surfaces since they present a regular network of steps. The dynamical behavior of adatoms in their vicinity is expected to be different from the behavior at extended terraces.

Section 5 provides examples of issues characteristic of diffusion on inhomogeneous surfaces. Specific processes occurring at the inhomogeneous surface are described, such as lattice mismatch effect, step and reconstruction diffusion anisotropy, and solitonic diffusion.

2 Interatomic Potential

A reliable energy model allowing to determine atomic configurations and to characterize microscopic diffusion processes has to be defined from an accurate scheme accounting for the electronic structure. Due to the huge computational effort, the achievement of atomic configurations from ab initio electronic calculations is a prohibitive task for inhomogeneous solids and surface systems containing thousands of atoms. Equally challenging is to follow the evolution of atomic processes on time scales of a few nanoseconds. A viable alternative is to build a semi-empirical description of the electronic structure based on many-body potentials which can be used in numerical simulations like molecular dynamics or Monte Carlo. To such families belong potentials such as the *Embedded Atom method* (EAM) [1], the effective medium theory (EMT) [2], the second moment approximation tight-binding [3], and other recipes most adapted to covalent bonding [4] or oxide materials [5–7]. The situation becomes more intricate when two different types of bonding are involved (metal/semiconductor, metal/oxide). In this particular case, a strategy consists in deriving hybrid potentials from ab initio calculations performed on various fixed configurations [8].

In this work we shall consider surface systems based on transition metals. We stress that the methodology and the considerations developed herein can be applied to any material provided a realistic many-body description is available. The following section deals with our theoretical framework and highlights some essential methodological steps.

2.1 Many-Body Potential in the Second Moment Approximation (SMA) for Transition Metals

The expression of the tight-binding Hamiltonian for a pure bulk metal can be written in the basis of atomic orbitals λ at sites i, $|i, \lambda\rangle$ [9]:

$$H = \sum_{i,\lambda} |i, \lambda\rangle \varepsilon_{\text{at}}^{\lambda} \langle i, \lambda| + \sum_{i,j\neq i,\lambda,\mu\neq\lambda} |j, \mu\rangle \beta_{ij}^{\lambda,\mu} \langle i, \lambda|, \tag{1}$$

which involves two types of parameters, the effective atomic levels $\varepsilon_{\text{at}}^{\lambda}$ and the hopping integrals $\beta_{ij}^{\lambda,\mu}$. From this Hamiltonian, one defines the local density of states (LDOS) at a site i by projecting on the atomic orbitals at this site the Green function:

$$n_i(E) = -\frac{Im}{\pi} \sum_{\lambda} \langle i, \lambda|(E\hat{I} - \hat{H})^{-1}|i, \lambda\rangle. \tag{2}$$

From the LDOS the total energy $E_{\text{tot}} = \sum_i E_i$ is calculated with site energies E_i given by [10]

$$E_{\text{tot}} = \sum_{\lambda} \int^{E_f} (E - \epsilon_{\text{at}}^{\lambda}) n_i^{\lambda}(E) dE + \frac{1}{2} \sum_{j\neq i} \int\int dr_1 dr_2 \frac{Q_i(r_1)Q_j(r_2)}{|r_1 - r_2|}, \tag{3}$$

where the first term is a negative, i.e., attractive, contribution due to the band formation from the atomic level and the second one a positive, i.e., repulsive, interaction between spheres with charge Q_i. The latter contribution is too weak to treat correctly the repulsive interaction between the ions and it has to be replaced by a pairwise empirical contribution. To make calculations tractable, the first term can be reduced to a simple analytical form by replacing the actual LDOS by an approximate one having the same second moment. Adding this expression to the repulsive contribution leads to a many-body contribution which cannot be written as a sum of pair interactions, but instead as the square root of such a sum [3]:

$$E_i = \sum_{j \neq i} A \exp\left(-p\left(\frac{r_{ij}}{r_0^{ij}} - 1\right)\right) - \sqrt{\sum_{j \neq i} \xi 2 \exp\left(-2q\left(\frac{r_{ij}}{r_0^{ij}} - 1\right)\right)}. \quad (4)$$

This expression contains four parameters A, ξ, p, and q. The exponent q characterizes the distance dependence of the hopping integral between atoms at sites i and j. ξ is an effective hopping integral and p is related to the bulk modulus of the metal. The fact that the parameters are keeping a straightforward physical significance makes these potentials very reliable for studying structural effects in an interpretable way.

When bimetallic systems are considered, we have to rewrite a new sum including mixed interactions and containing three sets of four parameters:

$$E_i = -\left\{\sum_{j, r_{ij} < r_c} \xi 2_{IJ} \exp\left[-2q_{IJ}\left(\frac{r_{ij}}{r_0^{IJ}} - 1\right)\right]\right\}^{1/2}$$
$$+ \sum_{j, r_{ij} < r_c} A_{IJ} \exp\left[-p_{IJ}\left(\frac{r_{ij}}{r_0^{IJ}} - 1\right)\right], \quad (5)$$

where I and J indicate each of the two species. r_0^{II} is the first neighbor distance in the metal I and $r_0^{IJ} = (r_0^{II} + r_0^{JJ})/2$. The interaction is set to zero beyond a cutoff radius r_c. In order to use the same cutoff radius for both metals, r_c has been fixed at the second-neighbor distance for the atom with the largest atomic radius. Beyond this distance, the potential approaches zero via a fifth-order polynomial so as to avoid discontinuities both in the energies and in the forces.

The two heteroepitaxial systems under investigation, Co/Pt(111) and Co/Au(111), can be taken as representative of growth under tensile stress. Besides the effects due to the size mismatch, different behaviors of bimetallic systems in epitaxy can be encountered depending on the surface energy difference between the two metals (wetting criterion) and/or the ordering or phase separation tendency.

The (Co,Pt) system is characterized by a strong size mismatch (−11% between the Co and Pt lattice parameters). For each one of the metal species, the set of parameters (ξ, A, p, q) is determined by fitting the potential to the universal equation of state expressing the variation of the cohesive energy with the interatomic distances [11]. This procedure requires the knowledge of the cohesive energy, the lattice parameter, and the bulk modulus [12]; other structural properties like elastic constants can also be used or checked. Fitting the universal equation leads to the values of parameters reported in Table 1. An undesirable drawback of this recipe is the severe underestimation of the surface energies' values, as shown by their comparison with the experimental values [13]:

Table 1 Parameters of SMA potentials for the couple (A,B)=(Co,Pt) and values of the main fitted quantities: cohesion energy [12], equilibrium parameter [12], surface energies [13], and solution energies of impurities [15]

α	β	$A_{\alpha\beta}$ (eV)	$p_{\alpha\beta}$	$\xi_{\alpha\beta}$ (eV)	$q_{\alpha\beta}$	E_{coh} (eV/atom)	r_0 (nm)	$\gamma_\alpha - \gamma_\beta$ (eV/atom)	E_{sol} α in β (eV/atom)	E_{sol} β in α (eV/atom)
Co	Co	0.189	8.80	1.907	2.96	−4.45	0.251			
Pt	Pt	0.242	11.14	2.506	3.68	−5.86	0.277			
Co	Pt	0.245	9.97	2.386	3.32		0.264	−0.16	−0.47	−0.65

- $\gamma_{Pt}^{cal} = 0.46\,\text{eV/at}$, $\gamma_{Co}^{cal} = 0.35\,\text{eV/at}$
- $\gamma_{Pt}^{exp} = 1.03\,\text{eV/at}$, $\gamma_{Co}^{exp} = 0.87\,\text{eV/at}$

This difficulty can be circumvented by the consideration that when working on issues typical of bimetallic surfaces (segregation, deposition) the role of the key quantity can be played by the difference between surface energies [14]. In the present case, this difference is well reproduced: $\Delta\gamma^{cal} = 0.11\,\text{eV/at}$ and $\Delta\gamma^{exp} = 0.15\,\text{eV/at}$.

The main requirement for the cross-interaction parameters is to account for the tendency to bulk ordering, i.e., favoring heteroatomic pairs. This is achieved by fitting to the experimental (negative) heats of dissolution of one impurity of Pt into bulk Co (resp. Co into bulk Pt). Such quantity can be obtained from the slopes of the mixing energies [15] in the dilute limits. Due to the large size mismatch between Co and Pt, the whole system (matrix+impurity) has to be relaxed during the fitting procedure. Having only two equations for four parameters, only two of them (A and ξ) are left free while p and q are taken as the arithmetic average between the pure metal values (see Table 1).

Turning to (Co,Au) the size mismatch is of the same order as for (Co,Pt) (−14% between Co and Au parameters). For each pure metal species, the parameters (ξ, A, p, q) are determined by fitting to the cohesive energy, the lattice parameter, the bulk modulus, and some elastic constants. In order to preserve the surface energy difference $\Delta\gamma_{Co,Au} = \gamma_{Co} - \gamma_{Au}$, an additional constraint has to be introduced in the fitting procedure. The main requirement for the mixed-interaction parameters is to reproduce the existence of a miscibility gap in the phase diagram of the bulk CoAu system [15]. This is achieved by fitting the positive heats of solution for single substitutional impurities in order to get a realistic representation of this miscibility gap. An iterative scheme combining fitting and relaxation is then used. The final set of parameters is reported in Table 2.

It is worth noticing that the second moment approximation is best adapted to close-packed materials where the cohesion is mainly driven by d-band electrons. For other metals where the structure is less compact like body-cubic centered (bcc) crystals the bonding has a more directional character. In this case the use of modified embedded atom method (MEAM) should be a better alternative [16].

Table 2 Parameters of SMA potentials for the couple (A,B)=(Co,Au) and values of the main fitted quantities: cohesion energy [12], equilibrium parameter [12], surface energies [13], and solution energies of impurities [15]

α	β	$A_{\alpha\beta}$ (eV)	$p_{\alpha\beta}$	$\xi_{\alpha\beta}$ (eV)	$q_{\alpha\beta}$	E_{coh} (eV/atom)	r_0 (nm)	$\gamma_\alpha - \gamma_\beta$ (111) surface (eV/atom)	E_{sol} α in β (eV/atom)	E_{sol} β in α (eV/atom)
Co	Co	0.106	10.87	1.597	2.36	−4.45	0.251			
Au	Au	0.189	10.40	1.744	3.87	−3.82	0.288			
Co	Au	0.141	10.63	1.614	3.11		0.270	+0.21	+0.50	+0.75

2.2 Useful Energy Criteria

Molecular dynamics simulations of diffusion on surfaces rely on the evolution of the total energy for different configurations taken by the system during the migration process. In order to compare the relative stabilities of different structures, it is worth establishing some additional definitions and criteria related to the energetics. We refer here to the surface energy, the adsorption energy, and the incorporation energy. The surface energy is defined as an energy excess due to the presence of a surface. If one considers a layer of N atoms limited by two surface planes, each containing n_s atoms and having an area $\mathcal{A}$, the surface energy can be obtained by subtracting the energy E_N^{bulk} of N atoms taken in a bulk to the energy E_N^L of the layer. The latter being limited by two surfaces, the obtained value has to be divided by two. The surface energy γ is then calculated as follows:

$$\gamma = \frac{1}{2\mathcal{A}}(E_N^L - E_N^{bulk}). \tag{6}$$

The units are mJ/m², although the surface energy is often expressed in eV/atom. In this case the area $\mathcal{A}$ is replaced by the number n_s of atoms in the surface plane.

When depositing n_s A atoms onto a B substrate with N_s atoms per plane, corresponding to a coverage $\theta = n_s/N_s$ $ML(=$ monolayer), one defines the adsorption energy per A atom as follows:

$$E_{ads}(\theta) = \frac{E_{tot}(A/B) - E_{tot}(B) - n_s\mu(A)}{n_s}, \tag{7}$$

where $E_{tot}(A/B)$ is the total energy (B substrate and n_s adsorbed A atoms) and $E_{tot}(B)$ is the energy of the bare substrate, which writes for a slab of k layers:

$$E_{tot}(B) = kN_s E_{coh}(B) + N_s E_{surf}(B). \tag{8}$$

$\mu(A)$ is the chemical potential of the vapor phase taken as the origin of energies. The unit of E_{ads} in such calculation is eV/atom.

In the same way one calculates the incorporation energy E_{inc} (eV/atom) of n_{inc} A atoms in a surface of a B material by replacing n_{inc} atoms of B by n_{inc} atoms of A so that the number of total atoms is the same both in the initial and in the final slab:

$$E_{inc}^{(AinB)} = E_{tot}(A \text{ in } B) + n_{inc} E_{coh}(B) - E_{init}(B). \tag{9}$$

3 Methods

3.1 Classical Molecular Dynamics

The molecular dynamics method was first introduced by Alder and Wainwright in the late 1950s to study the interactions of hard spheres [17, 18]. It consists in the integration of the equations of motion for each atom in the crystal. We used the velocity Verlet integration algorithm [19, 20]. It allows to compute the particles position $q_{i,\alpha}(t + \delta t)$ and momentum $p_{i,\alpha}(t + \delta t)$ at a time $t + \delta t$.

The algorithm is derived by writing the following Taylor expansions for the position and its temporal derivative:

$$q_{i,\alpha}(t + \delta t) = q_{i,\alpha}(t) + \frac{dq_{i,\alpha}(t)}{dt}\delta t + \frac{d^2 q_{i,\alpha}(t)}{dt2}\frac{\delta t2}{2}, \tag{10}$$

$$\frac{dq_{i,\alpha}(t + \delta t)}{dt} = \frac{dq_{i,\alpha}(t)}{dt} + \frac{d^2 q_{i,\alpha}(t)}{dt2}\delta t. \tag{11}$$

By combining them, the coordinate at $t + \delta t$ reads

$$q_{i,\alpha}(t + \delta t) = q_{i,\alpha}(t) + \left(\frac{dq_{i,\alpha}(t + \delta t)}{dt} + \frac{dq_{i,\alpha}(t)}{dt} \right)\frac{\delta t}{2}. \tag{12}$$

The temporal derivative is computed by considering the Hamiltonian of the system

$$H = \sum_{i,\alpha} \frac{p_{i,\alpha}2}{2m_i} + \sum_i \sum_{j \neq i} P_{ij}\left(r_{ij} \right) - \sum_i \sqrt{\sum_{j \neq i} Q_{ij}\left(r_{ij} \right)}, \tag{13}$$

with

$$P_{ij} = A_{ij} \exp\left(p_{ij}\left(\frac{r_{ij}}{r_0^{ij}} - 1\right)\right),$$

(14)

$$Q_{ij} = \xi 2_{ij} \exp\left(2q_{ij}\left(\frac{r_{ij}}{r_0^{ij}} - 1\right)\right),$$

(15)

$$r_{ij} = \sqrt{\sum \alpha \left(q_{j,\alpha} - q_{i,\alpha}\right) 2},$$

(16)

and the associated set of dynamical equations

$$\frac{dq_{i,\alpha}}{dt} = \frac{\partial H}{\partial p_{i,\alpha}},$$

(17)

$$\frac{dp_{i,\alpha}}{dt} = -\frac{\partial H}{\partial q_{i,\alpha}}.$$

(18)

Then, by using Eq. (17), Eq. (12) can be written as follows:

$$q_{i,\alpha}(t + \delta t) = q_{i,\alpha}(t) + \left(\frac{\partial H}{\partial p_{i,\alpha}}(t + \delta t) + \frac{\partial H}{\partial p_{i,\alpha}}(t)\right)\frac{\delta t}{2m_i},$$

(19)

$$= q_{i,\alpha}(t) + \left(p_{i,\alpha}(t + \delta t) + p_{i,\alpha}(t)\right)\frac{\delta t}{2m_i}.$$

(20)

By using a Taylor expansion of the momentum,

$$p_{i,\alpha}(t + \delta t) = p_{i,\alpha}(t) + \frac{dp_{i,\alpha}(t)}{dt}\delta t = p_{i,\alpha}(t) - \frac{\partial H(t)}{dq_{i,\alpha}}\delta t,$$

(21)

Equation (12) is written as follows:

$$q_{i,\alpha}(t + \delta t) = q_{i,\alpha}(t) + \left(p_{i,\alpha}(t) - \frac{\delta t}{2}\frac{\partial H}{\partial q_{i,\alpha}}(t)\right)\frac{\delta t}{m_i}.$$

(22)

A similar scheme is used to compute the momentum.

First, a Taylor expansion of the momentum and its temporal derivative is computed:

$$p_{i,\alpha}(t + \delta t) = p_{i,\alpha}(t) + \frac{dp_{i,\alpha}(t)}{dt}\delta t + \frac{d^2 p_{i,\alpha}(t)}{dt2}\frac{\delta t2}{2},$$

(23)

$$\frac{dp_{i,\alpha}(t + \delta t)}{dt} = \frac{dp_{i,\alpha}(t)}{dt} + \frac{d^2 p_{i,\alpha}(t)}{dt2}\delta t.$$

(24)

By combining them, we obtain

$$p_{i,\alpha}(t + \delta t) = p_{i,\alpha}(t) + \left(\frac{dp_{i,\alpha}(t + \delta t)}{dt} + \frac{dp_{i,\alpha}(t)}{dt} \right) \frac{\delta t}{2}, \tag{25}$$

and by using Eq. (18)

$$p_{i,\alpha}(t + \delta t) = p_{i,\alpha}(t) - \left(\frac{\partial H(t + \delta t)}{\partial q_{i,\alpha}} + \frac{\partial H(t)}{\partial q_{i,\alpha}} \right) \frac{\delta t}{2}. \tag{26}$$

It is worth noting that the spatial derivative of the Hamiltonian at $t + \delta t$ does not depend on the momentum of the particles at $t + \delta t$

$$\frac{\partial H}{\partial q_{k,\alpha}}(t + \delta t) = \sum_i \sum_{j \neq i} \left(P'_{ij}(r_{ij}) - \frac{1}{2} \frac{Q'_{ij}(r_{ij})}{\sqrt{\sum_{j \neq i} Q_{ij}(r_{ij})}} \right) \frac{\partial r_{ij}}{\partial q_{k,\alpha}}, \tag{27}$$

with

$$P'_{ij} = -\frac{p_{ij}}{r_0^{ij}} P_{ij}(r_{ij}), \tag{28}$$

$$Q'_{ij} = -\frac{2q_{ij}}{r_0^{ij}} Q_{ij}(r_{ij}). \tag{29}$$

Then, Eq. (26) fully determines the momentum of the particles at $t + \delta t$ from the position and the Hamiltonian at t.

r_{ij} reads

$$r_{ij} = \sum_\beta (q_{j,\beta} - q_{i,\beta}) 2, \tag{30}$$

and its spatial derivative reads

$$\frac{\partial r_{ij}}{\partial q_{k,\alpha}} = \frac{q_{j,\alpha} - q_{i,\alpha}}{r_{ij}} (\delta_{jk} - \delta_{ik}), \tag{31}$$

where the symbol δ_{jk} stands for the Kronecker delta.

Inserting Eq. (31) in Eq. (27) leads to

$$\frac{\partial H}{\partial q_{k,\alpha}} = \sum_{i \neq k} \left(2P'_{ik}(r_{ik}) - \frac{1}{2} Q'_{ik}(r_{ik}) \left(\frac{1}{\sqrt{\sum_{j \neq i} Q_{ij}(r_{ij})}} + \frac{1}{\sqrt{\sum_{j \neq k} Q_{kj}(r_{kj})}} \right) \right)$$

$$\frac{(q_{k,\alpha} - q_{i,\alpha})}{r_{ik}}. \tag{32}$$

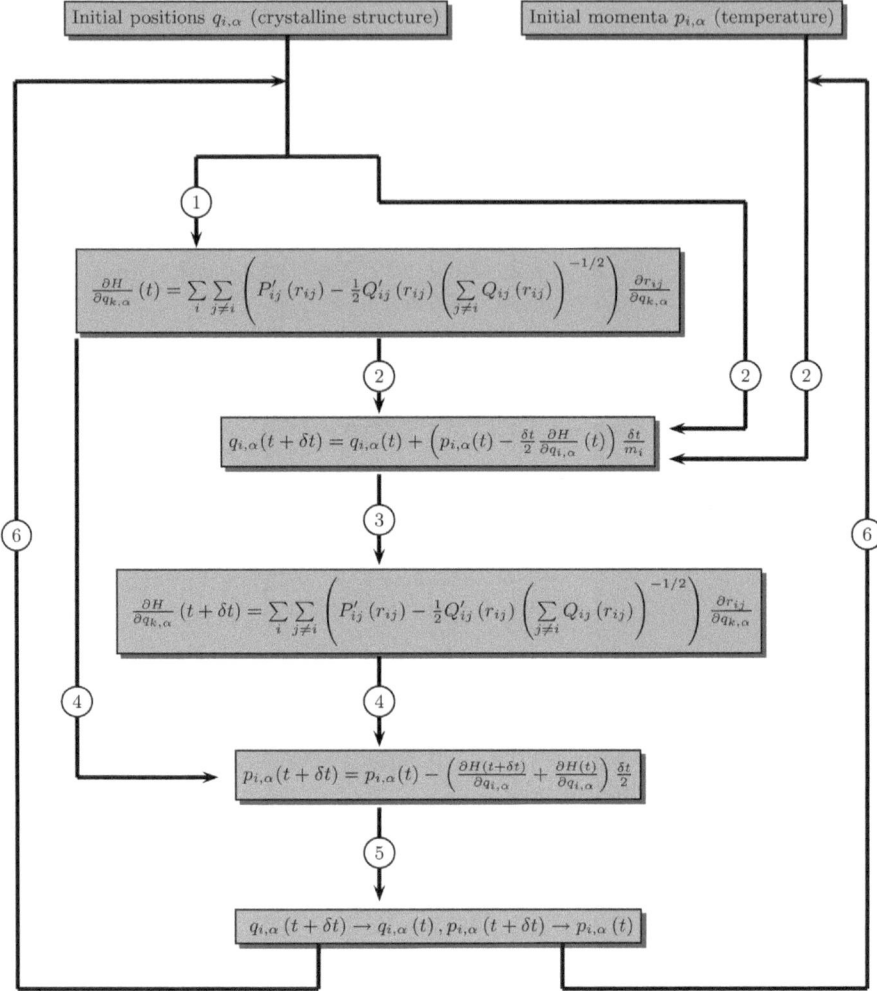

Fig. 1 Classical molecular dynamics algorithm

The molecular dynamics procedure is summarized in Fig. 1.

3.2 Quenched Molecular Dynamics

The quenched molecular dynamics procedure consists in a minimization of the total energy of the system, based on the idea that a minimum of the energy can be found by partially removing, at each time step, the kinetic contribution to the total energy [21]. The procedure amounts to extracting kinetic energy from the system by setting to zero the momentum $p_{i,\alpha}$ of an atom i anytime the scalar product of this momentum with the force is negative.

3.3 Nudged Elastic Band

The nudged elastic band (NEB) method is used to find reaction pathways or minimization energy paths (MEP) provided both the initial and final states are known [22–24]. The NEB algorithm works by linearly interpolating a set of images between the known initial and final states (as a "guess" of the MEP), and then it minimizes the energy of this string of images. Each "image" corresponds to a specific atomic configuration intermediate between the initial and the final state, playing the role of a snapshot along the reaction path. Thus, once the energy of this string of images has been minimized, the true MEP is found. The energy optimization of the string of images is performed by using the quenched molecular dynamics procedure and involves both the minimization of the force acting on the images and the minimization of the force acting on the atoms for each image.

4 The Inhomogeneous Surfaces

Inhomogeneous surfaces encompass reconstructed surfaces, vicinal ones, or surface alloys. These surfaces are characterized by a structural and/or a chemical inhomogeneity. Inhomogeneous surfaces are the result of mechanisms involving short-scale and long-scale rearrangements, as briefly described below.

4.1 Surface Mismatch Reconstruction

The first case we would like to consider is a stressed surface plane. In some cases, strain relief occurs through dislocation formation [25]. Then, due to the mutual long-range repulsion, the dislocations arrange into highly ordered periodic patterns [26]. A typical case of stressed surface plane is the (111) surface of transition metals. It is now well established that the ability of a transition metal surface to reconstruct depends on the surface stress [27, 28]. A reliable criterion to evaluate the surface stress in a metallic surface is the *surface mismatch* m_s [29],

$$m_s = \frac{r_{eq}^{surface} - r_{eq}^{bulk}}{r_{eq}^{bulk}}, \tag{33}$$

where r_{eq}^{bulk} and $r_{eq}^{surface}$ are, respectively, the interatomic equilibrium distance in the bulk and at the surface. For gold, due to the strong relativistic effects, characteristic of the $5d$ metals and the complete d-band filling, the equilibrium distances between the atoms at the surface are much smaller than in the bulk, leading to a surface mismatch m_s close to -3%. The relief of the existing strong tensile surface stress [29] leads to a reconstruction of the surface, the herringbone reconstruction of gold [30, 31]. Figure 2a, b shows an STM image of the topography of a clean reconstructed Au(111) surface and a simulation of the surface by using

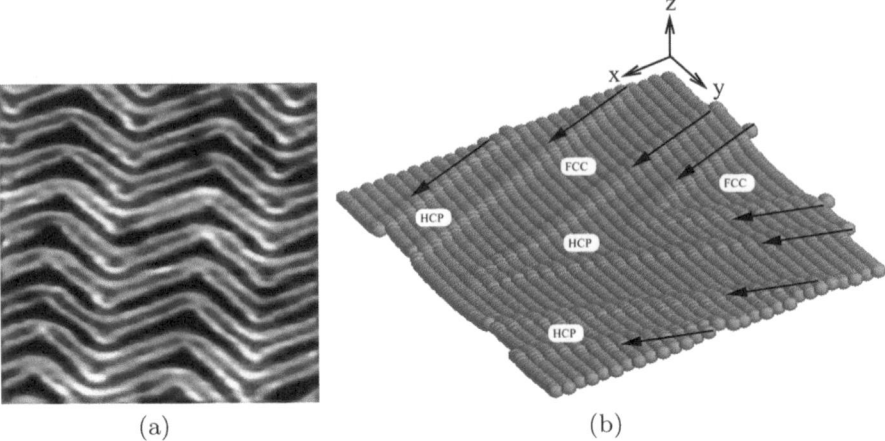

(a)   (b)

Fig. 2 (**a**) STM image of the herringbone reconstruction of Au(111). Image 70×70 nm [32]. (**b**) Geometry of the herringbone reconstruction of gold determined by using molecular dynamics. See Sect. 3 for more details about the method. Only the surface plane is displayed. The *black arrows* mark the discommensuration lines. The z-scale ($< 111 >$ direction perpendicular to the surface) is distended in order to emphasize the corrugation of the surface

molecular dynamics, respectively. Details about the method used to determine the reconstructed surface are given in Sect. 3. The present herringbone reconstruction consists of pair-wise arranged parallel stripes, the discommensuration lines (DL), running in a zigzag pattern. The mean distance between the lines of a pair is about 22 and 44 Å between the pairs. This arrangement is characterized by a strong structural inhomogeneity at the atomic scale. In Fig. 3 we visualize the map of the in-plane interatomic distances between the gold atoms belonging to the reconstructed surface. The corresponding distribution is given in Fig. 4. Details about the method used to determine the distribution of Fig. 4 are given in Sect. 3.

On the one hand, the surface mismatch forces the atoms of the surface to get closer in order to reach the surface equilibrium distance (2.82 Å). On the other hand, the underlying bulk lattice imposes the bulk equilibrium distance (2.88 Å) at the surface. The stablest configuration is a surface split into two main parts, respectively, of *fcc* and *hcp* stacking (Fig. 2b). In the fcc region, the surface atoms

Fig. 3 Mean in-plane interatomic $\bar{r}$ distances between the gold atoms belonging to the reconstructed surface

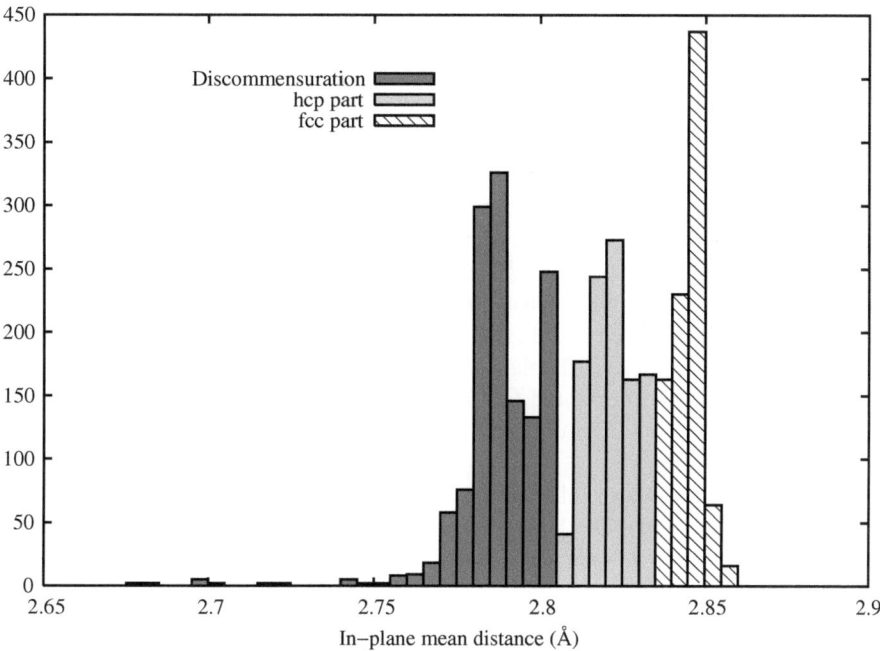

Fig. 4 Distribution of the in-plane interatomic distances between the gold atoms belonging to the reconstructed surface

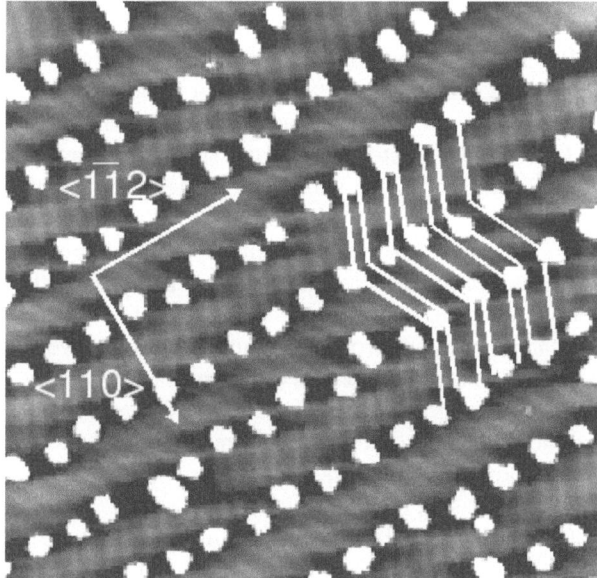

Fig. 5 STM image of self-organized Co clusters as obtained after deposition at 300 K. Zigzag DLs have been highlighted with *white lines*. Images 100×100 nm [33]

adopt the bulk lattice parameter whereas in the hcp region, they adopt the surface equilibrium distance. The DL acts as a transition region between the fcc and the hcp parts. Finally the DLs cross each other at specific points of the surface, the *kinks of the reconstruction*. The mean interatomic distance between the atoms of the surface in the DL is about 2.75 Å. The kinks form a network issued from the entanglement of the DLs and play a special role during the heteroepitaxial growth of many chemical species such as cobalt, iron, or nickel, since the nucleation takes place at these points. The mean interatomic distance at the kinks of the reconstruction is about 2.65 Å. Preferential nucleation is at the origin of the well-organized growth of nanostructures on the herringbone reconstruction of gold. A typical example is given in Fig. 5 where an STM picture of self-organized Co clusters grown at 300 K is displayed [33]. The clusters are two atomic layers in height, the amount of deposited Co corresponds to 0.26 atomic cobalt monolayer. Figure 5 shows a regular array of Co islands nucleated at the kink positions of the reconstruction.

Another interesting case of surface reconstruction is the (111) surface of platinum. The surface stress at the Pt(111) surface is smaller than the gold one. The associated surface mismatch m_s is only -2.5%. Then, there is no spontaneous reconstruction phase formation in normal conditions for Pt(111) as is the case for the Au(111) surface. However, in some specific conditions, as for instance after a sputtering of the surface [34] or in the presence of a supersaturated Pt gas-phase environment [35], a Pt(111) surface reconstruction has been reported.

4.2 Reconstruction Through Heteroepitaxy

While stress relief at pure surfaces may be responsible for reconstruction phenomena, it can be also responsible for the formation of network defects in an epitaxied atomic layer. This is the case for instance of copper deposited on Ru(0001) [36] or silver on Pt(111) [37, 38]. In heteroepitaxy the stress in the adsorbed layers stems from the lattice mismatch with the substrate. In the case of Ag/Pt(111) [37, 38], the first monolayer grows in a (1×1) structure, leading to an isotropic compression close to 9%. The spawned stress is then released in the second grown silver layer by the formation of metastable unidirectional phases, which transform into a trigonal network of crossing domain walls upon annealing. While Ag/Pt(111) has been taken as a famous example for growth under compressive strain, Co/Pt(111) is a system well representative of the relaxation mechanisms of the tensile strain in metal-on-metal heteroepitaxy. Using quenched molecular dynamics, it was found that the pseudomorphy is favorable in the first stages of the growth [39]. The adsorption energy corresponding to the continuous pseudomorphic (1×1) layer is 4.64 eV/atom. In comparison, the fully relaxed Co layer (1.23 ML) is less favorable with a value of 4.60 eV/atom. In this case, fitting the substrate by dilation has a lower energy cost than introducing unfavorable *on-top* positions while recovering

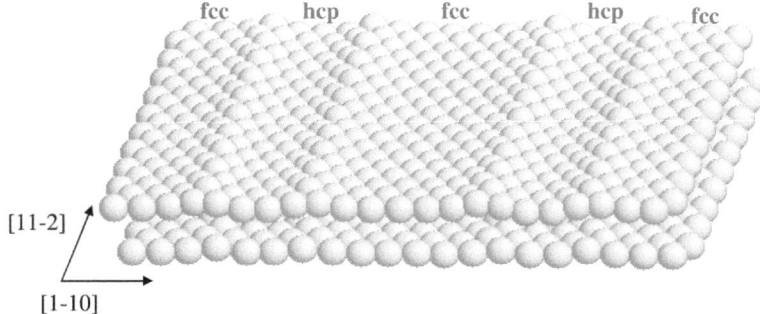

fcc   hcp   fcc   hcp   fcc

[11-2]

[1-10]

Fig. 6 Optimal relaxed structure for a Co layer onto Pt(111): 1D pseudoepitaxy. The *upper plane* is the Co deposit and the underlying layer is the first Pt plane of the Pt substrate slab. The scale has been magnified in the z-direction in order to better visualize the corrugation in the layer

bulk-like Co distances. The first layer structure can be optimized with respect to the pseudomorphy by slightly overfilling the substrate plane, i.e., by simulating layers with a coverage slightly higher than 1 ML. One way to build a more close-packed structure than the Pt(111) plane is obtained from the analogy with the reconstruction observed in the Au(111) surface which is attributed to the pure surface tensile strain. One gets in this way a pseudoepitaxy having an uniaxial character that features a 1D pseudomorphy along the [11-2] direction and a 1D pseudoepitaxy along the [1–10] direction. Different box sizes and/or fillings of the first Co layer have been tested within the relaxation procedure. The optimal value of the adsorption energy is $E_{ads} = -4.66\,\text{eV/atom}$ for a coverage of 1.10 ML. Figure 6 shows the corresponding relaxed superstructure. Upon the relaxation, the adatoms do not stay uniformly distributed on the surface and some corrugation appears in the layer. This is due to the formation of linear defective regions having a higher level in the z-direction $\delta z = 0.017\,\text{nm}$ and for Co atoms in bridge positions. These line defects separate regions in coherent epitaxy but with alternate fcc and hcp stacking with regard to the substrate. Two pairs of these lines are visualized in Fig. 6. In this case, the fcc regions are wider than the hcp regions situated between two close domain walls just as for the Au(111) reconstruction surface. Some discommensurations of this nature have been observed in scanning tunneling microscopy (STM) experiments of Lundgren et al. [40] for a Co layer grown onto Pt(111).

It appears that under tensile stress during heteroepitaxy, pseudomorphy is the most favorable situation in the first stages of the growth. This holds true despite the large expansion involved. The lower energy cost for this dilation is due to the very smooth variation of the interatomic potential above the equilibrium distance. The lattice dilation has a limited spatial extent meaning that the strain at the completion of the first layer is too large to get a perfect pseudomorphy. The strain induced by the large misfit is then partially released through the introduction of some contractive reconstruction discommensurations, separating the regions of coherent epitaxy. This mechanism allows an over-closepacking of the film similar to the 5*d* metal surface reconstructions like it was demonstrated precedently for Au(111) surfaces. In many systems presenting a large negative value of the misfit, the epitaxy is coherent in the first stages: Co/Pd(111) [41]; Cu/Ru(0001) [36]; Ni/Pt(111) [42]. The occurrence of relaxation by contractive discommensurations has been experimentally evidenced in the case of Co/Pt(111) [40] and Cu/Ru(0001) [36]. This scheme is the most generally valid for tensile strain relaxation in close-packed metallic interfaces. However, this behavior has a limit in terms of size mismatch. Indeed, it has been shown for Co/Au(111) epitaxy that Co is relaxing in the first layer toward its natural bulk parameter and this is due to a too strong size mismatch (–14% between Co and Au parameters) [43–45].

4.3 The Vicinal Surfaces

The vicinal surfaces are obtained by cutting a crystal along a plane making a small angle – typically lower than 10° – with a low index plane. Figure 7 shows the example of the Pt(997) vicinal surface where a periodic succession of terraces and steps of monoatomic height can be observed. The origin of inhomogeneity of vicinal surfaces is double. First, the coordination of the atoms belonging to the steps is lower than the one of the atoms inside the terraces, leading to different nucleation behavior of the adatoms at the step's edges and at the center of the terraces. Hence, nanowires can be obtained by step edge decoration [46–48].

Second, due to the steps, a side of the terraces is free, which allow the terraces to release (RELEASE...?) the surface stress. Figure 8 displays the mean in-plane interatomic distances between the platinum surface atoms in the neighborhood of a step edge. Distances are shorter at the edge compared to the bulk value (2.78 Å). Mainly three atomic rows are affected, and the distance ranges from 2.737 Å at the edge to 2.762 Å at the center of the terrace.

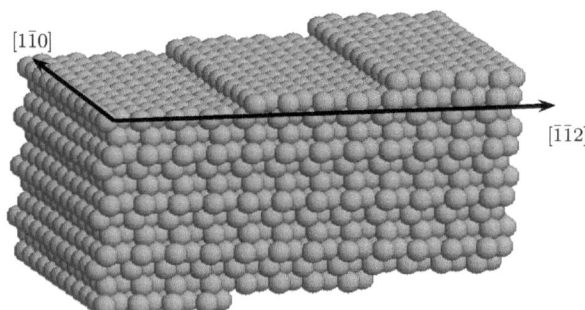

Fig. 7 Pt(997) vicinal surface in perspective view

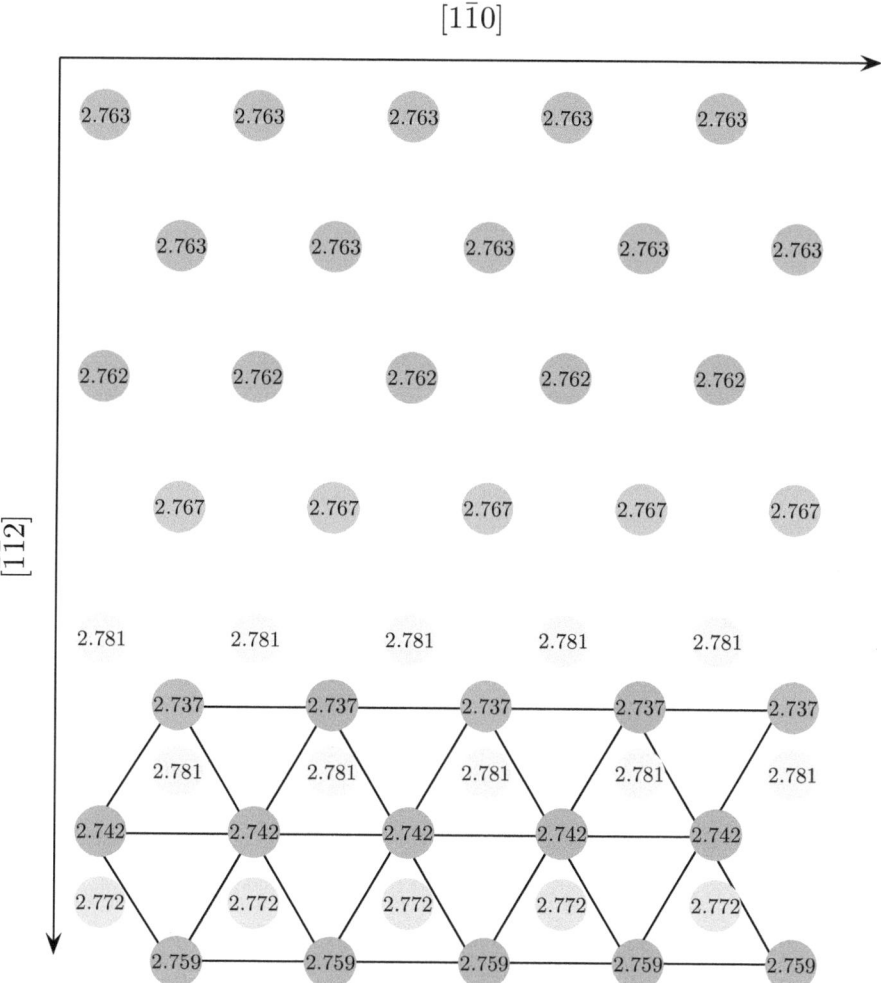

Fig. 8 Mean in-plane intertomic distances between platinum atoms on the Pt(997) vicinal surface. The *black lines* connect the platinum surface atoms of the upper terrace

5 Atomic Diffusion on Inhomogeneous Surfaces: Some Examples

5.1 Lattice Mismatch Effect in Atomic Migration During Heteroepitaxial Metal Growth

Lattice mismatch plays a determinant role in surface atomic diffusion in epitaxy, as shown by the example of atomic migration along steps during heteroepitaxial metal growth. One effect is the increasing anisotropy of diffusion along close-packed steps of adsorbed islands on (111) surfaces with increasing strain in the layer. This has been exemplified by results on homoepitaxy [49–55] and heteroepitaxy where the adlayer is under compressive strain [56]. In most of these investigations the barrier of diffusion along steps with (111) triangular microfacets (B steps) is greater than the barrier of diffusion along steps with (1000) square microfacets (A steps). The Co/Pt(111) system can be used as a model system to study diffusion along steps in the case of tensile strain [57]. Simulations have been performed within classical molecular dynamics (MD) applied to the diffusion along these two kinds of steps in the case of Co/Pt(111). MD is first used as a relaxation method by applying quenched molecular dynamics. In this way we can determine at 0 K both the equilibrium positions of atoms and the activation energies related to the motion of an adatom along a step. In a second step, constant energy molecular dynamics simulations are performed in order to link the obtained results to the actual trajectories of adatoms at temperatures of interest. The chosen system is a slab having a thickness of 12 planes in the z-direction and delimited by two (111) surfaces. On the side of the slab (x- and y-directions) periodic boundary conditions are applied. A smaller Co terrace has been added on the topmost Pt(111) plane where the Co atoms are in registry with the substrate following its fcc stacking. The Co terrace is limited by an A step in the lower part and by a B step in the upper part. To study the parallel motion we start with one adatom in the most stable fcc site on the side of each step, where the coordination is fivefold (there are three Pt neighbors underneath and two lateral Co ones). In order to determine the activation energies associated with the motion parallel to the steps, calculations have been performed at 0 K by quenched molecular dynamics. The Co adatom added on the side of each step is moved parallel to the step and potential energies are calculated at regular distances between two most stable sites. The adatom is left free to relax in all directions except in the direction parallel to the step. All other atoms of the slab are free to relax in the three space directions. Results are presented in Fig. 9 together with the main simulated atomic configurations labeled by numbers 1, 2, and 3.

The zero energy reference is taken at the most stable fcc site. The potential energy curves have similar appearance both presenting a secondary minimum, giving rise to two symmetrical saddle points. Migration along an A step gives a maximal variation of energy of $\Delta E(A) = 0.41$ eV. Within the same definition, migration along a B step gives a maximum at much lower energy with $\Delta E(B) = 0.18$ eV. Looking closer to the curves one has to pay attention to the secondary minimum. In both cases the maxima correspond to the crossing of a bridge position between two Pt substrate atoms. The secondary minimum corresponds to an effective fourfold position (see images labeled *Number* 2 in Fig. 9) where the Co adatom is linked to three underlying atoms plus one atom of the step border. It is clear for the B step that in this position the Co–Co distance is strongly reduced and gives a (secondary) minimum, since this leads to a distance of 0.226 nm closer to the Co–Co bulk one (0.250 nm) while much smaller than the Pt–Pt one (0.277 nm). We have identified here an inward displacement of the Co atom toward the step, $\Delta y = -0.037$ nm. Such a displacement is strictly forbidden in homoepitaxy where the motion is 1D giving a saddle point in the middle position between the two stable fcc sites [50, 54].

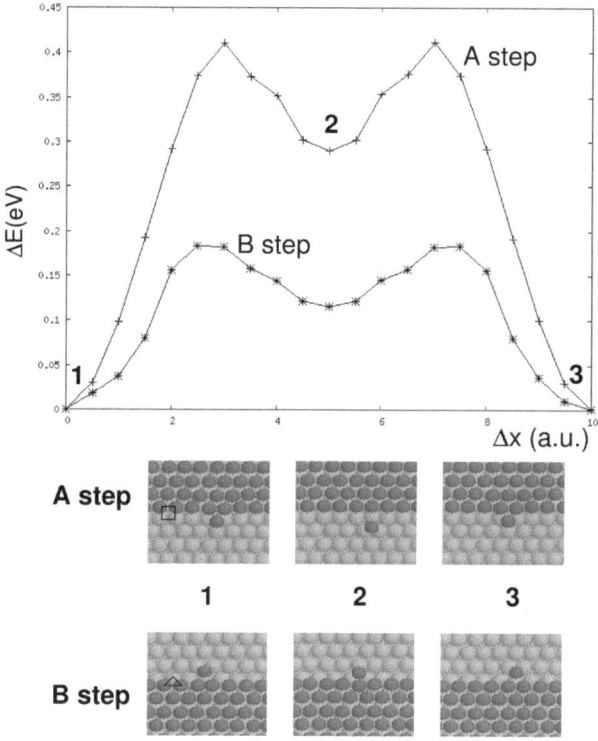

Fig. 9 Migration along steps: potential energy curves obtained for the displacement of a Co adatom starting from a fivefold coordinated site along A or B step and joining another equivalent site along the considered step. The x-axis corresponds to the displacement parallel to the step and is expressed in arbitrary units (a.u.). The images under the *curves* represent the key stages of this displacement. They are labeled with numbers which are also reported on the *curves* in order to locate the corresponding energy

In the case of the A step, the adatom moves outward from the step with regard to the stable fcc sites. A secondary minimum is also observed where the adatom recovers a coordination of strictly three (*Number* 2 in Fig. 9). This particular energy profile results also from a lattice mismatch effect. Indeed, for homoepitaxy along the A step, the displacement of the adatom is similar on a geometrical point of view but the energy maximum is obtained close to the hcp position where the adatom is the most distant from the step [50, 54]. As the energy for moving along A step is quite large with regard to migration along B step, one has to check the ability of an alternative path of lower cost before concluding on an actual anisotropy between migration along A and B steps. Another option for moving along the A step is through an exchange process where the adatom pushes out one of the step atoms. Simulating this motion leads to a saddle position represented in Fig. 9 corresponding to an activation energy of $\Delta E = 0.48$ eV, therefore higher than simple migration. This high energy comes from the fact that the two moving atoms have to turn around one underlying Pt atom of the substrate close to a much unfavorable on-top position. In order to link static activation energies to the actual relative mobility of the Co adatoms some complementary constant energy calculations have been performed. The trajectories of the two Co adatoms placed at each border are then followed in a same simulation of 2.4 ns at $T = 370$ K, this temperature being chosen in order to approach experimental conditions. They are plotted in Fig. 10.

As expected on the basis of static calculations, a much longer trajectory is obtained along the B step while along the A step no jumps are recorded within the same simulation time. Looking first at the Co atom moving along B step, the displacement occurs by hopping, with the adatom spending more time in stable five-fold sites where it undergoes small oscillations. Transitions toward the secondary, less stable site along the step are also observed and identified in the potential energy curves. In agreement with the potential energy profile, the atom spends less time in this site. The two oscillatory modes are also visible along the A step but the short span of the trajectory prevents from drawing clear-cut conclusions. Therefore, both static calculations and molecular dynamics provide evidence for a much larger

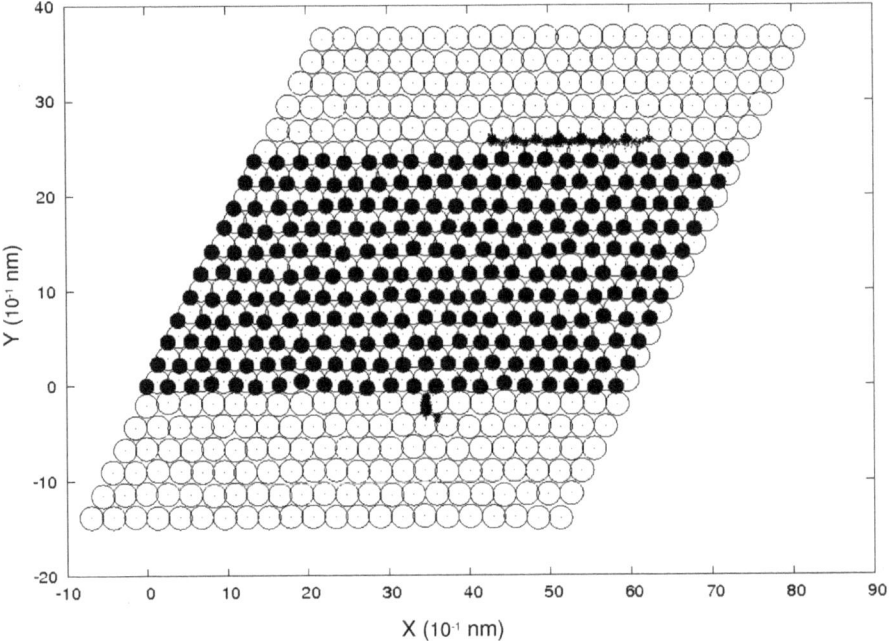

Fig. 10 Trajectories of a Co adatom, respectively, along A and B steps obtained for a simulation of 2.4 ns at 370 K. The Co atoms of the terrace are the *smaller dark circles* situated on *top* of the Pt(111) lattice represented with *empty circles*. The *lower part* of the terrace is an A step and the upper part is a B step. The trajectories of the Co atom are represented by *solid lines*

propensity to migrate along B steps than along A steps in the case of heteroepitaxial systems for which the adlayer has a smaller lattice constant relative to the substrate (tensile strain).

In order to obtain a general view for diffusion along steps as a function of lattice mismatch, we have compared calculated diffusion barriers obtained within different formalisms for various homoepitaxial systems and two heteroepitaxial cases (Co/Pt(111) and Ag/Pt(111)). These latter are representative of systems under tensile and compressive strain, respectively. Values of energy barriers are summarized in Table 3. Looking first at homoepitaxy, most of the systems present either no anisotropy or a much smaller one than in heteroepitaxy. Except for gold, the general tendency is to have slightly higher energy activation for diffusion along B steps. The same tendency, while much marked in terms of anisotropy, is obtained for the heteroepitaxial compressive system Ag/Pt(111). For this system, the energy barrier for migration along the B step is 2.1 times larger than the one for migration along the A step. The opposite behavior is obtained for Co/Pt(111), the energy barrier for migration along A step being 2.3 times larger than the one for migration along B step. This particular behavior is directly linked to the lattice mismatch between Co and Pt allowing a lower energy diffusion path along the B steps. In this case, the "smaller" Co atom undergoes an inward displacement toward the step during its displacement between two stable sites. This motion becomes possible only in the case of tensile strain where the adlayer has a smaller lattice constant than the substrate.

Table 3 Comparison of activation energies for migration along A and B steps for various metallic systems either calculated with semi-empirical methods or density functional theory (DFT) calculations, second moment approximation (SMA), effective medium theory (EMT), or embedded atom method (EAM). For DFT calculations we refer to either local density (LDA) or generalized gradient corrections (GGA) calculations. Corresponding references appear in the last column of the table

System	ΔE(eV) A step	ΔE(eV) B step	Method	References
Cu/Cu(111)				
HCP islands	0.25	0.31	EAM	[49]
FCC islands	0.34	0.30	EAM	[49]
Au/Au(111)	0.34	0.22	SMA	[50]
Ag/Ag(111)	0.25	0.29	SMA	[50]
	0.21	0.28	EAM	[51]
	0.22	0.30	EMT	[56]
Al/Al(111)	0.31	0.45	DFT-GGA	[52]
Pt/Pt(111)	0.45	0.40	EMT	[53]
	0.71	0.77	DFT-GGA	[54]
	0.84	0.90	DFT-LDA	[54]
Ag/Pt(111)	0.19	0.39	EMT	[56]
Co/Pt(111)	0.41	0.18	SMA	[57]

5.2 Co Adatom Diffusion Across a Stepped Pt(111) Surface

Stepped surfaces can be taken as the simplest examples of inhomogeneous surfaces. When preparing substrates as flat as possible for epitaxial growth a non-negligible density of atomic steps can remain, perturbing further diffusion through interlayer mass transport or simply acting as a trap for incoming atoms. This latter effect is also used to build in a controlled way regular networks of nanowires using a vicinal surface, i.e., a surface presenting a regular network of atomic steps [58]. To elucidate this point let us consider again the case of Co/Pt(111). The simulation slabs feature a thick Pt(111) substrate on top of which a six-row Pt terrace has been added. The terrace is bounded on the lateral sides by two straight steps of different symmetry, an A step (lower side of the terrace in Fig. 11) and a B step (upper side). Above the terrace we put an adatom which may move around, reach the border of one of the steps, and possibly descend. Once the adatom is on the border many different processes may occur: diffusion along the step, reflection back to the inner terrace, or descent to the lower terrace either by jump or by exchange. First considering step A,

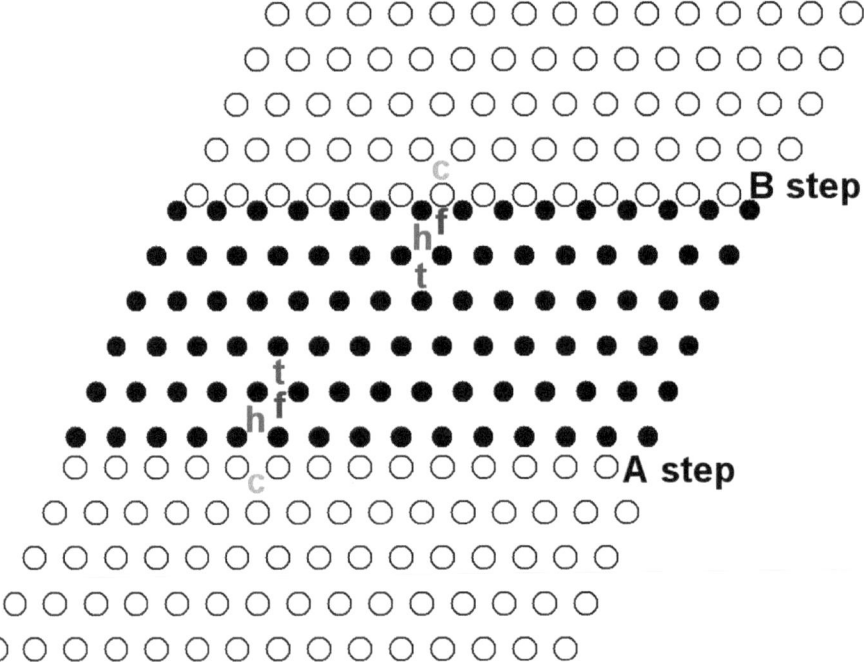

Fig. 11 *Top view* of the slab used in the simulation. The *open and full circles* represent the atoms of the *lower and upper terraces*, respectively. The terrace is bounded by an A step on its *lower side* and a B step on its *upper side*. *f* and *h* indicate fcc and hcp sites on the terrace borders; *c* indicates the nearest equilibrium site on the lower terrace (in the channel) and *t* the nearest equilibrium site in the inner terrace (terminology taken from [59])

the equilibrium sites just at the border of the upper terrace are hcp sites (as the site h in Fig. 11); their fcc neighbors (f site) are behind the atoms of the step. Once the adatom has reached an h site, several possibilities are let open. The adatom may jump directly down in the channel (to the site c just in front); it may exchange with an atom of the step, the latter being pushed toward one of the c sites; it may jump to an f site. If the adatom is on an f site, other processes may take place: it may jump to an h site, it may exchange with an atom of the step, or it may come back to the inner terrace jumping to the nearest t site. Along a B step similar processes can take place, the only difference being that the sites just at the border of the step are fcc ones.

The barriers for some of the above processes are reported in Table 4. Let us consider first migration on the upper terrace. The hcp and fcc positions just near the top (h and f) border have nearly equal site energies and are more stable than the other sites on the terrace. For both types of steps their site energies are lower by about 50.0 meV with regard to the other ternary sites on the terrace. For that reason, the barrier for coming back toward the terrace, or the reflection barrier ΔE_r, is increased by the same value with respect to the barrier for migration along the border. The barrier for migrating parallely to the steps border is the lowest one, suggesting that once the adatom has reached a border (A or B) it preferentially migrates along it. Then, as regards step descent, jump or exchange descent is proceeding from h sites at A step and f sites at B steps (as suggested by the energetics of these locations). For both types of steps, the jump-descent barriers ΔE_j are lower by a few tenths of electronvolts than exchange barriers. From these results step descent is expected to occur preferentially by jump. The anisotropy diffusion between steps is noticeable in the case of the exchange process.

An anisotropy in the case of the exchange process is similarly found in calculations (by embedded atom and effective-medium potentials) in Ag [60] and Pt [61], the barrier for exchange turning out to be considerably lower at step B than at A step. In the case of Ag, by means of SMA potentials, a strong anisotropy (0.35 and 0.21 eV) has been found for exchange from A and B steps, respectively [62], in agreement with [60]. For Au the anisotropy is much less marked whatever the considered process [59].

According to the static energy barriers, one may predict the qualitative following behavior at high T. ΔE_{edge} is much lower than the other barriers. Therefore, once the adatom has reached the border of one step it should diffuse along that step before being reflected back to the inner terrace. The probability for this process to occur

Table 4 Energy barriers for crossing Pt steps at a (111) surface for the following processes: exchange, ΔE_{ex}; jump, ΔE_j; reflection from the border toward inner terrace, ΔE_r; migration along the top border of the step, ΔE_{edge}. All data are in eV

Step	ΔE_{ex}	ΔE_j	ΔE_r	ΔE_{edge}
A	0.77	0.56	0.24	0.18
B	0.63	0.52	0.25	0.19

is low but much larger than the probability of descending to the lower level either by exchange or by jump. On the other hand, as the barriers for exchange are much higher than barriers for jump descent, one may expect that only jumps are occurring at A steps and few exchanges at B steps. In order to investigate whether those predictions are realistic, we have performed a series of constant-energy simulations at $T = 900$ K. In each simulation, the adatom starts in the middle of the upper terrace, then it diffuses, reaches the step, and finally descends. Each simulation is stopped when the adatom has descended to the lower terrace, either by jump or by exchange. Figure 12 represents a trajectory of a Co adatom diffusing onto the Pt terrace. It has been obtained after a simulation at $T = 900$ K corresponding to a diffusing time of 0.4 ns. Near the A step it nicely illustrates the Schwoebel effect [63] due to the fact that the adatom reaching the border of the terrace has to overcome a potential barrier effectively preventing the descent to the lower terrace. In this example, one exchange process has finally occurred at the B step between the adatom and a Pt atom of the step border.

Generally, as observed in Fig. 12, when the adatom reaches the border of an A step, it spends most of its time along this border with very few steps of descent and a multitude of jumps. On the contrary, when the adatom reaches the border of a B step it is rapidly incorporated into the step by an exchange. From a series of simulations at 900 K we have estimated the relative occurrence of each type of jump. At B steps 92% of descent events are exchanges. At A steps there is an equal frequency between jumps and exchanges. At B steps, it appears that the above predictions do not correspond to the actual high-T results where exchanges

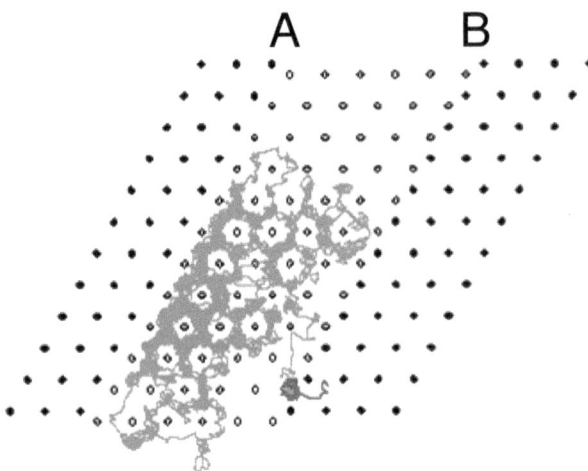

Fig. 12 Trajectory of a Co adatom initially placed on top of a Pt terrace. The Pt atoms of the terrace are the smaller empty circles situated on top of the Pt(111) lattice represented with *filled black circles*. The *left part* of the terrace is an A step and the *right part* is a B step. The trajectory of the Co atom is represented by a *gray solid line*. The *black solid line* corresponds to the Pt atom going out of the B step during the final exchange at this step

are clearly more likely to occur than reflections and jumps. In summary, there is an unexpectedly high frequency of exchanges with respect to that of reflections and jumps in the case of B steps. The apparent contradiction between static and dynamical results can be explained as follows. At high temperatures, the step atoms perform large-amplitude oscillations around their equilibrium sites. The vibrations are asymmetric in the direction perpendicular to the step, being easier for a step atom to oscillate outward than inward. This effect is much stronger when an adatom is on the border of the step. From a geometrical point of view, the outward motion is much easier for an atom coming out of the B step than for an atom coming out of an A step for which an on-top position has to be circumvented (see Fig. 11). The latter considerations are in favor of a high anisotropy of diffusion in step descent by exchange for heteroepitaxy of Co/Pt(111), this anisotropy being expected on the basis of static calculations. However, vibrational effects at stepped surfaces may play an important role and lead to diffusion phenomena that could not be predicted by static calculations. These pieces of evidence suggest that diffusion studies should be made through a combination of static and dynamic simulations. On a practical point of view, the simulations performed at high temperatures can help to identify the temperature regime of self-organization on a surface. For a vicinal substrate this strategy will allow to determine the range of temperatures where the formation of nanowires is possible without alloy formation by incorporation at steps.

5.3 Diffusion Anisotropy of Cobalt on the Herringbone Reconstruction of Gold

In what follows, we focus on the mass transport and the diffusion mechanism of cobalt adatoms on the herringbone reconstructed surface of Au(111). A typical diffusion path for Co atoms on such inhomogeneous surfaces is displayed in Fig. 13 where a snapshot of a simulation performed at 600 K for a time interval of 96 ps is presented. A strong diffusion anisotropy of the cobalt adatoms on both fcc and hcp parts of the reconstruction stands out as the most peculiar phenomenon. One notices a clear displacement of the cobalt adatoms toward the discommensuration lines, along a direction perpendicular to it. Conversely, escape of cobalt adatoms from the discommensuration lines to either the fcc or the hcp parts of the reconstruction occurs much less frequently. This highlights the role of the discommensuration lines as attractive traps for the cobalt adatoms.

A direct outcome of the attractive feature of the discommensuration lines is the cobalt depletion of the fcc and hcp parts of the reconstruction to the advantage of the discommensuration lines. The depletion effect is readily seen in Fig. 14, where the time evolution of the occupation ratio of the different parts of the reconstruction by the cobalt adatoms at 400 and at 600 K is presented. The diamond symbols give the *expected geometrical occupation ratios*. They represent the proportion of cobalt adatoms lying on the different parts of the reconstruction in the case of an uniform distribution. Figure 14 shows that irrespective of the temperature, the occupation

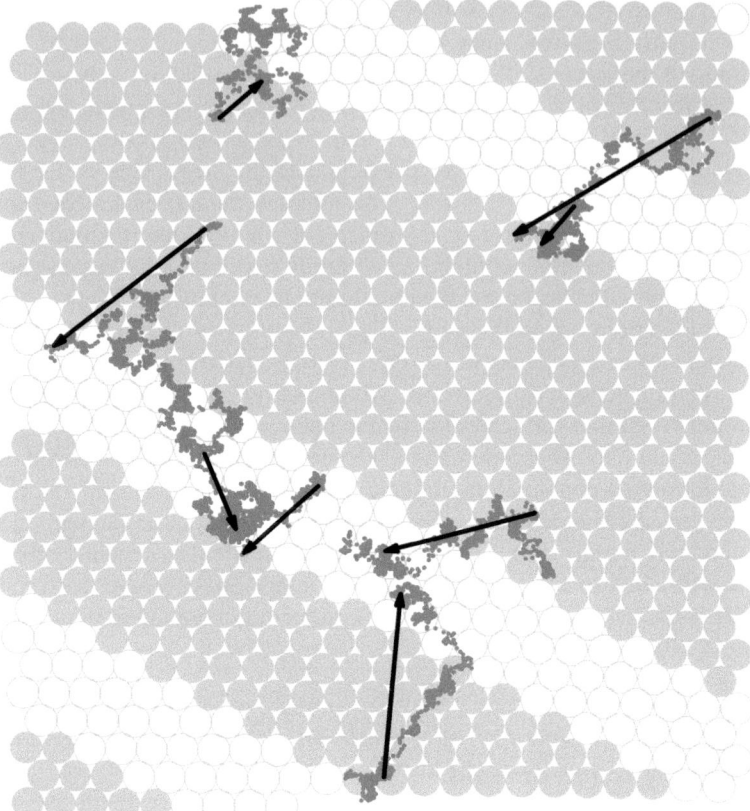

Fig. 13 Trajectories (*gray dots*) of Co adatoms on the herringbone reconstruction of Au(111) along a time interval of 96 ps at $T = 600$ K. The *unfilled circles* give the position of the gold surface atoms belonging to the discommensuration lines. The *black arrows* give the net displacement of the cobalt adatoms along the time interval

ratio of the discommensuration line reaches a stationary value much larger than the corresponding expected geometrical occupation ratio. This is an additional clue of the attractive character of the discommensuration lines at the expense of the other parts of the reconstruction.

It is worth noting the temperature dependence of the stationary occupation ratio, i.e., the higher the temperature, the larger the variation compared with the expected geometrical occupation ratio. Such a trend provides evidence for the thermodynamical origins of the depletion effect.

The origin of the diffusion anisotropy can be understood from Fig. 15 which displays a part of the surface with two discommensuration lines and a fcc part of the reconstruction. An adatom is located at the center of the triangle and the three possible directions to diffuse are indicated by the arrows. The main difference between the three directions $<11\bar{2}>$, $<1\bar{2}1>$, and $<\bar{2}11>$ lies on the stress state:

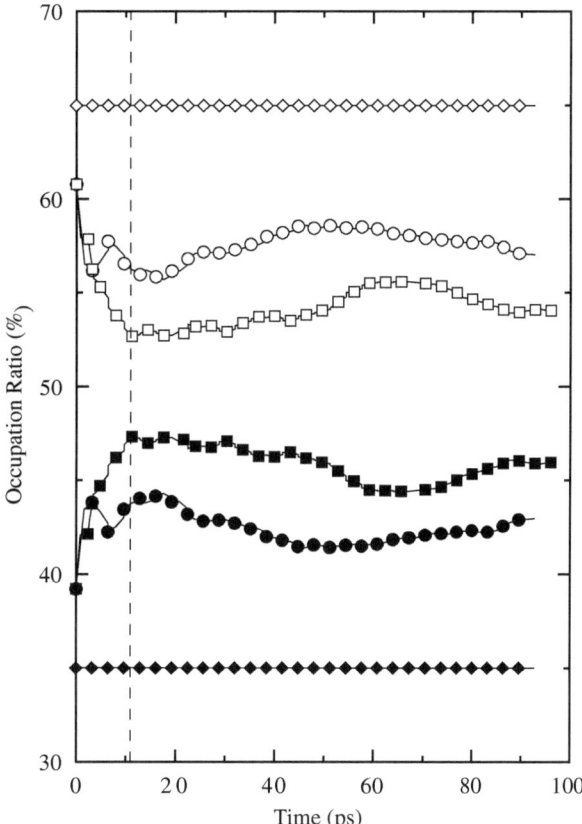

Fig. 14 Time evolution of the occupation ratio of the fcc+hcp parts of the reconstruction (*open symbols*) and of the DL (*full symbols*) at 600 K (•) and 400 K (□). The *diamond symbols* give the expected geometrical occupation ratio

while the stress along the direction $<\bar{2}11>$ perpendicular to the discommensuration lines is released due to the over-closepacking (see Sect. 4), the other $<11\bar{2}>$ and $<1\bar{2}1>$ directions are stretched. NEB calculations indicate an activation energy of 100 meV for moving along the directions perpendicular to the discommensuration lines whereas it is 200 meV for the directions parallel to the discommensuration lines [64]. This means that the movement along the $<11\bar{2}>$ and $<1\bar{2}1>$ directions is more demanding than along the $<\bar{2}11>$ direction. In the latter case, the presence of the adatom at the saddle point releases the tensile stress, thereby decreasing the energy barrier. Figure 15 shows that adatoms deposited on the fcc part of the reconstruction are constrained to move into a channel perpendicular to the discommensuration lines.

5.4 The Solitonic Diffusion

For a long time, the atomic diffusion by hopping, described in the previous section, has been considered the most common mechanism of mass transport on the surfaces.

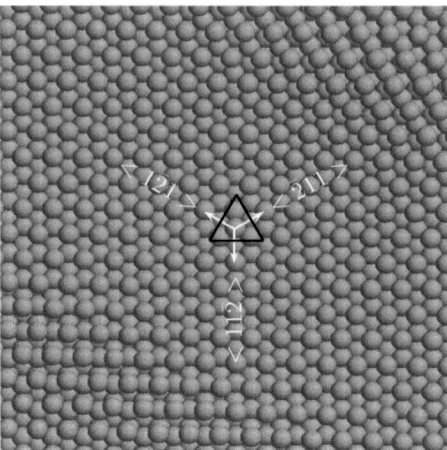

Fig. 15 Diffusion directions for an adatom located at the center of the fcc part of the reconstruction

However, on the basis of both experiment and theory, it has been discovered that a second major mechanism can occur, the atomic site exchange [65–67]. In such a mechanism, the adatom takes the place of a surface atom which in turn continues the diffusion. The occurrence of the exchange mechanism depends mainly on the chemical nature of the substrate and on its orientation. It has been proved that in the case of the $5d$ fcc metal surfaces such as gold, iridium, or platinum, the origin of the exchange mechanism has to be found in the tensile stress [67, 68]. This mechanism has been reported as a major one not only in the case of some $5d$ metallic surfaces [69, 70] but also in the case of some $3d$ ones [71, 72] and some $4d$ ones [73].

In Sect. 4, the strong structural inhomogeneity of the gold (111) surface has been described. In this section, we show that the surface stress inhomogeneity related to the structural inhomogeneity of the herringbone reconstruction of gold is at the origin of a long-range exchange mechanism.

Our approach involves the preparation of a Au(111) surface with adsorbed Co clusters [74]. Upon thermal activation, the Co clusters are allowed to burrow into the substrate, as shown on the STM topography of Fig. 16 obtained for a cobalt coverage $\theta = 3$ AL, before (Fig. 16a) and after (Fig. 16b) annealing the sample at 450 K for 1 min. As shown in the line scan of Fig. 16c, some clusters being originally two atomic layers in height appear as monolayers after annealing. These clusters have buried into the substrate by one atomic Co layer. This burrowing process is known as a "superexchange" mechanism and it has been observed in the case of Co into Cu(100) and Ag(100) and in the case of Ni clusters into the (001), (110), and (111) surface of gold as well [75–79]. During this process, the Co clusters remain intact as could be confirmed by following in situ their magnetic properties [75] Table 5.

Along with Co clusters burrowing, gold islands appear at some distance from the buried Co clusters. A detailed analysis shows that the amount of Au transferred to the surface corresponds to the amount of atoms displaced by the sunken Co clusters. The Au atoms ejected from the topmost atomic layer mainly form rims around

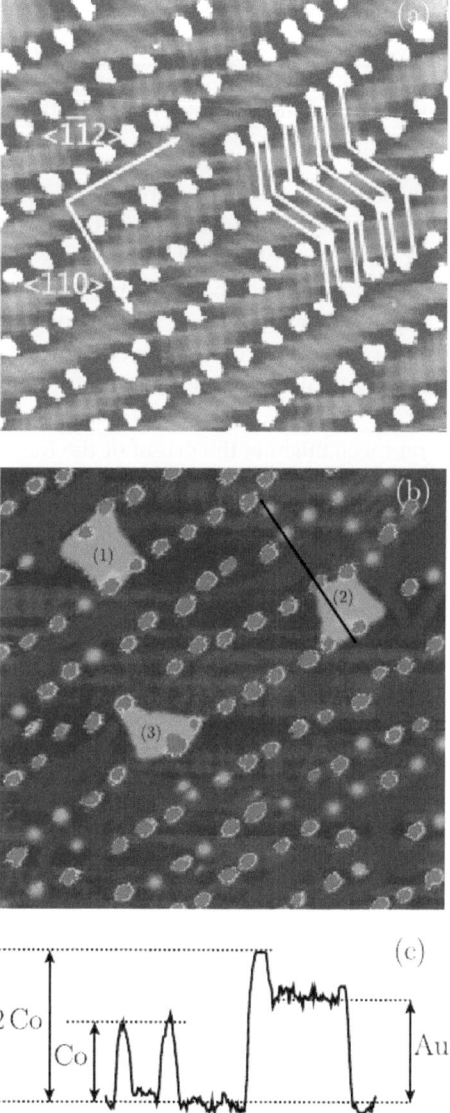

Fig. 16 (**a**) STM image of a self-organized Co cluster as obtained after deposition at 300 K. Zigzag DLs have been highlighted with *white lines*. (**b**) False color STM image of the Co cluster network after annealing at 450 K. *Dark gray* patches represent the Co bilayer clusters. *Light gray* patches are the sunken Co clusters. *Large bright patches* labeled (1), (2), and (3) are the gold rims. (**c**) The line scan confirms the presence of fully emerged and sunken Co clusters (4.0 and 2.0 Å above the surface, respectively) as well as monolayer high Au rims (2.36 Å). Images 100×100 nm (from [33])

Table 5 Conservation of cobalt and gold atoms upon thermally activated cobalt burrowing

Temperature (K)	455	500	550
Co buried (10^{-3} ML)	8.9 ± 0.5	8.5 ± 0.5	40 ± 1
Au transferred to surface (10^{-3} ML)	8.1 ± 2	10 ± 2	41 ± 2

the remaining, unperturbed bilayer Co clusters to which they attach exclusively (Fig. 16b). It is worth noting that the gold atoms ejected from the sunken Co clusters do not necessarily stick to the nearby Co bilayer clusters but are sometimes found several tens of nanometers away, as can be seen in Fig. 16b. This fact is highlighted by the histograms in Fig. 17, in which the distance between gold islands has been reported. The maximum probability of occurrence is reached for $\bar{r} = 70$ nm. The localization of the ejected gold atoms suggests long-range substrate-mediated mass transport.

What is the diffusion mechanism at the origin of the long-range mass transport? A plausible mechanism consists in a gold ejection close to a buried Co cluster, surface diffusion afterward, and a preferential nucleation at the Co cluster edges. However, kinetic Monte Carlo simulations performed on a rigid network consisting of 120 Au(111) reconstructed unit cells and corresponding to 240 Co clusters show that 92% of the ejected gold atoms stick to Co clusters located within a radius of 20 nm around the emitting island (Fig. 17). This percentage becomes even 100% for a radius of 25 nm, in contradiction with the experimental results described above. An additional clue ruling out the above mass transport mechanism is the high activation energy necessary to initiate this process.

Figure 19 displays the minimum energy path (MEP) corresponding to the ejection of a Au atom close to a 10 atoms Co cluster deposited on a DL and its propagation by means of conventional hopping. The computational slab (Fig. 18) used in our calculations consists in a crystal of $22 \times 22 \times 8$ Au atoms cut perpendicular to the <111> direction with a $\sqrt{3} \times 22$ reconstructed top layer of (22×23) Au atoms [80]. The MEP calculation shows that the ejection of a Au atom close

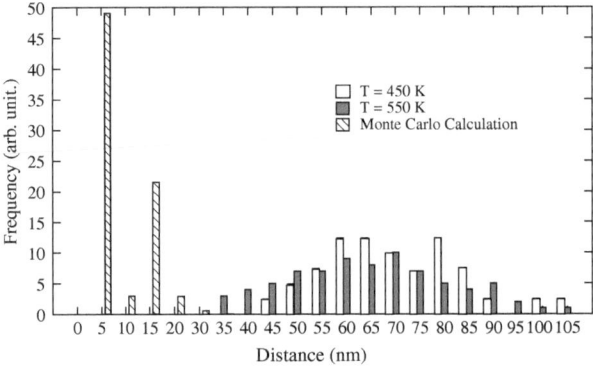

Fig. 17 Histograms of gold island–island distances after annealing at 450 and 550, respectively. For each temperature, the data have been taken from six different STM images

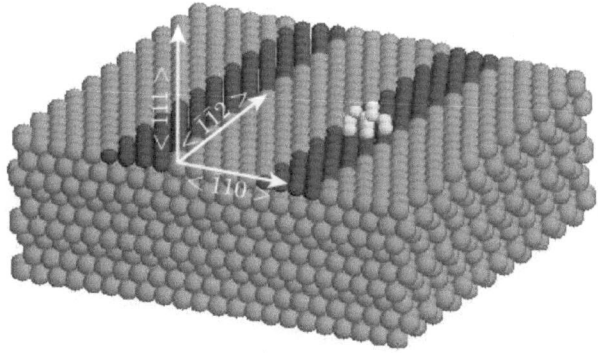

Fig. 18 Slab used in the simulations. Two discommensuration lines of highest corrugation run along the $< \bar{1}\bar{1}2 >$ direction. The DLs (*black atoms*) separate regions (*gray atoms*) of a regular fcc stacking from a smaller region with hcp faulted stackings. *White atoms* are the Co atoms (from [33])

to the Co cluster requires an activation energy of 950 meV [(1) in Fig. 19]. The large activation energy is due to the fact that a vacancy has to form in the gold surface plane before it can be filled with a Co atom [(2) in Fig. 19]. The ejected Au atom then propagates by conventional surface hopping with an activation energy of about 100 meV [(3), (4), and (5) in Fig. 19]. The large activation energy necessary for the Au atoms to escape from the edge of Co clusters makes this event highly improbable.

The existence of a second mechanism for the long-range mass transport is based on two observations. The first one is related to the reconstruction of the surface. As was described in Sect. 4, such reconstruction consists in an overclosed packing along

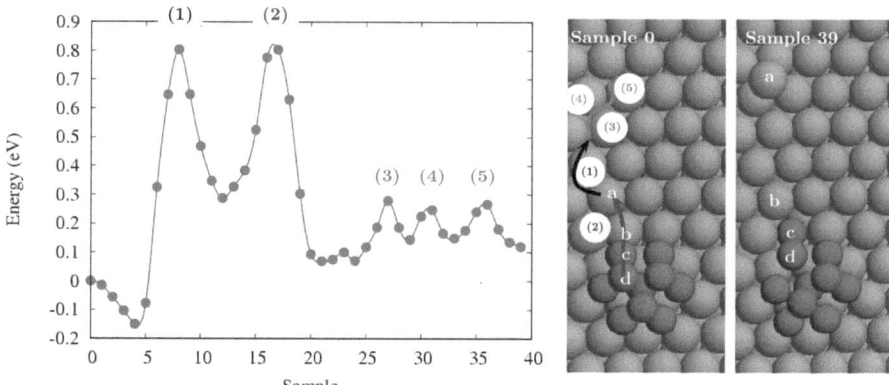

Fig. 19 Energy profile for an ejection of a Au atom close to a 10 atom Co cluster deposited on a DL, followed by its diffusion by means of conventional hoping: (1) ejection of the Au atom, (2) filling of the vacancy by a Co atom, and (3), (4), and (5) conventional surface hopping of the Au atom

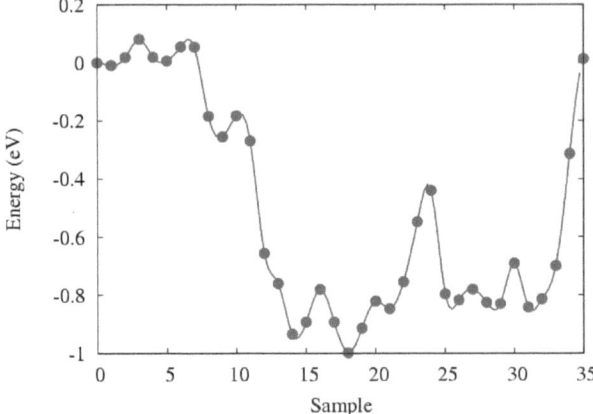

Fig. 20 Energy profile for an 1D soliton-like propagation mechanism along the $< \bar{1}\bar{1}2 >$ direction inside the discommensuration line

the direction $<\bar{1}10>$ in order to release the strain due to the surface mismatch [29]. However, there is no overclosed packing along the perpendicular $<\bar{1}\bar{1}2>$ direction (Fig. 18), inducing a strong tensile stress along this direction. Moreover, during the growth, the tensile stress is increased due to the growth of Co clusters and their influence on the surface. Then, a way for the system to release the tensile stress is the insertion of a small atom along the $<\bar{1}\bar{1}2>$ direction (Fig. 21). MEP calculations corresponding to this scenario were performed along the $<\bar{1}\bar{1}2>$ direction of the DL. In Fig. 20 a small activation barrier is noticeable, less than 100 meV, followed by an energy gain of about 1 eV and a series of smaller barriers. By analyzing the corresponding configuration in real space (Fig. 21), it was found that the first barrier (labeled (1) on Fig. 20) corresponds to the burrowing of the Co atom. This insertion process is the first step of a twosteps exchange process. A similar process has been found in the case of the burrowing of Ni clusters into a Au(111) substrate as well [78]. However, in the latter case, the gold ejection occurs close to the insertion point

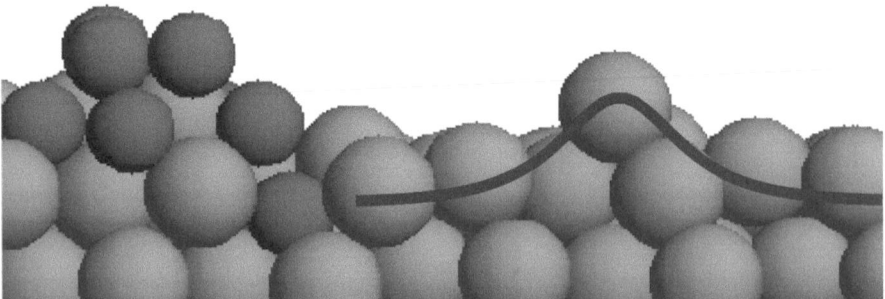

Fig. 21 Cut view through the Co cluster (*smallest atoms*). The strain induced by the Co insertion into the gold surface is given by the *line*

whereas in the present case, the cobalt insertion along the $<\bar{1}\bar{1}2>$ leads to the formation of a strain as shown in Fig. 21. Note that the value of the activation energy for inserting the cobalt adatom into the surface plane (100 meV) is much smaller than the one calculated in the case of Ni/Au(111) (580 meV) [78]. This is related to the additional tensile stress exerted by the cobalt cluster on the surface plane [68]. The smaller barriers of the MEP (Fig. 20) correspond to the propagation of this defect along the $<\bar{1}\bar{1}2>$ direction. Hence, the strain induced by the cobalt insertion can be identified as a traveling soliton wave, propagating along the $<\bar{1}\bar{1}1>$ direction. The wave form shown in Fig. 21 is approximatively described by the equations

$$x_n(t) = na - \frac{a}{2}\left(\tanh\left(\alpha\left(na - vt\right)\right) - 1\right), \tag{34}$$

$$z_n(t) = z_n^{(0)} + \frac{\Delta z}{\cosh\left(\alpha\left(na - vt\right)\right)}. \tag{35}$$

a and Δz are the maximum displacement of the atoms in the chain along, respectively, the $<\bar{1}\bar{1}2>$ direction and the $<111>$ ones. v is the velocity of the soliton wave and α is its width.

The propagation is strongly anisotropic. Indeed, an exchange along $<\bar{2}\bar{2}1>$, as calculated above, was found more favorable than an exchange in the perpendicular $<1\bar{1}0>$ direction, which clearly demonstrates that the propagation is channeled by the DLs.

Figure 22 displays the atomic displacement of the atoms involved in the process as a function of the sample number. A collective motion involving several atoms highlighted by the rings is observed. Once the soliton is formed, it propagates along a row of atoms belonging to the DL by small changes in the position of atoms, leading to the smaller oscillations seen in the energy curve in Fig. 20.

The 1D character of the propagation along an atomic row allows us to investigate the dynamics of the soliton in the Frenkel–Kontorova framework [81]. In this model,

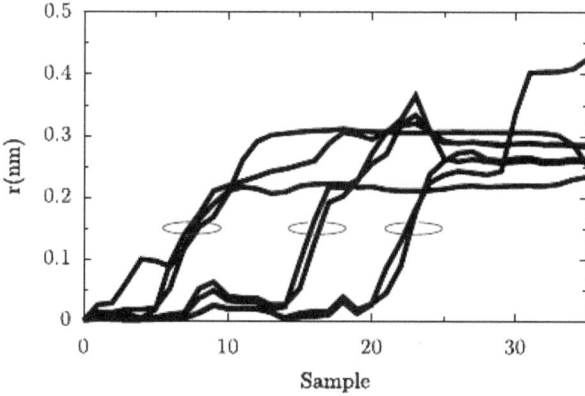

Fig. 22 Atomic displacements involved in the solitonic-exchange mechanism of surface diffusion. The *rings* highlight the concerted motion of the atoms (from [33])

a chain of $N + 1$ atoms, connected by springs, is assumed to be subjected to a sinusoidal substrate potential. Since the chain length $L = (N - 1) b$ (where b is the interatomic distance between atoms in the bulk) is held constant upon embedding an extra atom, the static energy of the chain is given by

$$H = \sum_{n=1}^{N} A \left(1 - \cos \frac{2\pi}{b} x_n\right) + \sum_{n=1}^{N-1} \frac{\gamma}{2} (x_{n+1} - x_n - a) 2$$
$$+ \lambda_1 x_1 + \lambda_2 \left(x_N - (N - 1) b + \lambda_3 \left(x_j - jb - \beta\right)\right), \tag{36}$$

where γ is the spring constant between neighboring atoms in the chain, a is the interatomic distance between atoms in the surface, and A is the amplitude of the harmonic potential exerted by the environment on the atoms of the embedded chain. λ_1 and λ_2 are the Lagrange multipliers which keep the position of the first and the last atoms of the chain fixed. λ_3 is an additional constraint which allows us to clamp the atom j at a given position $jb + \beta$. The Peierls barrier, governing the motion of the soliton wave, can then be estimated by using Hobart's procedure [82]. In this approach, the equilibrium equation $\partial H / \partial x_i = 0$ is used to adjust λ_1 and λ_2 in order to get the correct x_j and x_N positions. The Peierls barrier is then calculated from the total energy variation as a function of the displacement β of the soliton. The value of $A = 0.30\,\text{eV}$ has been chosen such that the solution of the equilibrium equations $\partial H / \partial x_i = 0$ matches the atomic displacements shown in Fig. 22. Estimating $\gamma = 21.7\,\text{eV/Å}^2$ from Rose's universal equation of states [83] and taking a typical chain length of $L = 120\,\text{nm}$ from a segment of a DL in Fig. 16a, a Peierls energy of 26 meV is calculated. The low migration barrier of the soliton confirms that this mechanism strongly competes with the conventional surface diffusion as found experimentally.

6 Conclusions

Looking at surfaces from the theoretical point of view by keeping in mind that these objects can be inhomogeneous has always been challenging due to the physical dimensions of the systems and of the related issues to be faced. This chapter has shown that the computational burden of atomic-scale calculations on extended surface "defects" (making surfaces at the same time very different from ideal ones, but much more realistic) can be circumvented by using reliable interatomic potentials and appropriate simulation probes. The first issue addressed in this chapter has been the one of the topology of a reconstruction, characterizing substrates such as the Pt(111) and the Au(111). We have shown that a reconstructed surface features peculiar regions, playing a crucial role in establishing the details of the nucleation and diffusion mechanisms. As a second issue, specific examples have been provided of migration mechanisms for atoms moving on the Pt(111) and Au(111) substrates. In both cases, combinations of simple hopping and exchange mechanisms have been underlined, enhancing the importance of both elementary steps to elucidate diffusion on inhomogeneous substrates. Finally, our report contains an example of long-range diffusion that can be explained by resorting to elastic effects and elementary atomic processes, fully consistent with experimental observations. What is the future of atomic-scale calculations in the area of surface diffusion and growth of metallic systems...? Even though the interatomic potential approach proved valuable, the explicit account of the electronic structure in defining the interaction appears unavoidable. Such applications are prohibitive due to the localization of d electrons (increasing the size of the basis set in plane wave calculations, for instance), the minimal size of the systems of interest, and the metallic nature of bonding. Recent attempts using methods scaling linearly with the size of the systems appear promising [84]. This holds in cases for which an extended dynamical simulation can be safely replaced by a less costly structural optimization with no sacrifice of atomic-scale understanding.

Acknowledgments Hervé Bulou has benefited greatly from the work of and discussions with Jean-Pierre Bucher. The authors are grateful to Mircea Rastei for providing STM image of the herringbone reconstruction of gold in Sect. 4.

References

1. S. M. Foiles, M. I. Baskes, and M. S. Daw, Phys. Rev. B **33**, 7983 (1986).
2. K. W. Jacobsen, J. K. Nørskov, and M. J. Puska, Phys. Rev. B **35**, 7423 (1987).
3. V. Rosato, M. Guillope, and B. Legrand, Phil. Mag. A **59**, 321 (1989).
4. J. Tersoff, Phys. Rev. B **39**, 5566 (1989).
5. J. Alvarez, J. A. Odriozola, Phys. Rev. B **58**, 6057 (1988).
6. C. Noguera, *Physique, chimie des surfaces d'oxyde* (Edition Eyrolles, Paris, 1995).
7. R. Tétot, A. Hallil, J. Creuze, and I. Braems, Europhys. Lett. **83**, 40001 (2008).
8. W. Vervisch, C. Mottet, and J. Goniakowski, Phys. Rev. B **65**, 245411 (2002).
9. J. Friedel, *The Physics of Metals*, ed. J. M. Ziman (Cambridge University Press, Cambridge, 1969), p. 340.
10. F. Ducastelle, *Order and Phase Stability and Alloys* (North Holland, Amsterdam, 1991).
11. H. Rose, J. R. Smith, and J. Ferrante, Phys. Rev. B **28**, 1835 (1983).
12. C. Kittel, *Introduction to Solid State Physics*, (New York, John Wiley and Sons, 1996).
13. F. R. de Boer, R. Boom, W. C. M. Mattens, A. R. Miedema, and A. K. Niessen, *Cohesion in Metals*, vol. 1 (North Holland, Amsterdam, 1988).
14. G. Tréglia, B. Legrand, and F. Ducastelle, Europhys. Lett. **7**, 575 (1988).
15. R. Hultgren, P. D. Desay, D. T. Hawkins, M. Gleiser, K. K. Kelley, and D. D. Wagman, *Selected Values of Thermodynamic Properties of the Elements* (American Society of Metals, Ohio, 1973).
16. B.-J. Lee, M. I. Baskes, H. Kim, and Y. Koo Cho, Phys. Rev. B **64**, 184102 (2001).
17. B. J. Alder and T. E. Wainwright, J. Chem. Phys. **27**, 1208 (1957).
18. B. J. Alder and T. E. Wainwright, J. Chem. Phys. **31**, 459 (1959).
19. L. Verlet, Phys. Rev. **159**, 98 (1967).
20. W. C. Swope, H. C. Andersen, P. H. Berens, and K. R. Wilson, J. Chem. Phys. **76**, 637 (1982).
21. B. Legrand, G. Tréglia, M. C. Desjonquères, and D. Spanjaard, J. Phys. C **19**, 4463 (1986).
22. H. Jónsson, G. Mills, and K. W. Jacobsen, *Classical and Quantum Dynamics in Condensed Phase Simulations*, eds. B. J. Berne, G. Ciccotti, D. F. Coker (World Scientific, Singapore, 1998), p. 385.

23. G. Henkelman and H. Jónsson, J. Chem. Phys. **113**, 9978 (2000).
24. G. Henkelman, B. P. Uberuaga, and H. Jónsson, J. Chem. Phys. **113**, 9901 (2000).
25. O. L. Alerhand, D. Vanderbilt, R. D. Meade, and J. D. Joannopoulos, Phys. Rev. Lett. **61**, 1973 (1988).
26. P. Zeppenfeld, M. Krzyzowski, Ch. Romainczyk, G. Comsa, and M. G. Lagally, Phys. Rev. Lett. **72**, 2737 (1994).
27. Ž. Crljen, P. Lazić, D. Šokčević, and R. Brako, Phys. Rev. B **68**, 195411 (2003).
28. Y. Wang, N. S. Hush, and J. R. Reimers, Phys. Rev. B **75**, 233416 (2007).
29. H. Bulou and C. Goyhenex, Phys. Rev. B **65**, 045407 (2002).
30. J. V. Barth, H. Brune, G. Ertl, and R. J. Behm, Phys. Rev. B **42**, 9307 (1990).
31. K. G. Huang, D. Gibbs, D. M. Zehner, A. R. Sandy, and S. G. J. Mochrie, Phys. Rev. Lett. **65**, 3313 (1990).
32. M. V. Rastei, PhD Thesis, Strasbourg University, France (2006).
33. H. Bulou and J.-P. Bucher, Phys. Rev. Lett. **96**, 076102 (2006).
34. Ch. Teichert, M. Hohage, Th. Michely, and G. Comsa, Phys. Rev. Lett. **72**, 1682 (1994).
35. M. Bott, M. Hohage, Th. Michely, and G. Comsa, Phys. Rev. Lett. **70**, 1489 (1993).
36. C. Günther, J. Vrijmoeth, R. Q. Hwang, and R. J. Behm, Phys. Rev. Lett. **74**, 754 (1995).
37. H. Brune, M. Giovannini, K. Bromann, and K. Kern, Nature **394**, 451 (1998).
38. K. Bromann, M. Giovannini, H. Brune, and K. Kern, Eur. Phys. J. D **9**, 25 (1999).
39. C. Goyhenex and G. Tréglia, Surf. Sci. **446**, 272 (2000).
40. E. Lundgren, B. Stanka, M. Schmid, and P. Varga, Phys. Rev. B **62**, 2843 (2000).
41. A. Atrei, G. Rovida, M. Torrini, U. Bardi, M. Gleeson, and C. J. Barnes, Surf. Sci. **372**, 91 (1997).
42. M. Sambi, E. Pin, and G. Granozzi, Surf. Sci. **340**, 215 (1995).
43. N. Marsot, R. Belkhou, H. Magnan, P. Le Févre, C. Guillot, and D. Chandesris, Phys. Rev. B **59**, 3135 (1999).
44. I. Chado, C. Goyhenex, H. Bulou, and J. P. Bucher, Phys. Rev. B **69**, 085413 (2004).
45. F. Leroy, G. Renaud, A. Letoublon, S. Rohart, Y. Girard, V. Repain, S. Rousset, A. Coati, and Y. Garreau, Phys. Rev. B **77**, 045430 (2008).
46. P. Gambardella, A. Dallmeyer, K. Maiti, M. C. Malagoli, W. Eberhardt, K. Kern, and C. Carbone, Nature **416**, 301 (2002).
47. P. Gambardella, A. Dallmeyer, K. Maiti, M. C. Malagoli, S. Rusponi, P. Ohresser, W. Eberhardt, C. Carbone, and K. Kern, Phys. Rev. Lett. **93**, 077203 (2004).
48. R. Cheng, K. Y. Guslienko, F. Y. Fradin, J. E. Pearson, H. F. Ding, D. Li, and S. D. Bader, Phys. Rev. B **72**, 0144409 (2005).
49. M. C. Marinica, C. Barreteau, D. Spanjaard, and M. C. Desjonquères, Phys. Rev. B **72**, 115402 (2005).
50. R. Ferrando and G. Tréglia, Phys. Rev. B **50**, 12104 (1994).
51. R. C. Nelson, T. L. Einstein, S. V. Khare, and P. J. Rous, Surf. Sci. **295**, 462 (1993).
52. A. Bogicevic, J. Strömquist, B. I. and Lundqvist, Phys. Rev. Lett. **81**, 637 (1998).
53. J. Jacobsen, K. W. Jacobsen, P. Stoltze, and J. K. Nørskov, Phys. Rev. Lett. **74**, 2295 (1995).
54. P. J. Feibelman, Phys. Rev. B **60**, 4972 (1999).
55. E. Cox, M. Li, P.-W. Chung, C. Gosh, T. S. Rahman, C. J. Jenks, J. W. Evans, and P. A. Thiel, Phys. Rev. B **71**, 115414 (2005).
56. H. Brune, H. Roder, K. Bromann, K. Kern, J. Jacobsen, P. Stoltze, K. Jacobsen, and J. Nørskov, Surf. Sci. **349**, L115 (1996).
57. C. Goyhenex, K. Farah, and A. Taobane, Surf. Sci. **601**, L132 (2007).
58. P. Gambardella, M. Blanc, H. Brune, K. Kuhnke, and K. Kern, Phys. Rev. B **61**, 2254 (2000).
59. R. Ferrando and G. Tréglia, Phys. Rev. Lett. **76**, 2109 (1996).
60. Y. Li and A. Depristo, Surf. Sci. **319**, 141 (1994).
61. M. Villarba and H. Jónsson, Surf. Sci. **17**, 15 (1994).
62. R. Ferrando and G. Tréglia, Surf. Sci. **377–379** 843 (1997).
63. R. L. Schwoebel, J. Appl. Phys. **40**, 614 (1969).

64. H. Bulou and C. Massobrio, J. Phys. Chem. C, **112**, 8743 (2008).
65. J. D. Wrigley and G. Ehrlich, Phys. Rev. Lett. **44**, 661 (1980).
66. P. J. Feibelman, Phys. Rev. Lett. **65**, 729 (1990).
67. B. D. Yu and M. Scheffler, Phys. Rev. B **56**, R15569 (1997).
68. H. Bulou and C. Massobrio, Phys. Rev. B **72**, 205427 (2005).
69. Ch. Chen and T. T. Tsong, Phys. Rev. Lett. **64**, 3147 (1990).
70. G. L. Kellogg and P. J. Feibelman, Phys. Rev. Lett. **64**, 3143 (1990).
71. L. Hansen, P. Stoltze, K. W. Jacobsen, and J. K. Nørskov, Phys. Rev. B **44**, 6523 (1991).
72. F. Montalenti and R. Ferrando, Phys. Rev. B **59**, 5881 (1999).
73. C. Massobrio and P. Fernandez, J. Chem. Phys. **102**, 605 (1995).
74. B. Voigtländer, G. Meyer, and N. M. Amer, Phys. Rev. B **44**, 10354 (1991).
75. S. Padovani, F. Scheurer, and J. P. Bucher, Europhys. Lett. **45**, 327 (1999).
76. J. Frantz and K. Nordlund, Phys. Rev. B **67**, 075415 (2003).
77. N. A. Levanov, V. S. Stepanyuk, W. Hergert, D. I. Bazhanov, P. H. Dederichs, A. Katsnelson, and C. Massobrio, Phys. Rev. B **61**, 2230 (2000).
78. D. N. Tafen and L. J. Lewis, Phys. Rev. B **77**, 075429 (2008).
79. C. G. Zimmermann, M. Yeadon, K. Nordlund, J. M. Gibson, R. S. Averback, U. Herr, and K. Samwer, Phys. Rev. Lett. **83**, 1163 (1999).
80. C. Goyhenex and H. Bulou, Phys. Rev. B **63**, 235404 (2001).
81. J. Frenkel and T. Kontorova, Zh. Eksp. Teor. Fiz. **8**, 89 (1938).
82. R. Hobart: J. Appl. Phys. **36**, 1944 (1965).
83. J. H. Rose, J. R. Smith, F. Guinea, and J. Ferrante, Phys. Rev. B **29**, 2963 (1984).
84. A. Jaafar and C. Goyhenex, Sol. Stat. Sci. (in press).

Electronic, Magnetic and Spectroscopic Properties of Vanadium, Chromium and Manganese Nanostructures

C. Demangeat and J.C. Parlebas

Abstract This chapter presents some aspects of the electronic, magnetic and spectroscopic properties of vanadium, chromium and manganese aggregates deposited on substrates or embedded in a host, mainly metallic. Those elements present a sizeable magnetic moment in their atomic form and non-ferromagnetic behaviour for the corresponding bulk materials. More precisely, vanadium, except in the atomic form and for very small cluster sizes (free or embedded), is magnetically dead. It presents a sizeable moment when in contact with strong ferromagnets like Fe, Co or Ni. Chromium presents, in the bulk form, a competition between classical anti-ferromagnetic behaviour and spin density wave (SDW). Those various behaviours are energetically almost degenerate in energy so that they can be both present. However, when the dimension shrinks, the SDW configuration does not survive. Moreover, when Cr atoms are in contact with Fe, Co or Ni a competition between the intrinsic non-ferromagnetic behaviour of Cr and the induced ferromagnetic polarization arising from the strong ferromagnet leads to complicated magnetic maps. Also the magnetic behaviour of manganese is non-trivial. The bulk magnetic behaviour is of complex anti-ferromagnetic type with some kind of non-collinear behaviour. Ferromagnetic behaviour of Mn aggregates generally seems unlikely, but a few Mn atoms may acquire a somewhat ferromagnetic configuration. However, without a constraint of collinearity, those Mn atoms follow non-collinear behaviour.

C. Demangeat (✉)
Institut de Physique et Chimie des Matériaux de Strasbourg, 23 rue du Loess, BP 43, F-67034 Strasbourg Cedex 2, France, claude.demangeat@ipcms.u-strasbg.fr

J.C. Parlebas
Institut de Physique et Chimie des Matériaux de Strasbourg, 23 rue du Loess, BP 43, F-67034 Strasbourg Cedex 2, France, jean-claude.parlebas@ipcms.u-strasbg.fr

Demangeat, C., Parlebas, J.C.: *Electronic, Magnetic and Spectroscopic Properties of Vanadium, Chromium and Manganese Nanostructures*. Lect. Notes Phys. **795**, 161–196 (2010)
DOI 10.1007/978-3-642-04650-6_6
© Springer-Verlag Berlin Heidelberg 2010

1 Introduction

The temperature tends to kill the magnetic behaviour of most of the magnetic components, an exception being described theoretically by Kondo [1]. The "paramagnetic Curie temperature" was defined after the very famous work of Pierre Curie [2] displaying a divergence of the magnetic susceptibility. Theoretical explanation was put forward by Langevin [3] within a model based on non-interacting individual magnets. This could explain the behaviour of paramagnetic systems, but not those where some directional magnetic ordering was present. Later on, Weiss [4] introduced the idea of "molecular field" in order to explain this "ferromagnetic behaviour". Shortly after, Weiss introduced the "experimental magneton", also called "Weiss magneton" [5]. Only integer numbers were allowed to represent the magnetism of Ni, Co and Fe so that for Ni the number 3 was attached (3 Weiss magnetons), 8 for Co and 11 for Fe. A few years later, Dirac [6] postulated the spin of the electron and opened the way to the Bohr magneton. Roughly, the Bohr magneton, which has a quantum mechanical origin, is equal to five "Weiss magnetons".

The quantum mechanical approach was able to explain some experimental results for which the classical theory was unable to give satisfactory answers. However, due to large number of electrons present in any condensed matter system, theoreticians introduced some semi-phenomenological Hamiltonians. The Ruderman–Kittel–Kasuya–Yosida (RKKY) [7, 8], originally proposed by Ruderman and Kittel [9] as a means of explaining unusually broad nuclear spin resonance lines, was extended by Kasuya [7] to explain indirect exchange coupling between localized d electron spins interacting via conduction electrons. Soon later, in order to explain the onset of magnetism for some systems when the temperature increases, Kondo [1] proposed a model in which, at low temperature, the magnetic moment of localized d electrons is screened by conduction electrons. Because those conduction electrons are less bound than localized d electrons, temperature effect tends to decouple both systems, so that, above a temperature called Kondo temperature, the magnetism of localized electrons do reappear. Hubbard model [10] is a good approximation for particles in a periodic potential. It can be considered as an improvement of the tight binding model. In the tight-binding approximation, electrons are viewed as occupying standard orbitals of their constituent atoms with "hopping integrals" or "transfer integral" between neighbouring atoms. Hubbard model includes the so-called onsite repulsion which stems from the Coulomb repulsion between electrons. The Haydock recursive method [11] based on the tri-diagonalization of Lanczos matrices [12] allowed the determination of the density of states in the direct space. Thus, the Hubbard Hamiltonian can be used to describe nanomaterials which do not present any essential periodicity.

Nowadays, most of the microscopic description of magnetic materials are based on density functional theory (DFT) following Kohn's approach [13, 14]. This approach is well documented in many textbooks so that there is no need to add here any new details. In general those DFT codes work in the reciprocal space, so that they essentially apply to materials presenting periodicity. For nanostructures, this periodicity is broken and in order to use those codes it is necessary to rely on supercells, periodically repeated. Let us note a recent book by Gross [15] focusing on the microscopic perspective of Surface Science. As pointed out by Gross, while most of the processes described via DFT are assumed to occur in the electronic ground state, the theoretical description of non-adiabatic phenomena has not reached the same level of maturity.

It is only in very rare cases that one can perform ab initio DFT calculations in real space. Frota-Pessoa [16] and Klautau & Frota-Pessoa [17] developed a real space linear muffin-tin-orbitals method in the atomic-sphere approximation (RS-LMTO-ASA). Within a screened Korriga–Kohn–Rostocker (KKR) method, Wildberger et al. [18] determined the Green function of an adsorbed cluster on a surface in terms of the surface Green function of pure substrate via a Dyson equation. Those authors extended their formalism in order to determine the non-collinear magnetism of nanoclusters on substrates.

On the one hand, DFT calculations are expected to be the best tool for the description of electronic nanostructures either free or deposited on a substrate. On the other hand, XMCD experiment is expected to be a versatile tool to discriminate between magnetic and non-magnetic behaviour (see for instance [19] and references therein). As already discussed by Dreysse and Demangeat [20], some transition elements like V, Rh and Pd, although non-magnetic in bulk phase, may present some kind of magnetization in the case of reduced geometry. This could be simply explained by a reduced coordination of transition metal atoms leading to an increase of the density of states at the Fermi level and consequently, via Stoner criterion, to the onset of ferromagnetism. Bansmann et al. [21] further pointed out a considerable increase of the magnetic anisotropy in those small clusters. Very recently Honolka et al. [22] questioned the onset of magnetism of Ru and Rh impurities and clusters deposited on non-magnetic substrates. They reported element-specific XMCD measurements of local magnetic moments of Ru and Rh adatoms and cluster ensembles deposited at 5 K on Ag and Pt surfaces. No magnetic moment was detected in the coverage range between 0.12 and 2.0 ML, independently of the magnitude of externally applied static magnetic fields. This is clearly at odds with most of the theoretical calculations displaying sizeable magnetic moments for those configurations. Honolka et al. [22] discussed the probable origin of those discrepancies: (i) a possible alloying of Rh, Ru with substrates; (ii) the absence of full relaxation and (iii) the neglect of many-body effects (Kondo or local spin fluctuations) in theoretical calculations. Also, De Siervo et al. [23] performed both experimental and theoretical studies of Pd ultrathin films on Ru(0001). No hysteretic MOKE loop was observed for Pd films on Ru(0001), as measured at 160 K. DFT calculations were done using the Quantum ESPRESSO package (Baroni et al.) within LDA or GGA approaches. Moreover, GGA predicts a ferromagnetic ground state for bulk Pd at odds with experimental results, so that GGA cannot be used for the determination of magnetic polarization of Pd on a Ru(0001) surface. For LDA, on the contrary, no magnetic moment is obtained, neither in fcc and hcp bulk Pd nor for Pd thin films on Ru(0001). Results by De Siervo et al. [23] confirm that GGA does not reproduce the experimental results of bulk Pd magnetization. Their claim that their LDA results are in agreement with their MOKE results, displaying non-magnetic behaviour for Pd on Ru(0001), is not really convincing because temperature effect may be important at 160 K. In that sense, the results of Honolka et al. [22], displaying no magnetization at 5 K, look more convincing. This chapter presents an overview of the electronic, magnetic and spectroscopic properties of a few transition metal-based nanostructures. In this short lecture note, it is difficult to also report about a new class of materials, namely those elements, the magnetism of which seems to be linked to "sp" electrons. Oxides like ZnO, TiO_2, HfO_2, which are non magnetic wide band semiconductors, could present high-temperature magnetism when oxygen defects are introduced [24]. Also ZnO doped with co-impurities displays anti-ferromagnetic behaviour [25]. A review on the "Origin of ferromagnetic response in diluted magnetic semiconductors and oxides" was written most recently by Dietl [26]. Also let us mention, among a great number of them, a paper by Schulthess et al. [27] concerning the first-principles calculation of the electronic structure of Mn-doped GaAs, GaP and GaN semiconductors. However, here the focus will be mainly on another timely subject which is the description of non-collinear (or unconstrained) magnetism. This non-collinear magnetism arises around topological defects like surface steps, skew dislocations or vacancies. The specific case of trimers will be discussed in full detail because it is the most simple case between frustration and non-collinear magnetism.

Section 2 is devoted to general trends of magnetism along the 3d transition metal series. Following the pioneering work of Blügel et al. [28], a considerable number of papers were devoted to the study of magnetism of thin films of transition metals deposited on substrates. See, for example, the review paper by Dreysse and Demangeat [20]. Also spectroscopic description of the magnetism can be found in the review by Binns et al. [29]. Most calculations were performed by taking account for periodicity in the direction perpendicular to the substrate. Then physicists considered not only films with one chemical component but also alloys. Nowadays the trend is clearly to sub-monolayer coverage, like clusters or chains, as shown in a relatively recent report by Vega et al. [30]. From the experimental side, the spin-polarized scanning tunnelling spectroscopy is now currently used to probe the spin polarization of individual atom adsorbed on various substrates [31].

Section 3 is devoted to vanadium nanostructures. While vanadium-free atom presents a magnetic moment of $3\mu_B$, bulk V is clearly non-magnetic. For dilute V atoms in noble metal or in some very specific case, like Au4V [32], magnetization is not completely killed. It has been argued that V surface is magnetic both experimentally and theoretically but serious doubts remain. It has also been argued that free-standing or adsorbed V clusters could be magnetic but no conclusive answer has been reached yet. We shall discuss those points in full detail by giving all the pro and counter arguments. It seems that V is magnetic only in contact with a magnetic system. It is generally admitted that strong induced magnetization is obtained for thin V nanostructures in contact with Ni, Co and Fe metals. Specific examples will be given, but in all considered cases the induced polarization is short ranged.

Section 4 deals with recent aspects of Cr nanostructures. Bulk Cr is known to present spin density waves (SDWs), but in thin films, the SDW has no space to develop so that it is clearly killed. Therefore, for most Cr nanostructures reported here, SDW is clearly absent. Thus magnetism oscillates between ferromagnetism when a thin Cr film is in contact with Fe, Co or Ni and anti-ferromagnetic-like behaviour. For thicker slabs, only Cr atoms at the interface with a strong ferromagnet are ferromagnetically polarized, whereas other Cr atoms exhibit either layered anti-ferromagnetic polarization or in-plane anti-ferromagnetic polarization. When the symmetry is broken, non-collinear magnetism takes place. A particular important aspect concerns a Cr trimer, either free standing or adsorbed. In all cases it looks non-collinear [33]. However, it has been recently argued by Kudasov and Uzdin [34] that a compact Cr trimer may have very complex magnetism. For an isocele trimer, the two atoms, which are close, actually present anti-ferromagnetic configuration whereas the remaining Cr seems more like an isolated atom and may be described by a Kondo model.

Section 5 is devoted to Mn nanostructures. Mn is a very important element because in its atomic form it presents a magnetic moment as high as $5\mu_B$. However, in its bulk form (see [35]) it is mainly non-collinear. It can become ferromagnetic in contact with a strong ferromagnet. Examples can be found in the review report by Demangeat and Parlebas [36]. Usually the induced polarization by Fe, Co or Ni metals is short ranged, so that the ferromagnetic behaviour of Mn is limited to Mn atoms in direct contact with a strong ferromagnet. Moreover, recent calculations by Hafner and Spisak [35] have shown that, for one monolayer of Mn in contact with Fe(001), some kind of non-collinearity stabilizes the system. For completeness let us mention the review report of Fukamichi et al. [37] concerned with magnetic and electrical properties of Mn alloys.

Section 6 presents some concluding remarks as well as a discussion about other trends in magnetism. The present content was mainly concerned with a description of V-; Cr- and Mn- based nanostructures at $T = 0$ K. At this temperature, the magnetization is frozen. However, at laboratory temperature experiment, other mechanisms can take place, so that it is necessary to mention temperature effects which are of utmost interest in the description of dilute magnetic semiconductors (DMS) [38]. Indeed these DMS appear to be the most versatile component for spin electronics. Also, spin dynamic aspects in confined magnetic structures [39] should be pointed out. The magnetization could be disturbed by an intense ultra short laser pulse or by a short magnetic field pulse that tips the magnetization out of its equilibrium position.

2 Trends of Magnetism Along the 3d Transition Metal Series

Since the work of Blügel et al. [28], concerning the magnetic properties of 3d transition metal monolayers on metal substrates, a considerable number of papers have been devoted to 3d nanoclusters deposited on substrates. Older work on that subject can be found in the review by Vega et al. [30]. Here mainly we focus on the most recent works treating small clusters [40], chains [41] and sub-monolayer coverages [42] on metallic substrates. More and more calculations are now considering unconstrained (or non-collinear) magnetization.

2.1 From Adatoms to Monolayers Through Chains

Lounis et al. [40] studied magnetic states of 3d atom clusters in and on Ni(001) surface. In a first step they restricted to collinear magnetization for adatoms, dimers and trimers. For 3d adatoms on Ni(001) Sc, Ti, V and Cr are anti-ferromagnetically (AF) coupled to the substrate whereas the couplings of Mn, Fe, Co and Ni are ferromagnetic (FM). Clearly the AF–FM transition occurs between Cr and Mn. For dimers, it is shown that, on the one hand, V atoms remain anti-ferromagnetically coupled to the substrate whereas Fe atoms stay ferromagnetically coupled. On the other hand, Cr and Mn dimers show magnetic frustration. Lounis et al. [40] also extended the full potential Korringa–Kohn–Rostocker Green function method to treat non-collinear magnetic nanostructures. Focusing on Cr and Mn dimers and trimers they obtained very different results between these two elements. For Cr dimers, the ground state stays collinear whereas it is non-collinear for Mn. Cr and Mn trimers present collinear ground states, but the Mn trimer also presents a metastable non-collinear state, a few meV above its collinear ground state. However, when going from a Ni(001) substrate to a Ni(111) one, non-collinear configurations for Cr and Mn are more likely to be considered [43].

Similarly, in mono-atomic 3d transition metal chains, Mokrousov et al. [41] investigated magnetic order and exchange interactions, within a full potential linearized augmented plane wave method (FLAPW), in its 1D and 2D formulations, as implemented within a FLEUR code. They investigated free-standing chains of V, Cr, Mn, Fe and Co, as well as deposited ones on unreconstructed (110) surfaces of Cu, Pd, Ag and NiAl. Actually, Cr and Mn chains show a transition from ferromagnetic coupling in free-standing chains to anti-ferromagnetic coupling on (110) surfaces of Pd, Ag and NiAl. For Fe and Co chains on NiAl(110), ferromagnetic and anti-ferromagnetic states differ by only 2 meV, suggesting the possibility of a more complex ground state.

2.2 Sub-monolayer Coverage of 3d Transition Metal Adatoms on Co(001)

In the sub-monolayer regime, Carillo-Cazares et al. [42] investigated the coverage-dependent magnetization of 3d transition metal adatoms on Co(001). The magnetic map is reported in Table 1 for 3d adatoms from Sc to Ni and for concentration x equal to 0.25 and 1.0 (see Fig. 1). The results for $x = 0.5$ are very similar to those of $x = 0.25$, so that we only report them in Fig. 2. From these results it is obvious that, for low 3d coverage, there is a drastic change of magnetic map between Cr and Mn. It is only for 3d = Ni, Co and Fe that the ferromagnetic coupling between 3d adatom and Co atoms is independent of 3d concentration. This is obvious because those three elements are bulk like ferromagnets. For all the other 3d elements a drastic modification of the magnetic polarization appears when the coverage is modified. This is also obvious because those elements are

Table 1 Magnetic moments per atom (in μ_B) for $3d_x/Co(001)$, with $x = 0.25$ and 1.00, in the ground state configuration. The letters a, b, c and d represent the sites A, B, C and D of the 3d transition-metal atoms and Co_s, Co_{s-1}, are the Co-substrate atoms. In the case of $x = 0.25$ ($x = 1.00$) we use four (two) inequivalent atoms per plane. For a monolayer coverage, the magnetic moments on A and B atoms are equal as well as those on C and D. Due to symmetry the magnetic moments on the Co atoms located underneath the $3d_x$-adatoms do not depend on their position in the plane

	Sc		Ti		V		Cr		Mn		Fe		Co		Ni	
x	0.25	1.00	0.25	1.00	0.25	1.00	0.25	1.00	0.25	1.00	0.25	1.00	0.25	1.00	0.25	1.00
3da	−0.50	0.15	−1.42	−0.40	−2.72	0.47	−3.91	2.94	4.06	3.22	3.16	2.85	2.07	1.86	0.91	0.73
3dc		0.15		−0.40		0.47		−2.87		−3.51		2.85		1.86		0.73
Co_s	1.40	0.88	1.45	0.98	1.54	1.17	1.63	1.36	1.66	1.15	1.71	1.58	1.77	1.63	1.79	1.68
Coa_{s-1}	1.69	1.77	1.66	1.78	1.68	1.79	1.67	1.76	1.66	1.75	1.66	1.74	1.67	1.72	1.66	1.70
Cob_{s-1}	1.65	1.77	1.66	1.78	1.65	1.79	1.65	1.76	1.66	1.75	1.66	1.74	1.67	1.72	1.66	1.70
Coc_{s-1}	1.64	1.77	1.65	1.78	1.65	1.79	1.62	1.74	1.62	1.79	1.59	1.74	1.56	1.72	1.55	1.70
Cod_{s-1}	1.64	1.77	1.65	1.78	1.65	1.79	1.65	1.74	1.66	1.79	1.66	1.74	1.67	1.72	1.67	1.70

Fig. 1 Positions of 3d adatoms in the unit cell, consisting of four sites labelled by A, B, C and D. For 3d$_{0.25}$ E$_{0.75}$ only the site A is occupied by a 3d atom, the others being empty (**a**). For 3d$_{0.5}$ E$_{0.5}$ the 3d atoms are located at sites A and B whereas sites C and D are empty (**b**). For 3d$_{0.75}$ E$_{0.25}$ the 3d atoms are located at sites A, B and C whereas site D is empty (**c**). Finally, for a complete over-layer of 3d atoms on Co(001), i.e. 3d$_{1.00}$, the A, B, C and D sites are occupied by 3d atoms with A and B of one type and C and D of another type (**d**). *Black and grey dots* represent 3d atoms whereas *white dots* are empty spaces (from [42])

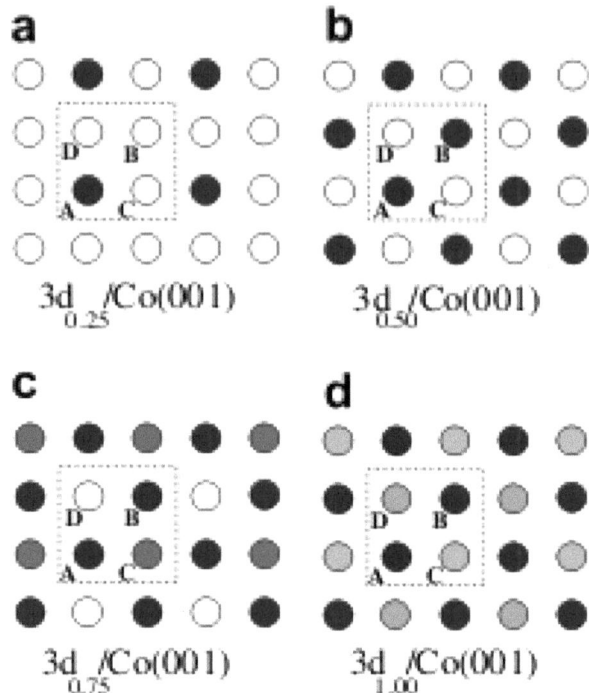

essentially of non-ferromagnetic type in the bulk form. For Sc, Ti and V which are non-magnetic in the bulk form the magnetization arises essentially from the proximity with Co ferromagnet. For Cr and Mn whose intrinsic bulk magnetization are rather complex, but mainly of non-ferromagnetic type, we can expect a

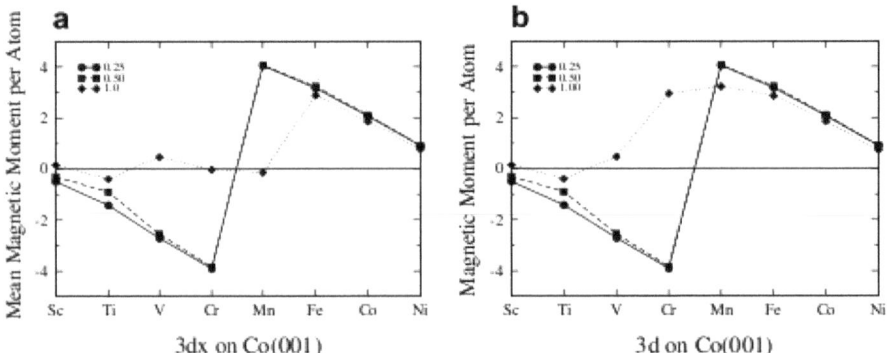

Fig. 2 (**a**) Magnetic map for 3d transition metal adatoms on Co(001) surface for concentration $x = 0.25$, 0.50 and 1.00. The *solid line* is for 0.25 coverage, the *dashed line* for 0.50 coverage, and the *dotted line* for 1.00 coverage. (**b**) Magnetic moment per atom (in μ_B), for atom in the position A, for concentration of $x = 0.25$, 0.5 and 1.00 (from [42])

Table 2 Magnetic moment per atom (in μ_B) for Crx/Co(001) and Mnx/Co(001), with $x = 0.25$, 0.50, 0.75 and 1.00. The letters a, b, c and d represent the sites A, B, C and D of Cr or Mn atoms and CoS, CoS-1 are Co atoms underneath Cr and Mn adatoms. In the case of $x = 0.25$, 0.50 and 0.75 ($x = 1.00$) we use four (two) inequivalent atoms per plane (from [42])

x	0.25	0.50	0.75	1.00
Cra (Mna)	−3.91 (4.06)	−3.84 (4.06)	−3.43 (3.64)	2.94 (3.22)
Crb (Mnb)		−3.84 (4.06)	−3.43 (3.64)	2.94 (3.22)
Crc (Mnc)			2.99 (−3.59)	−2.87 (−3.51)
Crd (Mnd)				−2.87 (−3.51)
Co$_s$	1.63 (1.66)	1.42 (1.56)	1.40 (1.35)	1.36 (1.15)
Coa$_{s-1}$	1.67 (1.66)	1.66 (1.66)	1.70 (1.72)	1.76 (1.75)
Cob$_{s-1}$	1.65 (1.66)	1.66 (1.66)	1.70 (1.72)	1.76 (1.75)
Coc$_{s-1}$	1.62 (1.62)	1.68 (1.68)	1.72 (1.70)	1.74 (1.79)
Cod$_{s-1}$	1.65 (1.66)	1.68 (1.68)	1.70 (1.70)	1.74 (1.79)

competition between the induced magnetization arising from the Co substrate and an intrinsic anti-ferromagnetic coupling present in the bulk (see Table 2). For Ti adatoms, the Ti–Co coupling is always of anti-ferromagnetic type: only the values of Ti magnetic moments decrease when the concentration of atoms in the over-layer increases. For Cr, an anti-ferromagnetic coupling between Cr and Co atoms is found in the low-coverage regime. For a complete Cr over-layer on Co(001), an in-plane anti-ferrimagnetic coupling is observed. However, the local values of Cr moments remain as high as $2.94\mu_B$ and $-2.87\mu_B$ contrary to the values observed for Sc, Ti and V, where a drastic decrease appears when going from the low-coverage to the mono-layer range. Besides, Izquierdo and Demangeat [44] pointed out the effect of crystal-lographic faces on the magnetic polarization of a Co monolayer either on Cr(001) or Cr(011) face. For Co on Cr(001) the Co monolayer is clearly of ferromagnetic type whereas for Co on Cr(011) an in-plane anti-ferromagnetic configuration is obtained. From that we can conclude that Co nanostructures could present non-ferromagnetic behaviour: this is reminiscent of a RKKY behaviour for a pair of impurities in noble metals. Also we can think of the Bethe curve displaying magnetic config-uration (ferromagnetic–anti-ferromagnetic) versus distance between magnetic ele-ments. For $x = 0.5$, we observe a moment of $3.84\mu_B$ on Cr atoms at A and B positions: the Cr–Co coupling is of anti-ferromagnetic type. When the concentration of Cr adatoms reaches $x = 0.75$, we see in Table 2 an anti-ferromagnetic coupling between Cr atoms at A and B positions and Co substrate as well as a ferromagnetic coupling between the Cr atom at C position and the Co substrate. For Mn, a com-pletely different behaviour is obtained. Now, for low Mn coverage ($x = 0.25, 0.50$), a ferromagnetic coupling between Mn and Co is obtained whereas for a complete Mn over-layer on Co(001) an in-plane antiferrimagnetic coupling is present. In order to shed some light on the variation of the magnetic behaviour with concentration, we report in Table 2 the variation of magnetic moments for both Cr and Mn adatoms for $x = 0.25$, 0.50, 0.75 and 1.0 coverages. For $x = 0.75$, a complex magnetic behaviour is present: contrary to Cr or Co, now the coupling between the Mn atoms on A and B sites is ferromagnetically coupled to the Co substrate whereas an anti-ferromagnetic coupling is present between Mn atom at C position and Co substrate. All these solutions reported here are "ground state" in the collinear constraint approach.

2.3 Outlook

In this section we discuss the magnetic map of very small clusters [40, 43], chains [41], periodic sub-monolayer coverage [42] of 3d transition metal elements on metallic substrates. Calculations by Lounis et al. were performed within an unconstrained approach of the spin direction. One main conclusion is that the magnetic map strongly depends on the crystallographic surface considered and on the number of nearest neighbours in the adsorbed nanostructure. Actually, V, Cr and Mn nanostructures have some tendency to develop non-collinear magnetism, especially when the substrate is non-magnetic. However, when the substrate is strongly magnetic this non-collinearity is somewhat quenched through the strong induced ferromagnetic polarization of the substrate. Thus, more detailed calculations are needed concerning this competition between anti-ferromagnetic coupling, mainly present in V, Cr and Mn, and the ferromagnetic induced polarization, arising from the substrate. In many cases this competition leads to frustration and non-collinearity [33].

Very recently, Tung and Guo [45] performed calculations of the electronic and magnetic properties of linear and "zigzag" atomic chains of 3d transition metals within a PAW–GGA approach. For all elements, a ferromagnetic state is obtained whereas the anti-ferromagnetic state is not always present. Those states are either stable or metastable. Giant magneto-lattice expansion is obtained for linear chains of V, Cr, Mn and Fe. The shape anisotropy energy is found to be comparable to the electronic one and always prefer an axial magnetization in both linear and zigzag chains.

3 Vanadium Nanostructures

As discussed by Dreysse and Demangeat [20], vanadium nanostructures were on the verge of magnetism. At a V(100) surface, this magnetism remains elusive [46]. However, Huttel et al. [47] evidenced a magnetic behaviour of vanadium embedded in copper through X-ray magnetic circular dichroism (XMCD). Very recently, within FLAPW, Hamad [48] reported on the determination of a magnetic map for very thin V films on Mo(001). The topmost V layers relax inward and an in-plane ferromagnetic ordering with appreciable magnetic moment is obtained on V overlayers. Moreover the Mo atoms, in contact with V, do present a small induced magnetic moment. The coupling between V and Mo atoms is of anti-ferromagnetic type. Also very recent calculations by Khmelevska et al. [32] have shown that a specific phase of Au4V do present some magnetism. The strong dependence of magnetic properties of Au4V upon the degree of chemical order has been the subject of intense experimental studies and controversial theoretical interpretations. Using a coherent potential approximation (CPA) method embodied in a Korringa–Kohn–Rostocker (KKR) calculation, Khmelevska et al. [32] performed first-principles calculation of Au4V, varying the degree of atomic chemical order from a disordered fcc alloy to a fully ordered Ni4Mo-type structure. To sum up, they showed that the complex behaviour is due to a combination of the following facts: (i) the existence of anti-ferromagnetic interactions on a geometrically frustrated lattice, (ii) highly non-trivial effects of chemical atomic order on the magnetic ordering and (iii) a dependence of the vanadium local moment formation on the local atomic environment.

Fritzsche et al. [49] used polarized neutron reflectometry to determine the absolute magnetic moment of uncovered and V-covered Fe films in the thickness range from 0.3 to 5.5 nm. The films were prepared by molecular beam epitaxy on a V(100) buffer layer grown on a MgO(100) crystal. The measurements on the V-covered Fe films revealed a reduction of $0.75\mu_B$ per Fe interface atom which is in agreement with some recent theoretical papers.

By a combination of element-specific X-ray resonant magnetic scattering (XRMS) experiments and model Hamiltonian calculations, Remhof et al. [50] showed that, upon hydrogen loading of a non-magnetic V spacer in Fe/V superlattices, the Fe magnetic moment becomes enhanced remotely, while the induced anti-parallel V moment at the Fe/V interface remains unaffected. This long range and remote control of Fe magnetic moments by hydrogen has been shown to be due to the following effect: a redistribution of d electrons between Fe and V, as a function of hydrogen concentration, leads to a shift of the d band relative to the Fermi level and thus to a change of exchange splitting.

Calleja et al. [51] performed a comparative study of the structure, magnetic properties and magnetic anisotropies of the VCo system formed in two different stacking sequences, namely, Co/V/MgO(100) and V/Co/MgO(100). These sequences gave rise to different Co crystalline structures (hcp versus fcc) and therefore to different magnetic behaviours. Within polarized neutron reflectivity and magnetization measurements, Baczewski et al. [52] determined the magnetic structure of epitaxial vanadium/gadolinium bilayers with different V thickness. The polarized neutron reflectivity results showed that in the fully magnetized state of V/Gd bilayer, about three to five monolayers of V became magnetic with a mean magnetic moment of around $0.8\mu_B$. The V slab was then anti-ferromagnetically aligned with the Gd layer. From magnetization measurements, Baczewski et al. [52] found an increase of the Curie temperature of a V/Gd system as compared to pure Gd.

In this section we focus on (i) the magnetization of V clusters supported on a Cu(111) surface [53] and (ii) the magnetization of V on Co(001) when going from sub-monolayer coverage to thin V films on Co(001) [54].

3.1 Non-collinear Magnetization of Supported V Clusters

Within a self-consistent non-collinear real space method, based on the Haydock recursion method [11, 53] determined the magnetization of very small V clusters on a Cu(111) surface. The method used is an extension of the collinear version of the real space (RS) LMTO-ASA method [16] to non-collinear arrangements of magnetic moments [53, 55]. The ground state for three V atoms forming an equilateral triangle presents highly non-collinear magnetic moments with an angle of 120° between them. When these bond lengths are modified, so that the frustration is avoided, Bergman et al. [53] found an anti-ferromagnetic ground state.

3.2 From V Sub-monolayer Coverage to V Thin Film on Co(001)

First we present results concerning a V sub-monolayer coverage, the geometrical configurations of which already have been reported in Fig. 1. For all V concentrations, from 0.25 to 0.75, a sizeable magnetic moment per V atom is always obtained. Moreover an anti-ferromagnetic polarization between V and Co atoms is predominantly obtained and the results are reported in Table 3. Let us comment these results. First, when V concentration is low (0.25 monolayer) the magnetic moment on each V atom is as high as $2.72\mu_B$ whereas the magnetic moments of Co surface atoms (in contact to V) are slightly diminished. This can be explained from the fact that V atoms are nearly isolated (at least far from each other) and Co atoms, being ferromagnetic, are inducing magnetization on V atoms. Second, when the concentration is 0.5 monolayer with the geometrical configuration displayed in Fig. 1b, the V–V distance is diminished. This leads to two effects: (i) a small decrease of V magnetic moment and (ii) also a small decrease of Co magnetic moment. For a V concentration of 0.75 monolayer, a strong decrease of V moments is observed. More precisely, V atoms at next nearest-neighbour positions, i.e. at A and B, display a sizeable decrease of their magnetic moments, as compared to the case of 0.5 monolayer coverage. Moreover the third V atom at C position (in the unit cell shown in Fig. 1c) presents a ferromagnetic coupling with the Co substrate and an anti-ferromagnetic coupling with its nearest-neighbouring V atoms at A and B positions.

In order to be complete, we also considered higher V coverages in order to see more precisely the effect of V–V distance on V magnetic moments. The mean magnetic moment for V atoms in terms of V coverage is reported in Fig. 3. For a complete V monolayer coverage a drastic modification of the distribution of V magnetic moments is found. For one V monolayer on Co(001) we considered an unit cell of successively one and two inequivalent V atoms in order to see if complex magnetic configuration could be stabilized. For one inequivalent V atom, per plane, both layered ferromagnetic and anti-ferromagnetic couplings can be obtained whereas for two inequivalent V atoms in-plane anti-ferromagnetic configuration can also be obtained.

Table 3 Magnetic moments (in μ_B) for three concentrations of V atoms on Co(001) slabs: (a) $V_{0.25}$ $E_{0.75}$ corresponding to configuration (a) of Fig. 1; $V_{0.50}$ $E_{0.50}$ corresponding to configuration (b) of Fig. 1. Magnetic moments on V atoms on sites A and B are equal, due to symmetry; and (c) $V_{0.75}$ $E_{0.25}$ on Co(001) fcc substrate. For (c) the magnetic moments on V atoms are found inequivalent. A, B, C and D are four sites of the unit cell; E is the empty space. Co4 (Co1) atoms are in the interface layer with V atoms (in the centre of the slab) (from [54]).

(a) $V_{0.25}$ Atom		(b) $V_{0.50}$ Atom		(c) $V_{0.75}$ Atom	
Va	−2.72	Va	−2.57	Va	−1.67
Eb	−0.01	Vb	−2.57	Vb	−1.67
Ec	−0.07	Ec	−0.12	Vc	0.86
Ed	−0.07	Ed	−0.12	Ed	0.11
Co4a	1.54		1.30		1.32
Co4b	1.54		1.30		1.32
Co4c	1.54		1.30		1.32
Co4d	1.54		1.30		1.32
Co3a	1.68		1.69		1.75
Co3b	1.64		1.69		1.75
Co3c	1.65		1.68		1.73
Co3d	1.65		1.68		1.73
Co2a	1.77		1.75		1.70
Co2b	1.77		1.75		1.70
Co2c	1.77		1.75		1.70
Co2d	1.77		1.75		1.70
Co1a	1.74		1.73		1.71
Co1b	1.74		1.73		1.71
Co1c	1.74		1.74		1.70
Co1d	1.74		1.74		1.70

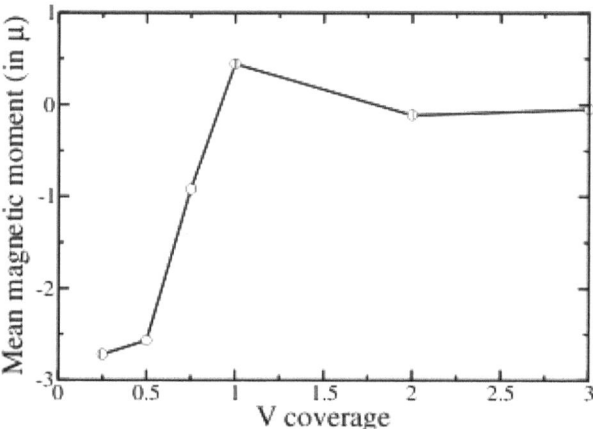

Fig. 3 Mean magnetic moment per V atom (in μ_B) and taken in function of V coverage. A drastic change of the direction of polarization is observed at full V coverage (from [54])

For V sub-monolayer coverage on Co(001), recent ab initio calculations by Carrillo-Cazares et al. [54] and Hong [57] obtained sizeable magnetic moments on V atoms with an anti-ferromagnetic coupling between V and Co atoms. The results obtained by Carrillo-Cazares et al. [54] were determined within the TB-LMTO code in ASA approximation. FLAPW approach was used by Hong [57] and the results obtained were very similar to those of Carrillo-Cazares et al. [54], except for a complete V monolayer on Co(001). Hong [57] got a small magnetic moment on V atoms, a small decrease of the Co moment at the V/Co interface as well as an anti-ferromagnetic coupling between V and Co atoms, whereas Carrillo-Cazares et al. [54] obtained a much larger moment on V atoms with a ferromagnetic coupling between V and Co atoms. The discrepancy seems to arise from the relaxation effect taken into account by Hong [57]. In their paper, Carrillo-Cazares et al. [54] used the same lattice parameter for Co and V: this is indeed not correct. In a more recent paper Carrillo-Cazares et al. [56] utilized the pseudopotential plane wave package of Baroni et al. [58] to perform additional calculations with relaxations along the z-axis. To solve the problem, Carrillo-Cazares et al. [56] followed the growth procedure of Huttel et al. [59], i.e. Co and V atoms deposited on Cu(001) substrates. Calculations were performed in the slab-geometry approach. For one and two complete V layers on Co(001), Carrillo-Cazares et al. [56] obtained (Table 4) an anti-ferromagnetic coupling at the Co/V interface in agreement with Hong [57]. In a second calculation Carrillo-Cazares et al. [56] tried to recover the result appearing in the last line of Table I of Huttel et al. [59], i.e. the system where 1.8 ML of V is grown on a Cu(100) substrate and covered by 7 ML of Co. For the calculation, Carrillo-Cazares et al. [56] took a slab of 7 layers of Cu(100) on which the V–Co layers were deposited. Table 5 reports the results of various geometries starting with a pure Cu film displaying no magnetism. When V layers are deposited on Cu no onset of magnetism is seen. Thus in Table 5, only the distances from the centre of the Cu slab are reported for pure Cu slab and for V over-layers on Cu(001) because no spin polarization has been obtained. However, when the two V monolayers, at the top of a Cu(001) film, are covered by one and two layers of Co, these Co atoms do present a sizeable magnetic moment in agreement with the XMCD

Table 4 Magnetic moments (in μ_B) and z atomic positions (in a.u.) with respect to the Co1 layer at the centre of the Co slab of 7 layers (noted by 0.00), for Vn layers ($n = 1, 2$) on Co(001), (a) one V monolayer on fcc Co(001) and (b) two V layers on fcc Co(001) (from [56])

| | (a) | | (b) | |
| | V/Co(001) | | 2V/Co(001) | |
	μ	z	μ	z
V2			−0.03	16.985
V1	−0.26	13.071	−0.20	13.162
Co4	1.34	9.670	1.35	9.606
Co3	1.78	6.521	1.75	6.481
Co2	1.77	3.229	1.76	3.214
Co1	1.77	0.000	1.76	0.000

Table 5 Magnetic moments (in μ_B) and z atomic positions (in a.u.) with respect to the Cu1 layer at the centre of the slab (noted by 0.00) for, respectively, (a) the fcc Cu(001) surface; (b) one V monolayer on fcc Cu(001) substrate; (c) two V layers on fcc Cu(001) substrate; (d) one Co monolayer on V2/Cu(001); and (e) 2Co layers on V2/Cu(001) (from [56])

| | (a) | (b) | (c) | (d) | | (e) | |
| | Cu(001) | V | 2V | Co/2V | | 2Co/2V | |
	z	z	z	μ	z	μ	z
Co2						1.91	25.013
Co1				1.33	21.734	1.49	22.002
V2			18.294	−0.31	18.418	−0.19	18.393
V1		14.385	14.487	−0.03	14.506	−0.07	14.481
Cu4	10.545	10.664	10.672	0.00	10.660	−0.00	10.643
Cu3	7.089	7.096	7.099	0.00	7.087	0.00	7.079
Cu2	3.536	3.555	3.550	0.00	3.543	0.00	3.539
Cu1	0.000	0.000	0.000	0.00	0.000	0.00	0.000

results of Huttel et al. [59]. Calculations also found an induced polarization on V sub-surface atoms but those values are at odds with those of Huttel et al. [59]: they are much smaller. However, an increase of V polarization can be obtained by considering an ordered Co–V ordered alloy at a Co/V interface. As shown in Table 6, a

Table 6 Magnetic moments (in μ_B) and z atomic positions (in a.u.) with respect to the centre of the slab (noted by 0.00) for (a) Co/CoV/V/Cu(001); (b) 2Co/CoV/V/Cu(001); and (c) 3Co/CoV/V/Cu(001) (from [56])

| | Co/VCo/V | | 2Co/CoV/V | | 3Co/CoV/V | | 4Co/CoV/V | |
	μ	z	μ	z	μ	z	μ	z
Co5a							1.91	31.193
Co5b							1.91	31.180
Co4a					1.92	27.887	1.72	28.094
Co4b					1.92	27.887	1.72	28.094
Co3a			2.01	24.729	1.81	24.823	1.82	24.805
Co3b			1.84	24.678	1.67	24.790	1.68	24.772
Co2a	1.17	21.447	1.27	21.600	1.16	21.466	1.17	21.544
Co2b	1.17	21.447	1.27	21.600	1.16	21.466	1.17	21.544
Co1	0.55	18.529	0.26	18.302	0.40	18.218	0.25	18.452
V2	−0.85	18.374	−0.63	18.417	−0.74	18.416	−0.69	18.302
V1b	0.10	14.589	0.05	14.527	0.02	14.506	0.01	14.541
V1a	0.10	14.589	0.05	14.527	0.02	14.506	0.01	14.541
Cu4a	0.01	10.768	0.00	10.722	0.00	10.699	0.00	10.732
Cu4b	0.00	10.772	0.00	10.713	0.00	10.688	0.00	10.723
Cu3a	0.00	7.157	0.00	7.132	0.00	7.124	0.00	7.123
Cu3b	0.00	7.157	0.00	7.132	0.00	7.124	0.00	7.123
Cu2a	0.00	3.575	0.00	3.565	0.00	3.565	0.00	3.559
Cu2b	0.00	3.575	0.00	3.565	0.00	3.565	0.00	3.559
Cu1a	0.00	0.000	0.00	0.000	0.00	0.000	0.00	0.000
Cu1b	0.00	0.000	0.00	0.000	0.00	0.000	0.00	0.000

polarization of about 0.7–0.8 μ_B can be obtained, as compared to $1.2\mu_B$, deduced by Huttel et al. [59].

3.3 Outlook

In this section we mainly discuss some attempt to describe the onset of non-collinear magnetization of V clusters supported on a Cu(111) surface, as well as the induced polarization of V atoms in contact with Co(001). For V compact trimer on Cu(111), it has been shown that the ground state is non-collinear. On the other hand, there is no doubt about a short-ranged induced polarization when those vanadium nanoclusters are deposited on a strong ferromagnet: this has been seen when V atoms are in contact with Co(001). Carrillo-Cazares et al. [54, 42, 56] did perform various ab initio calculations, in the collinear approach, but taking into account the relaxation. It is clearly shown that, for V nanoclusters, relaxation is fundamental for the determination of spin polarization. The reason is that V is at the verge of magnetism so that any external effect may induce (or destroy) the magnetization in those nanostructures. Nowadays, contrary to Cr and Mn for which magnetization is most generally present, the V nanostructures do not present (in general) any confirmed experimental proof of magnetization. It is only for very dilute V atoms in Cu and for a Au4V-specific phase that magnetization has been show to be present.

On the contrary, V atoms in contact with a strong ferromagnet do present induced polarization. This induced polarization is clearly short ranged, i.e. only nearest and somewhat next-nearest V atoms to the strong ferromagnet are clearly polarized. There are in fact two opposite effects for the induced polarization. The first one concerns the number of V atoms in contact with the ferromagnet: if the number of V adatoms is small then the induced polarization is strong [56, 54, 57]. However when the number of V atoms increases, then the induced polarization strongly decreases [54, 57]. Besides it was clearly shown that the effect of relaxation is essential to determine induced magnetic moments [56]. Thus it appears that in order to get a significant and valuable result, any ab initio calculation must be "non-collinear" within a cancellation of the forces acting on V atoms. Such calculation did not appear yet. Besides, when V atoms are in contact with a substrate, total energy calculation should also take into account a possible exchange between V atoms and substrate atoms.

4 Chromium Nanostructures

4.1 Introduction

From a historical point of view Overhauser [60] and Lomer [61] explained spin density waves (SDW) of bcc Cr through nesting mechanism. These SDW are observed experimentally, but density functional theory (DFT) calculation [62] predicts anti-ferromagnetic configuration. This apparent contradiction has been solved by Uzdin and Demangeat [63] within a Periodic Anderson Model (PAM) (see, for example, [64]) and confirmed by Vanhoof [65] within a full-potential augmented plane wave + local orbitals (APW + lo) method as implemented in the WIEN2k package. The claim is that "the ground state of Cr is really anti-ferromagnetic" and it is correctly predicted by DFT. However, this ground state is (so far) not yet observed experimentally, because there are quasi-particle excitations called "nodons" that can be populated already at very low temperatures. Each nodon corresponds to the introduction of one node (a lattice site with zero moment) in the anti-ferromagnetic Cr lattice. Such a node does not need to be localized, but can travel through the Cr lattice. These nodons have the following properties: (1) They can easily be created, but it is difficult to destroy them; once a nodon is created, it is much easier to displace it than to destroy it. (2) The nodon–nodon interaction is highly anisotropic: attractive for distances of a few monolayers along the [001] direction, weakly repulsive for all larger distances along this direction, strongly repulsive for distances smaller than the nearest-neighbour Cr–Cr distance in all directions perpendicular to [001], and attractive for all other distances in the latter directions. (3) When two nodons meet, they completely annihilate each other and locally the AF state is restored. With these features, the inevitable existence of SDW can be understood: Cr is AF at 0 K and as soon as the temperature rises above an unknown threshold, thermal fluctuations may excite a nodon. Once it is there, it is unlikely to be destroyed, and it starts travelling through the lattice. The number of nodons continuously increases, and they start to interact with each other. The attractive interaction in the (001) plane leads to the building of (001) planes of nodes. These planes repel each other along the [001] direction, and therefore try to maximize their mutual distance along this direction. Meanwhile, since the creation of nodes keeps going, ever more planes will be formed and the [001] distance between planes decreases, leading to occasional annihilations when planes come too close together. This evolution ends when formation and annihilation of planes are in equilibrium. Due to repulsive interaction, the system ends up with a collection of (001) node planes at regular distances from each other, which is clearly SDW. If temperature rises, the creation of nodes increases and extra node planes will be formed. These are now more densely packed so that more annihilations take place, and the net result is less node planes at larger distances from each other, in dynamic equilibrium with the formation of new nodes. This agrees with the experimental observation that the SDW period increases with temperature. In a final step, the nodon model was used in combination with ab initio calculations to study SDW doped with impurities [65].

Recently Kravtsov et al. [66] discussed the onset of SDW in Cr/V hetero-structures with different Cr thicknesses by using combined resistivity, neutron and synchrotron scattering measurements. They demonstrated that SDW behaviour is strongly affected by Cr Fermi surface nesting, finite size effects and proximity effects from neighbouring V layers. The magnetism on Cr layers in those Cr/V hetero-structures depends strongly on thickness of the considered Cr film. It is especially pointed out that Cr magnetization disappeared when its thickness is less than

100 Å. Very complex magnetic states are depicted in terms of Cr thickness: no magnetism for small thickness followed by a paramagnetic state when the thickness increases. Then commensurate SDW appears before the onset of incommensurate SDW in thick Cr layers. Effect of V is clearly disturbing, presenting a long-range nature which causes stabilization of longitudinal out-of-plane SDW.

Up to now we discussed the stability of SDW in Cr. These SDW are only present when the thickness of Cr slab is thick enough [63]. However, in the case of nanostructures investigated in the present section, the size is generally much too small so that SDW cannot develop. For free-standing clusters of 20–133 atoms at temperature between 60 and 100 K, Payne et al. [67] used a Stern–Gerlach deflection technique to study magnetism. What is striking is the observation of two magnetically distinguishable populations never observed in other transition metals. On the theoretical side, magnetization has been obtained for clusters with much smaller size so that comparison with experiment is presently lacking. Ohresser et al. [68] performed XMCD measurements of isolated-like Cr atoms deposited on Au(111). The magnetic moment, per Cr atom, extracted from a Brillouin function, is above $4\mu_B$ in agreement with fully relativistic electronic structure calculations, using the embedding technique within the KKR method. Also, Yayon et al. [69] used spin-polarized scanning tunnelling spectroscopy to observe the spin polarization state of individual Fe and Cr atoms adsorbed onto Co nanoislands.

It is difficult to discuss about Cr nanostructures without remembering the discovery of the giant magneto-resistance (GMR) by Baibich et al. [70] in Fe/Cr multilayers. Details on GMR can be found in a review paper by Coehoorn [71]. Baibich et al. [70] performed their experiments for Cr(001)-oriented thin films. For this orientation Fe/Cr multilayers clearly present two periods of oscillations (see [30] for details). By means of spin- and angle-resolved photoelectron spectroscopy, Dedkov [72] performed spin-resolved electronic structure measurements of thin Cr overlayers on top of a Fe(110) surface. The initial fast drop of photoelectron spin-polarization at the Fermi level was followed by weak oscillatory behaviour with a period of about 2 monolayers (ML). In the case of Fe/Cr(001) superlattices, the theoretical explanation of 2 ML oscillation relies on the anti-ferromagnetic coupling between nearest Fe–Cr and Cr–Cr monolayers. Thus, for a Cr spacer with an odd number of layers, the magnetic moments of Fe layers on both interfaces will order parallel and interlayer exchange coupling (IEC) will be ferromagnetic. If the Cr spacer has now an even number of layers, IEC will therefore be anti-ferromagnetic. Nothing like that for the (110)-oriented Fe substrate because, for this geometry, the (110) Cr monolayer has an in-plane anti-ferromagnetic configuration leading to a net magnetization of zero. Of course, for a small number of Cr layers adsorbed on Fe(110), an induced ferromagnetic component should be present.

In the next section, we will focus on small Cr clusters adsorbed on a substrate. Let us just mention, for completeness, very recent non-collinear calculations of unsupported Cr(111) monolayers [73] as well as unsupported small planar Cr clusters [74].

4.2 Cr Clusters on a Substrate

New phenomena discovered in low-dimensional magnetic structures have attracted considerable attention, as far as the connection between magnetism and dimensionality is concerned. Very interesting from this point of view is the evolution of Cr magnetic properties with the size of the considered sample. Bulk Cr has complex magnetic structure, including an incommensurate anti-ferromagnetic SDW. In epitaxial films, SDW can be manipulated by the choice of a magnetic cover, an interfacial roughness and a chromium film thickness [75]. For film thickness less than the SDW period, magnetic structure is determined by the interface region. In Fe/Cr multilayers with thin Cr spacer, interface defects, such as steps or pinholes, are responsible for non-collinear coupling of Fe magnetic moments through Cr layers [76]. Magnetism of small supported Cr clusters have special interest. They display non-collinear ordering due to the competition of exchange interactions between different atoms. The ratio between these interactions can be varied versus the interatomic distances in cluster by choosing a suitable substrate or by varying the conditions of the epitaxial growth.

Within STM experiment, Jamneala et al. [77] observed a narrow resonance at the Fermi level for a compact Cr trimer on Au(111) substrate. This appears to correspond to a sizeable Kondo temperature of about $50\,K$ much larger than the value of $6\,K$ corresponding to a single Cr impurity on Au(111). Kudasov and Uzdin [34] were the first to determine the ground state of this Cr trimer supported on a Au(111) surface by means of a variational approach to the Coqblin–Schrieffer Hamiltonian. The temperature of Kondo resonance formation for trimers was found much larger to that of a single Cr adatom in agreement with results of Jamneala et al. [77]. Following the work of Kudasov and Uzdin [34], let us notice interesting results obtained by Savkin et al. [78], Lazarovits et al. [79] and Aligia [80]. As pointed out by Aligia [80]: "The origin of this puzzling dependence of the Cr trimer TK with geometry is still unclear."

Gotsis et al. [81] carried out first-principles electronic structure calculation to study the structural, electronic and magnetic properties of monomer, dimer and trimer Cr clusters on a Au(111) surface. They used the projector augmented wave (PAW) method as implemented in the VASP code. The most favourable location for the single Cr adatom is the fcc hollow. The magnetic moment obtained is $3.93\mu_B$, which is a value smaller than that obtained by Ohresser et al. [68] within a fully relativistic KKR method. The compact dimer orders anti-ferromagnetically with a moment of $0.005\mu_B$. The compact triangular trimer displays non-collinear magnetism with angles of $120°$ between the direction of the various magnetic moments. The effect of spin–orbit is shown to be rather negligible on a local moment of about $3.16\mu_B$.

Later on, Bergman et al. [82], as well as Antal et al. [83], also determined electronic and magnetic properties of a compact Cr cluster on Au(111). Bergman et al. [82] considered various geometries for Cr on Au(111): on the surface; in the surface, and in the sub-surface. Clearly, as a consequence of the increasing numbers of Au atoms in the neighbourhood of Cr, its moment decreases from $4.31\mu_B$, for an adatom, to $3.85\mu_B$, in the surface and finally to $3.59\mu_B$, in the sub-surface. The result for Cr adatoms is very similar to that obtained by Gotsis et al. [81]. However, for a Cr dimer the results obtained are totally at odds with those of Gotsis et al. [81]. This is surely due to the unrelaxed approximation used by Bergman et al. [82]. These authors also considered the Cr trimer in linear and triangular geometries. For a linear chain, the coupling is found to be anti-ferromagnetic whereas, for the compact trimer, the coupling of the magnetic moments is non-collinear with $120°$ between moments of different atoms. Again, it is difficult to compare with the results of Gotsis et al. because of the absence of relaxation. Also Antal et al. [83] reported on the magnetic properties of a Cr trimer on Au(111) surface. The Cr atoms are situated at the hollow sites on the top of a fcc Au(111) surface, i.e. they neglected possible relaxations of the geometry. Stocks et al. [84] recently reviewed some of their recent work concerning first-principles calculation of the magnetic structure of surface and bulk nanostructures. Moreover, Bergman et al. [85] determined the magnetic structures of small clusters of Fe, Mn and Cr supported on a Cu(111) surface. Different geometries, such as triangles, pyramids and wires, are considered and the cluster sizes have been varied between two and ten atoms. Kawagoe et al. [86] fabricated high-density self-organized spiral terraces on Cr(001) films. Imaging of both topological and magnetic structures was realized at room temperature by spin-polarized scanning tunnelling spectroscopy.

4.3 Outlook

In the section concerning V nanostructures, most of the experiences do not point out spin polarization in their results. On the contrary, for Cr nanostructures, any experiment or calculation should take into account the inherent magnetization of the considered Cr nanostructures. The most recent results do take into account, not only the modulus of Cr magnetization but also its direction. Actually, since the work of Uzdin et al. [33] on Cr compact trimer on noble metals, more and more calculations do take into account the fact that a broken symmetry does lead to a non-collinear ground state. For example, magnetic moment distribution in Cr(001) and Cr(110) films with strips and random defects at surface have been calculated in the framework of a periodic Anderson model [87]. All Cr(001) films reveal an appreciable surface magnetic moment, the value of which depends on the type of defect. The film with flat Cr(110) surface does not exhibit any magnetization because of an equal number of atoms with opposite moments. Only well-ordered defects, like one or two atom strips, can give a noticeable magnetization in Cr(110) films. Figure 4 shows a schematic illustration of the anti-ferromagnetic non-collinear magnetic moment orientation for a Cr film with two atom strips. The maximal deviation of the surface moments on the substrate orientation is about $11°$. In all considered cases the energy of non-collinear configurations is very close to the collinear ones.

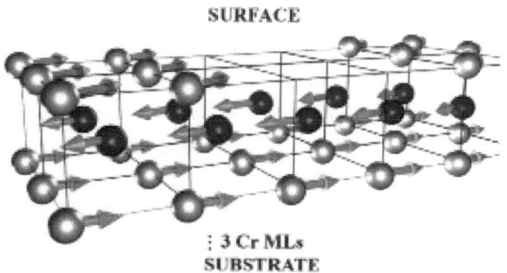

SURFACE

⋮ 3 Cr MLs
SUBSTRATE

Fig. 4 Schematic illustration of the non-collinear magnetic moments orientation (nearly anti-ferromagnetic) for a Cr film with two atom strips. The *dark spheres* indicate the atoms with nearly opposite orientations of moments (from [87])

5 Manganese Nanostructures

5.1 Introduction

As pointed out by Demangeat and Parlebas [36], manganese nanostructures have been the subject of a number of investigations because manganese is a unique element which exhibits a variety of unusual electronic and magnetic properties depending on its environment. Knickelbein [88] depicted a super-paramagnetic behaviour in Mn clusters within Stern–Gerlach experiments. Mejia-Lopez et al. [89] were able to explain this puzzling size dependence of magnetic properties via an effective spin Hamiltonian. In this section, first we shall focus on Mn films on substrates (Sect. 5.2). Also, we say a few words on Mn-based ordered alloys on ferromagnets. It seems worthy, after the recent result of Knickelbein [88], to write down a short paragraph on free-standing Mn clusters (Sect. 5.3). Another paragraph will be devoted on adsorbed nanoclusters deposited on a substrate (Sect. 5.4). At last, we shall give some recent spectroscopic results of Mn nanostructures on Ag(001) (Sect. 5.5). Finally, Sect. 5.6 will be devoted to an outlook upon the considered subject.

5.2 Thin Mn Films and Mn-Based Films Deposited on Substrates

Following the work of Hafner and Spisak [35] displaying non-collinear magnetism in a Mn monolayer on Fe(001), renewed interest was devoted, from an experimental point of view, on Mn films in contact with a strong ferromagnet. Lee et al. [90] observed an unexpected magnetic structure, with out-of-plane Mn moments perpendicular to those of Fe in Fe/Mn multilayers. For Mn on Fe(001) and within spin-polarized scanning tunnelling spectroscopy, with ring electrodes, Gao et al. [91] depicted a large reconstruction with a non-collinear spin structure. Besides Kohlhepp and De Jonge [92] stabilized a metastable expanded fct Mn on Co(001).

Also Biswas et al. [93] studied the growth and electronic structure of Mn on Al(111) within photoelectron spectroscopy and low-energy electron diffraction.

Within a full potential linearized augmented plane waves (FLAPW) code, Hong and Wu [94] determined the magnetic map and the magnetic anisotropy of a Mn monolayer on Nb(001). Total energy calculations revealed that a ferromagnetic state instead of $c(2 \times 2)$ anti-ferromagnetic state is a preferred state in Mn/Nb(001). A magnetic moment of $3.23 \mu_B$ was obtained on each Mn surface atom whereas an induced moment of $0.56 \mu_B$ was found on the sub-surface Nb atoms. Furthermore, Mn/Nb(001) has an in-plane easy axis because the negative shape anisotropy overwhelms the small positive magneto-crystalline anisotropy. Yamada et al. [95] investigated the magnetic structure on thin Mn films grown on a Fe(001) surface with a mono-atomic step. The spin-polarized scanning tunnelling microscopy/spectroscopy images display a change in the magnetic contrast when crossing one of those steps due to the change of Mn thickness. The width of the domain wall around the substrate steps does not depend on the thickness, at least for coverages up to seven Mn overlayers. This is due to the weakly defined magnetic coupling at the central Mn layers that decouple the surface from the interface to a large extent. These findings are described by parameterized self-consistent real space tight binding calculation in which the spin quantization axis is site dependent, thus allowing non-collinear magnetism. Also, Spisak and Hafner [96] reported non-collinear calculations performed within a VASP-code in a full relativistic mode, for Mn adlayer on Fe(100) surface. When the constraint of spin collinearity is dropped, a spin-flop state with a nearly perpendicular direction of Fe and Mn moments was found.

Besides those Mn thin films on ferromagnets, M'Passi-Mabiala et al. [97] investigated the ferromagnetic ground state for MnCo surface-ordered alloy on a Co(001) substrate. Within TB-LMTO-ASA, in a GGA approximation, this ordered alloy is found more stable as compared to a Mn overlayer on Co(001). Besides a ferromagnetic coupling is obtained (Table 7) in agreement with experimental results of Choi et al. [98]. Later on, Malonda-Boungou et al. [99] determined the magnetic

Table 7 Magnetic moments (in μ_B) for Mn–Co/Co(001) and difference of total energies per cell (in mRy) with a GGA–PW approximation. The ground state is indicated by 0 (from [97])

Input	Mn ↓ Co ↓ Mn ↓ Co ↑	Mn ↑ Co ↓ Mn ↑ Co ↑
Energy	36	0
Atom	Moments	Moments
Mn	−3.79	3.67
Co5	0.82	1.73
Co4b	1.33	1.59
Co4a	1.33	1.59
Co3b	1.77	1.76
Co3a	1.71	1.73
Co2b	1.71	1.70
Co2a	1.71	1.70
Co1b	1.70	1.70
Co1a	1.71	1.70

Table 8 Magnetic moments per atom (in μ_B) and differences of total energy per cell (DTEC) in mRy/cell for $(Mn_{0.5}Ni_{0.5})_2/Co(001)$. The input magnetic configurations noted "Input" $\uparrow$ and $\downarrow$ represent the ferromagnetic and anti-ferromagnetic couplings with Co substrate, respectively. The energy of the ground state is set to 0.0 mRy/cell (from [99])

Input	Mn1 $\uparrow$Mn2 $\downarrow$	Mn1 $\downarrow$Mn2 $\uparrow$	Mn1 $\uparrow$Mn2 $\uparrow$	Mn1 $\downarrow$Mn2 $\downarrow$
DTEC	0.0	17	38	71
Atom				
Mn2	−3.69	3.63	3.78	−3.80
Ni2	−0.12	0.12	0.64	−0.63
Mn1	2.90	−3.07	3.01	−3.01
Ni1	0.28	0.09	0.56	−0.32
Co4b	1.62	1.33	1.62	1.28
Co4a	1.55	1.34	1.62	1.38
Co3b	1.75	1.80	1.78	1.78
Co3a	1.71	1.70	1.72	1.69
Co2b	1.78	1.83	1.82	1.82
Co2a	1.83	1.83	1.82	1.83
Co1b	1.78	1.81	1.81	1.80
Co1a	1.76	1.78	1.77	1.77

map of Ni–Mn thin films on Co(001) and Co(111). For the $Ni_{0.5}Mn_{0.5}$ ordered alloy, one layer thick on Co(001), couplings are ferromagnetic between Mn and Ni, as well as for Mn and Ni, with the Co substrate. For the $Ni_{0.5}Mn_{0.5}$ ordered alloy, two layers thick, the surface is polarized ferromagnetically, but is coupled anti-ferromagnetically with the sub-surface and the Co atoms (Table 8). The mean magnetic moment of Mn versus the thickness of the surface-ordered alloy oscillates (Fig. 5) whereas the mean magnetic moment of Ni remains always small. Investigation of Ni–Mn monolayer on Co(111) versus Mn concentration displays a ferromagnetic polarization between Mn and Ni for small concentrations of Mn whereas an anti-ferromagnetic polarization between Mn atoms is present for high Mn concentrations.

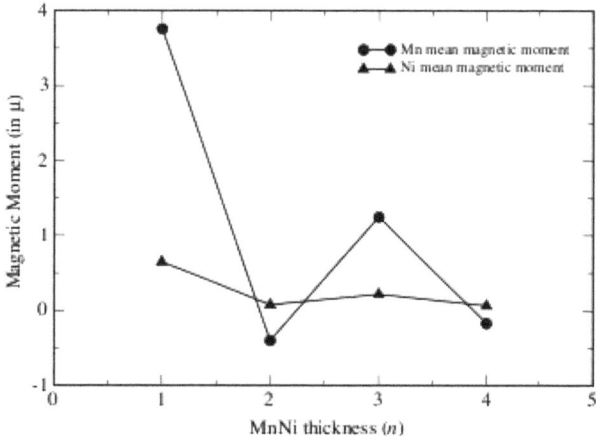

Fig. 5 Mean magnetic moment per atom (in μ_B) for $(Ni_{0.5}Mn_{0.5})n$ monolayers on Co(001), "n" is the thickness of MnNi over-layers, for Ni and Mn atoms as a function of the Mn–Ni thickness. *Line* with *circles (triangles)* represent the mean magnetic moment for Mn (Ni) atoms (from [99])

Since the work of Andrieu et al. [100], another point of current interest is the effect of oxygen contamination on the properties of Mn nanostructures deposited on a strong ferromagnet. Andrieu et al. [100] considered the effect of oxygen contamination on the magnetic properties of Mn grown on Fe(100) with a variable coverage between 0.1 and 1 ML and for different coverage rates of oxygen. Just after the sample preparation, a strong XMCD signal is found indicating sizeable magnetic moments on Mn atoms which are coupled ferromagnetically to the Fe ones. At this stage, the X-ray photoelectron spectroscopy (XPS) signal indicated low O contamination. The XMCD signal, 15 h later, has undergone a significant change due to a complete covering of the Mn layer by oxygen. Also, Yonamoto et al. [101] saw this oxygen-induced reversal of magnetic moments for Mn on Co(001). The effect of oxygen on the magnetic coupling between Mn on Fe(001) substrate and between Mn on Co substrate was determined, respectively, by Zenia et al. [102] and Pick and Demangeat [103]. Zenia et al. [102] performed ab initio density functional calculations on a Mn monolayer deposited on Fe(001) to which a top layer of oxygen is added. They used the SIESTA (Spanish Initiative for Electronic Structure of Thousands of Atoms) code developed by Soler et al. [104]. A clean Mn surface is found to be in a c(2 × 2) in-plane anti-ferromagnetic order with moments of $3.98\mu_B$ and $-4.39\,\mu_B$ on both inequivalents atoms, the sub-surface Fe moments being smaller than in the bulk case. Two metastable solutions with ferromagnetic polarization between Mn and Fe (p(1 × 1) ↑) and anti-ferromagnetic

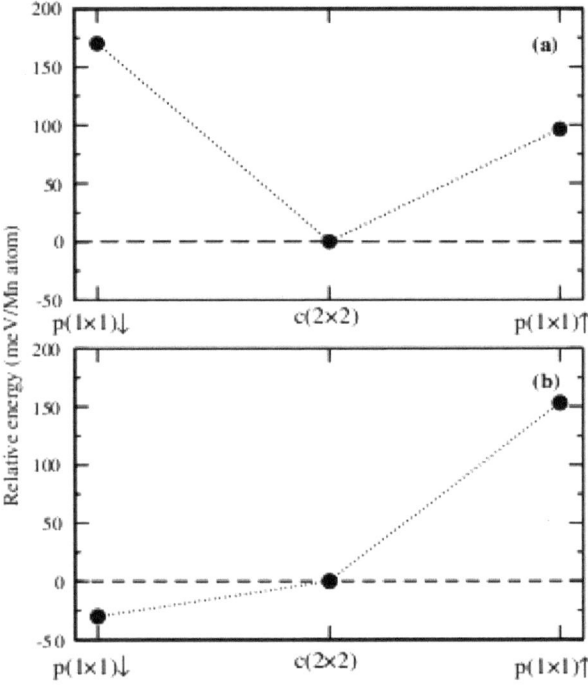

Fig. 6 Relative stability of three magnetic arrangements at Mn atoms in clean Mn/Fe(100) (**a**) and O/Mn/Fe(001) (**b**) systems. All the atoms are fixed to their ideal bcc positions (from [102])

polarization between Mn and Fe (p(1 × 1) ↓) are found some 100 meV above the ground state. The addition of oxygen modifies drastically the stability of the three solutions obtained without oxygen. Now the p(1 × 1) ↓ solution is the ground state (Fig. 6).

5.3 Non-collinear Free-Standing Mn Clusters

Soon after the Stern–Gerlach experiment of Knickelbein [88], displaying magnetization in Mn clusters, first-principle studies were performed on small Mn clusters. In the Stern–Gerlach experiment, the clusters in a beam passed through a gradient magnetic field that tries to orient the moments, as well as to deflect them. For Fe, Co or Ni clusters, the beam undergoes a net deflection upon application of the gradient fields. For Mn, however, the observed deflection profiles do not exhibit any net deflection but simply broaden with increasing gradient field. The absence of a net deflection could be accounted by a weak ferromagnetism or anti-ferromagnetic coupling or better a non-collinear arrangement. Within a projector augmented wave (PAW) formalism implanted in the Vienna Ab initio Simulation Package (VASP) code, Morisato et al. [105] performed an unconstrained magnetization of Mn clusters with five and six atoms. They showed that while the ground state of a Mn5 cluster had a collinear arrangement of spins, a Mn6 cluster was the smallest to exhibit a non-collinear ground state.

Just after the publication of the paper by Morisato et al. [105] and using a fully unconstrained magnetization option of SIESTA [104], as well as non local norm-conserving pseudo-potentials and a local spin density approximation (LSDA) for exchange and correlation, Longo et al. [106] performed DFT calculations of the structures, binding energies and magnetic moments of Mnn clusters ($n = 2$–7). They found that, for $n = 2$ the ferromagnetic solution is the ground state, and that for $n = 3, 4$, ferromagnetic and anti-ferromagnetic configurations are equiprobable. For $n = 5, 6$ and 7, there is a clear preference for anti-ferromagnetic ordering. However, for $n = 3$–7, they also found non-collinear solutions the total energies of which are more negative than those obtained for collinear constrained magnetization. This is clearly at odds with the result of Morisato et al. [105] displaying non-collinear ground state only for n greater than 5. No serious explanation was given concerning this very serious discrepancy. Morisato et al. [105], as well as Longo et al. [106], performed their ab initio calculation only for a small number of Mn atoms in the cluster. Therefore they do not give any explanation of the magnetic site dependence reported by Knickelbein et al. [88]. To make some link between calculations and experiment, Xie and Blackman [107] performed some semi-empirical non-collinear tight-binding calculation for Mn clusters containing more than 10 Mn atoms. Also, within the SIESTA-code, Mejia-Lopez et al. [89] did ab initio non-collinear calculations for Mn-clusters with less than nine atoms. For clusters with more Mn atoms they used an effective spin Hamiltonian. Within a gradient corrected DFT approach, Kabir et al. [108] performed a systematic investigation of electronic structure and emergence of non-collinear magnetism of pure Mnn clusters and Mnn clusters doped with As. The ground state for $n < 6$ is collinear and emergence of non-collinear ground states is seen for $n > 5$. This result is at odds with those of Longo et al. [106] who obtained non-collinear ground states as soon as the number of Mn atoms are greater than two.

5.4 Mn Chains and Compact Clusters on Substrates

Inelastic scanning tunnelling spectroscopy with STM was used by Hirjibehedin et al. [109] to obtain a direct probe of magnetic interactions between manganese atoms building linear chains. These chains, ranging from 1 to 10 atoms, were assembled by atomic manipulation on one monolayer high insulating islands of CuN deposited on Cu(001). Comparing the spin excitation spectra with magnetic properties predicted by an Heisenberg Hamiltonian, restricted to Mn–Mn nearest-neighbour exchange interactions, they obtained the strength of magnetic coupling J between individual spins per Mn. They showed that this coupling is strongly dependent on the deposition sites of the Mn chains: the value of the J factor goes from one to two when depositing Mn on top of N or on top of Cu. Barral et al. [110] performed WIEN2k calculations for different arrangements of Mn chains deposited on CuN/Cu(001). The results of Barral et al. [110] suggested that for a given configuration the interaction was of super exchange type through N–Mn molecular orbitals, while for another one, it was given by a combination of super exchange and indirect type interactions through Cu substrate. Costa et al. [111] showed that the proximity of two magnetic adatoms attached to the walls of carbon nanotubes may induce the formation of non-collinear alignment of their magnetizations. This effect is the result of a competition between direct and indirect contributions to exchange couplings which become comparable when magnetic adatoms are not too far apart from each other. Moreover, the ability to control the indirect exchange coupling through a careful selection of the Fermi energy of nanotubes opens the road to the possibility of controlling the magnetization of nanotube-based systems with magnetic dimers.

5.5 Spectroscopic Properties of Mn Nanostructures on Ag(001)

The study of magnetically stabilized surfaces has attracted much experimental and theoretical interest over the past decade because of its great importance in understanding the interfacial magnetism for this type of system in addition to its relevance as far as magneto-resistive devices are concerned. For the study of spontaneous two-dimensional itinerant magnetism, the well-known prototype system is a 3d transition metal monolayer (ML) adsorbed on a non-magnetic substrate.

Extending a previous study [112] and using a realistic impurity model that includes full atomic multiplet interaction as well as coupling to Mn 3d and Ag 4d bands, Taguchi et al. [113] performed a variety of calculations, i.e. core level

Fig. 7 Calculated 2p core level photoemission spectra for various Mn–Ag structures, along with experimental spectra from [114] (Mn impurity in Ag) and Schieffer et al. [115] (all others). The *short dashed line* represents the calculated background spectra. The labels M and S indicate the main peak and satellite, respectively (from [113]).

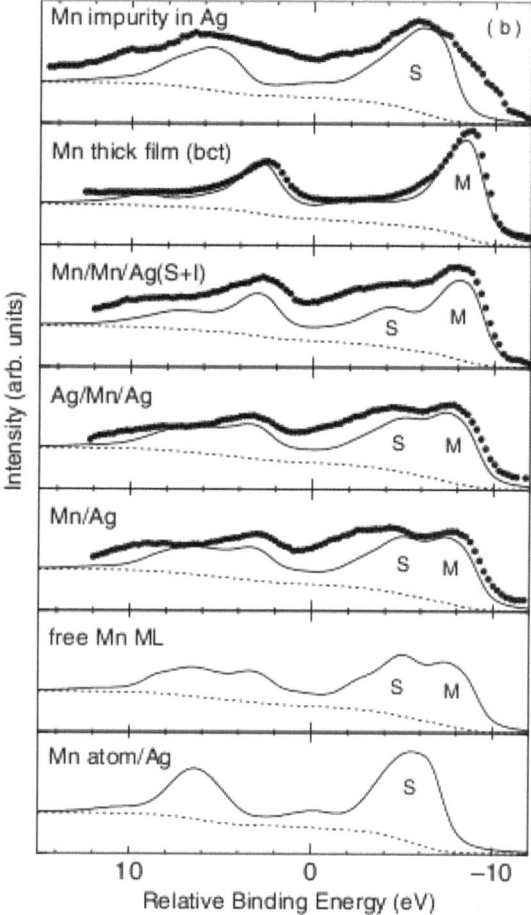

X-ray photoemission (c-XPS), X-ray absorption (XAS) and resonant X-ray emission (RXES) spectra at Mn L_{23} edge in Mn thin films on Ag(001) and related structures, namely an adsorbed Mn monolayer (ML) on Ag(001) [i.e. Mn/Ag(001)], a "buried" one [i.e. Ag/Mn/Ag(001)], an adsorbed bilayer Mn/Mn/Ag(001), a freestanding Mn ML, bulk body-centred-tetragonal (bct) Mn, a Mn impurity in Ag and a single Mn atom on Ag(001). The calculated 2p photoemission spectra (Fig. 7) well reproduce the experimental ones in the whole range of structures from Mn impurities in Ag and to bulk bct Mn [114, 115]. For comparison, the considered authors also calculated the spectra of a free Mn ML and that of a single Mn atom on Ag. Taguchi et al. [113] also calculated Mn 2p XAS with the same parameter values as those of 2p XPS. The results are shown in Fig. 8. The experimental spectrum is also shown by closed symbols, only for a Mn/Ag system, from [116]. The calculated result for Mn/Ag is in good agreement with the experiment. All systems have very similar spectral shape. This is because the charge

Fig. 8 Calculated 2p core level photo-absorption spectra for various Mn–Ag structures. The *closed symbol* represents the experimental 2p absorption spectra from [116]. The photon energies of the resonant X-ray emission spectra are marked by *vertical arrows* (from [113])

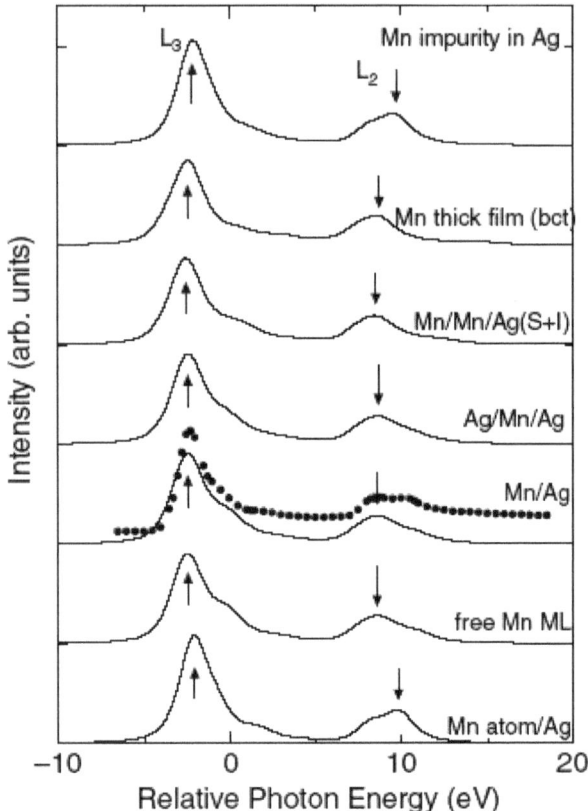

transfer effect is suppressed due to the screening of the core hole potential by 3d electrons.

Finally, Taguchi et al. [113] considered the resonant X-ray emission spectra with L_3 and L_2 excitation peaks. For simplicity, they neglected polarization effect of incident photons. The calculated results at L_3 and L_2 excitations are shown in the left and right panels of Fig. 9, respectively. At L_3 excitation, the elastic peak has a strong intensity for all systems and the spectral shape does not change very much. This is for the same reason as that for the case of 2p XAS (i.e. the ordering of the states does not change). For the RXES spectra with L_2 excitation (about 12 eV above L_3 edge), the intensity of the inelastic peak increases and is larger than that of the elastic peaks for all systems. Moreover, while the 2p XAS spectra at L_2 region does not change very much, Fig. 9 clearly shows that the RXES spectral shapes with L_2 excitation strongly depend on the system under consideration. Also, the calculated RXES spectra present satellite structures which are very sensitive to the hybridization strength.

Later on and still using an impurity model which includes full multiplet interaction and coupling to the Mn 3d and Ag 4d bands, Taguchi et al. [117] reported a

Fig. 9 Calculated
$2p \rightarrow 3d \rightarrow 2p$ resonant X-ray
emission at the photon
energies marked by *vertical
arrows* in the L_{23} spectrum of
Fig. 8. *Left part*: calculated
RXES results for L_3
excitation. *Right part*:
calculated RXES results for
L_2 excitation (from [113])

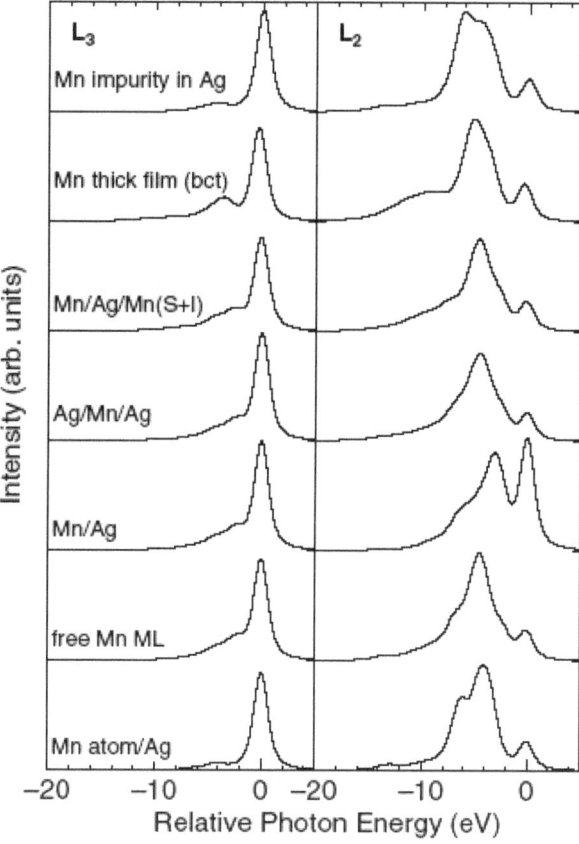

RXES study in the whole energy range of Mn L_{23} white lines for three prototypical
Mn/Ag(001) systems: (i) a Mn impurity in Ag, (ii) an adsorbed Mn monolayer on
Ag and (iii) a thick Mn film. The considered model allowed to investigate the interplay between on-site dd excitations and charge transfer screening from neighbouring
Mn and Ag atoms. For illustration, we plot here the calculated RXES spectra for
a Mn monolayer on Ag (Fig. 10). The most simple interpretation of these spectra
would be to superimpose charge transfer satellites originating from 3d 6 (Mn^+) onto
the impurity spectra with some weighting factor. This approach neglects, however,
configuration mixing in the final state, which strongly modifies the spectral shape,
especially the relative peak intensities. From a general point of view, the calculated RXES spectra depend strongly on the excitation energy. At L_3 excitation,
the spectra of all three systems are dominated by the elastic peak. For excitation
energies around L_2, and between L_3 and L_2, however, most of the spectral weight
comes from inelastic X-ray scattering. The line shape of these inelastic satellite
structures changes considerably between the three considered Mn/Ag systems, a
fact that may be attributed to changes in the bonding nature of the Mn d orbitals.

Fig. 10 (**a**) Absorption spectrum in the Mn L23 region with indication of the excitation energies used in the RXES calculation. X stands for Mn^+ and Ag. All RXES spectra have been rescaled to the same amplitude. (**b**) Calculated RXES spectra for a Mn monolayer on Ag (from [117])

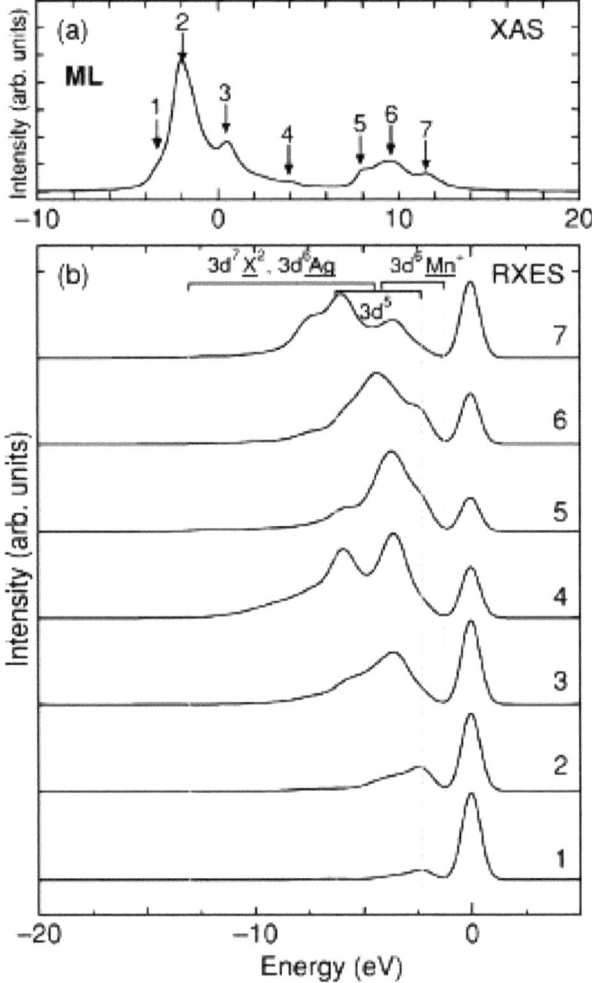

The system dependence of the RXES spectrum is thus found to be much stronger than that of the corresponding absorption spectrum. Taguchi et al.'s results [117] suggest that RXES in the Mn L_{23} region may be used as a sensitive probe of the local environment of Mn atoms.

5.6 Outlook

Nanomagnets frequently lack inversion symmetry because of the presence of surfaces so that, owing to the presence of spin–orbit interaction which connects the lattice with spin symmetry, this broken parity of the lattice gives rise to an additional interaction. This is the Dzyaloshinskii–Moriya interaction (DMI) described by Dzyaloshinskii [118] and Moriya [119]. Bode et al. [120] reported the occurrence of homo-chiral spin structures in a single atomic layer of Mn on W(110) substrate arising from the DMI and leading to a left rotating spin cycloid. More precisely chirality in nanoscale magnets may play a crucial role in spintronic devices. For instance, a spin-polarized current flowing through chiral magnetic structures will exert a spin-torque on the magnetic structure causing a variety of excitations or manipulations of the magnetization and giving rise to microwave emission, magnetization switching or magnetic motors [120].

6 Concluding Remarks

In this short lecture note we reported recent results concerning the electronic, magnetic and spectroscopic properties of V, Cr and Mn nanostructures. A great majority of present theoretical results were obtained at $T = 0\,\mathrm{K}$ and in static conditions. However, life is not so simple because most of the experimental results were obtained at non-zero temperature and spin dynamics cannot be excluded [39]. If the temperature is raised, the magnetic moment will decrease due to additional thermal disorder among the spins [121]. The ultimate limits of such dynamic process are achieved if the perturbation of the system is rapid enough to drive its internal degrees of freedom (spins as well as electronic and nuclear motion) out of equilibrium. Exploring these limits not only provides insights in elementary mechanisms governing spin dynamics but also contributes to development of novel ultrafast switching strategies for future magnetic memory devices [122].

In fact, the rapidly increasing information density required from magnetic data storage devices raises the question of fundamental limits in bit size and writing speed. Stamm et al. [123] used X-ray magnetic circular dichroism in ultrafast mode to obtain insight into spin relaxation in nickel on a femtosecond timescale, opening new horizons for research into spin dynamics with the highest resolution. Femtosecond laser pulses are the key to exploring non-equilibrium magnetism. Exciting a ferromagnetic thin film with such a pulse heats up the conduction electrons almost instantaneously, but it takes a finite time for the nuclear lattice and spin degree of freedom to adapt [122]. Since the pioneering work by Beaurepaire et al. [124] the underlying microscopic mechanisms have been heavily discussed. The debate led to the realization that XMCD, when performed with femtosecond X-ray pulses, provides a direct and well-defined view on spin dynamics [123].

Calculations of the magnetic properties of thin films were investigated by Jensen and Bennemann [121] as functions of temperature and atomic morphology. Special attention was paid to determine the influence of collective magnetic excitations and non-collinear magnetic structures at finite temperatures. Those points were studied within a Heisenberg model by application of a mean-field approximation as well as by a many-body Green's function theory. Also Buruzs et al. [125] performed ab initio calculations of the temperature-dependent magnetic anisotropy energy (MAE) of magnetic surfaces, interfaces or films by using an extension of the relativistic disordered local moments (RDLM) scheme to layered systems. The RDLM method was implemented within the relativistic screened KKR code for layered systems and applied to the study of temperature dependence of MAE of ferromagnetic Co films on Cu(001). Their results are in overall agreement with experiment, as far as it is found that the magnetization is oriented parallel to the surface for almost all temperatures below the Curie temperature, except for the two-monolayer system.

Let us conclude with the results displayed by Sipr et al. [126] about the influence of temperature on the systematics of magnetic moments of free Fe clusters. The focus of these authors is on free spherical Fe clusters with less than 100 atoms. As usual the interaction among individual magnetic moments is described by a classical Heisenberg Hamiltonian, with exchange coupling constants provided by spin-polarized relativistic KKR calculations [127]. The average magnetic moment for a given temperature is evaluated by a Monte Carlo method (see [128]). It is found that up to room temperature this average magnetic moment does not depend significantly on T. This implies that ground-state calculation should be able to reproduce the currently available experimental data. Thus the disagreement of current calculation with experiment should not be ascribed to the influence of finite temperature. In fact these discrepancies may come from the fact that experimental values have not been obtained directly, but inferred from measured deflections in a Stern–Gerlach magnet [126]. The magnetic moments can be derived from the experiment only after several assumptions about super-paramagnetic behaviour, single domain magnetization and spin relaxation.

Acknowledgments The authors are deeply indebted to Carlo Massobrio for encouraging them to contribute to LNP. Also, they are very much grateful to many collaborators on topics directly related to the present subject. Among them are A. Kotani, M. Taguchi, P. Krüger, N. Yartseva, V. Uzdin, H. Zenia, S. Bouarab, T.A. Carillo-Cazares, S. Meza-Aguilar, B.R. Malonda-Boungou, B. M'Passi-Mabiala, S. Pick, A. Vega and S. Lounis. At last the authors would like to thank Hervé Bulou for his valuable technical help in preparing this chapter.

References

1. J. Kondo, Prog. Theor. Phys. **32**, 37 (1964).
2. P. Curie, CRAS **115**, 805 (1892).
3. P. Langevin, CRAS **139**, 1204 (1904).
4. P. Weiss, CRAS **143**, 1136 (1906).
5. P. Weiss, CRAS **152**, 187 (1911).
6. P. A. M. Dirac, Proc. Roy. Soc. A **117**, 610 (1928).
7. T. Kasuya, Prog. Theor. Phys. **16**, 45 (1956).
8. K. Yoshida, Phys. Rev. **106**, 893 (1957).
9. M. A. Ruderman and C. Kittel, Phys. Rev. **96**, 99 (1954).
10. J. Hubbard, Proc. Roy. Soc. A **276**, 238 (1963).
11. R. Haydock, *Solid State Physics*, vol. 35, (Academic Press, New York, 1980).
12. C. Lanczos, J. Rest. Natl. Bur. Stand. **45**, 255 (1950).
13. P. Hohenberg and W. Kohn, Phys. Rev. 136, B **864** (1964).

14. W. Kohn and L. Sham, Phys. Rev. **140**, A113 (1965).
15. A. Gross, *Theoretical Surface Science* (Springer-Verlag, 2003).
16. S. Frota-Pessoa, Phys. Rev. B **46**, 14750 (1992).
17. A. B. Klautau and S. Frota-Pessoa, Surf. Sci. **579**, 27 (2005).
18. K. Wildberger, R. Zeller, and P. H. Dederichs, Phys. Rev. B **55**, 10074 (1997).
19. J. C. Parlebas, K. Asakura, A. Fujiwara, I. Harada, and A. Kotani, Phy. Rep. **431**, 1 (2006).
20. H. Dreysse and C. Demangeat, Surf. Sci. Rep. **28**, 65 (1997).
21. J. Bansmann, S. H. Baker, C. Binns, J. A. Blackman, J.-P. Bucher, J. Dorantes-Davila, V. Dupuis, L. Favre, D. Kechrakos, A. Kleibert, K.-H. Meiwes-Broer, G. M. Pastor, A. Perez, O. Toulemonde, K. N. Trohidou, J. Tuaillon and Y. Xie, Surf. Sci. Rep. **56**, 189 (2005).
22. J. Honolka, K. Kuhnke, L. Vitali, A. Enders, K. Kern, S. Gardonio, C. Carbone, S. R. Krishnakumar, P. Bencok, S. Stepanow, and P. Gambardella, Phys. Rev. B **76**, 144412 (2007).
23. A. De Siervo, E. De Biasi, F. Garcia, R. Landers, M. D. Martins, and W. A. A. Macedo, Phys. Rev. B **76**, 075432 (2007).
24. J. Osorio-Guillen, S. Lany, S. V. Barabash, and A. Zunger, Phys. Rev. B **75**, 184421 (2007).
25. P. Sati, C. Deparis, C. Morhain, S. Schäfer, and A. Stepanov, Phys. Rev. Lett. **98**, 137204 (2007).
26. T. Dietl, J. Phys.: Condens. Matter **19**, 165204 (2007).
27. T. C. Schulthess, W. M. Temmerman, Z. Szotek, A. Svane, and L. Petit, J. Phys.: Condens. Matter **19**, 165207 (2007).
28. S. Blügel, B. Drittler, R. Zeller, and P. H. Dederichs, Appl. Phys. Lett. A **49**, 547 (1989).
29. C. Binns, S. H. Baker, C. Demangeat, and J. C. Parlebas, Surf. Sci. Rep. **34**, 105 (1999).
30. A. Vega, J. C. Parlebas, and C. Demangeat, *Handbook of Magnetic Materials*, ed. K. H. J. Buschov vol. 15, p. 199 (Elsevier, Amsterdam, 2003).
31. R. Wiesendanger, Europhys. News **38/2**, 16 (2007) and references therein.
32. T. Khmelevska, S. Khmelevskyi, A. V. Ruban, and P. Mohn, Phys. Rev. B **76**, 054445 (2007).
33. S. Uzdin, V. Uzdin, and C. Demangeat, Europhys. Lett. **47**, 556 (1999a).
34. Y. B. Kudasov and V. Uzdin, Phys. Rev. Lett. **89**, 276802 (2002).
35. J. Hafner and D. Spisak, Phys. Rev. B **72**, 144420 (2005).
36. C. Demangeat and J. C. Parlebas, Rep. Prog. Phys. **65**, 1679 (2002).
37. K. Fukamichi, R. Y. Umetsu, A. Sakuma and C. Mitsumata, *Handbook of Magnetic Materials*, vol. 16, ed. K. H. J. Buschov (Elsevier, Amsterdam, 2000) p. 209
38. S. Ostanin, A. Ernst, L. M. Sandratskii, P. Bruno, M. Däne, I. D. Hughes, J. B. Staunton, W. Hergert, I. Mertig, and J. Kudrnovsky, Phys. Rev. Lett. **98**, 016101 (2007).
39. B. Hillebrands and K. Ounadjela, *Spin Dynamics in Confined Magnetic Structures I and II* (Springer, Berlin Hiedelberg 2002).
40. S. Lounis, Ph. Mavropoulos, P. H. Dederichs, and S. Blügel, Phys. Rev. B **72**, 224437 (2005) and references therein.
41. Y. Mokrousov, G. Bihlmayer, S. Blügel, and S. Heinze, Phys. Rev. B **75**, 104413 (2007).
42. T. A. Carrillo-Cazares, S. Meza-Aguilar, and C. Demangeat, Surf. Sci. **601**, 1763 (2007a).
43. S. Lounis, Ph. Mavropoulos, R. Zeller, P. H. Dederichs, and S. Blügel, Phys. Rev. B **75**, 174436 (2007).
44. J. Izquierdo and C. Demangeat, Phys. Rev. B **62**, 12287 (2000).
45. J. C. Tung and G. Y. Guo, Phys. Rev. B **76**, 094413 (2007).
46. D. Lacina, J. Yang, and J. L. Erskine, Phys. Rev. B **75**, 195423 (2007).
47. Y. Huttel, G. van der Laan, C. M. Teodorescu, P. Bencok, and S. S. Dhesi, Phys. Rev. B **67**, 052408 (2003a).
48. B. A. Hamad, Surf. Sci. **601**, 4944 (2007).
49. H. Fritzsche, Y. T. Liu, J. Hauschild, and H. Maletta, Phys. Rev. B **70**, 214406 (2004).
50. A. Remhof, G. Nowak, H. Zabel, M. Bjorck, M. Parnaste, B. Hjörvarsson, and V. Uzdin, Europhys. Lett. **79**, 37003 (2007).

51. J. F. Calleja, Y. Huttel, M. C. Contreras, E. Navarro, B. Presa, R. Mararranz, and A. Cebol-lada, J. Appl. Phys. **100**, 053917 (2006).
52. L. T. Baczewski, P. Pankowski, A. Wawro, K. Mergia, S. Messolaras, and F. Ott, Phys. Rev. B **74**, 075417 (2006)
53. A. Bergman, L. Nordström, A. B. Klautau, S. Frota-Pessoa, and O. Eriksson, Surf. Sci. **600**, 4838 (2006a).
54. T. A. Carrillo-Cazares, S. Meza-Aguilar, and C. Demangeat, Eur. Phys. J. B **48**, 249 (2005).
55. A. Bergman, L. Nordström, A. B. Klautau, S. Frota-Pessoa, and O. Eriksson, Phys. Rev. B **73**, 174434 (2006b).
56. T. A. Carrillo-Cazares, S. Meza-Aguilar, and C. Demangeat, Sol. St. Comm., **144**, 94 (2007b).
57. J. Hong, J. Magn. Magn. Mat. **303**, 191 (2006).
58. S. Baroni, A. Dal Corso, S. de Girancoli, and P. Gianozzi, http://www.pwscf.org (2007).
59. Y. Huttel, G. van der Laan, T. K. Johal, N. D. Telling, and P. Bencok, Phys. Rev. B **68**, 174405 (2003b).
60. A. W. Overhauser, Phys. Rev. **128**, 1437 (1962).
61. W. M. Lomer, Proc. Phys. Soc. **80**, 489 (1962).
62. R. Hafner, D. Spisak, R. Lorenz, and J. Hafner, Phys. Rev. B **65**, 184432 (2002).
63. V. Uzdin and C. Demangeat, J. Phys.: Condens. Matter **18**, 2717 (2006).
64. J. C. Parlebas, Phys. Stat. Sol. B **160**, 11 (1990).
65. V. Vanhoof, M. Rots, and S. Cottenier, Phys. Rev. B **80**, 184420 (2009).
66. E. Kravtsov, R. Brucas, B. Hjorvarsson, A. Hoser, A. Liebig, G. J. McIntyre, M. A. Milyaev, A. Nefedov, L. Paolasini, F. Radu, A. Remhof, V. V. Ustinov F. Yakhou, and H. Zabel, Phys. Rev. B **76**, 024421 (2007).
67. F. W. Payne, W. Jiang, and L. A. Bloomfield, Phys. Rev. Lett. **97**, 193401 (2006).
68. P. Ohresser, H. Bulou, S. S. Dhesi, C. Boeglin, B. Lazarovits, E. Gaudry, I. Chado, J. Faerber, and F. Scheurer, Phys. Rev. Lett. **95**, 195901 (2005).
69. Y. Yayon, V. W. Brar, L. Senapati, S. C. Erwin, and M. F. Crommie, Phys. Rev. Lett., **99**, 067202 (2007).
70. M. N. Baibich, J. M. Brote, A. Fert, Nguyen Van Dau, F. Petroff, P. Etienne, G. Creuzet, A. Friederich and J. Chazelas, Phys. Rev. Lett., **61**, 2472 (1988).
71. R. Coehoorn, *Handbook of Magnetic Materials*, vol. 15, ed. K. H. J. Buschow, (Elsevier, Amsterdam, 2003) p. 1.
72. Y. S. Dedkov, Eur. Phys. J. B **57**, 15 (2007).
73. S. Sharma, J. K. Dewhurst, C. Ambrosch-Draxl, S. Kurth, N. Helbig, S. Pittalis, S. Shallcross, L. Nordström, and E. K. U. Gross, Phys. Rev. Lett. **98**, 196405 (2007).
74. J. E. Peralta, G. E. Scuseria, and M. J. Frisch, Phys. Rev. B **75**, 125119 (2007).
75. P. Bodeker, A. Schreyer and H. Zabel, Phys. Rev. B **59**, 9408 (1999).
76. V. Uzdin, A. Mokrani, C. Demangeat, and N. S. Yartseva, J. Magn. Magn. Mat. **198–199**, 469 (1999b).
77. T. Jamneala, V. Madhavan, and M. F. Crommie, Phys. Rev. Lett. **87**, 256804 (2001).
78. V. V. Savkin, A. N. Rubtsov, M. I. Katsnelson, and A. I. Lichtenstein, Phys. Rev. Lett. **94**, 026402 (2005).
79. B. Lazarovits, P. Simon, G. Zarand, and L. Szunyogh, Phys. Rev. Lett. **95**, 077202 (2005).
80. A. A. Aligia, Phys. Rev. Lett., **96**, 096804 (2006).
81. H. J. Gotsis, N. Kioussis, and D. A. Papaconstantopoulos, Phys. Rev. B **73**, 014436 (2006).
82. A. Bergman, L. Nordström, A. B. Klautau, S. Frota-Pessoa, and O. Eriksson, J. Phys.: Condens. Matter **19**, 156226 (2007a).
83. A. Antal, L. Udvardi, B. Ujfalussy, B. Lazarovits, L. Szunyogh, and P. Weinberger, J. Magn. Magn. Mat. **316**, 118 (2007).
84. G. M. Stocks, M. Eisenbach, B. Ujfalussy, B. Lazarovits, L. Szunyogh, and P. Weinberger, Prog. Mat. Sci. **52**, 371 (2007).

85. A. Bergman, L. Nordström, A. B. Klautau, S. Frota-Pessoa, and O. Eriksson, Phys. Rev. B **75**, 224425 (2007b).
86. T. Kawagoe, Y. Iguchi, and S. Suga, J. Magn. Magn. Mat. **310**, 2201 (2007).
87. N. S. Yartseva, S. V. Yartsev, C. Demangeat, V. M. Uzdin, J. C. Parlebas, J. Mol. Struct.: THEOCHEM **777**, 29 (2006).
88. M. B. Knickelbein, Phys. Rev. Lett. **86**, 5255 (2001).
89. J. Mejia-Lopez, A. H. Romero, M. E. Garcia, and J. L. Moran-Lopez, Phys. Rev. B **74**, 140405 (2006).
90. S. J. Lee, J. P. Goff, G. J. McIntyre, R. C. C. Ward, S. Langridge, T. Charlton, R. Dalgliesh, and D. Mannix, Phys. Rev. Lett. **99**, 037204 (2007).
91. C. L. Gao, U. Schlickum, W. Wulfhekel, and J. Kirschner, Phys. Rev. Lett. **98**, 107203 (2007).
92. J. T. Kohlhepp and W. J. M. De Jonge, Phys. Rev. Lett. **96**, 237201 (2006).
93. C. Biswas, R. S. Dhaka, A. K. Shukla, and S. R. Barman, Surf. Sci. **601**, 609 (2007).
94. J. Hong and R. Q. Wu, Phys. Rev. B **73**, 094450 (2006).
95. T. K. Yamada, E. Martinez, A. Vega, R. Robles, D. Stoeffler, A. L. Vazquez de Parga, T. Mizoguchi, and H. van Kempen, Nanotechnology **18**, 235702 (2007).
96. D. Spisak and J. Hafner, Surf. Sci. **601**, 4348 (2007).
97. B. M'Passi-Mabiala, S. Meza-Aguilar, and C. Demangeat, Phys. Rev. B **65**, 012414 (2002).
98. B. C. Choi, J. P. Bode, and J. A. C. Bland, Phys. Rev. B **58**, 5166 (1998).
99. B. R. Malonda-Boungou, B. M'Passi-Mabiala, S. Meza-Aguilar, and C. Demangeat, Surf. Sci. **600**, 1763 (2006).
100. S. Andrieu, E. Foy, H. Fischer, M. Alnot, F. Chevrier, G. Krill, and M. Piecuch, Phys. Rev. B **58**, 8210 (1998).
101. Y. Yonamoto, T. Yokoyama, K. Amemiya, D. Matsumara, and T. Ohta, Phys. Rev. B **63**, 214406 (2001).
102. H. Zenia, S. Bouarab, J. Ferrer, and C. Demangeat, Surf. Sci. **564**, 12 (2004).
103. S. Pick and C. Demangeat, Surf. Sci. **584**, 146 (2005).
104. J. M. Soler, E. Artacho, J. D. Gale, A. Garcia, J. Junquera, P. Ordejon, and D. Sanchez-Portal, J. Phys.: Condens. Matter **14**, 2745 (2002).
105. T. Morisato, S. N. Khanna, and Y. Kawazoe, Phys. Rev. B **72**, 014435 (2005).
106. R. C. Longo, E. G. Noya, and L. J. Gallego, Phys. Rev. B **72**, 174409 (2005).
107. Y. Xie and J. A. Blackman, Phys. Rev. B **73**, 214436 (2006).
108. M. Kabir, D. G. Kanhere and A. Mookerjee, Phys. Rev. B **75**, 214433 (2007).
109. C. F. Hirjibehedin, C. P. Lutz, and A. J. Heinrich, Science **312**, 1021 (2006).
110. M. A. Barral, R. Weht, G. Lozano, and A. M. Llois, Physica B **398**, 369 (2007).
111. A. T. Costa, C. G. Rocha, and M. S. Ferreira, Phys. Rev. B **76**, 085401 (2007).
112. P. Krüger and A. Kotani, Phys. Rev. B **68**, 035407 (2003).
113. M. Taguchi, P. Krüger, J. C. Parlebas, and A. Kotani, J. Phys. Soc. Jpn **73**, 1347 (2004).
114. H. Höchst, P. Steiner, and S. Hüfner, Z. Phys. B **38**, 201 (1980).
115. P. Schieffer, C. Krembel, M. C. Hanf, and G. Gewinner, J. Electron Spectrosc. Relat. Phenom. **104**, 127 (1999a).
116. P. Schieffer, M. H. Tuilier, C. Krembel, M. C. Hanf, G. Gewinner, D. Chandesris, H. Magnan, and K. Hricovini, J. Synchrotron Radiat. **6**, 784 (1999b).
117. M. Taguchi, P. Krüger, J. C. Parlebas, and A. Kotani, Phys. Rev. B **73**, 125404 (2006).
118. I. E. Dzyaloshinskii, Sov. Phys. JETP **5**, 1259 (1957).
119. T. Moriya, Phys. Rev. **120**, 91 (1960).
120. M. Bode, M. Heide, K. von Bergmann, P. Ferriani, S. Heinze, G. Bihlmayer, A. Kunetza, O. Pietzsch, S. Blugel, and R. Wiesendanger, Nature **447**, 190 (2007).
121. P. J. Jensen and K. H. Bennemann, Surf. Sci. Rep. **61**, 129 (2006).
122. B. Koopmans, Nat. Mater. **6**, 715 (2007).
123. C. Stamm, T. Kachel, N. Pontius, R. Mitzner, T. Quast, K. Holldack, S. Khan, C. Lupulescu, E. F. Aziz, M. Wietstruk, H. A. Durr, and W. Eberhardt, Nat. Mat. **6**, 740 (2007).
124. E. Beaurepaire, J.-C. Merle, A. Daunois, and J.-Y. Bigot, Phys. Rev. Lett. **76**, 4250 (1996).

125. A. Buruzs, P. Weinberger, L. Szunyogh, L. Udvardi, P. I. Chleboun, A. M. Fischer, and J. B. Staunton, Phys. Rev. B **76**, 064417 (2007).
126. O. Sipr, S. Polesya, J. Minar, and H. Ebert, J. Phys: Condens. Matter **19**, 446205 (2007).
127. H. Ebert, The Munich SPR-KKR Package, Version 2.1 http://olymp.cup.uni-muenchen.de/ak/ebert/SPRKKR (2002).
128. K. Binder, Rep. Prog. Phys. **60**, 487 (1997).

Electronic Structure and Magnetism of Double Perovskite Systems

D. Stoeffler

Abstract Double perovskite systems (like Sr_2XMoO_6 (X=Fe, Co)) presenting an half-metallic behavior have been recently extensively experimentally and theoretically studied in relation with their potential use as magnetic electrode in spintronic devices. The half-metallic property has first been theoretically predicted for Sr_2FeMoO_6 with the LSDA but this method does not provide satisfactory results when Fe antisites are considered. This is related to the high sensitivity of the gap into the majority band, being at the origin of the half-metallic property, to structural variations or to imperfections. The aim of the present work is to investigate how the enhancement of this gap affects the electronic structure and the magnetic properties of such kind of double perovskite system.

Using the full potential linearized augmented plane wave ab initio method, we investigate first these bulk double perovskites by comparing the results obtained with the GGA and GGA+U methods in order to discuss their magnetic configuration in relation with the experiments. We show that both methods lead to significantly different results and that a good agreement with experimental results – antiferromagnetic insulator for X=Co – can be obtained only when the GGA+U method is used. For X=Fe, we exhibit the role played by oxygen vacancies on the stabilization of a negative magnetic moment on the Fe antisite with preserved half-metallicity. We show that such a negative moment can be obtained only when an oxygen vacancy occurs in the direct neighborhood of the Fe antisite with the GGA+U method.

We investigate also the electronic structure of $Sr_2FeMoO_6/SrTiO_3$ (SFMO/ STO) multilayers. We examine more especially the role of the interface on the magnetic and transport properties of these multilayers taking into account a possible Fe deficiency at the interface. We show that bulk behavior is rapidly recovered due to the strong localization of the interfacial perturbation. For perfect interfaces, the whole structure is found half-metallic within the GGA+U method; the situation being ambiguous within the GGA method where SFMO is at the limit of being half-metallic depending on the structural deformation induced by the STO layer. This

D. Stoeffler (✉)
Institut de Physique et de Chimie des Matériaux de Strasbourg (UMR 7504 CNRS -ULP)
23 rue du Loess, BP 43, 67034 Strasbourg Cedex 2, France,
Daniel.Stoeffler@ipcms.u-strasbg.fr

Stoeffler, D.: *Electronic Structure and Magnetism of Double Perovskite Systems*. Lect. Notes
Phys. **795**, 197–226 (2010)
DOI 10.1007/978-3-642-04650-6_7 ⓒ Springer-Verlag Berlin Heidelberg 2010

leads us to the conclusion that such a system could be used as injection electrode and tunnel barrier in magnetic tunnel junctions with a fully spin polarized injected current. For Fe-deficient interfaces, we show that the interfacial densities of states are nearly unpolarized suggesting that this kind of imperfection has potentially a strong impact on the properties of the multilayers.

1 Introduction

The observation at room temperature of a large magnetoresistance in magnetic tunnel junctions [1] gave rise to an increase of the interest for these systems mainly due to their potential applications such as read heads, recording media, magnetic memories, or field sensors [2]. In most of these junctions, the magnetic electrodes are based on CoFe alloys but their spin polarization does not exceed 50% and leads therefore to a limited magnetoresistive signal. Many works focused on the increase of the sensitivity of these systems by investigating alternative materials that could be used as tunnel barrier [3] or magnetic electrode [4]. Another direction of investigation consists in replacing the metallic electrodes by magnetic oxide ones having a high magnetic polarization. Sr_2FeMoO_6 (SFMO) double perovskites are expected to be good candidates as magnetic electrode materials for such applications because they present a half-metallic character (a 100% theoretical polarization) and a high transition temperature $T_C = 415K$. This suggests a potential large polarization of the conduction electrons at room temperature [5], in contrast with earlier studied manganite compounds for which T_C is generally smaller than 350 K [6].

In the perfect crystalline structure of SFMO, Fe and Mo atoms are arranged on two equivalent body-centered tetragonal sublattices (within the $I4/mmm$ space group with the lattice parameters $a = 5.58\text{Å}$ and $c = 7.90\,\text{Å}$, found experimentally by X-ray diffraction [7]) and are connected through oxygen octahedra. Assuming double exchange mechanism for the interaction, the $Fe^{3+}(S = 5/2)$ and $Mo^{5+}(S = 1/2)$ magnetic moments are antiferromagnetically coupled leading to a theoretical magnetization value of 4 μ_B per formula unit. In this compound Kobayashi et al. [5] succeeded to measure a reasonable high magnetoresistance of 42% at low temperature (5 K), originating from the electron transport through oxygen-rich grain boundaries, and a saturation magnetization of 3 μ_B per formula unit. Such a low saturation magnetization was found as well by almost all authors and was ascribed mainly to the disorder between Fe and Mo cations. Indeed, due to the supposed antiferromagnetic interaction of double exchange type between the Fe and Mo sublattices of perfect SFMO, the magnetic moment of Fe antisites, i.e., for Fe cations situated on Mo (Fe) sites should find itself in an antiparallel configuration with the magnetic moment of Fe cations on the regular sites, thus reducing the saturation magnetization [8]. This model of magnetization reduction is supported by Monte Carlo calculations which indicate an antiparallel alignment between two Fe cations separated by an oxygen atom [9]. More recently, the magnetization reduction has been associated with the reduction of the magnetic moment of Fe and Mo when the environment of these atoms is altered. This model is in agreement with ab initio Linear Muffin-Tin Orbital with Atomic Sphere Approximation (LMTO-ASA) calculations which indicate that the Fe—O—Fe moments have a parallel orientation which cannot reduce the total magnetic moment [10, 11]. Nevertheless, this magnetization reduction was never calculated for both types of imperfections (oxygen vacancies and antisite defect), and the band structure was reported only for SFMO with a perfect structure or with antisite defects [10–12]. In this last case the half-metallic character of the compound was reported to disappear. This leads to the conclusion that antisite defects have to be avoided in order to preserve the half-metallic properties and to recover the high-saturation magnetization.

Recently, Sr_2CoMoO_6 (SCMO) double perovskites, the Co analogous to SFMO, have been synthesized using a soft chemistry method and have been characterized by neutron powder diffraction [13]. At room temperature, the crystal structure is found tetragonal (space group $I4/m$) with $a = 5.565$ Å and $c = 7.948$ Å and contains alternating CoO_6 and MoO_6 octahedra tilted by $7°$ in the basal ab plane (Fig. 1). The stoichiometric samples are found antiferromagnetic ($T_N = 37$ K) and insulating. The reduction of these samples, leading to oxygen-deficient perovskites with the same crystal structure, gives rise to ferromagnetic domains ($T_C = 350$–370 K) and to a dramatic increase of the conductivity related to a large component of itinerancy for down-spin Mo t_{2g} electrons. We will see that a good qualitative agreement between calculated and measured properties requires to introduce the Generalized Gradient Approximation with a Hubbard-like contribution (GGA+U) method. The comparison of calculated features obtained with the GGA and the GGA+U methods will then be one of our goals.

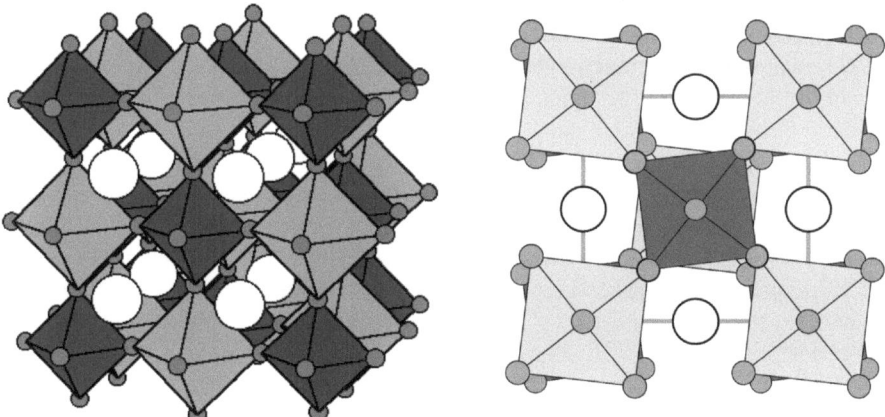

Fig. 1 Schematic representations of the Sr_2CoMoO_6 unit cell displaying tilted octahedra : (*left*) 3D view of the cell containing four formula units and (*right*) *top* view of the cell containing two formula units. The *white* (respectively gray) spheres correspond to Sr (respectively 0) atoms and the *light* (respectively dark) *gray* octahedra are *centered* on Co (respectively Mo) atoms

The present study of the SFMO system analyzes the importance of the oxygen vacancies and antisite defects perturbation from the point of view of the magnetization reduction and of the half-metallic character. We show clearly that the half-metallic character is preserved for structures containing nearly isolated oxygen vacancies, while it vanishes when antisite defects are present. Most interesting, the magnetization reduction is four times larger per oxygen vacancies ($2 \mu_B$/vacancy) than per antisite defect, in contrast with the common suggestion that a close saturation magnetization to $4 \mu_B$ per formula unit is needed in order to preserve good transport properties of SFMO. We examine also the stability of solutions with an anti-aligned magnetic moment on the Fe antisite and show that the solutions with all Fe magnetic moments parallel are always the most stable when a single antisite is considered. We exhibit the role played by oxygen vacancies on the stabilization of a negative magnetic moment on the Fe antisite with preserved half-metallicity. We show that such a negative moment can be obtained only when an oxygen vacancy occurs in the direct neighborhood of the Fe antisite with the GGA+U method.

Epitaxial SFMO thin films have been deposited on $SrTiO_3$ (STO) substrates and it has been shown that they present structural and magnetic properties similar to the bulk material [7, 14–18] confirming its potential. In order to investigate the polarized current outcoming from such SFMO electrodes, magnetic tunnel junctions using an ultrathin STO film as tunnel barrier between SFMO and Co thin layers have been grown [19]. In SFMO/STO/Co junctions, a clear positive magnetoresistive signal of 50% is obtained at low temperature yielding a negative spin polarization value of 85% for SFMO. This result shows that the SFMO/STO interface preserves a high polarization of the supposed half-metallic SFMO electrode. However, it is not clear if there is a lowering of the spin polarization due to the interface with the STO layer or if bulk SFMO has intrinsically this 85% spin polarization.

Experimentally, several studies were focused on $Sr_2FeMoO_6/SrTiO_3$ (SFMO/STO) magnetic tunnel junction but the expected magnetoresistance was not obtained even for well-prepared samples [20]. Most of these studies were unable to explain clearly this disappointing result. A depolarization of the current at the interface was finally supposed to be the most reasonable explanation. Recently, by combining X-ray magnetic circular dichroism and X-ray photoemission spectroscopy, it has been shown that the surface magnetic moment of SFMO is anomalously weak and is consistent with a lack of Fe at the surface [20]. The absence of magnetoresistance is consequently ascribed to Fe-deficient surfaces and interfaces of the SFMO layer where Fe atoms are replaced by Mo atoms and the polarization is strongly reduced over a significant SFMO thickness.

The present work also investigates the electronic structure and the magnetic properties of SFMO/STO superlattices taking into account such interfacial Fe deficiency. The aim is to examine the occurrence of such an interfacial spin polarization lowering using a first-principle method for determining the electronic structure taking electronic correlation into account in the density functionnal theory within the GGA+U approach (GGA including the semi-empirical Hubbard contribution).

2 Methodology

The band structure for the considered systems is calculated with the full potential augmented plane wave formalism (FLAPW) in the FLEUR implementation [21] taking core, semi-core, and valence states into account. This method is largely described in the literature and has been chosen for the present study because hyperfine fields can be obtained in correlation with nuclear magnetic resonance and Moessbauer measurements [7].

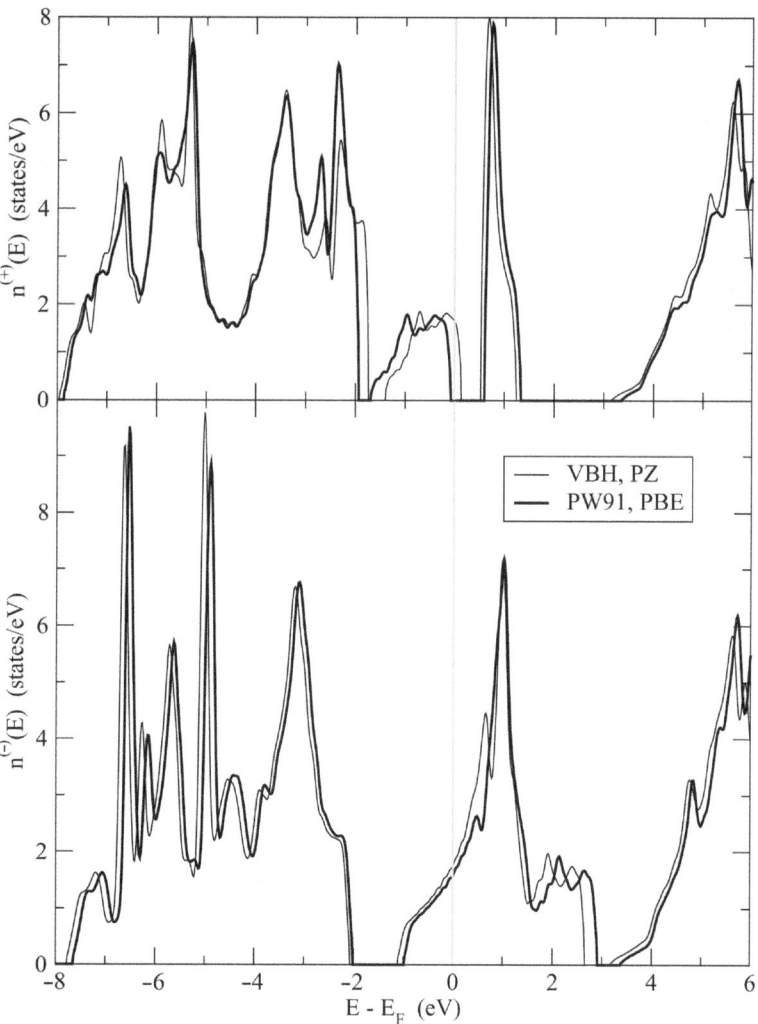

Fig. 2 Densities of states (DOS) for Sr_2FeMoO_6 in the experimental unit cell obtained with various expressions for the exchange correlation contribution taking (*thick line*) or not (*thin line*) gradient corrections into account. The vertical *gray line* corresponds to the Fermi level (E_F) and $n^{(+)}$ ($n^{(-)}$) corresponds the up-spin (respectively, down-spin) DOS

Using the local density approximation (LDA) with the Van Barth–Hedin (VBH) [22] or the Perdrew–Zhang (PZ) [23] expression for the exchange and correlation term, SFMO is not found half metallic for the bulk $I4/mmm$ experimental structure (Fig. 2), the band gap ranging from 0.13 to 0.52 eV. The overestimation of the metallic character by LDA approaches is a well-known problem which can be solved nicely by taking exciton states into account [24] or by correcting the electron–electron interaction terms in a semi-empirical way like in the LDA+U approach, [25, 26]. Within the GGA, the Fermi level is found to fall into the band gap of the majority spin density of states (DOS) (Fig. 2) whatever the Perdrew–Whang (PW91) [27] or the Perdrew–Burke–Ernzerhof (PBE) [28] expression is used, the band gap ranging from −0.08 to 0.60 eV. The GGA is consequently more satisfactory because it solves slightly the problem of the overestimation of the metallic character and will be used as reference in all this work.

More generally, in such oxide systems, the band gap is usually found too small with the GGA method. In this work, the GGA+U method is used in order to get a band structure which is expected to be more correct. Even if values of U for Fe and Co are approximately known, it is essential to check that the result is not sensitive to small variations of this parameter. Moreover, as we will discuss later, the GGA+U method will also be used in order to play with the band gaps and even introducing metallic to half metallic or to insulating character transitions.

3 Bulk SCMO and SFMO

3.1 Bulk SCMO

In order to examine the stability of the experimental structure, we should relax the cell parameters (a, c) and the internal degrees of freedom like the z_1 position of the oxygen atom $(O1)$ linking Co and Mo into the c direction and the position (x_2, y_2) of the O atom $(O2)$ in the basal plane (see Table 1). However, with the used method, such a calculation requires too much computer time. Nevertheless, aiming to investigate the role of the tilted octahedra, we have done a partial relaxation varying only a and c into the $I4/mmm$ symmetry (where $z_1 = x_2 = y_2 = 1/4$). The comparison between this relaxed $I4/mmm$ structure and the experimental one will give us strong indications about the role of the tilting (mainly because the experimental structure will finally be the most stable one).

Table 1 Internal atomic position for all inequivalent sites for the two considered structures

$(I4/mmm)/(I4/m)$	site	x	y	z
Sr	$4d$	0/0	0.5/0.5	0.25/0.25
Co	$2a$	0/0	0/0	0/0
Mo	$2b$	0/0	0/0	0.5/0.5
O1	$4e$	0/0	0/0	0.2500/0.2589
O2	$8h$	0.2500/0.2895	0.2500/0.2296	0/0

3.1.1 SCMO in a $I4/mmm$ Structure

We have determined the values of a and c minimizing the total energy by interpolating the values calculated for a set of (a, c) points on a grid. The energy miminum is obtained for $a = 5.61$ Å and $c = 7.93$ Å. It is not surprising to find a slightly larger ab basal plane parameter than the experimental value because the octahedra tilting affects more especially in-plane Co–O–Mo bonds and leads to a reduction of the Co–Mo distance.

The densities of states (DOS) for the ferromagnetic (FM) solution within GGA, displayed in Fig. 3, are very similar to the ones obtained for SFMO [29] : the spin-up DOS exhibits a gap (ranging from –0.1 to 0.7 eV) around the Fermi level (E_F) separating mainly Co occupied states from mainly Mo t_{2g} unoccupied states, whereas the down-spin DOS shows a metallic behavior. As a consequence of this half-metallic character, the total moment is equal to 3 μ_B per formula unit (f.u.). For the antiferromagnetic (AFM) solution, built by considering a unit cell containing 2 f.u. with opposite magnetic moments on the two Co sites (the up- and down-spin DOS are identical), we found finite DOS at E_F indicating metallic behavior. However, the FM solution is found to be the most stable by 117 meV per 2 f.u. cell. Consequently, the SCMO is found half metallic and ferromagnetically ordered for this $I4/mmm$ structure with the GGA method.

As we will see later, the energy difference between antiferromagnetic (AFM) and ferromagnetic (FM) solutions ($\Delta E_{AFM-FM} = E_{AFM} - E_{FM}$) is found small (a few meV/cell) when the GGA+U method is used; that can make the choice of the Hubbard parameter U and the Hund's-rule exchange parameter J of the Hubbard Hamiltonian very questionable. In the literature, usual values for Co are $U_{Co} = 5$ eV and $J_{Co} = 0.89$ eV which makes U_{Co} particularly large as compared to the resulting total energy difference. In order to investigate the role of these two parameters, we have varied U and J over a reasonable range of values (see Table 2) and found very smooth variations of ΔE_{AFM-FM} (its absolute value increases even when U and J are reduced) without change of sign. This is a clear indication of the reliability of these calculations and we will use $U_{Co} = 5$ eV and $J_{Co} = 0.89$ eV for all other GGA+U calculations in this work [30].

With the GGA+U method and for the FM solution, giving also a total moment of 3 μ_B/f.u., both spin DOS (Fig. 3) present a gap around E_F: the gap into the down-spin DOS coming from the splitting of the Fe and Mo states (the Fe states are shifted toward lower energy, whereas the Mo ones are shifted toward higher energy). This illustrates the major contribution of the GGA+U method: it shifts occupied states to lower energy and unoccupied states to higher energy resulting usually in the occurrence of new gaps or the enlargement of gaps around E_F. A similar result is obtained for the AFM solution presenting a small gap (from –0.11 to 0.11 eV). However, contrary to the results of the GGA method, the AFM solution is found to be more stable by 20 meV/cell than the FM one. Consequently, the SCMO is found insulating and antiferromagnetically ordered for the $I4/mmm$ structure with the GGA+U method, this result being clearly more satisfactory when compared to the experiments.

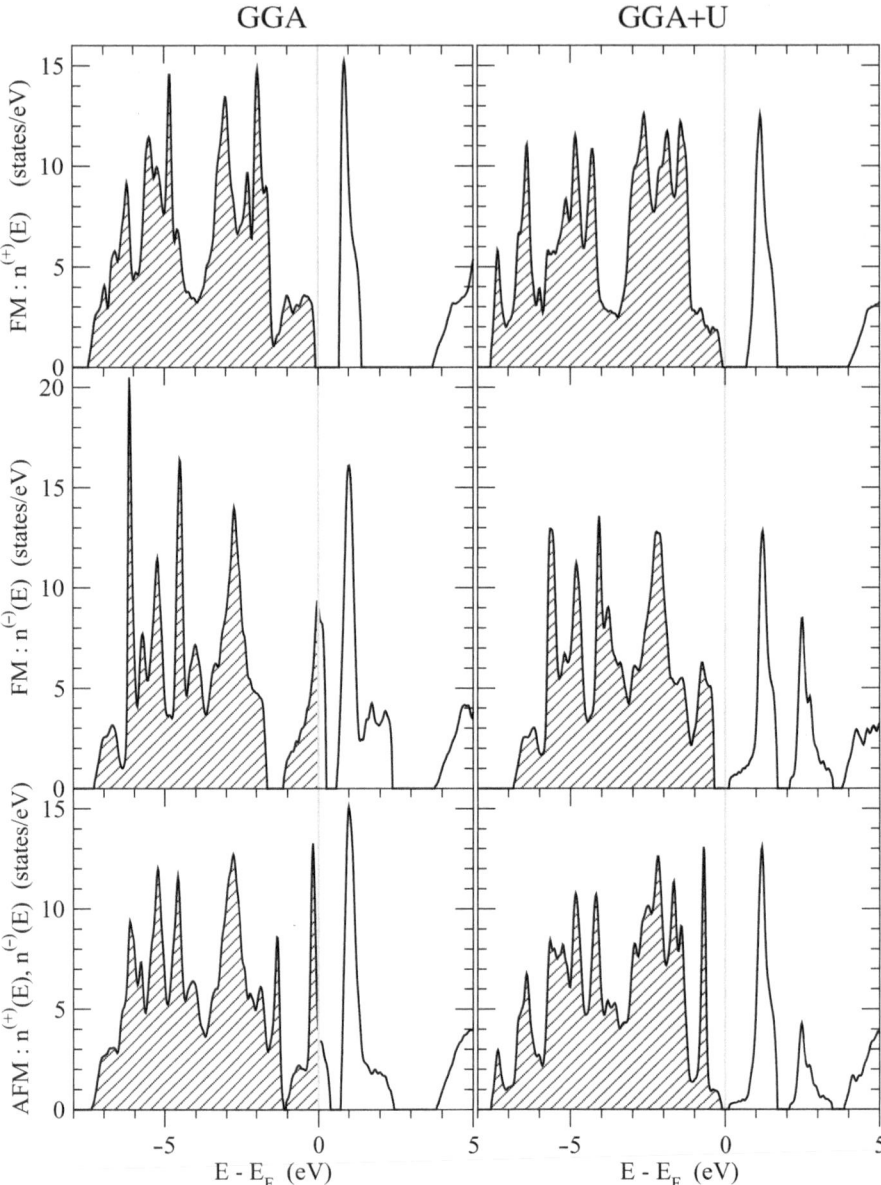

Fig. 3 Densities of states (DOS) for Sr_2CoMoO_6 in the $I4/mmm$ unit cell obtained by minimization of the total energy for ferromagnetic (FM) and antiferromagnetic (AFM) solutions using the GGA (*left*) and the GGA+U (*right*) method. The *vertical gray line* corresponds to the Fermi level (E_F) and $n^{(+)}$ ($n^{(-)}$) corresponds the up-spin (respectively, down-spin) DOS

Table 2 Energy difference between ferromagnetic (FM) and antiferromagnetic (AFM) solutions in meV per cell (a cell corresponds to 2 f.u.) for the two structures considered for various U and J values of the GGA+U method (for comparison, with the GGA method we get $\Delta E_{AFM-FM} =$ 117/142 meV/cell)

ΔE_{AFM-FM} (meV/cell) ($I4/mmm$)/($I4/m$)	$J = 0.78$ eV	$J = 0.89$ eV	$J = 1.00$ eV
$U = 3$ eV	−34/−14	−32/−13	−32/−18
$U = 4$ eV	−28/−10	−26/−8	−21/−3
$U = 5$ eV	−22/−8	−20/−7	−18/−6

3.1.2 Experimental $I4/m$ Structure

This structure introduces only limited modifications relatively to the previously considered $I4/mmm$ one: the neighborhood of each Co or Mo atom remains the same in terms of O coordination and only the volume of the octahedra is slightly altered (the O—Co and O—Mo distances vary by +/−3.6% in the ab plane). Consequently, we do not expect the main structures in the DOS to be affected. However, because the four in-plane ab Co—O—Mo bonds are no more rectilinear, the interaction between Co and Mo states can be significantly changed affecting directly the gap between occupied Co and unoccupied Mo states. This is exactly what we obtain.

With the GGA method (Fig. 4), the DOS are very similar to the previous ones (Fig. 3) and exhibit more narrow structures separated by larger gaps (for example, the gap around E_F in the spin-up DOS for the FM solution ranges from −0.35 to 1.05 eV and is nearly doubled) indicating clearly more tight Co—O—Mo bonds. The AFM solution is found to be also less stable than the FM one by 142 meV/cell. Surprisingly, this FM solution is 457 meV/cell less stable into the $I4/m$ crystal structure than the corresponding FM solution into the $I4/mmm$ crystal structure. This difference can be hardly overcome by internal relaxations. Consequently, the disagreement with the experimental results becomes more pronounced with the GGA method: the $I4/mmm$ crystal structure is the most stable and ferromagnetism is favored to antiferromagnetism.

With the GGA+U method (Fig. 4), an insulating behavior is obtained for both FM and AFM solutions with enlarged gaps as compared to the ones of Fig. 3 and the AFM solution is also the most stable one by 7 meV/cell. The major new result of this calculation is that, for the AFM solution, the $I4/m$ crystal structure is 160 meV/cell more stable than the $I4/mmm$ one (for the FM solution). Even if this result does not prove that the present $I4/m$ structure is the most stable one, this shows clearly that internal relaxations have to be taken into account. Consequently, we found a complete agreement with the experiment using the GGA+U method: the experimental $I4/m$ crystal structure is more stable than all other $I4/mmm$ ones and the SCMO is an antiferromagnetic insulator. In the following, we consider only the case of GGA+U calculations for the discussion.

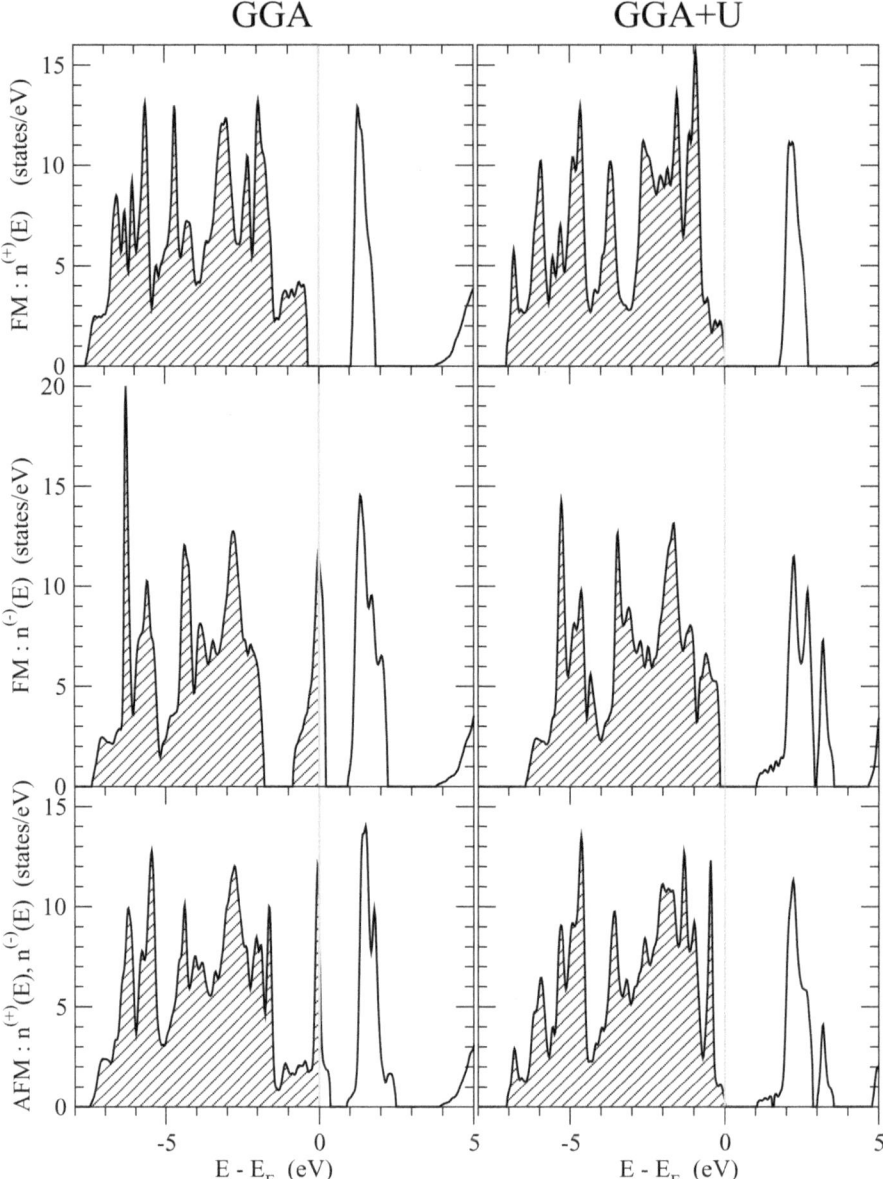

Fig. 4 Densities of states (DOS) for Sr_2CoMoO_6 in the $I4/m$ unit cell for ferromagnetic (FM) and antiferromagnetic (AFM) solutions using the GGA (*left*) and the GGA+U (*right*) method. The *vertical gray line* corresponds to the Fermi level (E_F) and $n^{(+)}$ ($n^{(-)}$) corresponds to the up-spin (respectively down-spin) DOS

3.1.3 Oxygen-Deficient Case

Experimentally, it has been shown that ferromagnetism is recovered and that the resistivity is decreased for the oxygen-deficient samples. Using a superexchange approach as guide, this can be understood by considering that the removal of oxygen atoms adds two electrons to the delocalized charge per O vacancy and that a fraction of this additional electronic charge is gained by the Mo atoms in the vicinity of the vacancy. Consequently, $4d$ electrons become available on these Mo sites carrying a negative local magnetic moment and resulting in a ferromagnetic coupling between Co moments mediated by the Mo one like in SFMO.

It has been shown experimentally that the structure of oxygen-deficient samples is very similar to the stoichiometric ones [13]. Consequently, we assume that an oxygen vacancy (O^*) induces no structural relaxations and our cell is built starting from the same $I4/m$ cell and removing one O atom of a Co—O—Mo link along the c direction corresponding to a lower symmetry space group. We have done the calculation for $Sr_4Co_2Mo_2O_{11}$, corresponding to a much higher O vacancy concentration than the experimental one, and we get the FM solution to be more stable than the AFM one with both GGA ($\Delta E_{FM-AFM} = -95$ meV/cell) and GGA+U ($\Delta E_{FM-AFM} = -76$ meV/cell) methods. As expected, we recover clearly Mo

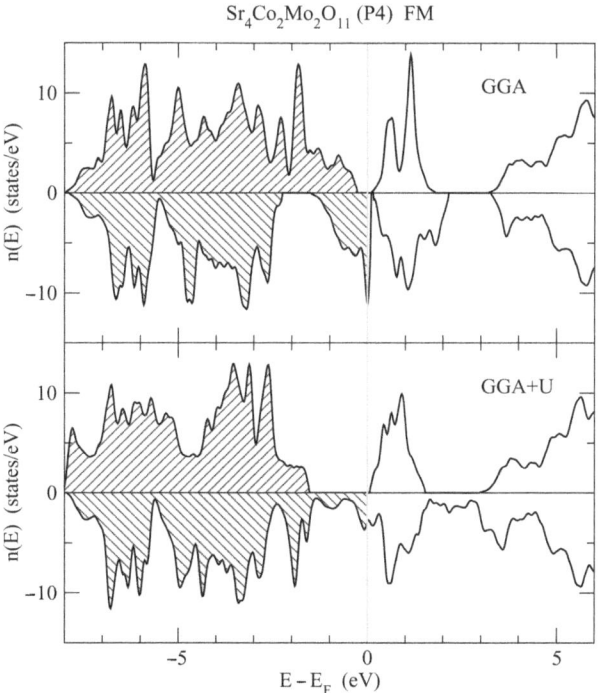

Fig. 5 Densities of states (DOS) for $Sr_4Co_2Mo_2O_{11}$ in the $I4/m$ unit cell with tilted octahedra for the ferromagnetic solution using the GGA and the GGA+U methods. The *vertical gray line* corresponds to the Fermi level (E_F), the down-spin DOS is displayed with negative values

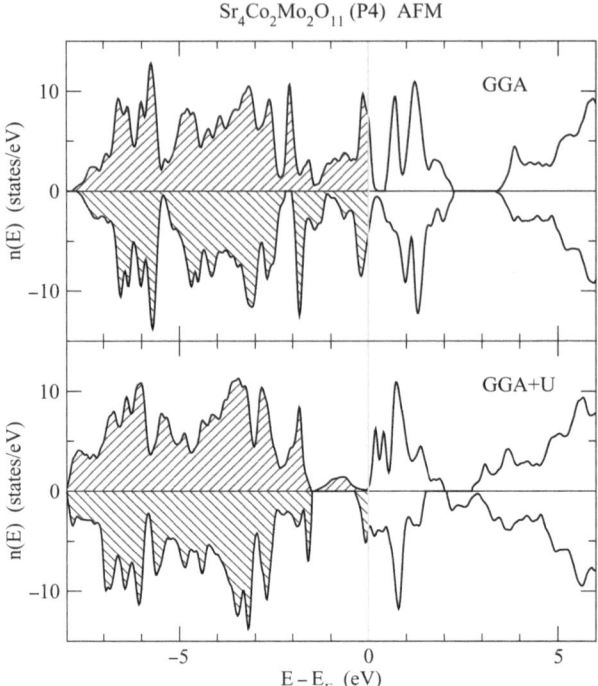

Fig. 6 Densities of states (DOS) for $Sr_4Co_2Mo_2O_{11}$ in the $I4/m$ unit cell with tilted octahedra for the antiferromagnetic solution using the GGA and the GGA+U methods. The *vertical gray line* corresponds to the Fermi level (E_F), the down-spin DOS is displayed with negative values

states around the Fermi level and the magnetic moment on the Mo site having an O vacancy as neighbor is found large ($-0.45\ \mu_B$ for FM and $-0.26\ \mu_B$ for AFM). Finally, the DOS (Fig. 5 and 6) show clearly that the stabilization of a FM solution corresponds to a half-metallic electronic structure in perfect agreement with the observed increase of the conductivity for this situation.

3.2 Anti-aligned Fe Magnetic Moments in Bulk SFMO

From the point of view of the band structure, the main change when using the GGA+U method relative to GGA is a lowering (respectively an increase) of the energy of occupied (unoccupied) states resulting in an enhancement of the main gap in the majority spin band from 0.6 to 1.7 eV. However, for both methods, the Fermi level lies approximately 0.5 eV (0.55 for GGA and 0.48 for GGA+U) below the bottom of the conduction band in reasonable agreement with the measured half-gap of 0.7 eV [31]. This gap enhancement has a direct consequence on the predicted transport properties of epitaxial SFMO films deposited on $SrTiO_3$ (STO) [17]. Indeed, using GGA, a possible transition from metallic to half-metallic SFMO has been predicted when the cell is constrained by epitaxy on STO; SFMO being metallic for c smaller than 7.900 Å $+ 23.614(a - 5.524$ Å$)$. This separation line crosses the experimental $c(a)$ curve for $a = 5.54$ Å and $c = 8.16$ Å corresponding approximately to a deposited SFMO thickness of 38 nm. On the contrary, with GGA+U, SFMO remains half metallic in the range of a and c values explored. Spin-resolved photoemission experiment as a function of the deposited SFMO thickness could consequently allows to prove the reliability of the use of GGA+U for such kind of materials.

For the perfect SFMO case, the considered experimental crystalline structure is tetragonal within the $I4/mmm$ space group, with the lattice parameters $a = 5.58$ Å and $c = 7.90$ Å, found experimentally by X-ray diffraction [7]. For cases with imperfections, the same lattice parameters have been used but, depending on the position of the imperfection site, a lower symmetry space group has been used. Three situations were considered : (i) SFMO with a perfect crystalline structure, (ii) SFMO with a single imperfection (Mo or Fe antisite defect, oxygen vacancy), and (iii) SFMO with two imperfections (Mo or Fe antisite defect and an oxygen vacancy, two oxygen vacancies), in order to detect any variation which could lead to a reduction of the total magnetic moment as the one observed. We assume that no structural relaxation occurs when imperfections are taken into account. This assumption is supported by the large total energy differences obtained during this work which would be only slightly affected by relaxations.

Up to now, only Fe or the Mo intersite defects have been considered [10, 12] leading to the conclusion that the half-metallic character is highly sensitive to such kind of imperfections. This is not extremely surprising because several Fe—O—Mo bonds are concerned by such defects. For example, in the antisite considered in this section (Fig. 7), half of the X—O—Y bonds into the unit cell become X—O—X (X,Y = Fe or Mo). On the contrary, one oxygen vacancy concerns only one X—O—Y bond and has consequently a more limited effect on the concerned atoms. This is the basic reason for which the electronic structure should be less affected by the introduction of an oxygen vacancy than by an antisite defect.

The essential role played by the delocalized electrons has been discussed recently [10]. It is exhibited that the antiferromagnetic coupling between Fe and Mo magnetic moments originates mainly from a kinetic energy mechanism due to the polarization of the delocalized states. The introduction of an oxygen vacancy, which corresponds to removing one O^{2-} atom, should alter significantly the amount of delocalised electrons and affect (i) the strength of the Fe—Mo coupling and (ii) the magnitude of the localized moments. A close look to the charge shows that an oxygen muffin tin sphere (which radius is equal to 1.4 Bohr) contains 6.5 electrons. Consequently, by removing one oxygen atom per cell, 1.5 electrons delocalized or localized on other atoms are also removed. The calculation shows that these 1.5 electrons are nearly entirely removed from the interstitial charge. This leads us to the conclusion that an oxygen vacancy affects weakly the charge carried by the atoms. The spin density around the central Mo atom (having the O vacancy in its neighborhood) is significantly more altered by the oxygen vacancy than around the corresponding Fe atom which bond to this Mo atom is removed. More explicitly,

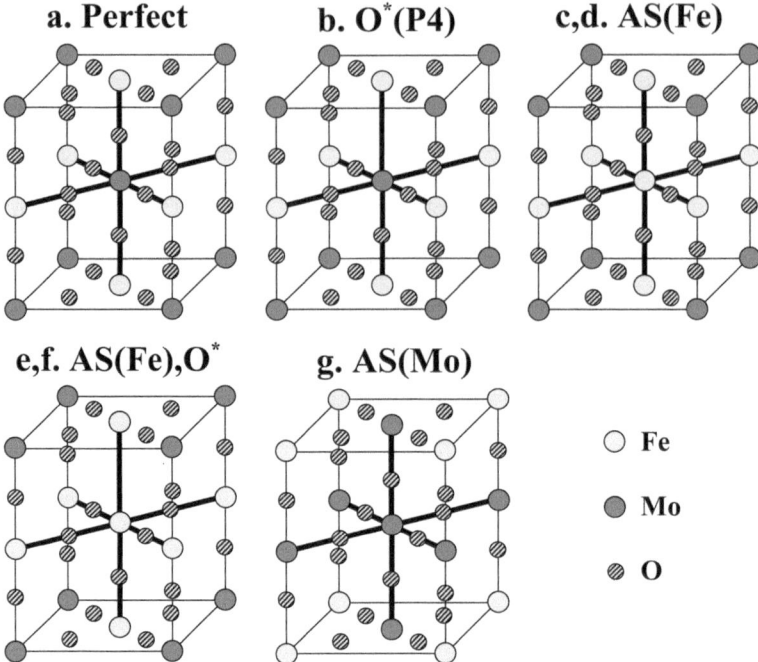

Fig. 7 Schematic representation of the unit cells used in this work for SFMO: perfect case (**a**), with one oxygen vacancy (**b**), for an Fe antisite (**c**, **d**), for an Fe antisite with one oxygen vacancy (**e**, **f**), and for a Mo antisite (**g**). The Sr atoms are not shown and the antisite is at the *center* of the cell

the tetragonal symmetry of the spin density around this Mo atom is completely lost. We conclude that most of the 1.5 electrons given by the oxygen atoms are delocalized and not transfered to other atoms : this confirms that they play certainly an important role on the Fe—Mo bond.

The calculations lead us to the conclusion that the observed magnetization cannot be explained by antisite imperfections because the reduction is too small for parallel moments and the solution with anti-aligned Fe moments is highly unstable. If we assume that the loss of magnetization finds its origin into imperfections of the SFMO phase, the occurrence of oxygen vacancies is clearly a candidate accounting for it and cannot be neglected.

In order to examine a possible magnetization reduction resulting from a combination of oxygen vacancies and Fe antisites, we have studied a particular case having one oxygen vacancy on the octahedra surrounding the Fe antisite (Fig. 7f). This case can be reasonably supposed to be probable because vacancies are well known to favor the atomic mobility inside compounds and it is reasonable to assume that a vacancy and an antisite form a pair of neighbors.

According to the previous section, for all following calculations we use now the experimental lattice parameters $a = 5.58$ Å and $c = 7.90$ Å [7]. For this crystalline structure, when one O^* is introduced, the gap remains but, due to the broadening of the unoccupied Mo t_{2g} states – resulting from the splitting of the peak into two peaks – it is reduced by 0.45 eV. As a consequence of the preserved half-metallic property, the total magnetic moment into the cell is reduced by 2 μ_B (from 8 μ_B to 6 μ_B) because five (resp. 3) electrons are removed from the majority (minority) spin band. Nearly all eight electrons removed with the vacancy come from the muffin tin sphere of the removed O atom and from the interstitial volume which become highly polarized ($-0.6\ (GGA)/-0.8\ (GGA+U)\ \mu_B$/cell) as compared to perfect SFMO for which it is nearly unpolarized. This leads us to the conclusion that O^* may have a strong impact on the magnetic couplings into the whole cell.

In the cell considered here for simulating Fe or Mo antisite, one of the three Fe or Mo atoms is on an antisite (AS) which gives a high AS concentration comparable to the highest value experimentally explored [32]. In this experimental work, the Mo NMR spectra have been found to be not affected when the AS concentration is varied from 28 to 3% leading to the conclusion that the transferred hyperfine field (HF) on Mo sites can be neglected and that the measured HF is given only by the local Mo magnetic moment. Band structure calculations allow to determine explicitly these local magnetic moments for varying situations: for instance, for $Sr_4FeMo_3O_{12}$ (SFMO[Mo AS]), the 3 inequivalent Mo atoms carry a local moment of -0.18, 0.04, and 0.26 μ_B showing clearly that this local moment is highly affected by imperfections. Whatever the used method is, $Sr_4FeMo_3O_{12}$ (SFMO[Fe AS]) and SFMO[Mo AS] are found metallic (Figs. 8 and 9) and, consequently, the solutions with all Fe moments aligned are always the most stable. For the particular case of SFMO[Fe AS], the energy difference between Ferrimagnetic (Fi) solutions with opposite local moment on the AS is found strongly reduced when using GGA+U ($\Delta E(Fi - F) = 78$ meV/AS) as compared to GGA ($\Delta E(Fi - F) = 710$ meV/AS). However, a total energy difference of 78 meV/AS remains a large value and we believe that relaxations would not change the sign of this difference. Consequently, even if we relax the atomic degrees of freedom, the F solution is expected to remain the most stable and the Fi one remains hard to stabilize. This result confirms that stable Fi solutions involving only AS can be hardly obtained by band structure calculations.

As previously discussed, oxygen vacancies may affect the magnetic couplings between Fe atoms. By combining AS and O^* imperfections, in a $Sr_4Fe_3MoO_{11}$ ($SFMO[Fe\ AS, O^*]$) cell, we get a most stable Fi and half-metallic solution with the GGA+U method ($\Delta E(Fi - F) = -222$ meV/AS), whereas it remains F and metallic with the GGA ($\Delta E(Fi - F) = 360$ meV/AS). Consequently, we conclude that the occurrence of an oxygen vacancy at the octahedra of an Fe AS is nearly the only way to get a stable negative moment on the Fe AS. Moreover, because the half-metallic property is preserved in such a solution, we get also an agreement between the double exchange model (DEM) and the band structure calculations concerning the stabilization of a negative magnetic moment on the AS. Because the Fe AS without an oxygen vacancy as neighbor carry a positive moment, only the fraction of Fe AS forming a pair with an oxygen vacancy give rise to a magnetization reduction: our work points out that the relevant parameter for studying the occurrence of the

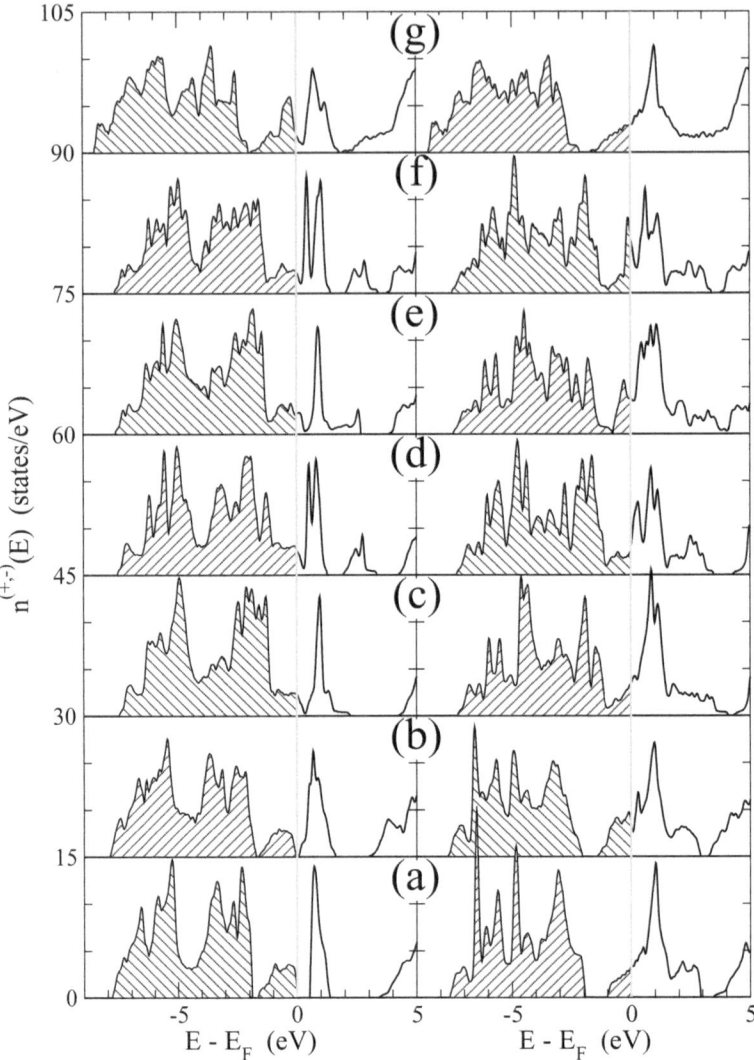

Fig. 8 Majority (*left*) and minority (*right*) spin total densities of states obtained with the GGA method for SFMO for all seven configurations considered. Each *curve* has been offset by 15 states/eV per formula unit (f.u.) relatively to the previous one and the *vertical line* corresponds to the Fermi level

Fi solution is not the concentration c_{AS} of Fe AS but the concentration c_{AS,O^*} of Fe AS and O^* forming a pair of neighbors. If the occurrence of Fe AS results from Fe atoms in excess, each single AS adds 1.5 μ_B to the total magnetization, whereas each AS and O^* pair removes 8 μ_B. If it results from exchange between Fe and Mo atoms, each single AS adds 1.15 μ_B to the total magnetization, whereas each AS and O^* pair removes only 3.6 μ_B. Experimentally, a slope around -8 μ_B per AS is found corresponding to excess Fe AS and O^* pairs. However, this would mean that

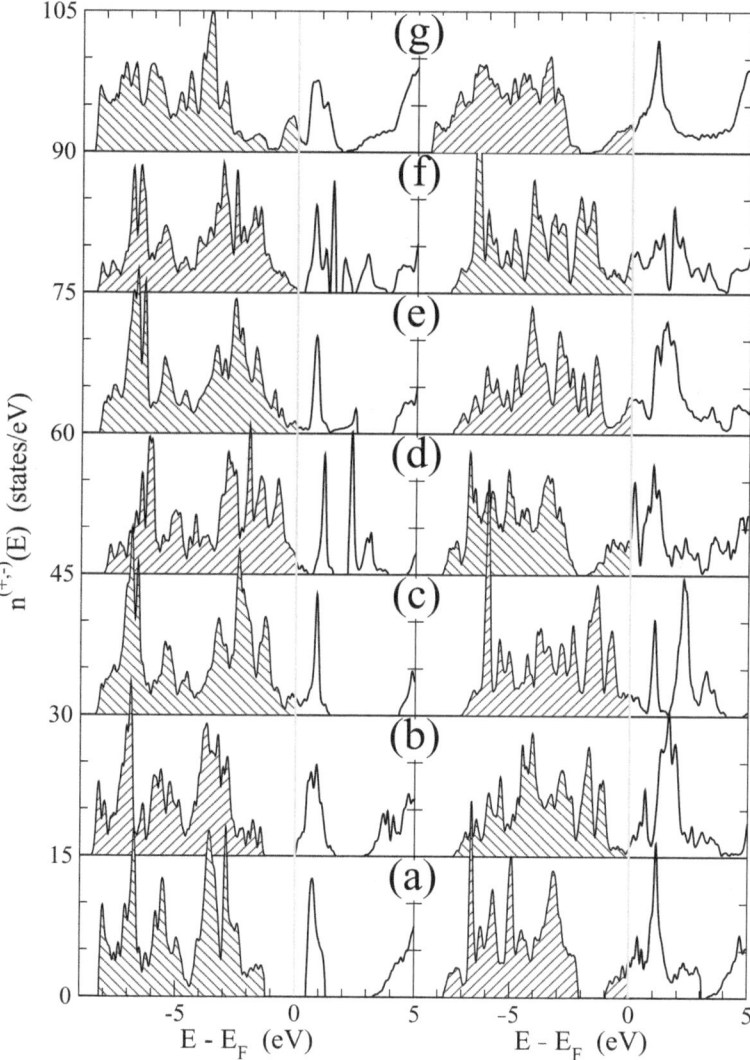

Fig. 9 Majority (*left*) and minority (*right*) spin total densities of states obtained with the GGA+U method for SFMO for all seven configurations considered. Each *curve* has been offset by 15 states/eV per formula unit (f.u.) relatively to the previous one and the *vertical line* corresponds to the Fermi level

each Fe AS form a pair with an oxygen vacancy. A priori, it seems difficult to have $c_{AS} = c_{AS,O^*}$ and, experimentally, no appreciable deviations to the nominal oxygen content have been found [33]. But this becomes more reasonable if we consider that $c_{AS} = 25\%$ corresponds to a O^* concentration of $c_{AS}/12 = 2\%$. Consequently, only a few percent of oxygen vacancies are required and, considering that they enhance the atomic mobility of the Fe and Mo atoms, it is highly probable that they form pairs with most antisites.

3.3 Comparison with an Exchange-Based Model (EBM)

Usually, to explain the magnetic and the electronic properties of SFMO double perovskites, models based on double exchange coupling of two successive large and positive Fe moments to a common small and negative Mo moments are used : Fe^{3+} $(3d^5)$ ions carrying a 5 μ_B magnetic moment are negatively coupled to Mo^{5+} $(4d^1)$ ions carrying a 1 μ_B magnetic moment via a ferrimagnetic super-exchange mechanism involving the O atom in between the Fe and Mo atoms (Fig. 10).

This explains well the F order obtained on the Fe sublattice by considering Fe—O—Mo—O—Fe chains where two successive Fe moments are negatively coupled to a common Mo moment. It explains also the half-metallic property giving (i) a completely filled Fe 3d $(t_{2g}$ and $e_g)$ up-spin band split by super-exchange from the empty Mo 4d (t_{2g}) up-spin band by a gap and (ii) a partially filled Mo 4d (t_{2g}) down-spin band. Consequently, the single conduction electron into the down-spin 4d shell of Mo is at the origin of the half metallic and F states. Our explicit band structure calculations using the GGA+U method agree with the conclusions of the EBM (Ferromagnetic half metal with a total magnetic moment of 4 μ_B/f.u.) but not with the considered charge occupation. Indeed, by integrating the charge density into the muffin tin spheres, we get an occupation of 24.6 electrons for Fe carrying a moment of 4 μ_B and an occupation of 38.7 electrons for Mo carrying a moment of -0.4 μ_B; the configuration corresponding much more to $Fe^{2.1+}$ $(3d^{5.9})$ and $Mo^{3.5+}$ $(4d^{2.5})$ ions. Consequently, Fe—Mo hybridized states have to be considered: the gap into the up-spin DOS splits hybridized (mainly Fe(e_g)) fully occupied states from hybridized (mainly Mo(t_{2g})) unoccupied states and the partially occupied down-spin metallic band consists also in hybridized (Fe(t_{2g}) and Mo(t_{2g})) states.

Considering SCMO, an insulating AF solution is found the most stable with our calculations. This agrees with the DEM only if we assume that we have Co^{2+} $(3d^7)$ ions carrying a 3 μ_B magnetic moment and Mo^{6+} $(4d^0)$ ions carrying no magnetic moment. With such a configuration, no d electrons are available on the Mo site giving the insulating behavior and no double exchange mechanism can take place for coupling the Co magnetic moments. Indeed, only an antiferromagnetic super-exchange mechanism can occur involving two Co moments coupled through two directly linked O atoms of the octahedra surrounding the Mo atom situated at the corner of the Co—Mo—Co bond making a right angle. Instead of having the

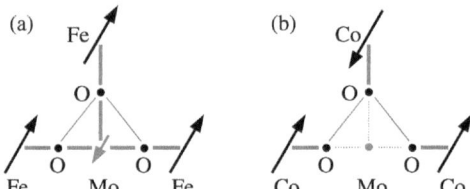

Fig. 10 Sketch of the bonds into (**a**) Sr_2FeMoO_6 and (**b**) Sr_2CoMoO_6. The strength of the interatomic interaction is given by the style of the *line* representing the bond where the strongest is the *thick solid line*, the weakest is the *thin dashed line* and the medium is the *thin solid line*

magnetic coupling dominated by Co—O—Mo—O—Co chains, like in SFMO, the coupling is dominated by Co—O—O—Co ones. Consequently, the two Co moments are negatively coupled giving rise to an AF order. Like for SFMO, this model does not agree with our explicit band structure calculations (using the GGA+U method) when considering local moments and charge occupations. We get an occupation of 25.8 electrons for Co carrying a moment of $\pm 2.7\mu_B$ and an occupation of 38.9 electrons for Mo carrying a moment of $\pm 0.02\mu_B$; the configuration corresponding much more to $Co^{1.9+}$ $(3d^{7.1})$ and $Mo^{3.6+}$ $(4d^{2.4})$ ions. However, a close look to the DOS shows that these occupied Mo states are very deep in energy (around -5 eV below the Fermi level) and there are effectively no Mo states available above -2 eV even if the configuration is not $4d^0$.

We conclude that BS calculations and the EBM lead to the same conclusion when the charge occupations obtained from the BS calculations allow the EBM to be applied.

4 Multilayers with SFMO

4.1 SFMO/STO Perfect Multilayers

4.1.1 Unit Cells

In the [001] direction, SFMO presents an alternance of $FeMoO_4$ and Sr_2O_2 atomic layers, whereas STO presents an alternance of TiO_2 and SrO atomic layers. Using a $c(2 \times 2)$ cell for STO, by doubling its in-plane cell, we get an alternance of Ti_2O_4 and Sr_2O_2 atomic layers similar to the one of SFMO. In order to have the smallest possible SFMO/STO total cell and to insure that the inner atomic plane of each SFMO and STO layer has a bulk-like environment, the total cell is consequently built by juxtaposition of a SFMO layer consisting of a $FeMoO_4/Sr_2O_2/FeMoO_4/Sr_2O_2/FeMoO_4$ stacking (denoted by $SFMO_5$) and of a STO layer consisting of a $Sr_2O_2/Ti_2O_4/Sr_2O_2/Ti_2O_4/Sr_2O_2$ (denoted by STO_5) directly in contact (see Fig. 11); the denomination of the total cell corresponds then to $SFMO_5/STO_5$. Even if the bulk in-plane lattice parameters for SFMO and STO are very similar ($a_{SFMO} = 5.57$ Å, $a_{STO}\sqrt{2} = 5.52$ Å), the tetragonal distortion resulting from the adjustment of one in plane parameter to the other can play a significant role on the half-metallic character of SFMO. Consequently, we consider the two extreme situations, i.e., using the in-plane parameter of SFMO or STO for the whole stack with a tetragonal distortion applied, respectively, on the STO or SFMO layer in order to preserve the cell volume. These cells will be used for comparing the results obtained with the GGA and the GGA+U methods.

A larger cell will be used for the investigation of the Fe interfacial deficient case where the interfacial Fe atoms are replaced by Mo atoms; the $SFMO_5$ layer being replaced by a $AMoO_4/Sr_2O_2/SFMO_5/Sr_2O_2/AMoO_4$ layer with $A = Fe$ (perfect interface) or $A = Mo$ (Fe-deficient interface) denoted, respectively, by $FMO/SFMO_7/FMO$ or $MMO/SFMO_7/MMO$.

Fig. 11 Schematic representation of half of the $AMoO_4/Sr_2O_2/SFMO_5/Sr_2O_2/AMoO_4/STO_5$ cell containing all nonequivalent atomic planes labeled from a. to h. as in Fig. 12, 13 and 16

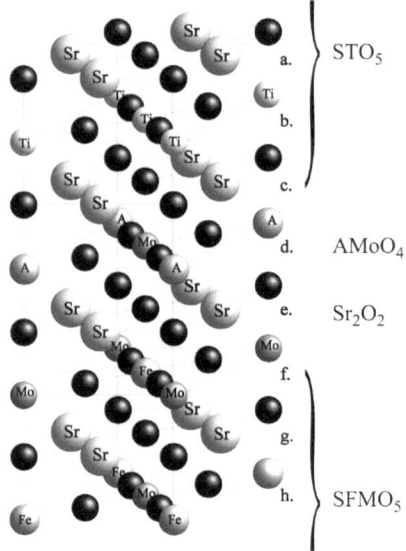

4.1.2 SFMO/STO Multilayers : GGA Method

With this method, as anticipated from bulk atomic layer projected densities of states (ALPDOS), the whole STO/SFMO superlattice is weakly half metallic and a band energy gap, ranging from 0 to 0.3 eV, remains for all ALPDOS (Fig. 12). The STO/SFMO interface has a very limited impact onto the electronic structure of the SFMO even on the interfacial $FeMoO_4$ atomic layer. This is also reflected in the magnetic moments profile which shows nearly no variation when considering Fe or Mo atoms from the interfacial and from the central atomic planes $M_{Fe} = 3.78$, 3.77 μ_B and $M_{Mo} = -0.29$, -0.29 μ_B, respectively. On the contrary, SFMO-induced states appear into a large part of the bulk-like band gap of the majority spin density of states of the STO layer as a consequence of the occurrence of a large number of Fe states for this range of energy. Consequently, if we consider a SFMO/STO/CoFe tri-layer where the magnetizations of SFMO and CoFe layers are aligned, the majority spin transmission should be highly asymmetric in small applied voltage (V). When injecting electrons from SFMO to CoFe toward the STO Barrier (for positive V), the majority spin transmission should be very small because no states are available from the SFMO electrode. On the contrary, when injecting from CoFe (for negative V), the majority spin transmission should rapidly increase because a large number of states are directly available below the Fermi level. A similar behavior should also be obtained for a SFMO/STO/SFMO trilayer but with opposite magnetizations of the two SFMO layers.

This transmission asymmetry in applied voltage for one spin channel should be reflected into the total current voltage characteristic and could be a direct proof that the electronic structure of such oxide layers is correctly or not correctly described by the GGA method.

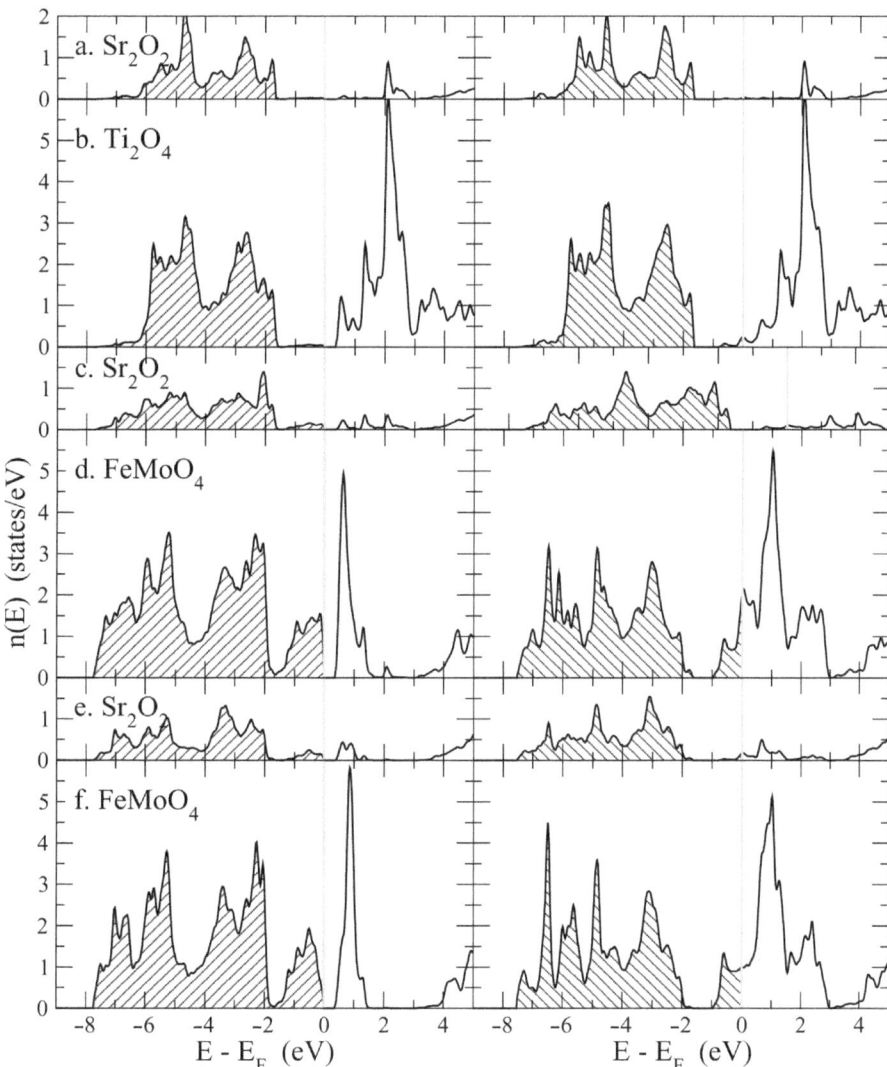

Fig. 12 ALPDOS for $SFMO_5/STO_5$ constrained to the in-plane cell parameter of SFMO (see text) (**a**) Sr_2O_2 inner STO atomic layer (AL), (**b**) Ti_2O_4 AL, (**c**) Sr_2O_2 interfacial AL (**d**) $FeMoO_4$ interfacial AL, (**e**) Sr_2O_2 AL, and (**f**) $FeMoO_4$ inner AL obtained with the GGA method. The *vertical gray line* corresponds to the Fermi level (E_F), *left* and *right* panels correspond, respectively, to the up-spin and down-spin ALPDOS

4.1.3 SFMO/STO Multilayers : GGA+U Method

With this method, the SFMO/STO superlattice is really entirely half metallic: as displayed in Fig. 13, there are no states available around the Fermi level in the majority spin density of states and the energy gaps of both STO and SFMO layer are preserved as a consequence of the very similar gaps, the gap of bulk STO ranges from −1.54 to 0.32 eV. Only electrons of the minority spin band can flow through this superlattice which corresponds to the parallel configuration of the magnetization of the two electrodes. For the antiparallel configuration, the gap into the majority band of the electrode with positive magnetization will present some states induced by the minority states of the next electrode with negative magnetization through the thin STO layer and the other way round. Consequently, SFMO is no more strictly half metallic but the current is certainly extremely weak as compared to the one in the parallel configuration because each electrode acts as an insulator in its bulk for one or the other spin channel. The interfacial $FeMoO_4$ ALPDOS (Fig. 13d) is found very similar to the one of the most central $FeMoO_4$ atomic layer (which can be considered as bulk-like). Again, this very limited impact of the interface on the properties of the SFMO layer is also reflected in the magnetic moments profile which shows nearly no variation when considering Fe or Mo atoms from the interfacial and from the central atomic planes (M_{Fe} = 3.97, 3.99 μ_B and M_{Mo} = −0.39, −0.39 μ_B, respectively). Consequently, the STO layer, terminated by the SrO atomic plane, has a weak impact on the electronic structure and the magnetic properties of the SFMO layer.

A high magnetorestistive signal is consequently expected for this case and the current voltage characteristic should be much less asymmetric than for the previous case. This shows clearly that "perfect" SFMO/STO multilayers are a priori good candidates for integration into spintronic devices when their electronic structure can be described with the GGA+U method.

4.2 The SFMO/STO/CFO MIS-Diode Like

One example of integration of the SFMO/STO interface into a magnetic tunnel junction consists in depositing a second magnetic oxide like $Fe(FeCo)O_4$ (CFO) onto STO. Figure 14 represents the evolution of the current as a function of the voltage at different temperatures between 20 and 290 K, for a junction of 50×50 μm^2 for such a junction. When the temperature is decreased, the current tends to become zero for negative voltage so that the junction behaves like a diode with a linear dependency of the current after the threshold voltage of 0.5 V [20]. In order to get an insight into the electronic structure of the three used materials, the bulk electronic structure has been determined for all three materials. Because the present discussion remains at a qualitative level, the variations of the lattice parameters due to strain are not taken into account.

The electronic structure of SFMO is the one previously obtained for a perfect unit cell. For STO, we use the usual perfect cubic cell with a = 3.905 Å. Our unit cell of CoFe2O4 corresponds to the cubic cell with a = 8.397 Å but the primitive

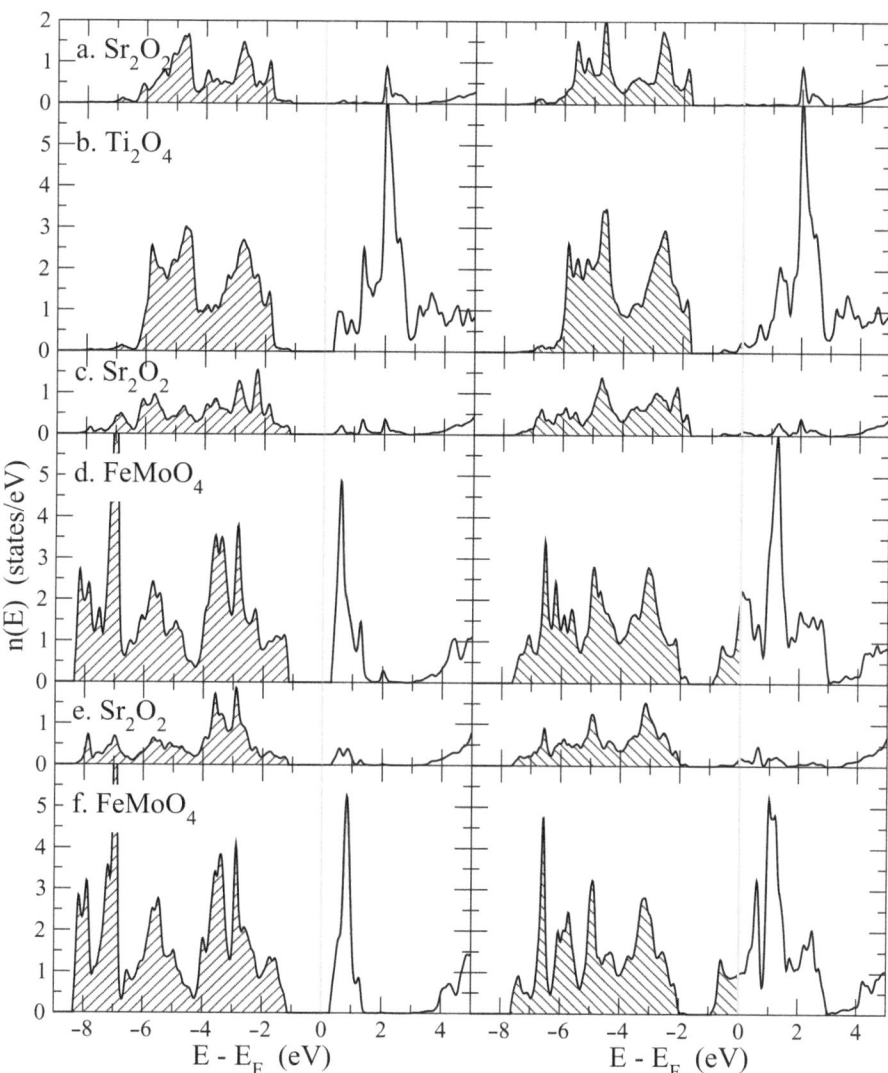

Fig. 13 ALPDOS for $SFMO_5/STO_5$ constrained to the in-plane cell parameter of SFMO (see text) (**a**) Sr_2O_2 inner STO atomic layer (AL), (**b**) Ti_2O_4 AL, (**c**) Sr_2O_2 interfacial AL (**d**) $FeMoO_4$ interfacial AL, (**e**) Sr_2O_2 AL, and (**f**) $FeMoO_4$ inner AL obtained with the GGA+U method. The *vertical gray line* corresponds to the Fermi level (E_F), *left* and *right* panels correspond, respectively, to the up-spin and down-spin ALPDOS

Fig. 14 Current vs. voltage
curves of a $50 \times 50\,\mu m^2$
junction for different
temperatures between 20 and
290 K

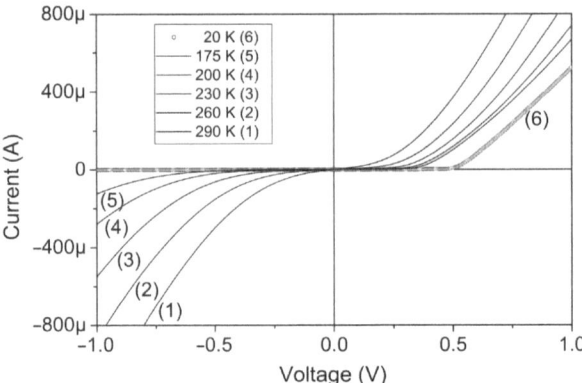

cell contains only two atoms of Co in octahedral sites, two atoms of Fe in octahedral sites, two atoms of Fe in tetrahedral sites, and eight atoms of O. Despite the fact that Co and Fe atoms have been found to alternate randomly on the same octahedral sites we use a highly symmetric-ordered cell where the Co atoms have only Fe neighbors on the octahedral sites sublattice in order to reduce the size of the primitive cell and to preserve the essential features of the electronic structure.

Figure 15 shows that STO is insulating with a gap of 1.8 eV and that CFO is a magnetic insulator with a majority spin gap of 0.65 eV and a minority spin gap of 2.0 eV. This figure exhibits the similarity of the system to a metal/insulator/ semi-conductor (MIS) structure at low temperature: minority spin electrons are injected from the metallic-like SFMO electrode into the minority spin band of CFO having a small gap through the insulating STO. The diode-like feature results from the shift of the CFO bands to lower energies, making available unoccupied minority spin states, for a positive applied voltage larger than approximately 0.4 V. When the applied voltage is small enough or negative, no states are available in the CFO electronic structure so that the current is very low. When the voltage is highly positive, the electronic structure of CFO is shifted toward negative energies so that there are unoccupied states available and therefore a current can be observed.

The use of such magnetic electrodes could also allow to control the diode I–V characteristic by applying a magnetic field. When the magnetization of the SFMO and CFO electrodes are parallel, we obtain the previously discussed behavior with a small threshold related to the small gap into the majority spin CFO band. By reversing only the magnetization of the SFMO layer, the electrons are injected into the minority spin band of CFO giving another I–V characteristic presenting a higher threshold and a different slope in the conducting regime.

4.3 SFMO/STO Interface with Fe Deficiency

As presented in the first paragraph of the present section, the unit cell is increased by considering a thicker SFMO layer so that two nonequivalent $FeMoO_4$ atomic lay-ers remain when interfacial Fe deficiency will be introduced and only the GGA+U

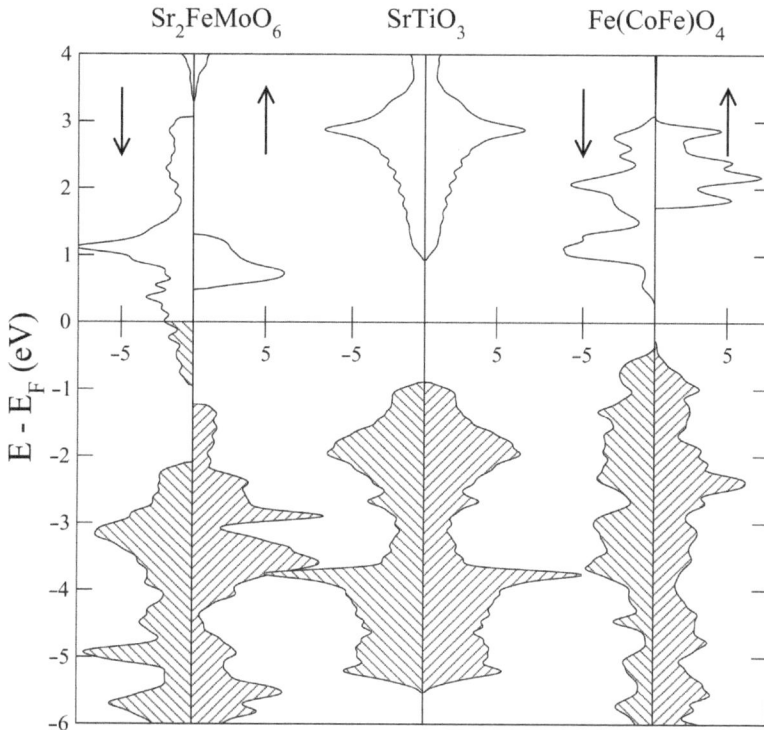

Fig. 15 Density of states for spin up and spin down of bulk (**a**) SFMO, (**b**) STO, and (**c**) CFO. The relative position of the STO and CFO bands to the SFMO one has been tentatively set assuming that the Fermi level falls in the *middle* of the gap of the total densities of states

method will be used. For the stoichiometric case, due to the very limited impact of the interface on the properties of the SFMO layer, the present results are very similar to the ones discussed in the previous section: the ALPDOS are very similar to the corresponding ones of Fig. 13 and the magnetic moments profile is very similar too (when considering Fe or Mo atoms from the interfacial to the central atomic planes $M_{Fe} = 3.99, 3.96, 3.97 \mu_B$ and $M_{Mo} = -0.40, -0.40, -0.37 \mu_B$ successively).

Replacing the interfacial Fe atoms by Mo atoms has clearly a strong impact on the global magnetic and transport properties (Fig. 16). Previous works on bulk SFMO with imperfections [29] have shown that (i) the half-metallic property is lost when Mo antisite is introduced by substituing one of the two Fe atom in a $Sr_4FeMo_3O_{12}$ cell by a Mo atom and (ii) the local moment on the Mo antisite is nearly opposite ($+0.26 \mu_B$) to the one on the regular Mo sites ($-0.39 \mu_B$). However, because this case corresponds to a bulk situation with a high concentration of Mo antisites (half of the Fe sites are occupied by Mo), the role played by such antisites is certainly overestimated.

As displayed by Fig. 16, the electronic structure of the SFMO/STO superlattice with Fe deficiency at the interface shows significant differences as compared to the "perfect" case. The half-metallic property is lost for the whole cell and the

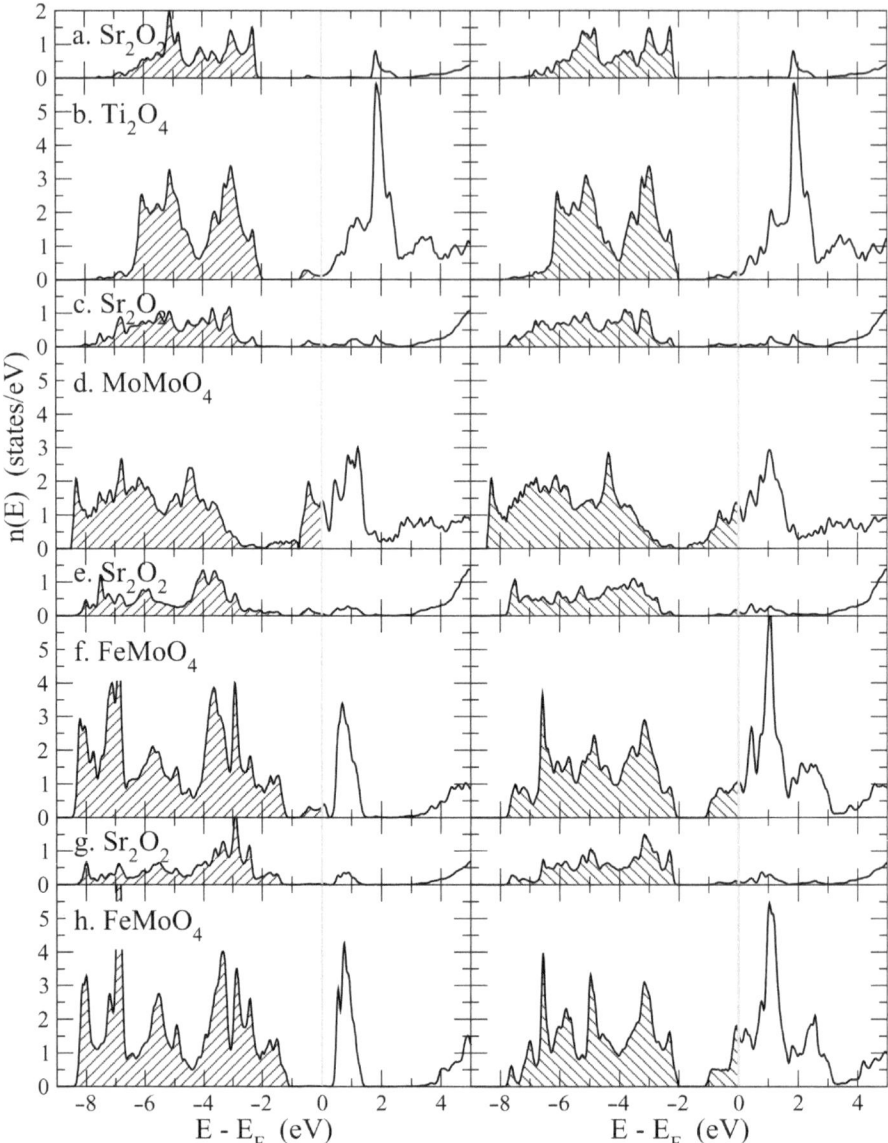

Fig. 16 ALPDOS for $SFMO_9/STO_5$ presenting Fe deficiency constrained to the in-plane cell parameter of SFMO (see text) (**a**) Sr_2O_2 inner STO atomic layer (AL), (**b**) Ti_2O_4 AL, (**c**) Sr_2O_2 interfacial AL (**d**) $MoMoO_4$ interfacial AL, (**e**) Sr_2O_2 AL, (**f**) $FeMoO_4$ AL, (**g**) Sr_2O_2 AL and (**h**) $FeMoO_4$ inner AL obtained with the GGA+U method. The *vertical gray line* corresponds to the Fermi level (E_F), *left* and *right* panels correspond, respectively, to the up-spin and down-spin ALPDOS

spin polarization of the ALPDOS at the Fermi level, defined by $P = (n^{(+)}(E_F) - n^{(-)}(E_F))/(n^{(+)}(E_F)+n^{(-)}(E_F))$, even vanishes completely on the interfacial atomic layer (from the a to the h atomic layer of Fig. 17, P is equal to -0.33, -0.21, $+0.01$, $+0.04$, -0.37, -0.45, -0.54, -0.99). Consequently, for this case, the interface is weakly polarized from the point of view of the spin polarization but also from the point of view of the magnetization; the local moment on the two interfacial Mo atoms being equal to 0.15 μ_B on the antisite and to 0.04 μ_B on the regular site. As clearly exhibited by Fig. 16f, the interfacial perturbation of the density of states extents up to the first $FeMoO_4$ atomic layer: its majority spin density of states presents a peak at the Fermi level being at the origin of the strong reduction of P. For this ALPDOS, as compared to the bulk-like central $FeMoO_4$ atomic layer for which the gap ranges from -1.20 to $0.35\,eV$, the gap into the majority spin band is clearly partially filled by states ranging from -0.75 to $0.30\,eV$: the ALPDOS corresponding to these states presents two peaks which can also be found into the majority-spin density of states of the Mo_2O_4 interfacial atomic layer showing clearly that they result from the extension of the interfacial states due to the Mo antisite. A similar feature is obtained into the Ti_2O_4 density of states of Fig. 17b showing that this interfacial states extend on both sides of the interface over five atomic layers. Consequently, the SFMO/STO interface becomes clearly spin unpolarized when an Fe deficiency is introduced but SFMO recovers rapidly its half-metallic feature outside this interface.

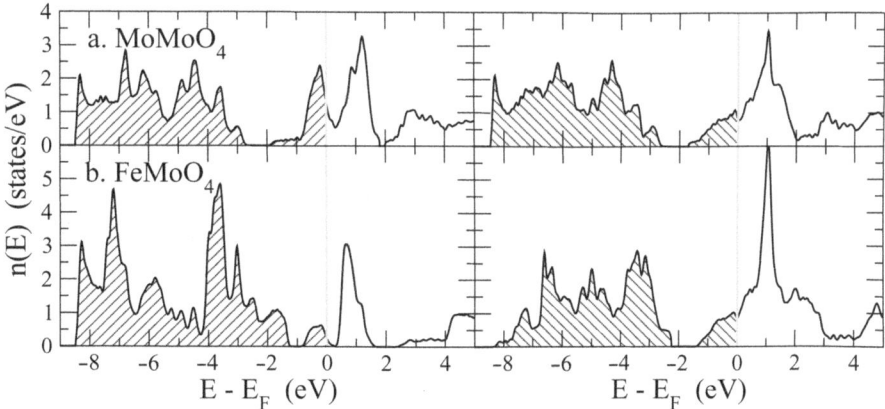

Fig. 17 ALPDOS for bulk $Sr_4FeMo_3O_{12}$ (**a**) $MoMoO_2$ atomic layer (AL) and (**b**) $FeMoO_4$ AL obtained with the GGA+U method. The *vertical gray line* corresponds to the Fermi level (E_F), *left* and *right* panels correspond, respectively, to the up-spin and down-spin ALPDOS

4.4 Discussion

The present band structure calculations show that a lack of Fe atoms at the SFMO/ STO interface results in a nearly unpolarized interface in terms of local magnetic moments (the Mo interfacial atoms carry small magnetic moments) and in terms of polarization P which is found nearly equal to zero. This result is in agreement with the experimental one of [20] which explains the absence of tunnel magnetoresistance signal in terms of an Fe-deficient SFMO surface and interface of SFMO with STO.

By comparing the ALPDOS displayed in Fig. 16d, f to the equivalent ALPDOS in a $Sr_4FeMo_3O_{12}$ cell (Fig. 17), it appears clearly that most of the main features of the density of states are similar confirming that the Ti_2O_4 atomic layer has a limited impact on the electronic structure of the Mo_2O_4 interfacial layer. However, from the viewpoint of the spin polarization, the Fe-deficient interface cancels completely P, whereas it is only reduced in the bulk $Sr_4FeMo_3O_{12}$ cell where $P(Mo_2O_4) = +0.24$ and $P(FeMoO_4) = -0.54$. Consequently, when coming from bulk SFMO, the fully polarized current becomes progressively weakly polarized when flowing through the bulk Fe-deficient layer (which is modeled here by $Sr_4FeMo_3O_{12}$) and becomes finally completely unpolarized when reaching the interface. Similarly, for a current flowing into the other direction, after tunneling through STO, it becomes unpolarized at the interface and the resulting current becomes unsensitive to the direction of the magnetization in the second SFMO layer. If it is assumed that an intrinsic mechanism (like Fe segregation or evaporation) is at the origin of this interfacial Fe deficiency, this can explain why no tunnel magnetoresistance is observed by most of the groups working on this system and why there has been only one positive result with $P = -0.85$ by Bibes et al. [19]. When considering that only a fraction λ of the interface present this Fe deficiency, the resulting spin polarization can be written as: $P = (\lambda n^{(+)} - (1 - \lambda)n^{(-)} - \lambda n'^{(-)})/(\lambda n'^{(+)} + (1 - \lambda)n^{(-)} + \lambda n'^{(-)})$, where n and n' are, respectively, the interfacial ALPDOS at the Fermi level for the perfect and the Fe-deficient cases. In order to get the measured P value of -0.85 with the present calculated values of the density of states, λ has to be around 0.1 which corresponds to only 10% of the interface presenting Fe deficiency. This confirms that Bibes et al. [19], using an improved three-step process growth for the SFMO layer and fabricating nanometer-size tunnel junctions, have effectively reached a high degree of quality of the SFMO/STO interface.

5 Conclusion

To conclude briefly (each section having its own conclusion), this work shows that electronic structure calculations give essential information on the magnetic and transport properties of multilayers build by alternating magnetic and nonmagnetic oxide layers. It shows also that, for the considered systems, the comparison between experiments and calculations requires to take imperfections into account by building less symmetric and/or larger cells resulting in a significative increase of the computation time. Moreover, for the oxide systems considered here, the GGA+U method is required to get a satisfactory agreement with the available experiments.

References

1. J. S. Moodera, L. R. Kinder, T. M. Wong, R. Meservey, Phys. Rev. Lett. **74**, 3273 (1995)
2. G. A. Prinz, Science **282**, 1660 (1999)
3. T. Dimopoulos, G. Gieres, S. Colis, J. Wecker, Y. Luo, K. Samwer, Appl. Phys. Lett. **83**, 3338 (2003)
4. J. M. De Teresa, A. Barthelemy, A. Fert, J. P. Contour, F. Montaigne, P. Seneor, Science **286**, 507 (1999)
5. K. I. Kobayashi, T. Kimura, H. Sawada, K. Terakura, Y. Tokura, Nature **395**, 677 (1998)
6. A. Gupta, J. Z. Sun, J. Magn. Magn. Mater. **200**, 24 (1999)
7. S. Colis, D. Stoeffler, C. Mény, T. Fix, C. Leuvrey, G. Pourroy, A. Dinia, P. Panissod, J. Appl. Phys. **98**, 033905 (2005)
8. L. Balcells, J. Navarro, M. Bibes, A. Roig, B. Martinez, J. Fontcuberta, Appl. Phys. Lett. **78**, 781 (2001)
9. A. S. Ogale, S. B. Ogale, R. Ramesh, T. Venkatesan, Appl. Phys. Lett. **75**, 537 (1999)
10. T. Saha-Dasgupta, D. D. Sarma, Phys. Rev. B **64**, 064408 (2001)
11. D. D. Sarma, Phys. Rev. Lett. **85**, 2549 (2000)
12. I. V. Solovyev, Phys. Rev. B **65**, 144446 (2002)
13. M. C. Viola, M. J. Martinez-Lope, J. A. Alonso, P. Valasco, J. L. Martinez, J. C. Pedregosa, R. E. Carbonio, M. T. Fernandez-Di, Chem. Mater. **14**, 812 (2002).
14. W. Westerburg, D. Reisinger, G. Jakob, Phys. Rev. B **62**, R767 (2000).
15. T. Manako, M. Izumi, Y. Konishi, K. Kobayashi, M. Kawasaki, Y. Tokura Appl. Phys. Lett. **74**, 2215 (1999).
16. S. R. Shinde, S. B. Ogale, R. L. Greene, T. Venkatesan Ken Tsoi, S.-W. Cheong, A. J. Millis, J. Appl. Phys. **93**, 1605 (2003).
17. T. Fix, D. Stoeffler, S. Colis, C. Ulhaq, G. Versini, J. P. Vola, F. Huber, A. Dinia, J. Appl. Phys. **98**, 023712 (2005).
18. S. Wang, H. Pan, X. Zhang, G. Lian, G. Xiong, Appl. Phys. Lett. **88**, 121912 (2006).
19. M. Bibes, K. Bouzehouane, A. Barthelemy, M. Besse, S. Fusil, M. Bowen, P. Seneor, J. Carrey, V. Cros, A. Vaurès, J.-P. Contour, A. Fert, Appl. Phys. Lett. **83**, 2629 (2003).
20. T. Fix, D. Stoeffler, Y. Henry, S. Colis, A. Dinia, T. Dimopoulos, L. Bar, J. Wecker, J. Appl. Phys. **99**, 08J107 (2006).
21. FLEUR is an implementation of the Full Potential Linearized Augmented Plane Wave method freely available at *http://www.flapw.de* funded by the European Research Network Ψ_k and managed by Prof. S. Bluegel.
22. U. von Barth, L. J. Hedin, J. Phys. C **5**, 1629 (1972).
23. J. P. Perdew, A. Zunger, Phys. Rev. B **23**, 5048 (1981).
24. L. Hedin, Phys. Rev. **139**, A796–A823 (1965).
25. V. I. Anisimov, J. Zaanen, O. K. Andersen, Phys. Rev. B **44**, 943–954 (1991).
26. V. I. Anisimov, F. Aryasetiawan, A. I. Lichtenstein, J. Phys. Condens. Matter **9**, 767–808 (1997)
27. Y. Wang, J. P. Perdew, Phys. Rev. B **44**, 13298 (1991).
28. J. P. Perdew, K. Burke, M. Ernzerhof, Phys. Rev. Lett. **77**, 3865 (1996).
29. D. Stoeffler, S. Colis, J. Phys. Condens. Matter **17**, 41, 6415–6424 (2005).
30. As a test, we have also done a calculation with $U_{Mo} = 3\ eV$ but, compared to the results obtained with $U_{Mo} = 0\ eV$, the densities of states and the energy difference between FM and AFM solutions were nearly not affected.
31. J. Navarro, J. Fontcuberta, M. Izquierdo, J. Avila, M. C. Asensio, Phys. Rev. B **69**, 115101 (2004)
32. M. Wojcik, E. Jedryka, S. Nadolski, J. Navarro, D. Rubi, J. Fontcuberta Phys. Rev. B **69**, 100407(R) (2004)
33. C. Frontera, D. Rubi, J. Navarro, J. L. García-Muñoz, J. Fontcuberta, C. Ritter Phys. Rev. B **68**, 012412 (2003)

34. T. Saitoh, M. Nakatake, A. Kakizaki, H. Nakajima, O. Morimoto, Sh. Xu, Y. Moritomo, N. Hamada, Y. Aiura, Phys. Rev. B **66**, 035112 (2002).
35. S. Ray, P. Mahadevan, A. Kumar, D. D. Sarma, R. Cimino, M. Pedio, L. Ferrari, A. Pesci, Phys. Rev. B **67**, 085109 (2003).

Effect of Spin–Orbit Coupling on the Magnetic Properties of Materials: Theory

M. Alouani, N. Baadji, S. Abdelouahed, O. Bengone, and H. Dreyssé

Abstract This contribution concerning the effect of spin–orbit coupling on the magnetic properties of materials is divided into two chapters. In the first chapter we review the method based on the density functional theory (DFT) within the local density approximation (LDA) used to compute the electronic structure, the magnetic anisotropy, the x-ray absorption spectra, and the x-ray magnetic circular dichroism. We give the major approximations used to derive the Kohn–Sham equations with or without the Hubbard interaction for correlated orbitals. We give also a brief introduction to the generalized gradient approximation (GGA). We then provide a solution of the latter equations using the full-potential linear augmented plane wave (FLAPW) basis set and discuss the so-called LDA+U method, where the Hubbard U is included for localized orbitals. We show how the relativistic effects, such as the spin–orbit coupling, can be introduced into band structure calculations and show their effect on magnetism, i.e., magnetic anisotropy energy (MAE), magneto-optical properties, and x-ray magnetic circular dichroism (XMCD). Then we show a brief derivation of the force theorem for the calculation of the magnetic anisotropy as well as a description of its application to the MAE calculations and show the details of the calculation of the XMCD matrix elements in the electric–dipole approximation.

M. Alouani (✉)
Institut de Physique et Chimie des Matériaux de Strasbourg, 23 Rue du Loess, BP 43, F-67034 Strasbourg Cedex 2, France

N. Baadji
Institut de Physique et Chimie des Matériaux de Strasbourg, 23 Rue du Loess, BP 43, F-67034 Strasbourg Cedex 2, France

S. Abdelouahed
Institut de Physique et Chimie des Matériaux de Strasbourg, 23 Rue du Loess, BP 43, F-67034 Strasbourg Cedex 2, France

O. Bengone
Institut de Physique et Chimie des Matériaux de Strasbourg, 23 Rue du Loess, BP 43, F-67034 Strasbourg Cedex 2, France

H. Dreyssé
Institut de Physique et Chimie des Matériaux de Strasbourg, 23 Rue du Loess, BP 43, F-67034 Strasbourg Cedex 2, France

Alouani, M. et al.: *Effect of Spin–Orbit Coupling on the Magnetic Properties of Materials: Theory*. Lect. Notes Phys. **795**, 227–308 (2010)
DOI 10.1007/978-3-642-04650-6_8

The second chapter of this contribution includes some applications of the method to the computation of the electronic, magnetism, and spectroscopic properties of spintronics materials. In particular , we investigate the electronic structure and x-ray magnetic circular dichroism (XMCD) of Sr_2FeMoO_6 (SFMO for short) and other useful ferromagnetic half-metals with 100% spin polarization, materials useful for spin injection. In particular, we show that the spin–orbit coupling reduces the spin polarization while the intra-site electronic correlations tend to increase it. For example, SFMO is found to be a half-metallic ferrimagnet with a gap in the spin-up channel. The calculated spin magnetic moments on Fe and Mo sites confirm the ferromagnetic ordering and settle the controversy existing between the earlier experimental works. The orbital magnetism at the Fe and Mo sites agrees quite well with the recent experimental XMCD measurements. The computed $L_{2,3}$ XMCD at the Fe and the Mo sites compares fairly well with experiment. The XMCD sum rule computed spin and orbital magnetic moments are in good agreement with the values obtained from the direct self-consistent calculations. In the last application, we focus on the GGA+U treatment of the electronic and magnetic structure of Gd and Gd-related compounds, such as GdN and $GdFe_2$. We compare the calculated density of states to the experimental photoemission and inverse photoemission spectra (XPS and BIS) and determine the Fermi surface with and without the Hubbard U and spin–orbit coupling. The GGA+U is found to be the most appropriate for treating the $4f$ Gd electrons. We have investigated the bulk properties and calculated the XMCD spectra at the $L_{2,3}$ edges at the Gd site of GdN. The agreement of the calculated spectra with experiment is the indication of the relevance of the XMCD formalism within the one-electron picture. The results also show that the ground-state electronic structure of GdN is that of a half-metal. Finally our computational method is used to determine the magnetic anisotropy aspect of the Gd and its compounds GdN and $GdFe_2$. Using the force theorem, we have calculated the MAE of Gd, GdN, and $GdFe_2$ for different directions of the magnetization. Indeed, owing to the nil spin–orbit interaction of the $4f$ half-filled shell, the force theorem is expected to be efficient for Gd and Gd compounds MAE calculations. This theorem allows a considerable computational effort gain since the spin–orbit coupling could be calculated only for one self-consistent iteration. Once again, the GGA+U method is found to be the most adequate approach for the force theorem calculations of the Gd MAE. The GGA and GGA-core model treatments of the $4f$ states have led to a wrong MAE. It turns out that the electronic properties and the magnetic properties of $4f$ systems are tightly related, and the $4f$ electrons play a crucial role in the computed magnetic anisotropy. Although the Gd MAE is found to be similar to that of a typical $3d$ transition metal like hcp Co, the GdN and $GdFe_2$ cubic crystals MAEs are found to be different from that of a pure $3d$ cubic material like fcc Ni.

1 Introduction

In the last few years, with the advance in the computer technology, it has become clear that a correct solution of the Kohn–Sham equations [1] is needed to produce accurate results concerning the electronic structure and magnetic properties of materials. Accordingly, ab initio methods have been extended to treat correctly the effective potential of solids, the spin polarization, and the relativistic effects [2–5]. Thus, new methods have been used to compute the magnetic properties of materials [3, 5–7]. In particular, important advances have been made in the determination of the magnetocrystalline anisotropy [8–10], non-collinear magnetism [11], the x-ray magnetic circular dichroism [5, 10], and magneto-optics [12]. To calculate accurately these properties both the spin–orbit coupling and the spin polarization have to be incorporated in band structure methods.

On the other hand, the experimental development of the x-ray magnetic circular dichroism spectroscopy [13, 14] (XMCD), together with the discovery of the sum rules which permit the determination of the spin and orbital moments from the integrated XMCD spectra [15, 16], has made this spectroscopy an interesting tool for studying magnetic properties of materials. This is so because XMCD can probe the magnetic properties of any specific atom and orbital of magnetic materials. XMCD can be also used to explore the magneto-crystalline anisotropy (MCA) by determining the orbital moment anisotropy. On the theoretical level, Bruno connected the orbital moment anisotropy to the MCA in the special case of the $3d$ transition metals [17] in which there are no holes in the spin-up band and where the crystalline field parameter is much smaller than the spin–orbit coupling. Later, van der Laan showed that the magnetocrystalline anisotropy is directly related to the anisotropic part of the spin–orbit coupling rather than to the orbital moment [18]. A general relation that strictly relates the MCA to the anisotropy of the orbital moments is still lacking, however.

These advances in XMCD spectroscopy enabled further interest in ab initio description of magnetism. Freeman and coworkers developed a slab LAPW method to study the XMCD of transition metals and their surfaces [3]. The method was used to check the validity of the XMCD sum rules, and it was found that the orbital moment obtained from the sum rule is within 10% from the direct calculation and the spin moment is much worse and can be off by upto 50% in the case of Ni(001) surface. Using a different formalism based on the full-potential relativistic linear muffin-tin orbital method, Alouani, Wills, and Wilkins study the XMCD of Fe nitrides [5, 19]. They showed that the XMCD intensity is directly proportional to the spin magnetic moment and that the spin and orbital moments obtained from the XMCD sum rules are in good agreement with the direct calculation. Later, this formalism was used by Galanakis et al. to study the transition metal binary alloys and the so-called Heusler alloys [10]. The same method was also extended to study the magneto-optical properties of materials [12]. Using multiple scattering theory Ankudinov and Rehr studied the XMCD in Gd [20] and Brouder, Alouani, and Bennemann studied the K-edge of Fe [21].

This chapter is structured as follows. We first discuss and give an overview of new development in methodology to compute the electronic structure problem. We then discuss and outline the founder ideas and the formulation of the DFT together with relativistic effects in a whole section about the density functional method. The main features of the FLAPW method which is used here to compute XMCD spectra and magnetic anisotropy energy are presented. Emphasis is put on the description of the spin–orbit coupling relativistic effect within the FLAPW method; a brief derivation of the force theorem as well as a description for its application to the MAE calculations is provided. A whole section is devoted to another consequence of the spin–orbit coupling, that of the XMCD and a special attention is given to the description and the derivation of the XMCD formalism as a magneto-optical effect.

2 Methodological Developments

Based on the work of Callaway [22], Ebert implemented the relativistic effects in ab initio methods to explore the magnetic properties of the spin-polarized electrons. In particular, Ebert implemented a fully relativistic linear muffin-tin method in the atomic-sphere approximation [2], and later extended it to determine the magneto-optical properties and the x-ray magnetic dichroism of transition metals and their binary alloys [4]. This allowed him to make the first calculation of the magnetic x-ray dichroism of the Fe K-edge which was found in good agreement with the experimental results of Schütz et al. [14]. Later, Halilov and Uspenskii used a similar method to determine the effect of the spin–orbit coupling and the spin polarization on the optical conductivity tensor of $3d$ ferromagnetic transition metals [23, 24]. However, the agreement with experiment was limited probably due to their use of the atomic-sphere approximation. During the same period Kübler and coworkers [25] used the augmented spherical wave method (ASW) in the atomic-sphere approximation to study the magneto-optical Kerr effect of Fe and Ni. The agreement with experiment was much better than in case of Halilov and Uspenskii calculation [23, 24] due to the better calculation of the matrix elements involved in the interband transitions. The ASW method was later used to determine the magneto-optical spectra of uranium compounds [25] (US, USe, and UTe). While the diagonal conductivity tensor elements were well reproduced, the off-diagonal elements and the Kerr angle did not agree well with the experimental data. The relativistic LMTO-ASA was also used by Oppeneer and Antonov to study the magneto-optical properties of the so-called Heusler alloys and good agreement with the experimental data was achieved [26].

Usually spin–orbit coupling is included in the band structure calculation using perturbation theory [2]. In some cases the full Dirac Hamiltonian is solved self-consistently. It has been argued that the latter description is more appropriate since the basis set has no restriction, especially when dealing with heavy elements where spin–orbit coupling is very strong [4]. Nonetheless, perturbation theory is more widely used in the literature [5, 20, 21].

3 Density Functional Theory

Calculating electronic and magnetic properties of materials is not a simple task even in the ground state. Materials are composed of atoms held together by chemical bonds insured by the valence electrons. Involving so many particles gives rise to a complex many-body problem. One of the early proposed simplification to deal with such complexity is the Born–Oppenheimer approximation.

3.1 The Born–Oppenheimer Approximation

The Born–Oppenheimer approximation [27] (BOA) consists of dividing the total solid-state problem into two parts: the motion of the electrons in a stationary lattice and that of the ions in a uniform space charge of electrons. The total Hamiltonian H which represents the total energy of a *realistic* system can be written as

$$H = H_e + H_I + H_{e-I}, \tag{1}$$

where H_e, H_I, and H_{e-I} are the electrons, the ions, and the electrons–ions interaction parts of the Hamiltonian respectively.

To understand the full meaning of the BOA we consider a system of nuclei described by coordinates $\mathbf{R}_1, ..., \mathbf{R}_N \equiv \mathbf{R}$ and momenta $\mathbf{P}_1, ..., \mathbf{P}_N \equiv \mathbf{P}$ and masses $M_1, ..., M_N$ and the electrons described by coordinates $\mathbf{r}_1, ..., \mathbf{r}_{N_e} \equiv \mathbf{r}$ and momenta $\mathbf{p}_1, ..., \mathbf{p}_{N_e} \equiv \mathbf{p}$ and spin variables, $s_1, ..., s_{N_e} \equiv s$. The Hamiltonian of the system is given by

$$H = \sum_{I=1}^{N} \frac{\mathbf{P}_I^2}{2M_I} + \sum_{i=1}^{N_e} \frac{\mathbf{p}_i^2}{2m} + \sum_{i>j} \frac{e^2}{|\mathbf{r}_i - \mathbf{r}_j|} + \sum_{I>J} \frac{Z_I Z_J e^2}{|\mathbf{R}_I - \mathbf{R}_J|} - \sum_{i,I} \frac{Z_I e^2}{|\mathbf{R}_I - \mathbf{r}_i|}$$

$$\equiv T_N + T_e + V_{ee}(\mathbf{r}) + V_{NN}(\mathbf{R}) + V_{eN}(\mathbf{r}, \mathbf{R}), \tag{2}$$

where m is the mass of the electron, and $Z_I e$ is the charge on the nucleus, T_N, T_e, V_{ee}, V_{NN}, and V_{eN} represent, respectively, the nuclear and electron kinetic energy operators and electron–electron, nuclear–nuclear, and electron–nuclear interaction potential operators, respectively. This Hamiltonian is very general and can describe any material. The solution of this Hamiltonian would predict any physical or chemical property of the material we are studying. This problem cannot be solved, so many approximations should be made in order to hope for some reasonable description of the properties of materials. The formal solution of this full many-body problem will amount in solving the following Schrödinger equation:

$$[T_N + T_e + V_{ee}(\mathbf{r}) + V_{NN}(\mathbf{R}) + V_{eN}(\mathbf{r}, \mathbf{R})] \, \Psi(\mathbf{x}, \mathbf{R}) = E\Psi(\mathbf{x}, \mathbf{R}), \tag{3}$$

where $\mathbf{x} \equiv (\mathbf{r}, s)$ represents the electron positions and spin variables, and $\Psi(\mathbf{x}, \mathbf{R})$ is an eigenfunction of H wih eigenvalue E.

In order to make progress, we use the Born–Oppenheimer approximation (BOA) by noticing that because the electrons are lighter than the nuclei by three orders of magnitude there is a strong separation of timescales between the electronic and the nuclear motion. In Eq. (3) we assume the separation of the degrees of freedom of the electrons and that of the nucleus and use the following form:

$$\Psi(\mathbf{x}, \mathbf{R}) = \phi(\mathbf{x}, \mathbf{R})\chi(\mathbf{R}), \tag{4}$$

where $\chi(\mathbf{R})$ is a nuclear wave function and $\phi(\mathbf{x}, \mathbf{R})$ an electronic wave function that depends parametrically on the nuclear positions $(\mathbf{R})$. This BOA is justified by the fact that nuclei are several thousand times heavier than electrons due to the fact that a proton or a neutron is about 1,836 times more massive than an electron. The electrons follow the nuclear motion adiabatically, i.e., they are dragged along with the nuclei with almost no relaxation time. This is a reasonable approximation, because the non-adiabatic effects that do not allow the electrons to follow in this *instantaneous* manner are usually small. In almost all materials, the adiabatic separation between electrons and nuclei is a good approximation:

$$\left[T_{\mathrm{e}} + V_{\mathrm{ee}}(\mathbf{r}) + V_{\mathrm{eN}}(\mathbf{r}, \mathbf{R})\right] \phi(\mathbf{x}, \mathbf{R}) = E(\mathbf{R})\phi(\mathbf{x}, \mathbf{R}). \tag{5}$$

Equation (5) is an electronic eigenvalue equation for an electronic Hamiltonian which will yield a set of normalized eigenfunctions, $\phi_n(\mathbf{x}, \mathbf{R})$, and eigenvalues, $E_n(\mathbf{R})$, which depend parametrically on the nuclear positions. For each solution, there will be a nuclear eigenvalue equation:

$$\left[T_{\mathrm{N}} + V_{\mathrm{NN}}(\mathbf{R}) + E_n(\mathbf{R})\right] \chi(\mathbf{R}) = E\chi(\mathbf{R}). \tag{6}$$

Moreover, each electronic eigenvalue, $E_n(\mathbf{R})$, will give rise to an electronic surface, and these surfaces are known as Born–Oppenheimer surfaces (BOS). Thus, the internuclear potential for each electronic surface is given by $V_{\mathrm{NN}}(\mathbf{R}) + E_n(\mathbf{R})$. On each Born–Oppenheimer surface, the nuclear eigenvalue problem can be solved, which yields a set of levels. The Born–Oppenheimer surfaces are surfaces on which the nuclear dynamics is described by a time-dependent Schrödinger equation for the time-dependent nuclear wave function $\chi(\mathbf{R}, t)$:

$$\left[T_{\mathrm{N}} + V_{\mathrm{NN}}(\mathbf{R}) + E_n(\mathbf{R})\right] \chi(\mathbf{R}, t) = i\hbar\frac{\partial}{\partial t}\chi(\mathbf{R}, t). \tag{7}$$

Equation (7) tells us that the electrons respond instantaneously to the nuclear motion; therefore, it is sufficient to obtain a set of instantaneous electronic eigenvalues and eigenfunctions for every nuclear configuration and therefore we obtain the parametric dependence of $\phi_n(\mathbf{x}, \mathbf{R})$ and $E_n(\mathbf{R})$. The eigenvalues give a set of independent potential surfaces on which the nuclear wave function can evolve. It is possible that for some materials these BOS can couple by the non-adiabatic effects, contained in the terms that have been neglected that might couple the electron degrees of freedom to those of the nucleus.

An important assumption of the BOA is that there are no excitations of the electrons among the various BOS. Such excitations constitute non-adiabatic effects which are, therefore, neglected. As an example of a condition in which this approximation is valid, consider a system at temperature T; if the electrons are in their ground state $E_0(\mathbf{R})$ then, if $E_1(\mathbf{R})$ denotes the first excitates state, there will be no excitations to this state if

$$|E_1(\mathbf{R}) - E_0(\mathbf{R})| \gg kT, \tag{8}$$

for all nuclear configurations. Without complete determination of these BOS, it is not possible to know whether this condition will be satisfied or not. There could be regions where the surfaces approach each other with an energy spacing close to kT. If the system visits such nuclear configurations, then the BOA will break down.

In many cases, non-adiabatic effects can be neglected, and we may consider motion only on the ground electronic surface described by

$$[T_{\mathrm{e}} + V_{\mathrm{ee}}(\mathbf{r}) + V_{\mathrm{eN}}(\mathbf{r}, \mathbf{R})] \, \phi(\mathbf{x}, \mathbf{R}) = E(\mathbf{R})\phi(\mathbf{x}, \mathbf{R}),$$

$$[T_{\mathrm{N}} + E(\mathbf{R}) + V_{\mathrm{NN}}(\mathbf{R})] \, \chi(\mathbf{R}, t) = i\hbar\frac{\partial}{\partial t}\chi(\mathbf{R}, t). \tag{9}$$

Moreover, if nuclear quantum effects can be neglected, it can be shown that the nuclei are described by the classical Hamilton–Jacobi equations with

$$H_{\mathrm{N}}(\mathbf{P}, \mathbf{R}) = \sum_{I=1}^{N} \frac{\mathbf{P}_I^2}{2M_I} + V_{\mathrm{NN}}(\mathbf{R}) + E(\mathbf{R}). \tag{10}$$

With the ground-state total energy given by $\mathcal{E} = E(R) + V_{\mathrm{NN}}(\mathbf{R})$.

$$\dot{\mathbf{R}}_I = \frac{\mathbf{P}_I}{M_I},$$

$$\dot{\mathbf{P}}_I = -\nabla_I\mathcal{E}(\mathbf{R}). \tag{11}$$

The force on the atoms $-\nabla_I\mathcal{E}(\mathbf{R})$ contains a term from the nuclear–nuclear repulsion and a term from the derivative of the electronic eigenvalue $E(\mathbf{R})$. Using the Hellman–Feynman theorem, the latter term can be expressed as

$$\nabla_I E(\mathbf{R}) = \langle\phi(\mathbf{R})|\nabla_I H_{\mathrm{e}}(\mathbf{R})|\phi(\mathbf{R})\rangle. \tag{12}$$

Equations (11) and (12) form the theoretical basis of ab initio molecular dynamics approaches. The practical implementation of the ab initio molecular dynamics method requires an algorithm for the numerical solution of Eq. (11) with forces obtained from Eq. (12) at each step of the calculation. Moreover, since an exact solution for the ground-state electronic wave function, $|\phi(\mathbf{R})\rangle$, and eigenvalue, $E(\mathbf{R})$, are not available, in general, it is necessary to introduce a method for obtaining these quantities. At this stage the wave function $\phi(\mathbf{r}_1, s_1, \ldots \mathbf{r}_N, s_N)$ depends on the coordinates of the N electrons and their spins.[1] However, since the Hamiltonian is the observable accounting for the measurable total energy, according to quantum mechanic principles, the eigenfunctions ϕ have to be written as an expansion in terms of a complete set of basis functions. This is the case, for example, for the configuration interaction (CI) method where the ground-state wave function is a linear combination of Slater-wave functions or the Hartree–Fock mean-field approximation which has offered the simplest approach to handle the N electrons problem and where the wave function is expressed as a single Slater determinant.

3.2 The Hartree–Fock Approximation

We focus now on the motion of the electrons, as described in Eq. (5). We consider an *electron gas* which is embedded in a homogeneous, positively charged medium (jellium medium) or in a rigid lattice of positively charged ions. Even with a jellium medium this problem is very difficult to solve because of the complexity of the electron–electron interaction. In the absence of this interaction, the many-body problem would decouple into one-body problems which describe the motion of an electron in an effective potential (*the one-electron approximation*). In this case the electron Hamiltonian (5) becomes[2]

$$H = -\sum_k \frac{1}{2}\nabla_k^2 + \sum_k V_{\text{ext}}(\mathbf{r}_k) + \frac{1}{2}\sum_{k,k'} \frac{1}{|\mathbf{r}_k - \mathbf{r}_{k'}|} = \sum_k H_k + \sum_{k,k'} H_{k,k'}. \quad (13)$$

According to the variational principle, those φ_k which minimize E represent the best set of functions for the ground state. For the Pauli principle to apply, the wave function should be written as a Slater determinant

$$\phi = (N!)^{-1/2} \begin{vmatrix} \varphi_1(\mathbf{x}_1) & \cdots & \varphi_N(\mathbf{x}_1) \\ \vdots & & \vdots \\ \varphi_1(\mathbf{x}_N) & \cdots & \varphi_N(\mathbf{x}_N) \end{vmatrix}, \quad (14)$$

where the x_N coordinates stand for both the spatial coordinates r_N and the spin coordinates χ_N. The normalizing factor $(N!)^{-1/2}$ accounts for the indistinguishability of the electrons since there are $N!$ possible ways of distributing N electrons

---

[1] The wave function depends also on the coordinates of atoms, but as it is seen above, these coordinates appear only as parameters in Eq. (5).
[2] We use atomic units $\hbar = m = e^2 = 1$.

at the N positions $\mathbf{r}_1, \cdot, \mathbf{r}_N$. The fermionic character of the electrons is therefore insured by the antisymmetric wave function (14). With the wave function (14), we can again calculate the expectation value $E = \langle \phi | H | \phi \rangle$. To include spin-polarized systems, we define the one-electron wave function for each electron spin as $\varphi_{k\sigma}(\mathbf{x}) = \varphi_{k\sigma}(\mathbf{r})\chi_\sigma$, where $\chi_+ = \begin{pmatrix} 1 \\ 0 \end{pmatrix}$ and $\chi_- = \begin{pmatrix} 0 \\ 1 \end{pmatrix}$. The total energy is then given by

$$
E = \sum_\sigma \sum_k^{N_\sigma} \int \varphi_{k\sigma}^*(\mathbf{r}) \left(\frac{1}{2} \nabla_k^2 + V_{\text{ext}}(\mathbf{r}) \right) \varphi_{k\sigma}(\mathbf{r}) \mathrm{d}^3 r + \frac{1}{2} \int \frac{\varrho(\mathbf{r}_1)\varrho(\mathbf{r}_2)}{|\mathbf{r}_1 - \mathbf{r}_2|} \mathrm{d}^3 r_1 \mathrm{d}^3 r_2
$$
$$
- \frac{1}{2} \sum_\sigma \int \frac{\varrho_\sigma(\mathbf{r}_1, \mathbf{r}_2)\varrho_\sigma(\mathbf{r}_2, \mathbf{r}_1)}{|\mathbf{r}_1 - \mathbf{r}_2|} \mathrm{d}^3 r_1 \mathrm{d}^3 r_2,
\tag{15}
$$

where $\varrho_\sigma(\mathbf{r}_1, \mathbf{r}_2) = \sum_{k=1}^{N_\sigma} \varphi_{k\sigma}(\mathbf{r}_2)\varphi_{k\sigma}(\mathbf{r}_1)$ and $\varrho(\mathbf{r}) = \sum_\sigma \varrho_\sigma(\mathbf{r})$, and $\varrho_\sigma(\mathbf{r})$ is the trace of $\varrho_\sigma(\mathbf{r}_1, \mathbf{r}_2)$. The integration here includes a summation over the spin variable σ. We further note that in the absence of spin–orbit coupling, every wave function can be written as the product of a space function and a spin function. The last term on the left-hand side of Eq. (15) leaves us with just a summation over electrons with the same spin, because the orthogonality of the spin functions causes the other spin terms to disappear. We therefore vary Eq. (15) for any $\varphi_{k\sigma}^*$ or $\varphi_{k\sigma}$ and equate the variation to zero:

$$
\delta/\phi^*_{\,k\sigma}[E - \sum_{k\sigma} E_{k\sigma}(\langle \varphi_{k\sigma} | \varphi_{k\sigma} \rangle - 1)] = 0.
\tag{16}
$$

We get

$$
\left[-\frac{\nabla^2}{2} + V_{\text{ext}}(\mathbf{r}) + \int \frac{\varrho(\mathbf{r}')}{|\mathbf{r} - \mathbf{r}'|} \mathrm{d}^3 r' \right] \varphi_{k\sigma}(\mathbf{r})
$$
$$
- \int \frac{\varrho_\sigma(\mathbf{r}, \mathbf{r}')}{|\mathbf{r} - \mathbf{r}'|} \varphi_{k\sigma}(\mathbf{r}')\mathrm{d}^3 r' = E_{k\sigma} \varphi_{k\sigma}(\mathbf{r}),
\tag{17}
$$

where $E_{k\sigma}$ are Lagrange parameters fulfilling condition (16), and we use $\mathbf{r}$ for the coordinates of the electron under consideration. This is the *Hartree–Fock* equation [28, 29]. The Schrödinger equation for the many-electron problem is thus split up into one-electron wave equations. While the Hartree equation was easy to interpret, the newly added third term on the left-hand side of Eq. (17) has no classical analogue. It is called the *exchange interaction*. There is an equation of the same form for each of the different one-electron functions, and these equations must be solved simultaneously. For a single atom this can be done by a method of successive approximations, until self-consistency of the required degree of accuracy is reached. In metals, the problem is too complicated and cruder approximations must be used. Electrons repel one another, so that they do not move independently but in such a way as to avoid each other as far as possible. Such correlations among the electrons' motions, or positions, are called *Coulomb correlations*. In the Hartree method, Coulomb correlations are completely ignored, each electron being supposed to move in the *average* charge distribution of the other electrons. The total wave function is a single product of one-electron functions, so that the probability of a given configuration depends only upon the one-electron functions and not directly upon the distances between pairs of electrons. The Hartree–Fock method again neglects proper Coulomb correlations, but includes correlations of another kind. These are correlations among the positions of electrons with parallel spins only (the exchange interaction of the Eq. (17)) and are due not to the Coulomb force but to the Pauli principle, as embodied in the use of a determinantal wave function. The exchange potential can be rewritten as an electrostatic potential which is due to a fictitious nonlocal exchange charge density:

$$n_{k\sigma}(\mathbf{r}_1, \mathbf{r}_2) = \frac{\varphi_{k\sigma}(\mathbf{r}_2)\varrho_\sigma(\mathbf{r}_1, \mathbf{r}_1)}{\varphi_{k\sigma}(\mathbf{r}_1)},$$

so that the exchange term becomes

$$-\int \frac{\varrho_\sigma(\mathbf{r}, \mathbf{r}')}{|\mathbf{r} - \mathbf{r}'|}\varphi_{k\sigma}(\mathbf{r}')\mathrm{d}^3 r' = -\int \frac{n_{k\sigma}(\mathbf{r}, \mathbf{r}')}{|\mathbf{r} - \mathbf{r}'|}\mathrm{d}^3 r' \varphi_{k\sigma}(\mathbf{r}).$$

This nonlocal exchange density integrates exactly to one $\int n_{k\sigma}(\mathbf{r}_1, \mathbf{r}_2)\mathrm{d}^3 r_1 = 1$. This implies that each electron of a given spin is surrounded by exactly one exchange hole originating from the polarization of the medium due to the same-spin electron repulsion. This exchange hole sum rule will turn out to be very important for the determination of the density functional exchange and correlation potentials. The correlations, associated with the exchange, are known rather under the name of the exchange–correlation potential in the density functional theory (DFT) formalism, which will be approached in more detail in the next section. The Hartree–Fock method becomes well known among chemists because it verifies the Koopman's Theorem for the ionization energies $I_{k\sigma}$ of any state $k\sigma$,

$$I_{k\sigma} = E(N - 1_{k\sigma}) - E(N) = -E_{k\sigma},$$

and the results agree very well with experimental data. However, electronic structure of materials and energy band gaps obtained within this approximation are in very bad agreement with experiments. Therefore most of electronic structure calculations are obtained with methods that go beyond this approximation to the many-body problem.

One of the early attempts to estimate the electron–electron interaction in solids and introduce the charge-dependent total energy in solids is that of the Thomas–Fermi model [30, 31], the Hartree–Fock approximation, and the X-α method of Slater [32]. The extension of these ideas, which have given rise to a revolution in the parameter-free ab initio description of complex electronic structure, is known as DFT. This was established by Hohenberg and Kohn [33] and Kohn and Sham [1] and will be reviewed next.

3.3 The Hohenberg–Kohn Theorems

The finding of Hohenberg and Kohn for non-magnetic systems with a non-degenerate ground state is based on two theorems [33].

Theorem 1 *The external potential v, and hence the total energy of a system, is a unique functional of the ground state electron density $n(\mathbf{r})$.*

Theorem 2 *The exact ground state electron density minimizes the total energy functional $E[n(\mathbf{r})]$.*

A brief demonstration is provided in the Hohenberg–Kohn paper [33]. In their paper, the Hamiltonian H is defined as $H = T + V + W$, for which T represents the kinetic energy of the system, V the interaction of the electrons with an external potential, and W the electron–electron interaction. The solution of this Hamiltonian is the many-body wave function $\phi(\mathbf{r}_1, \mathbf{r}_2, \ldots \mathbf{r}_N)$, and we have

$$H\phi = E\phi. \tag{18}$$

The electron density can be calculated from

$$n(\mathbf{r}) = \left\langle \Psi \left| \sum_{i=1}^{N} \delta(\mathbf{r} - \mathbf{r}_i) \right| \Psi \right\rangle \tag{19}$$

The extension of these theorems to the spin-polarized systems can be done by including an external magnetic field, $B(\mathbf{r})$, so that the Hamiltonian becomes $H = T + U + W$, where $U = \int v(\mathbf{r})n(\mathbf{r}) - \mathbf{B}(\mathbf{r}).\mathbf{m}(\mathbf{r})\mathrm{d}^3 r$.

Using the variational principle (in the same way as it was used to demonstrate Theorem 1), one can show that the ground-state energy is a unique functional of the electron and magnetization density ($n(\mathbf{r})$ and $m(\mathbf{r})$). Using the theorems above to get a practical scheme to use DFT in describing solids, Kohn and Sham [1] have shown that instead of solving the many-body equation (18), it suffices to solve an effective one-particle equation.

3.4 The Kohn–Sham Equations

An important step on the way to finding an applicable approximation of the Hohenberg–Kohn functional energy is the Kohn–Sham equations [1]. The main idea is to split the many-body equation (18) into an effective one-particle equation

$$\left[\frac{-\nabla^2}{2} + V_{\text{eff}}(\mathbf{r})\right]\psi_i(\mathbf{r}) = \epsilon_i\psi_i(\mathbf{r}), \tag{20}$$

where the effective potential $V_{\text{eff}}(\mathbf{r})$ has the form

$$V_{\text{eff}}(\mathbf{r}) = V_{\text{ext}}(\mathbf{r}) + \int \frac{n(\mathbf{r'})}{|\mathbf{r} - \mathbf{r'}|}\mathrm{d}^3r' + V_{\text{xc}}(\mathbf{r}), \tag{21}$$

where the first term is the external potential generated by the nuclei, the second the electrostatic potential, and the last the exchange–correlation potential supposed to include all many-body effects. The density is now constructed using

$$n(\mathbf{r}) = \sum_{i=1}^{N} |\psi_i(\mathbf{r})|^2, \tag{22}$$

where the sum runs over all occupied states.

The set of Equations (20), (21), and (22) represents the Kohn–Sham equations. The Kohn–Sham equation given by Eq. (20) can be viewed as a Schrödinger-like equation in which the external potential is replaced by the effective potential (21), which depends on the electron density. The electron density itself depends on the one-particle states ψ_i. The Kohn–Sham equations need therefore to be solved in a self-consistent manner. The total energy functional $E[n(\mathbf{r})]$ expressed in terms of the one-particle energies ϵ_i (the Fock eigenvalues) has the form

$$E[n(\mathbf{r})] = T_0[n(\mathbf{r})] + \int n(\mathbf{r})V_{\text{ext}}(\mathbf{r})\mathrm{d}^3r + \frac{1}{2}\int\int \frac{n(\mathbf{r}).n(\mathbf{r'})}{|\mathbf{r} - \mathbf{r'}|}\mathrm{d}^3r\mathrm{d}^3r' + E_{\text{xc}}[n(\mathbf{r})], \tag{23}$$

where $T_0[n(\mathbf{r})]$ accounts for independent-electron kinetic energy. This kinetic energy can be expressed in terms of the one-particle energies ϵ_i as

$$T_0[n(\mathbf{r})] = \sum_i f_i\epsilon_i - \int V_{\text{eff}}(\mathbf{r})n(\mathbf{r})\mathrm{d}^3r, \tag{24}$$

where f_i is the Fermi distribution function (one for occupied states and zero for empty states). Using Eq. (24), the total energy functional can be rewritten as

$$E[n(\mathbf{r})] = \sum_i f_i \epsilon_i - \frac{1}{2} \int \int \frac{n(\mathbf{r}).n(\mathbf{r}')}{|\mathbf{r} - \mathbf{r}'|} d^3r d^3r' - \int V_{\mathrm{xc}}(\mathbf{r}) n(\mathbf{r}) d^3r + E_{xc}[n(\mathbf{r})].$$

(25)

The exact exchange–correlation potential V_{xc} and functional $E_{\mathrm{xc}}[n(\mathbf{r})]$ are, however, not known and further approximations are needed for the solution of the electronic structure problem.

3.5 The Local Density Approximation

Since the first three terms on the right-hand side of Eq. (23) represent most of the total energy can be calculated numerically, the remaining complexity of the fully interacting system is mapped into the problem of finding the exchange and correlation functional. The most common and widely used approximation of the exchange–correlation functional is the so-called local density approximation (LDA) where the exchange–correlation energy is approximated by that of a homogeneous uniform electron gas,

$$E_{\mathrm{xc}}[n(\mathbf{r})] = \int \epsilon_{\mathrm{xc}}^{\mathrm{hom}}[n(\mathbf{r})] n(\mathbf{r}) d^3r,$$

(26)

where $\epsilon_{\mathrm{xc}}^{\mathrm{hom}}$ is the sum of the exchange and the correlation energy of the uniform electron gas of density $n(\mathbf{r})$. The exchange energy can be calculated analytically and the correlation energy has been parametrized and calculated to a great accuracy by means of a quantum Monte Carlo method [34]. The exchange–correlation potential $V_{\mathrm{xc}}^{\mathrm{LDA}}(\mathbf{r})$ is the functional derivative of $E_{\mathrm{xc}}^{\mathrm{LDA}}$, which can be written as

$$V_{\mathrm{xc}}(\mathbf{r}) = \epsilon_{\mathrm{xc}}[n(\mathbf{r})] + n(\mathbf{r}) \frac{\partial(\epsilon_{\mathrm{xc}}[n(\mathbf{r})])}{\partial n(\mathbf{r})}.$$

(27)

The most early parametrization attempts of the exchange–correlation energy ϵ_{xc} are those of Barth and Hedin [35]:

$$\epsilon_{xc}(n_\uparrow, n_\downarrow) = \epsilon_{xc}^P(r_s) + [\epsilon_{xc}^F(r_s) - \epsilon_{xc}^P(r_s)] f(n_\uparrow, n_\downarrow),$$

(28)

where

$$f(n_\uparrow, n_\downarrow) = [(2n_\uparrow/n)^{4/3} + (2n_\downarrow/n)^{4/3} - 2]/(2^{4/3} - 2),$$

(29)

$n_\uparrow$ and $n_\downarrow$ represent, respectively, the spin-up and spin-down components of the total charge n ($n = n_\uparrow + n_\downarrow$), and r_s is defined by

$$(4/3)\pi r_s^3 = 1/n.$$

(30)

The paramagnetic, P, and ferromagnetic, F, exchange–correlation energies in Eq. (28) are given by

$$\epsilon_{xc}^i = \epsilon_{xc}^i(r_s) - c_i G(r_s/r_i), i = P, F \tag{31}$$

where $\epsilon_x^P = -0.91633/r_s$, $\epsilon_x^F = 2^{1/3}\epsilon_x^P$,

$$G(x) = (1 + x^3)\ln(1 + 1/x) - x^2 + x/2 - 1/3, \tag{32}$$

and c_P, c_F, r_P, r_F were chosen by fitting Eq. (28) to ϵ_{xc} for the homogeneous electron gas. The resulting parameters [35] are

$$c_P = 0.045, r_P = 21, c_F = c_P/2, r_F = 2^{4/3}r_P. \tag{33}$$

The most commonly used parametrization is that of Moruzzi et al. [36]. The corresponding parameters are

$$c_P = 0.0504, r_P = 30, c_F = 0.0254, r_F = 75. \tag{34}$$

According to Eqs. (27) and (28) the resulting potential takes the form:

$$\begin{aligned}
V_{xc}^\sigma = \quad & [4/3\epsilon_x^P(r_s) + \gamma(\epsilon_c^F(r_s) - \epsilon_c^P(r_s))](2n_\sigma/n)^{1/3} \\
& +\mu_c^P(r_s) - \gamma(\epsilon_c^F(r_s) - \epsilon_c^P(r_s)) \\
& +[\mu_c^F(r_s) - \mu_c^P(r_s) - 4/3(\epsilon_c^F(r_s) - \epsilon_c^P(r_s)]f(n_\uparrow, n_\downarrow),
\end{aligned} \tag{35}$$

where

$$\begin{aligned}
\mu_c^P(r_s) &= -c_P \ln(1 + r_s/r_P), \\
\mu_c^F(r_s) &= -c_F \ln(1 + r_s/r_F), \\
\gamma &= 4/3(2^{1/3} - 1).
\end{aligned} \tag{36}$$

This potential is referred to as the LDA exchange–correlation potential in the rest of the chapter.

Although the local density approximation is rather simple and expected to be valid only for homogeneous cases, it turns out that it usually works remarkably well even for inhomogeneous cases. However, for solids LDA very often gives too small equilibrium volumes ($\sim$3%) due to overbinding. A simple improvement to the LDA that corrects the lattice parameter is based on the generalized gradient approximation (GGA).

3.6 The Generalized Gradient Approximation

Even though the LDA has been successfully applied to systems with varying charge density, it is rather valid for systems with nearly constant charge density. In order to understand the effect of the charge density variation in terms of the exchange–correlation interaction many attempts have been made so far. One of these attempts has given rise to the so-called generalized gradient approximation (GGA), where not only the density itself enters in the exchange–correlation energy but also its local gradient. The most successful one is the one suggested by Perdew and Wang (PW91) [37] and its simpler form by Perdew, Burke, and Enzerhof (PBE) [38]. We focus here on the latter one, which will be henceforth referred to as the GGA exchange–correlation potential.

The exchange–correlation energy now has the form:

$$E_{xc}^{GGA} = \int n(\mathbf{r})\epsilon_{xc}^{hom}(n(\mathbf{r}), |\nabla n|)d^3r, \tag{37}$$

which can be expressed as [38]

$$E_{xc}^{GGA} = \int f(n_\uparrow, n_\downarrow, \nabla n_\uparrow, \nabla n_\downarrow)d^3r. \tag{38}$$

The simplified scheme of the PBE approximation consists of evaluating separately the correlation and the exchange energy as follows:

$$E_c^{GGA} = \int [\epsilon_c^{unif} + H(r_s, \xi, t)]d^3r, \tag{39}$$

where r_s is the Seitz radius (as defined by Eq. (30)), ξ is the relative spin polarization, and $t = |\nabla n|/2\phi(\xi)k_s n$ is a dimensionless density gradient. Here $\phi(\xi) = [(1 + \xi)^{2/3} + (1 - \xi)^{2/3}]/2$ is a spin scaling factor and $k_s = \sqrt{(4k_F/\pi a_0)}$ is the Thomas–Fermi screening wave number. The constructed H function has the form

$$H = 2\gamma\phi^3 \ln\left\{1 + \frac{\beta}{\gamma}t^2\left[\frac{1 + At^2}{1 + At^2 + A^2t^4}\right]\right\}, \tag{40}$$

where

$$A = \frac{\beta}{\gamma}\left[\exp(-\epsilon_c^{unif}/(\gamma\phi^3 e^2/a_0)) - 1\right]^{-1}. \tag{41}$$

$\beta = 0.066725$ and $\gamma = (1 - \ln 2)/\pi^2$. The exchange energy functional obeys the relationship

$$E_x[n_\uparrow, n_\downarrow] = (E_x[2n_\uparrow] + E_x[2n_\downarrow])/2, \tag{42}$$

where

$$E_x = \int n\epsilon_x(n)F_x(s)\mathrm{d}^3r, \tag{43}$$

and

$$F_x(s) = 1 + \kappa - \kappa/(1 + \mu s^2/\kappa), \tag{44}$$

and where $s = |\nabla n|/2k_F n$ is another dimensionless density gradient, $\kappa = 0.804$ and $\mu = 0.21951$.

It is worth mentioning here that like the LDA, the GGA obeys the exchange and correlation hole density sum rules, first derived for the LDA [39]:

$$\int n_x(\mathbf{r}, \mathbf{r}')\mathrm{d}^3r' = -1, \tag{45}$$

$$\int n_c(\mathbf{r}, \mathbf{r}')\mathrm{d}^3r' = 0, \tag{46}$$

and the negativity condition of the exchange hole:

$$n_x(\mathbf{r}, \mathbf{r}') \leq 0, \tag{47}$$

where $\mathbf{r}' = \mathbf{r} + \mathbf{u}$ and $n_x(\mathbf{r}, \mathbf{r} + \mathbf{u})$, $n_c(\mathbf{r}, \mathbf{r} + \mathbf{u})$ are, respectively, the exchange and the correlation hole density of radius $\mathbf{u}$ surrounding the electron at $\mathbf{r}$ according to the exchange energy definition of Gunnarson and Lundqvist [40]: the exchange–correlation energy is the electrostatic interaction of each electron at $\mathbf{r}$ with the density $n_{xc}(\mathbf{r}, \mathbf{r} + \mathbf{u}) = n_x + n_c$ at $\mathbf{r} + \mathbf{u}$ of the exchange–correlation hole which surrounds it.

Figure 1 illustrates the difference between the exchange–correlation potential calculated using the GGA and the LDA. As it can be seen from the figure, although both approximations lead to small differences for different radii (because each of them is satisfying the same sum rules), this difference is locally perceptible (varying from 0.01 to 0.1 Htr). We have to mention here that, compared to the LDA, the GGA leads to better structural properties, i.e., it gives lattice parameters in better agreement with experiments. However, both the LDA and the GGA potentials suffer from electron self-interaction. Perdew and Zunger [41] self-interaction correction consists in proposing an exchange–correlation potential parametrization so that the sum of the self-interaction from the Coulomb interaction and from the exchange–correlation tends to cancel each other:

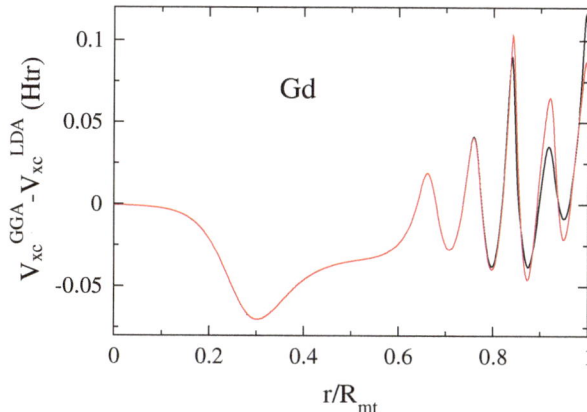

Fig. 1 The difference between the GGA exchange–correlation and that of the LDA up to the muffin-tin radii for gadolinium metal Gd. The spin-up part (in *black*) and the spin-down part (in *red*)

$$U\left[n_{l,\sigma}\right] + E_{\text{Coulomb}}\left[n_{l,\sigma}\right] = 0, \tag{48}$$

where l is the orbital quantum number and σ the spin. Although this approximation has led to improved total energy and charge density of light atoms and a number of monovalent metallic atoms compared to the Hartree–Fock one, it has not been able to provide a satisfactory result for molecules and solids with localized orbitals. A powerful alternative is the so-called LDA(GGA)+U method, which allows us not only to keep the LDA(GGA) potential but also to add the intra-atomic Coulomb interaction, particularly inavoidable for strongly localized and correlated electrons systems. The LDA(GGA)+U method are more efficient in removing the self-interaction of localized orbitals.

3.7 The LDA(GGA)+U Method

The LDA+U method [42], which is a generalization of the Hubbard model [43–46], is aimed to include the intra-atomic Coulomb interaction U in a mean-field (MF) Hartree–Fock-like manner. The original idea of this method is to replace the LDA exchange and correlation functional for localized orbitals by an intra-atomic Coulomb potential. This leads to the partitioning of the electronic system into two subsystems: the first subsystem of localized electrons $d(f)$ is treated like in the Hartree–Fock approximation and has therefore no self-interaction; the second subsystem is constituted of the rest of the electrons which are delocalized electrons and is treated within the LSDA. Because of the involved localized orbitals (d or f) it would be technically practical to use atomic-like orbitals as basis functions. The linearized muffin-tin orbital method (LMTO) in the atomic-sphere approximation [42] (ASA) or its full-potential version [47] has been the first method within which the LDA+U method was implemented. We present here the LDA+U implementation within the FLAPW method as it has been described by Shick et al. [48] without supplying details about the FLAPW method (the FLAPW method will be discussed in the next section). The variational LDA+U total-energy functional takes the form

$$E^{\text{LDA}+U}\left[n_\sigma\right] = E^{\text{LDA}}\left[n_\sigma\right] + E^{ee}\left[n_\sigma\right] - E^{dc}\left[n_\sigma\right],
\tag{49}$$

where $E^{\text{LDA}}\left[n_\sigma\right]$ is the standard LDA total energy functional, $E^{ee}\left[n_\sigma\right]$ is the electron–electron interaction energy of the correlated orbitals.

$$E^{\text{ee}} = \frac{1}{2} \sum_{\sigma,\sigma'} \sum_{m_1,m_2,m_3,m_4} n^\sigma_{m_1,m_2} n^{\sigma'}_{m_3,m_4}$$
$$\times \left[\langle m_1, m_3 | V^{\text{ee}} | m_2, m_4 \rangle - \langle m_1, m_3 | V^{\text{ee}} | m_4, m_2 \rangle \delta_{\sigma,\sigma'} \right],
\tag{50}$$

which can be also written as [47]

$$E^{\text{ee}} = \frac{1}{2} \sum_{\sigma} \sum_{m_1,m_2,m_3,m_4} \langle m_1, m_3 | V^{\text{ee}} | m_2, m_4 \rangle n^\sigma_{m_1,m_2} n^{-\sigma}_{m_3,m_4}$$
$$+ \left[\langle m_1, m_3 | V^{\text{ee}} | m_2, m_4 \rangle - \langle m_1, m_3 | V^{\text{ee}} | m_4, m_2 \rangle \right] n^\sigma_{m_1,m_2} n^\sigma_{m_3,m_4},
\tag{51}$$

where the occupation matrix of some localized orbitals is defined as

$$n^\sigma_{m_2,m_1} = \sum_{i\{occ\}} \langle \psi^\sigma_i | m_1, \sigma \rangle \langle m_2, \sigma | \psi^\sigma_i \rangle,
\tag{52}$$

with ψ^σ_i the FLAPW wave function.

$V^{\text{ee}} = 1/|r_1 - r_2|$ is the interaction potential between two electrons. The interaction term between localized orbitals $\langle m_1, m_3 | V^{\text{ee}} | m_2, m_4 \rangle$ can be given a simple interpretation when $(m_1 = m_2)$ and $(m_3 = m_4)$. In this case $\langle m_1, m_3 | V^{\text{ee}} | m_1, m_3 \rangle$ corresponds to the Coulomb interaction between localized orbitals with occupation $n^\sigma_{m_1,m_1}$ and $n^\sigma_{m_3,m_3}$. In the case where $(m_1 = m_4)$ and $(m_3 = m_2)$, the term $\langle m_1, m_3 | V^{\text{ee}} | m_3, m_1 \rangle$ represents the exchange energy of the same previous localized orbitals $n^\sigma_{m_1,m_1}$, and $n^{-\sigma}_{m_3,m_3}$ with electrons of opposite spin.

In terms of U and J matrices one obtains [49]

$$E^{\text{ee}} = \frac{1}{2} \sum_{\sigma} \sum_{m_1,m_2,m_3,m_4} n^\sigma_{m_1,m_2} n^{-\sigma}_{m_3,m_4} U_{m_1,m_2,m_3,m_4}$$
$$+ \left(U_{m_1,m_2,m_3,m_4} - J_{m_1,m_2,m_3,m_4} \right) n^\sigma_{m_1,m_2} n^\sigma_{m_3,m_4}.
\tag{53}$$

$E^{dc}[n_\sigma]$ is the double counting term to subtract for the interaction already accounted for, albeit incorrectly, in LDA (GGA):

$$E^{dc} = \frac{U}{2}n(n-1) - \frac{J}{2}\sum_\sigma n^\sigma(n^\sigma - 1), \tag{54}$$

where $n^\sigma = Tr(n^\sigma_{m_1,m_2})$ and $n = \sum_\sigma n^\sigma$.

The LDA+U potential which corresponds to the modified $E^{\text{LDA}+U}$ functional can be expressed as

$$V^{\text{LDA}+U} = \sum_\sigma \sum_{m_1,m_2} |m_1,\sigma\rangle V^\sigma_{m_1,m_2} \langle m_2,\sigma|, \tag{55}$$

where the potential matrix elements $V^\sigma_{m_1,m_2}$ are defined as

$$V^\sigma_{m_1,m_2} = \frac{\partial E^{\text{LDA}+U}}{\partial n^\sigma_{m,m'}} = \frac{\partial E^{ee}}{\partial n^\sigma_{m,m'}} - \frac{\partial E^{dc}}{\partial n^\sigma_{m,m'}}. \tag{56}$$

Using Eqs. (49), (50), and (54), Eq. (56) can be expressed as

$$\begin{aligned}
V^\sigma_{m_1,m_2} = &\sum_\sigma \sum_{m_3,m_4} \langle m_1, m_3|V^{ee}|m_2,m_4\rangle n^{-\sigma}_{m_3,m_4} \\
&+ (\langle m_1, m_3|V^{ee}|m_2,m_4\rangle - \langle m_1, m_3|V^{ee}|m_4,m_2\rangle)n^\sigma_{m_3,m_4} \\
&- \delta_{m_1,m_2}U(n - \frac{1}{2}) + \delta_{m_1,m_2}J(n^\sigma - \frac{1}{2}).
\end{aligned} \tag{57}$$

According to Eq. (55) the expected value of $V^{\text{LDA}+U}$ is then

$$\langle \psi^\sigma_i|V^{\text{LDA}+U}|\psi^\sigma_i\rangle = \sum_{m_1,m_2} V^\sigma_{m_1,m_2} n^\sigma_{m_2,m_1}. \tag{58}$$

With the help of the variational principal, one can minimize the LDA+U total energy functional (Eq. 49) with respect to ψ^σ_i:

$$\left[\nabla^2 + V^\sigma_{\text{LDA}}(\mathbf{r})\right]\psi^\sigma_i(\mathbf{r}) + \sum_{m_1,m_2} V^\sigma_{m_1,m_2} \frac{\delta n^\sigma_{m_1,m_2}}{\delta\psi^\sigma_i} = e^\sigma_i \psi^\sigma_i(\mathbf{r}). \tag{59}$$

This set of equations is that of the Kohn–Sham equations with an additional term accounting for the U (LDA+U) correction.

It is worth noticing that the present derivation of the LDA+U method (to which we will refer to from now on as the LDA+U method) is rotationally invariant and therefore independent of the quantization axis and hence of the crystal orientation.

Because LDA+U approach aimed at improving electron–electron interaction within localized orbitals, it is natural to suppose atomic-like character of these orbitals with separation between angular and radial variables. Therefore the electron–electron interaction term will read

$$\langle m_1, m_2 | V^{ee} | m_2, m_4 \rangle = \sum_k a_k(m_1, m_2, m_3, m_4) F_k,$$

$$a_k(m_1, m_2, m_3, m_4) = \frac{4\pi}{2k+1} \sum_{q=-k}^{k} \langle lm_1 | Y_{kq} | lm_2 \rangle \langle lm_3 | Y_{kq}^* | lm_4 \rangle, \qquad (60)$$

with separation between radial Slater integrals and angular a_k integrals. In this expression, F_k are the Slater integrals, $|l, m\rangle$ are $d(f)$ spherical harmonics, and a_k are related to the Gaunt coefficients through the complex spherical harmonics.

The on-site Coulomb and exchange interactions U, J are identified with the averaged Coulomb and exchange interactions:

$$U = \frac{1}{(2l+1)^2} \sum_{m_1, m_3} \langle m_1, m_3 | V^{ee} | m_1, m_3 \rangle,$$

$$J = U - \frac{1}{2l(2l+1)} \sum_{m_1, m_3} [\langle m_1, m_3 | V^{ee} | m_1, m_3 \rangle - \langle m_1, m_3 | V^{ee} | m_3, m_1 \rangle$$

$$= \frac{1}{2l(2l+1)} \sum_{m_1 \neq m_3, m_3} \langle m_1, m_3 | V^{ee} | m_3, m_1 \rangle. \qquad (61)$$

Although this atomic formulation is appropriate and reliable to incorporate these intra-atomic interactions, the electron–electron interaction (60) is *unscreened* and is therefore overestimated.

Some attempts have been already made to compute U and J interactions using ab initio approaches. The results obtained within the constrained LDA [50] calculations have shown the difficulty of simulating the screening effect for $3d$ and $4f$ system in solids and led to a too strong (for $3d$ metals) and to a too small (for $4f$ metals) intra-atomic interactions compared to that provided by experiment. Recent developments in LDA+U approach, in particular in the calculation of the U and J values by Pickett and Cococcioni [51, 52], suggested a constrained LDA approach combined with linear response theory, where screening could be treated more carefully. The obtained values for $3d$ metals are smaller and more consistent with experiment.

It turns out that the most realistic way to get an estimation of these intra-atomic interactions is to make use of the experimental spectra such as XPS (X-ray photoemission spectroscopy) and BIS (Bremsstrahlung isochromat spectroscopy) spectra to extract the U and J interactions and use them as input *parameters* for LDA+U calculations.

Since the U and J parameters (the *screened* interactions) are known, the Slater integrals can be calculated using Eq. (61);

for d orbitals (l=2):

$$U = F_0,$$
$$J = (F_2 + F_4)/14; F_4/F_2 = 5/8, \qquad (62)$$

for f orbitals (l=3):

$$U = F_0,$$
$$J = (286F_2 + 195F_4 + 250F_6)/6435;$$
$$F_2/F_4 = 675/451; F_2/F_6 = 2025/1001. \qquad (63)$$

Even though the conception of the LDA+U scheme parametrization (using U and J as parameters) makes the ab initio DFT calculations lose its parameter-free character, this method has provided a better understanding (compared to the LDA) of the electronic structure of transition-metal oxides and Mott–Hubbard insulators such as NiO and CoO [53–56]. We will show in the following chapters that the LDA+U approach is also appropriate for describing electronic structure of correlated $4f$ rare-earth metals [57–59].

4 Relativistic Effects

4.1 Importance of Relativistic Effects

Non-relativistic or semi-relativistic Kohn and Sham density functional theory [1] is extensively used to determine the band structure of materials, but it is insufficient in explaining many properties related to optics and magnetism. Its local spin density approximation (LSDA) which can calculate the magnetic properties can determine neither the magnetic anisotropy nor the optical threshold of gold. As we can see from Fig. 2, at the X high symmetry point in the Brillouin zone, the energy goes from its non-relativistic value of -1.8 to -1.1 eV when the spin–orbit coupling (SOC) is included, i.e., the highest occupied band gets much closer to the Fermi level when the SOC is included. Therefore, without the SOC, the LSDA is incapable of explaining the yellow color of gold [60]. The LSDA is also incapable of explaining the Faraday effect, the Kerr or the x-ray circular magnetic dichroism. All the above phenomena are due to relativistic effects, and precisely to the SOC. When the SOC is included in the Hamiltonian the spin and orbital number are no longer principle quantum numbers and only the total angular momentum is a good quantum number. In the case of heavy metals, like gold, where the SOC is large, d-valence bands are split into total angular momentum of 5/2 and 3/2. If the SOC is not very large, like in the case of $3d$ transition metals, then valence bands are not much affected; however, the SOC produces a small orbital moment (the orbital moment is no longer

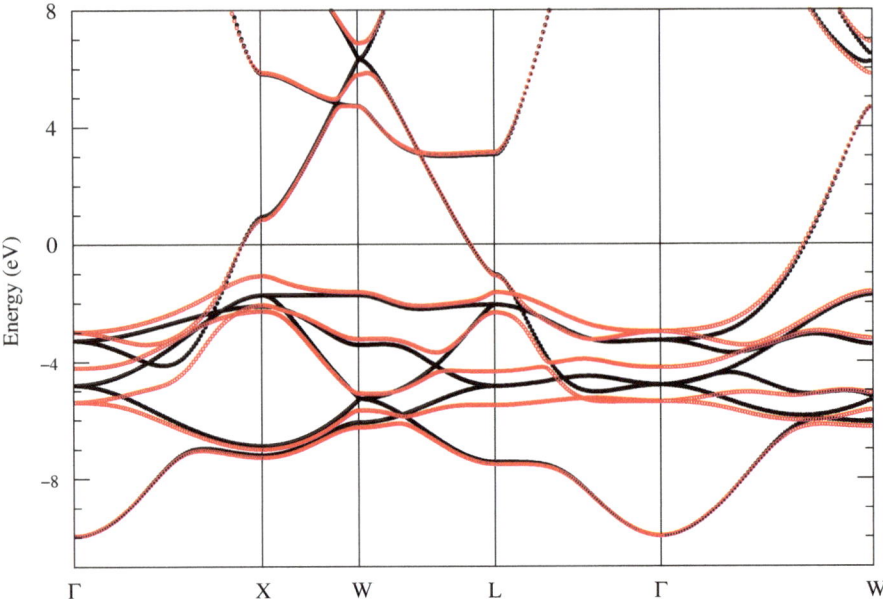

Fig. 2 Band structure of gold along some high symmetry directions including spin–orbit coupling (*red lines*) and without it (*black lines*)

quenched). The SOC term, as it can be seen below, is proportional to the scalar product of the spin moment and orbital moment, and thus aligns the spin magnetic moment with respect to the crystal. This will explain the easy and hard axes.

To describe the relativistic effects we start from the Dirac equation using relativistic electrodynamics. For light atoms, we will treat relativistic effects using perturbation theory. The standard technique is the use of the Foldy–Wouthuysen [61] transformation which decouple the large and small components of the Dirac wave function.

4.2 Dirac Equation

We will first derive the Dirac equation of a free particle and then that of a charged particle in an external potential. The Klein–Gordan (K–G) equation for a free particle can be written as[3]

$$P^{\mu} P_{\mu} |\Psi\rangle = m^2 |\Psi\rangle \tag{64}$$

---

[3] We use the atomic units $\hbar = c = 1$ and Einstein notation $A^{\mu} B_{\mu} = A_0 B_0 - \boldsymbol{A}.\boldsymbol{B}$ and $\underline{A}$ to designate the quadrivector $\underline{A} = (A_0, \boldsymbol{A})$

where $P^\mu P_\mu = p_0^2 - |\boldsymbol{p}|^2 = E^2 - |\boldsymbol{p}|^2$. Equation (64) is just the definition of the total energy of a relativistic particle:

$$E = \sqrt{|\boldsymbol{p}|^2 + m^2}$$

Using the spin-position representation, the correspondence principle gives

$$\left(\partial^\mu \partial_\mu + m^2\right) \Psi(\underline{r}) = 0 = \left(\frac{\partial^2}{\partial t^2} - \nabla^2 + m^2\right)\Psi(\underline{r}). \tag{65}$$

The problem with the K–G equation is that it doesn't have a probabilistic interpretation [62].[4] To find an equation with a probabilistic interpretation, Dirac replaced the K–G equation by an equation with first-order derivative in space and time, analogous to Schrödinger's equation. This last equation is covariant and of first order with respect to the momentum operator:

$$\not{P} |\Psi\rangle = m |\Psi\rangle = \gamma^\mu P_\mu |\Psi\rangle, \tag{66}$$

where $\not{P}$ is the slash momentum operator using the Feynman slash notation and where the matrices γ_μ are related to the Pauli matrices.

One has to find the energy of the relativistic free particle ($E^2 = P_0^2 = P^2 + m^2$). This implies that the matrices γ^μ verify the following equality:

$$(\gamma^\mu \gamma^\nu + \gamma^\nu \gamma^\mu) = 2g^{\mu\nu}, \tag{67}$$

where $g^{\mu\nu}$ is the Lorentz metric tensor, whose signature in a Galilean reference frame is $(+ - - -)$. We can write the Dirac equation, using the spin-position representation as

$$\left(i\gamma^\mu \partial_\mu - m\right) \Psi(\underline{r}) = 0. \tag{68}$$

We find an analogous equation to that of Schrödinger:

$$i\partial_0 \Psi(\underline{r}) = (\gamma^0 \gamma^i \partial_i + m\gamma^0)\Psi(\underline{r}) = (\boldsymbol{\alpha}.\boldsymbol{P} + m\beta)\Psi(\underline{r}). \tag{69}$$

---

[4] Since this equation is of second order and that $\Psi(\underline{r})$ and its temporal derivative are independent, it is not sure whether the first component of the current quadrivector J^μ obtained from this equation is positive definite at a later time. The first component of the current quadrivector J^μ is given by

$$J^\mu(\underline{r}) = \frac{1}{2mi\hbar}(\Psi^*(\underline{r})\partial^\mu \Psi(\underline{r}) - (\partial^\mu \Psi^*(\underline{r}))\Psi(\underline{r})).$$

The Dirac Hamiltonian H_D is then defined by

$$H_D = \boldsymbol{\alpha}.\boldsymbol{P} + m\beta, \tag{70}$$

$$\alpha^i = \gamma^0 \gamma^i,$$
$$\beta = \gamma^0. \tag{71}$$

For the Dirac Hamiltonian H_D to have a probabilistic interpretation, it has to be Hermitian. In the spin-position representation of a particle of spin 1/2, the wave function $|\Psi\rangle$ is a quadrispinor, having two spinors analogous to that of Pauli $|\varphi\rangle$ of a particle, whereas $|\chi\rangle$ is that of the antiparticle[5]

$$|\Psi\rangle = \begin{pmatrix} |\varphi\rangle \\ |\chi\rangle \end{pmatrix} \tag{72}$$

In this representation the matrices γ^μ can be written as

$$\gamma^0 = \begin{pmatrix} 1 & 0 \\ 0 & -1 \end{pmatrix},$$
$$\gamma^1 = \begin{pmatrix} 0 & \sigma_x \\ \sigma_x & 0 \end{pmatrix},$$
$$\gamma^2 = \begin{pmatrix} 0 & \sigma_y \\ \sigma_y & 0 \end{pmatrix}, \tag{73}$$
$$\gamma^3 = \begin{pmatrix} 0 & \sigma_z \\ \sigma_z & 0 \end{pmatrix},$$

and the Dirac equation can be written as

$$(i\partial_0 - m)\varphi(\underline{r}) = \boldsymbol{\sigma}.\boldsymbol{p}\chi(\underline{r}),$$
$$(-i\partial_0 - m)\chi(\underline{r}) = \boldsymbol{\sigma}.\boldsymbol{p}\varphi(\underline{r}). \tag{74}$$

As before, we write the Dirac equation for a charged particle in external potential, by replacing the quadrivector energy–momentum by the momentum quadrivector:

$$\pi^\mu = P^\mu - eA^\mu, \tag{75}$$

---

[5] In the Dirac equation the energy has no lower limit, where it was necessary to introduce the notion of antiparticle

where A^μ is the quadrivector potential. The Dirac Hamiltonian becomes

$$H_D = \boldsymbol{\alpha}.(\boldsymbol{P} - e\boldsymbol{A}) + m\beta + eV. \tag{76}$$

As for the free-particle equation within the spin-position representation, we find

$$\begin{aligned}
(i\partial_0 - m - eV)\varphi(\underline{r}) &= \boldsymbol{\sigma}.(\boldsymbol{p} - e\mathbf{A}(\underline{r}))\chi(\underline{r}), \\
(-i\partial_0 - m + eV)\chi(\underline{r}) &= \boldsymbol{\sigma}.(\boldsymbol{p} - e\mathbf{A}(\underline{r}))\varphi(\underline{r}).
\end{aligned} \tag{77}$$

4.3 The Foldy–Wouthuysen Transformation and the Spin–Orbit Coupling

As we have seen in Eqs. (74) and (77), the two components of the wave function are coupled. This is due to the fact that the γ^μ matrices are not diagonals in the spin-position representation and consequently the Hamiltonian is not bloc diagonal in this representation. We search for a representation where the Hamiltonian is diagonal [61]. For that Foldy–Wouthuysen used a unitary transformation U:

$$U(\underline{r}) = e^{iS(\underline{r})}, \tag{78}$$

where S is Hermitian.

It can be shown then that the Foldy–Wouthuysen Hamiltonian becomes

$$\begin{aligned}
H_{\text{FW}} = {}&\beta\left(m + \frac{(p-eA)^2}{2m}\right) + eV - \frac{e}{2m}\beta\boldsymbol{\sigma}.\boldsymbol{B} - \frac{ie}{8m^2}\boldsymbol{\sigma}.(\boldsymbol{\nabla}\wedge\boldsymbol{E}) \\
&- \frac{e}{4m^2}\boldsymbol{\sigma}.(\boldsymbol{E}\wedge\boldsymbol{p}) - \frac{e}{8m^2}\boldsymbol{\nabla}.\boldsymbol{E} + o\left(\frac{1}{m^2}\right).
\end{aligned} \tag{79}$$

If the potential has a spherical symmetry, we obtain

$$\nabla V(\boldsymbol{r}) = \frac{1}{r}\frac{\partial V}{\partial r}\mathbf{r}, \tag{80}$$

and as a consequence

$$-\frac{e}{4m^2}\boldsymbol{\sigma}.(\boldsymbol{E}\wedge\boldsymbol{p}) = \frac{e}{4m^2}\frac{1}{r}\frac{dV}{dr}\boldsymbol{\sigma}.\boldsymbol{L}. \tag{81}$$

We find the SOC term and the Darwin correction term. Using the fact that $\mathbf{E} = -\nabla V$, the Foldy–Wouthuysen Hamiltonian becomes

$$H_{\text{FW}} = \beta\left(m + \frac{(p-eA)^2}{2m}\right) + eV - \frac{e}{2m}\boldsymbol{\sigma}.\boldsymbol{B} - \frac{e}{4m^2}\boldsymbol{\sigma}\frac{dV}{dr}.\boldsymbol{L} - \frac{e}{8m^2}\Delta V. \tag{82}$$

Notice, since the Foldy–Wouthuysen Hamiltonian contains the SOC term, neither the spin nor the momentum is a constant of motion since

$$[H_{\text{FW}}, L_z] = -[H_{\text{FW}}, S_z] = i\zeta(r)\left(\sigma_x L_y - \sigma_y L_x\right) \neq 0, \tag{83}$$

with

$$\zeta(r) = \frac{e}{4m^2}\frac{1}{r}\frac{dV}{dr}.$$

This Hamiltonian, however, commutes with the sum $J_z = \sigma_z + L_z$; consequently the ensemble $\{H, J^2, J_z\}$ constitutes a complete ensemble of commuting operators and possess, therefore, a common eigenvector which we define as $|JM\rangle$. Using the addition of angular momentum theorem we find

$$|JM\rangle = \sum_{m=\pm\frac{1}{2}} \langle L, M - m; S, m \mid JM\rangle |L, M - m; S, m\rangle, \tag{84}$$

where $\langle L, M - m; S, m \mid JM\rangle$ are the Clebsch–Gordan coefficients. We can also define the K operator as

$$K = \beta(\boldsymbol{\sigma}.\boldsymbol{L} + 1) \tag{85}$$

which commutes with H, J^2, and J_z and with the inversion operator $P = i\gamma_0$. We find that

$$K^2 = J^2 + \frac{1}{4}. \tag{86}$$

The eigenvalues of K^2 are given by

$$\kappa = \pm\left(j + \frac{1}{2}\right). \tag{87}$$

Dirac equation for a spin-dependent potential: The treatment of the magnetic materials within a completely relativistic theory has been discussed in the literature [63, 4]. Analogous to the non-relativistic theory, we can describe the fundamental state of many relativistic electrons. The approach leads to a current functional theory where the quadrivector current J^μ is the central quantity. The corresponding Hamiltonian has the following form:

$$H = \boldsymbol{\alpha}.\left[\boldsymbol{p} + \boldsymbol{A}_H(j_\mu) + \boldsymbol{A}_{\text{xc}}(j_\mu)\right] + \frac{1}{2}(\beta - I) + V_H + V_{\text{xc}}. \tag{88}$$

The H, xc indices represent the Hartree and the exchange–correlation to the the scalar potential $V(r)$ and vector potential $\boldsymbol{A}$. Because of the numerical difficulties, an alternative approach was suggested, analogous to the local spin density approximation (LSDA) for the non-relativistic theory. We assume an interaction of the spins of the electrons with a hypothetical field of magnetic or exchange–correlation origin:

$$H = -i\boldsymbol{\alpha}.\nabla + \frac{1}{2}(\beta - I) + V(\boldsymbol{r}), \tag{89}$$

with a periodic potential V of the following form:

$$V(\boldsymbol{r}) = V_H(\boldsymbol{r}) + V_{xc}(\boldsymbol{r}) + V_{\text{spin}}(\boldsymbol{r}), \tag{90}$$

with

$$V_{\text{spin}}(\boldsymbol{r}) = \beta\boldsymbol{\sigma}.\left(\boldsymbol{B}_{\text{ext}} + \frac{\partial E_{xc}}{\partial m(\boldsymbol{r})}\right). \tag{91}$$

4.4 Solution of the Dirac Equation

4.4.1 Wave Equation in Polar Coordinates

To determine the wave equation in polar coordinates, we consider a stationary state of energy W in a central potential $V(\boldsymbol{r})$ and we transform the term of the kinetic energy $\boldsymbol{\alpha}.\boldsymbol{p}$. For this we will use the following identity [62]:

$$\nabla = \boldsymbol{u}_r(\boldsymbol{u}_r.\nabla) - \boldsymbol{u}_r \wedge (\boldsymbol{u}_r \wedge \nabla) = \boldsymbol{u}_r\frac{\partial}{\partial r} - i\frac{\boldsymbol{u}_r}{r} \wedge \boldsymbol{L}, \tag{92}$$

where $\boldsymbol{L}$ is the angular momentum, $\boldsymbol{u}_r$ the radial unitary vector. From this equation, the kinetic energy operator becomes

$$\boldsymbol{\alpha}.\boldsymbol{p} = -\alpha_r\frac{\partial}{\partial r} + i\frac{\alpha_r}{r}\boldsymbol{\sigma}.\boldsymbol{L}. \tag{93}$$

This equation can be substituted in the wave equation and using the K operator defined by Eq. (85) we obtain

$$W\Psi = H\Psi = \left[i\gamma^5\sigma_r\left(\frac{\partial}{\partial r} + \frac{1}{r} - \frac{\beta}{r}K\right) + V + \beta\right]\Psi, \tag{94}$$

with $\gamma^5 = i\gamma^0\gamma^1\gamma^2\gamma^3$. Thus we obtain a wave equation in polar coordinates.

From the fact that J^2, J_z, and K commutes with $V(r)$, these three operators commute with H. We will be interested by a representationwhich diagonalizes these three operators and H. The eigenvalues of J^2, J_z, and K are, respectively, $j(j+1)$, μ, and κ. As we have mention it in Eq. (72) Ψ can be written as:

$$\Psi(\boldsymbol{r}) = \begin{pmatrix} \psi^\mu(\boldsymbol{r}) \\ \psi^l(\boldsymbol{r}) \end{pmatrix}. \tag{95}$$

We have

$$(\boldsymbol{\sigma}.\boldsymbol{L}+1)\psi^\mu = -\kappa\psi^\mu, \tag{96}$$

$$(\boldsymbol{\sigma}.\boldsymbol{L}+1)\psi^l = \kappa\psi^l, \tag{97}$$

$$J^2\psi^{\mu,l} = j(j+1)\psi^{\mu,l}, \tag{98}$$

$$J_z\psi^{\mu,l} = \mu\psi^{\mu,l}, \tag{99}$$

where $\psi^{\mu,l}$ are a two component spinors and are, respectively, proportional to χ^μ_κ and $\chi^\mu_{-\kappa}$. We therefore can write

$$\Psi_\Lambda = \Psi^\mu_\kappa = \begin{pmatrix} g(r)\chi^\mu_\kappa \\ if(r)\chi^\mu_{-\kappa} \end{pmatrix}, \tag{100}$$

where

$$\chi^\mu_\kappa = \sum_{m_s} C^{\kappa,\mu}_{l,\mu-m_s;S,m_s} Y^{\mu-m_s}_l \chi_{m_s}, \tag{101}$$

and $C^{\kappa,\mu}_{l,\mu-m_s;S,m_s}$ are the Clebsch–Gordan coefficients and χ_{m_s} Pauli spinors, and $g(r)$ and $f(r)$ are the radial functions which depend on κ. The phase i is there to make f and g explicitly real. We obtain the following equations:

$$(W-V-1)g(r)\chi^\mu_\kappa = \left[-\left(\frac{\partial f(r)}{\partial r} + \frac{f(r)}{r}\right) + \kappa\frac{f(r)}{r}\right]\chi^\mu_\kappa, \tag{102}$$

$$(W-V-1)f(r)\chi^\mu_{-\kappa} = \left[\frac{\partial g(r)}{\partial r} + \frac{g(r)}{r} + \kappa\frac{g(r)}{r}\right]\chi^\mu_{-\kappa}, \tag{103}$$

where $\sigma_r\chi^\mu_\kappa = -\chi^\mu_{-\kappa}$ is used. These last equations give the final radial equations:

$$\frac{\partial g(r)}{\partial r} = (W-V+1)f(r) - (\kappa+1)\frac{g(r)}{r}, \tag{104}$$

$$\frac{\partial f(r)}{\partial r} = (\kappa - 1)\frac{g(r)}{r} - (W - V - 1)g(r). \tag{105}$$

It often helpful to use $\xi_1 = rg(r)$ and $\xi_2 = rf(r)$. We obtain the equivalent following radial equations:

$$\frac{d}{dr}\begin{pmatrix} \xi_1 \\ \xi_2 \end{pmatrix} = \begin{pmatrix} -\frac{\kappa}{r} & W + 1 - V \\ -(W - 1 - V) & \frac{\kappa}{r} \end{pmatrix}\begin{pmatrix} \xi_1 \\ \xi_2 \end{pmatrix}, \tag{106}$$

$$\left\{\frac{d^2}{dr^2} + \frac{dV/dr}{W - V + 1}\frac{d}{dr} + \left[(W - V)^2 - 1 - \frac{\kappa(\kappa + 1)}{r^2} + \frac{\kappa}{r}\frac{dV/dr}{W - V + 1}\right]\right\}\xi_1(r) = 0. \tag{107}$$

These equations are solved numerically, and we obtain the solution of the Dirac equation in the spherical region. In the first appendix we will give some details about the use of spin–orbit coupling in the FLAW formalism.

5 The FLAPW Method

5.1 Introduction

Before introducing the full-potential linear augmented plane wave (FLAPW) method, we would like to give a brief overview of ab initio methods. Several methods have been developed to solve the Kohn–Sham (KS) equations. The idea of dividing the space into spheres centered at each atom site, the so-called muffin-tin (MT) regions or augmentation region, and the remaining interstitial region was already proposed by Slater [64–66] before the KS equations. The concept of this division for a periodic potential corresponds to the Augmented Plane Wave (APW) technique. Soon after, this concept has been adopted by the Korringa [67], Kohn, and Rostoker [68] (KKR) to develop the so-called Green's function KKR method. The APW method, as all the others MT orbital-based methods, has known some deficiencies. The most problematic is that of the non linearity of the eigenvalue equations with respect to the energy. Other methods, such as the orthogonalized plane wave (OPW) method [69] and the linear combination of atomic orbitals (LCAO) method [7], which are quite similar to the APW method,[6] have been successful due to their accurate calculations of particular crystals. The applications of the OPW method, however, have been limited primarily to nearly free-electron (NFE) crystals. The reasons for that can be summarized in two points. The first one is that this method requires the electrons in the crystal to be separated into core and itinerant electrons,

---

[6] Terrell has shown that the APW method gives nearly the same results as the OPW method for the Be metal [70].

and all the non-overlapping atomic states with the neighboring lattice site states are considered as core states, so that the d-states, for example, will be considered as such.[7] The second one is that the OPW method is more difficult to apply to heavy elements since they have more core electron states. Therefore orthogonalizing a plane wave function[8] to these states requires more efforts. The complications of the OPW method had stimulated, at that time, the development of the actually used pseudopotential methods.

In the APW method, all that is required is the total electronic charge density based on atomic self-consistent calculations.

Some years later Andersen [71] succeeded in linearizing these eigenvalue equations within the same muffin-tin (MT) model, which thus has given rise to both the linear muffin-tin orbitals (LMTO) method and the linear augmented plane wave (LAPW) method.

One of the commonly used methods to solve the Kohn–Sham equations is to use some kind of basis set to represent the Bloch wave functions. A suitable basis-set choice suggested by Bloch's theorem is a sum of plane waves. They have several advantages: the implementation of the plane waves-based methods is rather straightforward because of their simplicity; they are orthogonal and diagonal in momentum. The only problem which arises from this representation is that it requires so many plane waves to account for the fast varying electron wave functions near the core. To overcome this problem with only a few basis functions, one can use a basis set which contains radial wave functions to describe the oscillations near the core. This is the suggested fundamental idea by Slater [64] for the augmented plane wave (APW) method.

5.2 The APW Concept

Within the APW approach, space is divided into spheres centered at each atom site (the MT spheres), and the remaining region is the interstitial region (IR). These regions can be seen in Fig. 3. Inside the MT spheres the potential is of spherical symmetry, and the interstitial potential is constant. The single-particle wave function $\psi_n(\mathbf{k}, \mathbf{r})$, which describes the physics within such environment, is therefore expressed in terms of the following basis functions:

---

[7] The problem which we would like to notice here is that though the d-states are relatively narrow and do not overlap with the other states, they are still far from being considered as frozen core states.

[8] The OPW basis functions are constructed by orthogonalizing plane waves to the core states. The resulting OPW's have nodal character in the core region but are essentially plane waves in the outer part.

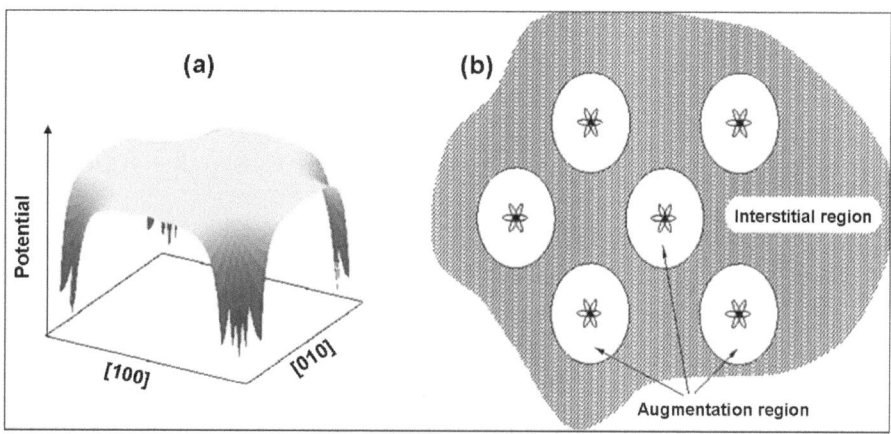

Fig. 3 (**a**) A typical form of a based APW potential, (**b**) The representation of space into MT and interstitial region

$$\varphi_{\mathbf{G}}(\mathbf{k}, \mathbf{r}) = \begin{cases} e^{i(\mathbf{G}+\mathbf{k})\mathbf{r}} & \text{IR} \\ \sum_{lm} A_{lm}^{\mu\mathbf{G}}(\mathbf{k}) u_l(r) Y_{lm}(\hat{\mathbf{r}}) & \text{MT} \quad \mu. \end{cases} \tag{108}$$

Thus, the wave function takes the form

$$\psi_n(\mathbf{k}, \mathbf{r}) = \sum_{\mathbf{G}} C_{n\mathbf{k}}^{\mathbf{G}} \varphi_{\mathbf{G}}(\mathbf{k}, \mathbf{r})$$

$$= \begin{cases} \sum_{\mathbf{G}} C_{n\mathbf{k}}^{\mathbf{G}} e^{i(\mathbf{G}+\mathbf{k})\mathbf{r}} & \text{IR} \\ \sum_{\mathbf{G}} \sum_{lm} C_{n\mathbf{k}}^{\mathbf{G}} A_{lm}^{\mu}(\mathbf{k}+\mathbf{G}) u_l(r) Y_{lm}(\hat{\mathbf{r}}) & \text{MT}, \end{cases} \tag{109}$$

where $\mathbf{k}$ is the Bloch wave vector, $\mathbf{G}$ is the reciprocal lattice vector, l and m are the angular quantum numbers, and u_l is the radial solution of the Schrödinger equation for a given energy E_l solved for the MT sphere located at τ_μ in the unit cell:

$$\left\{ -\frac{\hbar^2}{2m} \frac{\partial^2}{\partial r^2} + \frac{\hbar^2}{2m} \frac{l(l+1)}{r^2} + V(r) - E_l \right\} r u_l(r) = 0, \tag{110}$$

where $V(r)$ is the spherical component of the potential. Since the u_l functions account for the regular solutions, the basis functions inside the spheres should form a completely orthogonal basis set and the u_l functions should be orthonormal. Using the Rayleigh expression

$$e^{i(\mathbf{k}+\mathbf{G})\mathbf{r}} = 4\pi \sum_{lm} i^l j_l(|\mathbf{k}+\mathbf{G}||\mathbf{r}|) Y_{lm}^*(\widehat{\mathbf{k}+\mathbf{G}}) Y_{lm}(\hat{\mathbf{r}}), \tag{111}$$

and the continuity of the wave functions at the boundary of the MT spheres determine the $A_{lm}^{\mu}(\mathbf{k}+\mathbf{G})$ coefficients according to

$$A_{lm}^{\mu}(\mathbf{k}+\mathbf{G}) = \frac{4\pi i^l}{u_l(R_{\mu})} \exp(i(\mathbf{k}+\mathbf{G})\tau_{\mu}) j_l(|\mathbf{k}+\mathbf{G}||\mathbf{R}_{\mu}|) Y_{lm}^{*}(\widehat{\mathbf{k}+\mathbf{G}}), \qquad (112)$$

where R_{μ} is the MT radius of sphere μ.

The eigenvalue problem has the following form:

$$\hat{H}\psi_n(\mathbf{k},\mathbf{r}) = \varepsilon_{n\mathbf{k}}\psi_{\mathbf{n}}(\mathbf{k},\mathbf{r}), \qquad (113)$$

where n is the band index.

Even though plane waves form an orthogonal basis set, the APW functions do not. The plane waves in the interstitial region are non-orthogonal, because the MT regions are cut-out and, therefore, the integration over r space (in terms of which the orthogonality is defined) is not carried out over the whole unit cell but only over the interstitial region. An additional contribution comes from the MT regions, this is the so-called *augmented* contribution, which somehow, makes the plane waves coupled to the MT functions $(u_l(r)Y_{lm}(\hat{r}))$.

Due to the non-orthogonality of the basis functions the overlap matrix $\mathbf{S}$

$$S^{\mathbf{G},\mathbf{G}'}(\mathbf{k}) = \int \varphi_{\mathbf{G}'}(\mathbf{k},\mathbf{r})\varphi_{\mathbf{G}}(\mathbf{k},\mathbf{r})\mathrm{d}^3r, \qquad (114)$$

is not diagonal.

Using the wave function expansion (109), the eigenvalue problem (113) can be rewritten in its generalized form as:

$$(H(\mathbf{k}) - \varepsilon_{\nu\mathbf{k}}S(\mathbf{k}))C_{\nu\mathbf{k}} = 0 \quad \forall \mathbf{k} \in \text{BZ}. \qquad (115)$$

Within the APW method, the E_l parameters are mapped to the *real*-band energies $\varepsilon_{\nu\mathbf{k}}$; thus the u_l solutions become the functions of these band energies $u_l(r,\varepsilon_{\nu\mathbf{k}})$, and Eq. (115) is therefore nonlinear in energy,[9] so it can no longer be determined by a simple diagonalization. One way of solving this problem is to fix the energy E_l and scan over $\mathbf{k}$ to find the solution $u_l(\varepsilon_{\nu\mathbf{k}})$, which corresponds to the optimal shape of the band energies $\varepsilon_{\nu\mathbf{k}}$, instead of diagonalizing a matrix to find all the bands at a given $\mathbf{k}$. The Slater's formulation of the secular equation is, thus, computationally much more demanding than an ordinary linear one.

Another limitation of the APW method (known as the *asymptote* problem) is that of the zero value of $u_l(R)$ at the MT boundary in Eq. (112). The $A_{lm}^{\mu G}$'s are no longer finite, and the radial function and the plane wave become decoupled. Further details about the APW method can be found in the book by Loucks [72].

---

[9] The Hamiltonian matrix H depends not only on $\mathbf{k}$ but also on $\varepsilon_{\nu\mathbf{k}}$, H($\varepsilon_{\nu\mathbf{k}}$).

These problems are circumvented within the LAPW method proposed by Andersen [71]. The following section is devoted to the discussion of the main features of this method which will be necessary to follow the calculation of the XMCD matrix elements. Further details of the methods can be found in the article of Bluegel and Bihlmayer [73].

5.3 The LAPW Method

The basic idea of the linearized version of the APW (LAPW) is to expand the u_l functions into a Taylor-series around the E_l energy parameter

$$u_l(\varepsilon, r) = u_l(E_l, r) + \dot{u}_l(E_l, r)(\varepsilon - E_l) + O[(\varepsilon - E_l)^2], \qquad (116)$$

where the $\dot{u}_l$ denotes the energy derivative of u_l, $\partial u_l(\varepsilon, r)/\partial \epsilon$, and $O[(\varepsilon - E_l)^2]$ denotes errors that are quadratic in energy. Therefore, according to the variational principle the error in the calculated band energies is of order $(\varepsilon - E_l)^4$. Because of this high order, the linearization works well even over a rather broad energy region.

With this linearization, the explicit form of the basis functions is now as following:

$$\varphi_G(\mathbf{k}, \mathbf{r}) = \begin{cases} e^{i(\mathbf{G}+\mathbf{k})\mathbf{r}} & \text{IR} \\ \sum_{lm}(A_{lm}^{\mu G}(\mathbf{k})u_l(r) + B_{lm}^{\mu G}(\mathbf{k})\dot{u}_l(r))Y_{lm}(\hat{\mathbf{r}}) & \text{MT} \quad \mu. \end{cases} \qquad (117)$$

The values of the coefficients $A_{lm}^{\mu G}(\mathbf{k})$ and $B_{lm}^{\mu G}(\mathbf{k})$ are determined by insuring the continuity of the basis functions and their derivatives at the MT boundary (a detailed description of these coefficients will be provided in the following sections). The energy dependence of the Hamiltonian is therefore removed, which reduces the energy search given by Eq. (115) to a standard eigenvalue problem of linear algebra. This is a direct consequence of the disappearance of the discontinuity in the basis function derivatives (encountered in the APW method).

Taking the energy derivative of Eq. (110):

$$\left\{-\frac{\hbar^2}{2m}\frac{\partial^2}{\partial r^2} + \frac{\hbar^2}{2m}\frac{l(l+1)}{r^2} + V(r) - E_l\right\}r\dot{u}_l(r) = ru_l(r), \qquad (118)$$

where the $\dot{u}_l$ can be calculated as a solution of a Schrödinger-like equation. The E_l is fixed energy for each MT orbital and is chosen to minimize the errors due to the linearization of the wave function. For that this energy is chosen at the center of gravity of the occupied canonical band [74].

Since it is no longer necessary to set the energy parameters equal to the band energies, the latter can be determined by a single diagonalization of the Hamiltonian matrix (Eq. 115).

In order to simplify the calculation of the elements of the Hamiltonian matrix, the normalization of u_l is required

$$\int_0^{R_{\mathrm{MT}}} u_l^2(r)r^2\mathrm{d}r = 1, \tag{119}$$

which implies that the energy derivatives of u_l, $\dot{u}_l(r)$ are orthogonal to the radial functions, i.e.,

$$\int_0^{R_{\mathrm{MT}}} u_l(r)\dot{u}_l(r)r^2\mathrm{d}r = 0. \tag{120}$$

Once the u_l and $\dot{u}_l$ are made orthogonal, the basis functions inside the spheres form a completely orthogonal basis set, since the angular functions $Y_{lm}(\hat{r})$ are also orthogonal. However, the LAPW functions are, in general, not orthogonal to the core states, which are treated separately in the LAPW method.

In some materials the high-lying core states, the so-called semicore states, pose a problem to LAPW calculations: they are too delocalized to be described as core states and too deep in energy to be described as valence or conduction states. One of the strategies to overcome this problem is the use of local orbitals [75]. The local orbitals are an extension to the FLAPW basis that can be used to improve the representation of the semicore states. The extra basis functions are completely localized inside the MT spheres, and their values and derivatives fall to zero at the MT radii.[10] This can be achieved via a linear combination including three radial functions. The standard FLAPW functions u_l and $\dot{u}_l$ plus a further radial function $u_{l_{lo}}$, where l_{lo} is the quantum number l for local orbitals. This new radial function is constructed in the same way as u_l, but with different energy parameter $E_{l_{lo}}$. A detailed discussion of these problems can be found in the book by Singh [76] and a good review is given in [73].

Given that the LAPW basis set offers enough variational freedom, its extension to non-spherical potentials could be done with little difficulty. This leads then to the full-potential linearized augmented plane wave method (FLAPW).

6 The FLAPW Concept

In the full-potential LAPW method (FLAPW) [77, 78] the full-potential and charge density without are calculated without any shape approximation in the interstitial region and inside the muffin-tins:

---

[10] That is why no additional boundary conditions has to be satisfied.

$$V(\mathbf{r}) = \begin{cases} \sum_{\mathbf{G}} V_I^{\mathbf{G}} e^{i\mathbf{Gr}} & \text{IR} \\ \sum_{lm} V_{MT}^{l,m}(r) Y_{lm}(\hat{\mathbf{r}}) & \text{MT,} \end{cases} \tag{121}$$

the charge density, $\rho(\mathbf{r})$, is represented in the same way as the potential:

$$\rho(\mathbf{r}) = \begin{cases} \sum_{\mathbf{G}} \rho_I^{\mathbf{G}} e^{i\mathbf{Gr}} & \text{IR} \\ \sum_{lm} \rho_{MT}^{l,m}(r) Y_{lm}(\hat{\mathbf{r}}) & \text{MT} \quad \mu \end{cases} \tag{122}$$

We have to mention here that though the potential is nearly constant in the interstitial region in most materials, it is not necessarily true for the open systems or for systems with small packing factor.

It turns out that on contrary to the methods using the atomic-sphere approximation (ASA) [71], the FLAPW method accounts for the most realistic potential and leads, therefore, to a realistic distribution of the charge density within the whole space. In other words, within the FLAPW scheme the charge density is sensitive to the slightest variation of the potential in the whole space.

As other density functional theory-based codes, the first-principles FLAPW (Fleur code [79]) method is implemented according to a typical self-consistent loop (Fig. 4). We provide in the following sections the features of the main steps of a bulk calculation.

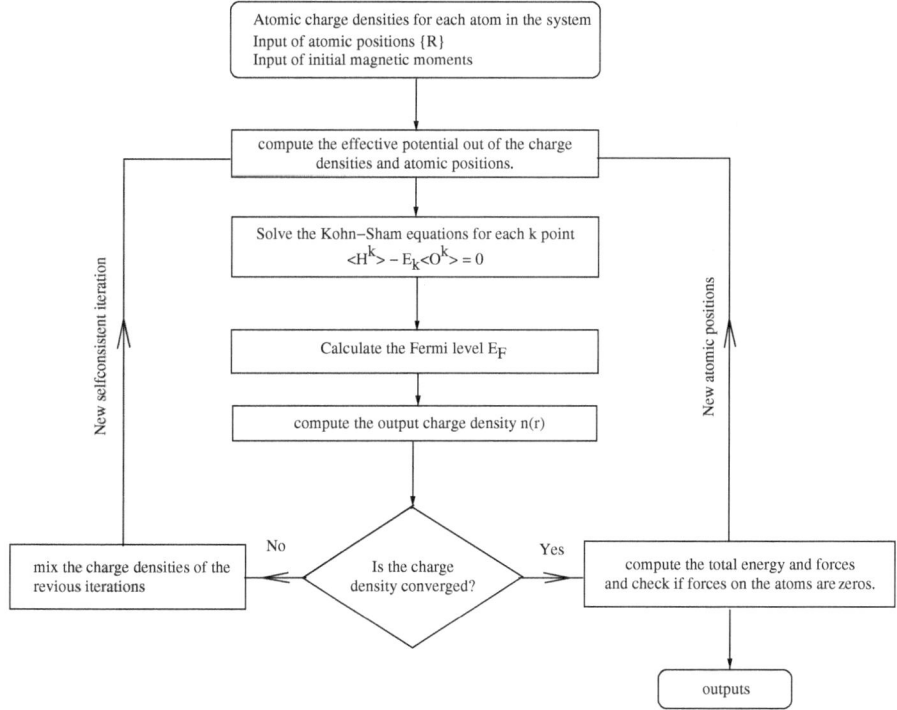

Fig. 4 Typical loop structure of a first-principles code based on density functional theory

6.1 Construction of the Potential

The total potential consists of two parts, the Coulomb potential and the exchange–correlation potential. The Coulomb potential is composed of the Hartree potential $V_H(\mathbf{r})$ and the external potential of the nuclei $V_i(\mathbf{r})$:

$$V_c(\mathbf{r}) = V_H(\mathbf{r}) + V_i(\mathbf{r}). \tag{123}$$

Once an initial charge density $n^0(\mathbf{r})$ (atomic charge) and atom positions $(\mathbf{R})$ are given, the Hartree potential can be determined from the charge density via the Poisson equation:

$$\triangle V_H(\mathbf{r}) = 4\pi n(\mathbf{r}), \tag{124}$$

in real space the solution of Eq. (124) is given by

$$V_H(\mathbf{r}) = \int \frac{n(\mathbf{r})}{|\mathbf{r} - \mathbf{r}'|} d^3 r. \tag{125}$$

In reciprocal space, however, the Poisson equation is diagonal:

$$V_H(\mathbf{G}) = 4\pi \mathbf{G}^2. \tag{126}$$

Therefore, and because of the representation of the charge density and the potential in the interstitial region, the solution of the Poisson equation in reciprocal space appears to be convenient. However, due to the rather localized core and valence states the charge density changes on a very small length scale near the nuclei (the MT region). Thus, the plane wave expansion of n converges slowly, and a direct use of Eq. (126) is impractical. The pseudocharge method [80] is used to circumvent this difficulty.

The problem of determining the exchange–correlation potential is quite different from that of the Coulomb potential, V_{xc}^σ is a local quantity and depends only on $n_\uparrow(\mathbf{r})$ and $n_\downarrow(\mathbf{r})$ at the same position $\mathbf{r}$. Thus, the MT and the interstitial region can be treated independently. Furthermore, V_{xc}^σ and ϵ_{xc}^σ are nonlinear functions of $n_\uparrow$ and $n_\downarrow$ and have to be calculated in real space. First, $n_\uparrow$ and $n_\downarrow$ are transformed to the real space, where V_{xc}^σ and ϵ_{xc}^σ are calculated,[11] and then back-transformed. The potential V_{xc}^σ is then added to the Coulomb potential, yielding the spin-dependent potential $V_\uparrow$ and $V_\downarrow$, whereas ϵ_{xc}^σ is needed for the determination of the total energy.

---

[11] As it was explained in Sects. (3.5) and 3.6, V_{xc}^σ and ϵ_{xc}^σ are calculated using either the LDA or the GGA.

6.2 The LDA(GGA)+U Approach Within the FLAPW

The LDA(GGA)+U implementation within the FLAPW method follows the same logic as explained in Sect. 3.7. The variational LDA(GGA)+U Schrödinger equations are those of Eq. (59):

$$\left[\nabla^2 + V_{\text{LDA}}^{\sigma}(\mathbf{r})\right] \psi_{\mathbf{k},\nu}^{\sigma}(\mathbf{r}) + \sum_{m_1,m_2} V_{m_1,m_2}^{\sigma} \frac{\delta n_{m_1,m_2}^{\sigma}}{\delta \psi_{\mathbf{k},\nu}^{\sigma}} = \varepsilon_{\mathbf{k},\nu}^{\sigma} \psi_{\mathbf{k},\nu}^{\sigma}(\mathbf{r}), \qquad (127)$$

where $V_{\text{LDA}}^{\sigma}(\mathbf{r})$ is the LSDA or the GGA potential calculated using the LDA(GGA)+U charge density:

$$n_{m_1 m_2}^{\sigma} = \sum_{\mathbf{k},\nu} w(\nu,\mathbf{k}) \left[A_{lm_1}(\mathbf{k}) A_{lm_2}^{*}(\mathbf{k}) + \langle \dot{u}_l^{\sigma} | \dot{u}_l^{\sigma} \rangle B_{lm_1}(\mathbf{k}) B_{lm_2}^{*}(\mathbf{k}) \right], \qquad (128)$$

where $A_{lm}(\mathbf{k}) = \langle u_l^{\sigma} Y_{lm} | \psi_{\mathbf{k},\nu}^{\sigma} \rangle$ and $B_{lm}(\mathbf{k}) = \langle \dot{u}_l^{\sigma} Y_{lm} | \psi_{\mathbf{k},\nu}^{\sigma} \rangle$ and (cf. Eq. (57)):

$$V_{m_1,m_2}^{\sigma} = \sum_{m_3,m_4}^{\sigma'} (\langle m_1, m_3 | V^{\text{ee}} | m_2, m_4 \rangle - \langle m_1, m_3 | V^{\text{ee}} | m_2, m_4 \rangle \delta_{\sigma,\sigma'}) n_{m_3,m_4}^{-\sigma'}$$
$$- \delta_{m_1,m_2} U \left(n - \frac{1}{2} \right) + \delta_{m_1,m_2} J \left(n^{\sigma} - \frac{1}{2} \right). \qquad (129)$$

The last term of the variational Hamiltonian is calculated from Eq. (128):

$$\frac{\delta n_{m_1,m_2}^{\sigma}}{\delta \psi_{\mathbf{k},\nu}^{\sigma}} = \langle \psi_{\mathbf{k},\nu}^{\sigma} | u_l^{\sigma} Y_{lm_2} \rangle u_l^{\sigma} Y_{lm_1} + \langle \dot{u}_l^{\sigma} | \dot{u}_l^{\sigma} \rangle \langle \psi_{\mathbf{k},\nu}^{\sigma} | \dot{u}_l^{\sigma} Y_{lm_2} \rangle \dot{u}_l^{\sigma} Y_{lm_1}$$
$$= \left[|u_l^{\sigma} Y_{lm_1} \rangle \langle u_l^{\sigma} Y_{lm_2}| + \langle \dot{u}_l^{\sigma} | \dot{u}_l^{\sigma} \rangle | \dot{u}_l^{\sigma} Y_{lm_1} \rangle \langle \dot{u}_l^{\sigma} Y_{lm_2}| \right] \psi_{\mathbf{k},\nu}^{\sigma}. \qquad (130)$$

The U intra-atomic Coulomb interaction and the J exchange interaction can be calculated according to Eq. (61) within an unscreened atomic formulation. Within the Fleur implementation [79], these interactions are considered as parameters, and they are usually extracted from experimental results.

7 Spin–Orbit Coupling and Magnetic Anisotropy

In a ferromagnetic material, below the Curie temperature, the total energy depends on the orientation of the magnetization. This is usually what is meant by the magnetocrystalline anisotropy in the literature. Two principle magnetic mechanisms are responsible for such phenomena. One is of classical origin and relies in the multipole (mainly dipole) interaction between the moments localized at lattice points [81–83]. The second one is related to the orientation of the spin axis and is of pure relativistic character appearing only when the spin–orbit interaction is taken into account [84]. The spin–orbit coupling (SOC) provides the mechanism that couples the spin moment to the crystal generating thereby a dependence of the energy from the spin axis.

7.1 The Kohn–Sham–Dirac Equation

Relativistic effects are important for the correct description of core or valence electrons. Both core and valence electrons have finite wave functions near the nucleus, where the kinetic energy is large. This kinetic energy becomes more significant for heavier elements and compounds. Additionally, only relativistic effects, in particular the spin–orbit coupling, introduce a link between spatial and spin coordinates. Thus, information about the orientation of spins relative to the lattice can only be gained if relativity is taken into account. For fully relativistic description of the electronic structure, all relativistic effects, i.e., mass-velocity, Darwin-term, spin–orbit coupling, have to be taken into account [85]. However, in many applications an approximation is used, where the spin–orbit coupling is neglected. This approximation is called the scalar relativistic approximation. It consists in including the spin–orbit interaction additionally,[12] either self-consistently or with the use of Andersen's force theorem [86].

In a relativistic density functional theory, the Kohn–Sham equations have the form of a single-particle Dirac equation

$$\{c\boldsymbol{\alpha}.\mathbf{p} + (\beta - 1)mc^2 + V_{eff}(\mathbf{r})\}\boldsymbol{\Psi} = E\boldsymbol{\Psi}, \tag{131}$$

$$\boldsymbol{\alpha} = \left(\begin{pmatrix} 0 & \sigma_x \\ \sigma_x & 0 \end{pmatrix}, \begin{pmatrix} 0 & \sigma_y \\ \sigma_y & 0 \end{pmatrix}, \begin{pmatrix} 0 & \sigma_z \\ \sigma_z & 0 \end{pmatrix}\right)^{tr} = \begin{pmatrix} 0 & \boldsymbol{\sigma} \\ \boldsymbol{\sigma} & 0 \end{pmatrix}, \tag{132}$$

$$\beta = \begin{pmatrix} \mathbf{I}_2 & 0 \\ 0 & -\mathbf{I}_2 \end{pmatrix}. \tag{133}$$

σ_x, σ_y, and σ_z are the three components of the Pauli matrix vector $\boldsymbol{\sigma}$, $\mathbf{p}$ is the momentum operator, and $\mathbf{I}_n$ is the $(n \times n)$ unit matrix. V_{eff} is the effective potential that contains electron–nucleon Coulomb potential, Hartree potential, and exchange–correlation potential. In the case of spin polarization, V_{eff} is spin dependent. Finally, $\boldsymbol{\Psi}$ is the relativistic four component wave function. The straightforward way to solve this problem would be to expand each of the four components of $\boldsymbol{\Psi}$ in terms of the FLAPW basis. However, if all four components were treated with the same accuracy,

---

[12] This is known as the second variational scheme.

this would result in a basis set which contains four times as many functions as in the non-relativistic (non-magnetic) case. Since the numerical effort of the Hamiltonian diagonalization scales with the dimension of the matrix to the third power, the computing time needed for the diagonalization increases by a factor of 64.

The FLAPW implementation within the Fleur code [79] introduces some approximations to make relativistic calculations more efficient. One of these approximations is the scalar relativistic approximations, which has been suggested by Koelling and Harmon [87], where the spin–orbit term is neglected, and spin and spatial coordinates become decoupled. Hence, the Hamiltonian matrix reduces to two matrices of half the size, which can be diagonalized separately. This saves a factor of four in computing time. The scalar relativistic approximation will be discussed in more detailed in the next section. It should be noted that relativistic effects are only significant close to the nucleus, where the kinetic energy is large. It is therefore reasonable to treat the interstitial region non-relativistically. Thus, merely within the muffin-tin spheres the electrons are treated relativistically. And only the large component of Ψ is matched to the non-relativistic wave functions at the boundary between the muffin-tins and the interstitial region, because the small component is already negligible at this distance from the nucleus. The small component is attached to the large component and cannot be varied independently. However, this is somewhat a sensible approximation for two reasons: First even inside the muffin-tin sphere the large component is still much bigger than the small component and plays an important role and second the two components are determined by solving the scalar relativistic equations for the spherically averaged potential. Therefore, they are very well suited to describe the wave functions.

Hence, the size of the basis set and the Hamiltonian matrix remain the same as in non-relativistic calculations, but the problem has to be solved twice, once for each direction of spin. This numerical effort is equal to that needed in spin-polarized non-relativistic calculations.

7.2 The Scalar Relativistic Approximation

As it was pointed out in the previous section, the electrons are only treated relativistically inside the muffin-tin spheres. Thus the first problem that has to be addressed is the construction of the radial function. This is done by solving the scalar relativistic equation, including only the spherically averaged part of the potential. The starting point is the following Dirac equation:

$$\{c\boldsymbol{\alpha}.\mathbf{p} + (\beta - 1)mc^2 + V(\mathbf{r})\}\Psi = E\Psi. \tag{134}$$

The solution of Eq. (134) is discussed in many textbooks, e.g., E.M. Rose [62]. Due to the spin–orbit coupling m_l and m_s are not good numbers any more, and they have to be replaced by the quantum numbers κ and μ (or j and μ), which are eigenvalues of the operators K and the z-component of the total angular momentum. The operator K is defined by

$$K = \beta(\boldsymbol{\sigma}.\mathbf{l} + 1).$$

(135)

The solution of Eq. (134) has the following form

$$\Psi = \Psi_{\kappa\mu} = \begin{pmatrix} g_\kappa(r)\chi_{\kappa\mu} \\ if_\kappa(r)\chi_{-\kappa\mu} \end{pmatrix},$$

(136)

where $g_\kappa(r)$ is the large component, $f_\kappa(r)$ is the small component, $\chi_{\kappa\mu}$ and $\chi_{-\kappa\mu}$ are spin angular functions, which are eigenfunctions of $\mathbf{j}^2$, j_z, K with eigenvalues $j(j+1), \mu, \kappa$, respectively. The spin angular functions can be expanded into a sum of products of spherical harmonics and Pauli spinors, where the expansion coefficients are the Clebsch–Gordan coefficients. The radial functions have to satisfy the set of coupled equations:

$$\begin{pmatrix} -\frac{\kappa+1}{r} - \frac{\partial}{\partial r} & 2Mc \\ \frac{1}{c}(V(r) - E) & \frac{\kappa-1}{r} - \frac{\partial}{\partial r} \end{pmatrix} \begin{pmatrix} g_\kappa(r) \\ f_\kappa(r) \end{pmatrix} = 0,$$

(137)

with

$$M = m + \frac{1}{2c^2}(E - V(r)).$$

(138)

To derive the scalar relativistic approximation, Koelling and Harmon [87] have introduced the following transformation:

$$\begin{pmatrix} g_\kappa(r) \\ \phi_\kappa(r) \end{pmatrix} = \begin{pmatrix} 1 & 0 \\ \frac{1}{2Mc}\frac{\kappa+1}{r} & 1 \end{pmatrix} \begin{pmatrix} g_\kappa(r) \\ f_\kappa(r) \end{pmatrix}.$$

(139)

Using this transformation, Eq. (137) becomes

$$\begin{pmatrix} -\frac{\partial}{\partial r} & 2Mc \\ \frac{1}{2Mc}\frac{l(l+1)}{r^2} + \frac{1}{c}(V(r) - E) + \frac{\kappa+1}{r}\frac{M'}{2M^2c} & -\frac{2}{r} - \frac{\partial}{\partial r} \end{pmatrix} \begin{pmatrix} g_\kappa(r) \\ \phi_\kappa(r) \end{pmatrix} = 0,$$

(140)

where M' denotes the radial derivative of M ($\frac{\partial M}{\partial r}$), and the identity $\kappa(\kappa+1) = l(l+1)$ has been used. Since κ is the eigenvalue of $K = \beta(\boldsymbol{\sigma}.\mathbf{l} + 1)$, the term $\frac{(\kappa+1)M'}{2M^2cr}$ can be identified as the spin–orbit term. This term is dropped in the scalar relativistic approximation, because it is the only one, that causes coupling of spin-up and spin-down contributions.

The radial functions $g_l(r)$ and $\phi_l(r)$ (the index κ has been replaced by l) can now be calculated from the following differential equations:

$$\frac{\partial}{\partial r} g_l(r) = 2Mc\phi_l(r), \tag{141}$$

$$\frac{\partial}{\partial r} \phi_l(r) = \left(\frac{1}{2Mc} \frac{l(l+1)}{r^2} + \frac{1}{c}(V(r) - E) \right) g_l(r) - \frac{2}{r}\phi_l(r). \tag{142}$$

The energy derivative of these equations yields straightforwardly a set of equations for $\dot{g}_l(r)$ and $\dot{\phi}_l(r)$, which are the relativistic analog of $\dot{u}_l(r)$. For numerical reasons the functions $g_l(r)$ and $\phi_l(r)$ are replaced by $p_l(r) = rg_l(r)$ and $q_l(r) = cr\phi_l(r)$.

7.3 The Spin–Orbit Coupling Implementation Within the FLAPW

In the present Fleur code implementation [79] of the FLAPW method the relativistic radial wave functions are normalized according to

$$\left\langle \begin{pmatrix} g_l(r) \\ \phi_l(r) \end{pmatrix} \middle| \begin{pmatrix} g_l(r) \\ \phi_l(r) \end{pmatrix} \right\rangle = \int_0^{R_{MT}} (g_l^2(r) + \phi_l^2(r))r^2 dr = 1. \tag{143}$$

The energy derivatives of the radial functions have to be made orthogonal to the radial functions:

$$\left\langle \begin{pmatrix} g_l(r) \\ \phi_l(r) \end{pmatrix} \middle| \begin{pmatrix} \dot{g}_l(r) \\ \dot{\phi}_l(r) \end{pmatrix} \right\rangle = 0. \tag{144}$$

So that the scalar relativistic FLAPW basis set takes the form

$$\varphi_{\mathbf{G}}(\mathbf{k}, \mathbf{r}) = \begin{cases} e^{i(\mathbf{G}+\mathbf{k})\mathbf{r}} & \text{IR} \\ \sum_{lm} \left(A_{lm}^{\mu\mathbf{G}}(\mathbf{k}) \begin{pmatrix} g_l(r) \\ \phi_l(r) \end{pmatrix} + B_{lm}^{\mu\mathbf{G}}(\mathbf{k}) \begin{pmatrix} \dot{g}_l(r) \\ \dot{\phi}_l(r) \end{pmatrix} \right) Y_{lm}(\hat{\mathbf{r}}) & \text{MT} \quad \mu, \end{cases} \tag{145}$$

which is too similar to that of a non-relativistic basis set (Eq. 117).

Note that the Pauli spinors have been omitted, since the spin-up and spin-down problems are solved independently within the scalar relativistic approximation. Ignoring the spin–orbit coupling term in Eq. (140) the scalar relativistic Hamiltonian including only the spherically averaged part of the potential can be expressed as

$$H_{sp} \begin{pmatrix} g_l(r) \\ \phi_l(r) \end{pmatrix} = E \begin{pmatrix} g_l(r) \\ \phi_l(r) \end{pmatrix}, \tag{146}$$

with

$$H_{sp} = \begin{pmatrix} \frac{1}{2M} \frac{l(l+1)}{r^2} + V(r) & -\frac{2c}{r} - c\frac{\partial}{\partial r} \\ c\frac{\partial}{\partial r} & -2mc^2 + V(r) \end{pmatrix}. \tag{147}$$

Thus, the Hamiltonian will be set up and diagonalized in a manner similar to that of non-relativistic one.

In a *second* step, the spin–orbit coupling is calculated according to the following relation:

$$\hat{V}_{so}(r) = \frac{1}{2m^2c^2} \frac{\hbar}{2} \frac{1}{r} \frac{dV}{dr} \mathbf{L}.\boldsymbol{\sigma} = \begin{pmatrix} \hat{V}_{so}^{\uparrow\uparrow} & \hat{V}_{so}^{\uparrow\downarrow} \\ \hat{V}_{so}^{\downarrow\uparrow} & \hat{V}_{so}^{\downarrow\downarrow} \end{pmatrix}. \tag{148}$$

Therefore, the spin–orbit coupling of the two-spin channels is related to the *unperturbed* potential[13] via the angular momentum operator $\mathbf{L}$ and the Pauli spin matrix $\boldsymbol{\sigma}$.

The 2 x 2 matrix form is written in spinor basis. The two-spin directions are denoted with $\uparrow$ and $\downarrow$. The derivation of the spin–orbit coupling angular part $\mathbf{L}.\boldsymbol{\sigma}$ is supplied in Appendix.

Finally the scalar relativistic Hamiltonian matrix elements will be constructed as

$$H_{v,v'}^{\sigma,\sigma'}(\mathbf{k}) = \varepsilon_v(\mathbf{k})\delta_{v,v'}\delta_{\sigma,\sigma'} + \langle \psi_v(\mathbf{k}, \mathbf{r}) | \hat{V}_{so} | \psi_{v'}(\mathbf{k}, \mathbf{r}) \rangle, \tag{149}$$

where the corresponding eigenfunctions are of the form

$$\boldsymbol{\Psi}_n(\mathbf{k}, \mathbf{r}) = \sum_{v,\sigma} a_{v',v}^{\sigma} \psi_v(\mathbf{k}, \mathbf{r}), \tag{150}$$

where $\psi_v(\mathbf{k}, \mathbf{r})$ and $\varepsilon_v(\mathbf{k})$ are the eigenfunctions and the eigenvalues of the Hamiltonian (147) calculated without spin–orbit coupling, and n, v are the band index. As it can easily seen from Eq. (150) the n index should be twice that of v because of the summation is carried out over both spins. This leads to a spin *mixing*[14] which makes this latter not a good quantum number.

We now derive the angular part of the spin–orbit coupling $\mathbf{L}.\boldsymbol{\sigma}$, for more details see the Appendix. In order to account for the appropriate geometry of this spin–orbit operator we shall remind the reader that the quantization axis is conventionally the z-axis. Therefore, one should *rotate*[15] the SOC operator toward the z-axis to get insight onto the involved z-components of the spin orbital and magnetic moments. The rotation operation of the SOC is given by

$$[\mathbf{L}.\boldsymbol{\sigma}]_z = R(\mathbf{L}.\boldsymbol{\sigma})R^+, \tag{151}$$

---

[13] This is the spherical potential of Eq. (148) calculated without including the spin–orbit interaction.

[14] This effect results in a lifting of the degeneracy and can be observed in the band structure of typical magnetic metals.

[15] Rotating the SOC operator from the local frame to the global frame is equivalent to rotating the system of reference from the global frame to the local frame.

where R is the rotation matrix operator [88]:

$$R = \begin{pmatrix} \cos(\frac{\theta}{2})e^{-i\frac{\phi}{2}} & \sin(\frac{\theta}{2})e^{i\frac{\phi}{2}} \\ -\sin(\frac{\theta}{2})e^{-i\frac{\phi}{2}} & \cos(\frac{\theta}{2})e^{i\frac{\phi}{2}} \end{pmatrix}, \tag{152}$$

where θ and ϕ are the polar angles. Writing the spin operator σ in terms of the Pauli matrices σ_x, σ_y, and σ_z

$$\sigma_x = \begin{pmatrix} 0 & 1 \\ 1 & 0 \end{pmatrix}, \sigma_y = \begin{pmatrix} 0 & -i \\ i & 0 \end{pmatrix}, \sigma_z = \begin{pmatrix} 1 & 0 \\ 0 & -1 \end{pmatrix}, \tag{153}$$

the spin–orbit operator takes the form

$$\mathbf{L}.\sigma = \begin{pmatrix} l_z & l^- \\ l^+ & -l_z \end{pmatrix}, \tag{154}$$

where l^- and l^+ are the angular momentum operator defined as

$$l^- = l_x - il_y, l^+ = l_x + il_y. \tag{155}$$

Substituting Eqs. (154) and (152) into Eq. (151) leads to

$$[\mathbf{L}.\sigma]_z = \begin{pmatrix} [\cos(\theta)l_z + \frac{1}{2}\sin(\theta)(e^{-i\phi}l^- + e^{i\phi}l^+)] & [\cos^2(\frac{\theta}{2})e^{-i\phi}l^- - \sin^2(\frac{\theta}{2})e^{i\phi}l^+ - \sin(\theta)l_z] \\ [-\sin^2(\frac{\theta}{2})e^{-i\phi}l^- + \cos^2(\frac{\theta}{2})e^{i\phi}l^+ - \sin(\theta)l_z] & -[\cos(\theta)l_z + \frac{1}{2}\sin(\theta)(e^{-i\phi}l^- + e^{i\phi}l^+)] \end{pmatrix}. \tag{156}$$

This is the formula we have adopted during our XMCD and magnetic anisotropy investigations.

7.4 Force Theorem Determination of the Magnetic Anisotropy

One of the interesting aspect of the magnetism is that of the magnetic anisotropy. Indeed this anisotropy result from a complex interplay of the crystal and the magnetic degree of freedom. This interplay is a direct consequence of the spin–orbit coupling [84]. In 3d magnetic materials, for example, the magnetocrystalline anisotropy energy (MAE) is found to be of about some μeV [89, 90] for bulk and up to some meV [17, 91] for surfaces and thin films. According to Bruno's [17] and van der Laan's [92] models these small values of the MAE are a direct consequence of the tiny effect of the spin–orbit coupling. Given that the spin–orbit coupling is small compared to the rest of the contributions to the Hamiltonian, this coupling can be treated as a perturbation in the same way that is explained in the previous section. In this respect, because of the computational effort saving gained by the force theorem [86] many computational investigations [93–95] have been performed to satisfactorily explain the corresponding experimental MAE results [96, 97]. These investigations have allowed a better understanding of the MAE of magnetic $3d$-based materials. However, magnetic $4f$ material anisotropy is only rarely studied. In order to get insight into the magnetic anisotropy in $4f$ rare-earth magnetic metals we have chosen to work with the gadolinium (Gd) materials. This choice was motivated by the interesting magnetic properties of Gd, especially its high spin magnetic moment.

The MAE is defined as the difference in energy

$$\text{MAE} \equiv E(\text{hardaxis}) - \text{E(easyaxis)}. \tag{157}$$

One can also define an anisotropy energy with respect to the angle θ between a reference axis and the magnetization direction, $E_A(\theta)$ as

$$E_A(\theta) \equiv E(\theta) - E(\text{ref.axis}), \tag{158}$$

where ref. axis indicates the axis chosen as reference (typically the easy axis or a symmetry axis of the crystal) and θ is the angle measured from it. The anisotropy energy can also be expanded as

$$E_A(\theta) = K_1 \sin\theta + K_2 \sin^4(\theta) + (K_3 + K_4 \cos\phi) \sin^6\theta + \ldots \tag{159}$$

where K_i are the anisotropy constants, which are increasingly small.

7.5 The Force Theorem

The fact that the spin–orbit interaction can be introduced as a perturbation to scalar relativistic systems can be exploited in order to speed up the evaluation of the MAE. The way to do so is given by the force theorem for band structure calculations [86]. Let us consider an unperturbed system[16] with its total energy given by Eq. (23):

$$E = T_0[n(\mathbf{r})] + \int n(\mathbf{r})\upsilon(\mathbf{r})d^3r + \frac{1}{2}\int\int \frac{n(\mathbf{r}).n(\mathbf{r}')}{|\mathbf{r} - \mathbf{r}'|}d^3rd^3r' + E_{xc}[n(\mathbf{r})]. \tag{160}$$

By *switching on* a perturbation, one introduces a change in the total energy, to the first order in the charge density, δn, equal to

---

[16] The perturbation will be, for our purposes, the SOC, but, the theorem is far more general.

$$\delta E = \delta T_0[n(\mathbf{r})] + \int \delta n(\mathbf{r}) \upsilon(\mathbf{r}) \mathrm{d}^3 r$$

$$+ \int \int \frac{n(\mathbf{r}).\delta n(\mathbf{r}')}{|\mathbf{r} - \mathbf{r}'|} \mathrm{d}^3 r \mathrm{d}^3 r' + \int V_{\mathrm{xc}}[n(\mathbf{r})] \delta n(\mathbf{r}) \mathrm{d}^3 r + O(\delta n^2) \quad (161)$$

$$\equiv \delta T_0[n(\mathbf{r})] + \int V(\mathbf{r}) \delta n(\mathbf{r}) \mathrm{d}^3 r + O(\delta n^2),$$

where the change in the nuclei has been disregarded and the identities

$$E_{\mathrm{xc}} \equiv \int \epsilon_{\mathrm{xc}}[n(\mathbf{r})] n(\mathbf{r}) \mathrm{d}^3 r, \quad (162)$$

$$\frac{\delta E_{\mathrm{xc}}}{\delta n(\mathbf{r})} = V_{\mathrm{xc}}(\mathbf{r}) = n(\mathbf{r}) \frac{\delta \epsilon_{\mathrm{xc}}[n(\mathbf{r})]}{\delta n(\mathbf{r})} + \epsilon_{\mathrm{xc}}[n(\mathbf{r})], \quad (163)$$

$$V = \upsilon(\mathbf{r}) + V_H + V_{\mathrm{xc}}, \quad (164)$$

have been used. The last equality is that of the total Kohn–Sham potential of Eq. (21). According to Eq. (24) the kinetic energy can be rewritten in the form

$$T_0[n(\mathbf{r})] = \sum_i \epsilon_i - \int V(\mathbf{r}) n(\mathbf{r}) \mathrm{d}^3 r, \quad (165)$$

therefore, its change is (also to the first order in the charge density change)

$$\delta T_0[n(\mathbf{r})] = \delta \sum_i \epsilon_i - \int \delta V(\mathbf{r}) n(\mathbf{r}) \mathrm{d}^3 r - \int V(\mathbf{r}) \delta n(\mathbf{r}) \mathrm{d}^3 r. \quad (166)$$

Thus, if the potential is kept frozen, a substitution of Eq. (166) in Eq. (161) yields

$$\delta E = \delta \sum_i \epsilon_i, \quad (167)$$

which is the force theorem we wanted to derive and is valid to order $O(\delta n)$. The reason why we wanted to show here the derivation of Eq. (167) is that it is interesting to see that some changes in the single contributions to the total energy are not zero but they partially cancel each other to first order.

Since the change in the total energy in a frozen potential is equal to just the change in the eigenvalue sum, one calculates this latter, less computationally demanding quantity,[17] in order to obtain the former. A large number of evaluations

---

[17] Using the force theorem, a self-consistent calculations is performed without including the SOC. Since all one needs is the difference of the eigenvalues sum for two magnetization directions, one iteration would be sufficient to introduce the spin–orbit interaction.

of MAE via the force theorem in various elements and compounds have been carried out in the past 20 years [93, 98–100], showing that contributions of order $O(\delta n^2)$ are most often negligible and that the change in the eigenvalue sum is very close to the total energy change.

7.6 The Peculiar MAE of Gd

Gd metal is in the middle of the rare-earth (RE) series and its f-shell is half filled. This means that, in a Russel–Saunders (RS) scheme, no orbital moment is to be expected from the f-electron shell. Because of the sphericity of the $4f$-shell, one expects no crystal electric field (CEF) contribution to the anisotropy and indeed the MAE of Gd ($\sim$35 μ eV/atom) is two order of magnitude smaller than the MAE of other RE metals ($\sim$meV/atom). The relevant question one may ask is what is the origin of the observed MAE of Gd? We found that this conduction band MAE is completely driven by the SOC band structure anisotropy. Our calculated anisotropy is found to be in excellent agreement with experiment [101] and can be explained by Bruno's model [17], according to which for spherical shell with no orbital moment (a half-filled $4f$-shell, for example) the magnetic anisotropy stems from the spin–orbit anisotropy. The force theorem investigations of this work have shown that the Gd MAE stems from an interplay between the dipole interaction of the large localized $4f$ spin moments and the SOC conduction band MAE. This MAE was found to explain well the observed anisotropy energy [101], $E_A(\theta)$. These calculations will be discussed in Part II of this presentation.

8 X-ray Magnetic Circular Dichroism

8.1 History

Since the x-ray discovery by Röntgen [102] on 1895 a considerable attention and effort have been devoted to the use of the x-ray in different research areas. Some years after the finding of Röntgen, Bassler brought to light the polarization character of this light within the experimental work of his thesis: *Polarization of x-rays evidenced with secondary radiation* [103]. To go further in the understanding of this x-ray's properties many experiments were set up to observe the interaction of light with magnetic materials [104] or non-magnetic materials with an external magnetic field, e.g., aluminum, carbon, copper, iron, nickel, platinum, zinc, and silver [105]. The influence of the magnetism on x-ray absorption has then been investigated measuring the difference of the absorption (or the cross-section) rate between two different orientations of the magnetization. Unfortunately the tiny amount of this absorption difference rate has made the results of Bassler [103] questionable if not doubtful. The measurements recorded using the equipment of that time does not have a sufficient sensitivity [106] and have given rise to a long controversial debate [107–110]. In 1983, G. Schütz and her colleagues [111] concentrated their efforts on the use of circularly polarized x-rays in order to elucidate the influence of the magnetic state of iron on x-ray absorption spectra. Again, the sensitivity of the experimental setup was not high enough to detect in this energy range any spin-dependent contribution to x-ray absorption. One year later, the attempt of Keller and Stern [112], despite of the use of a synchrotron radiation, has failed to reveal the dichroism of Gd in $Gd_{18}Fe_{82}$ alloy because of the circular polarization rate of the incident x-rays was only 5%. Shortly later, the existence of x-ray magnetic circular dichroism (XMCD) was proved experimentally by Schütz et al. at the Fe K-edge in an iron foil [113] and at the L edges of Gd in $Gd_3Fe_5O_{12}$ [114].[18]

The first theoretical investigation of XMCD was performed by Erskine and Stern [116]. Unfortunately, very few people paid attention to their band structure calculation of XMCD at the $M_{2,3}$ absorption edges of ferromagnetic nickel. The most important finding from a theoretical point of view which marked the beginning of modern days for XMCD is that of Thole and coworkers [15, 16]. They considered a single-ion electric–dipole transitions model and derived a magneto-optical sum rules relating separately, integrated intensities of XMCD spectra to the ground-state orbital [15] and spin [16] moments. These sum rules provided experimentalists with a powerful tool to analyze XMCD spectra and to extract magnetic moments magnitudes and directions, with the full benefit of the element and orbital selectivity of x-ray absorption spectroscopy.

8.2 Theory

We have devoted this section to our implementation of XMCD calculations within the Fleur code. However, before presenting the theoretical background of this implementation it would be interesting to remind previous attempts to model, simulate, and understand experimental dichroic x-ray absorption spectra.

The first theoretical investigations of XMCD are those of Thole et al. [117] who implemented an atomic multiplet approach [118]. This approach is based on an empirical atomic calculation. In addition to the absence of the hybridization effect (atomic) this method (as all the others empirical methods) relies on the experimental spectra. Calculations applying this method to the $3d^94f^{n+1}$ multiplets of the $M_{4,5}$ edges of Lantanides are summarized in the paper of Thole et al. [119]. Some years later Chen et al. [120] made use of the Erskine and Stern model[19] [116] for their experimental $L_{2,3}$ edges spectra of nickel. The disagreement between the measured branching ratio and that predicted by the model has been ascribed to the change of spin-dependent unoccupied density of states near the Fermi level caused by the

---

[18] A more detailed story of the XMCD can be found in the section entitled *X-ray Magnetic Circular Dichroism: Historical Perspective And Recent Highlights* by Andrei Rogalev et al. [115].

[19] According to this model the large spin–orbit coupling of the core states and its small value for valence states should allow us to treat these valence states without spin–orbit coupling.

spin–orbit coupling effect. A year later the same group [121] published results of a
tight-binding analysis in which they presented an attempt to include the spin–orbit
coupling for d valence states. The valence spin–orbit ξ and exchange splitting Δ_{ex}
parameters extracted from numerical experiments are found to be respectively larger
and smaller than those of the ground state to achieve an optimal agreement between
the simulated and experimental spectra. Later Smith et al. [122] included properly
the spin–orbit coupling within a tight-binding scheme. The results for nickel are
not too different from those of the previous calculation [121] but the parameters
(ξ, Δ_{ex}) found for iron revealed the sensitivity of the XMCD spectra on the unfilled
d band width. The discrepancies between the calculated and the experimental
parameters were imputed to many-body effects, e.g., since the core hole is created,
the $3d$ valence electrons will see a stronger attractive core potential and the spatial
extent of their orbitals will contract. Consequently, relativistic effects such as the
spin–orbit coupling will be stronger, and the exchange interaction among the first
neighbors will be weaker.

The development of x-ray spectroscopy experiments probing the magnetic properties of a large variety of magnetic rare-earth materials and the growing interest of
the scientific community toward their applications in media storage, strong magnets
and the emerging field of spintronics have stimulated our XMCD calculations for
these materials. The discovery of XMCD sum rules may be a powerful tool for
understanding and characterizing magnetic properties.

In order to study the strongly localized magnetism of rare-earth metals, we have
implemented the XMCD absorption within the dipolar approximation using polar
geometry.[20] Before providing the corresponding theoretical background we will
briefly discuss two much-earlier magneto-optical (MO) effects which are, to some
extent, related to XMCD.

When the linearly polarized light beam penetrates a magnetized sample, the light
will become elliptically polarized upon transmission as well as reflection. No matter
whether the magnetization is present spontaneously or induced by an external magnetic field, these phenomena are called the Faraday [123] and Kerr [124] effects.

The quantum mechanical understanding of the Kerr MO effect began as early as
1932 when Hulme [125] ascribed Kerr effect to the spin–orbit coupling (SOC).

The interaction of the electromagnetic radiation with a magnetic medium is
described classically by Maxwell's equations [126]:

$$\nabla \times \mathbf{E} + \frac{\partial \mathbf{B}}{\partial t} = 0, \tag{168}$$

$$\nabla.\mathbf{B} = 0, \tag{169}$$

$$\nabla \times \mathbf{H} - \frac{\partial \mathbf{D}}{\partial t} = \mathbf{J}, \tag{170}$$

$$\nabla.\mathbf{D} = \rho, \tag{171}$$

---

[20] This configuration corresponds to the case where the magnetization direction is parallel to the
wave vector of the x-ray beam.

where $\mathbf{D}$ is the electric displacement, which is related to the total electric field $\mathbf{E}$ caused in part by the polarization $\mathbf{P}$ of the medium:

$$\mathbf{D} = \epsilon_0 \mathbf{E} + \mathbf{P} = (1 + \chi_e)\epsilon_0 \mathbf{E} = \epsilon \mathbf{E}, \tag{172}$$

and $\mathbf{B}$ is the magnetic induction, which is related to the macroscopic magnetic field $\mathbf{H}$ resulting from the magnetization $\mathbf{M}$:

$$\mathbf{B} = \mu_0(\mathbf{H} + \mathbf{M}) = (1 + \chi_m)\mu_0 \mathbf{H} = \mu \mathbf{H}, \tag{173}$$

where ϵ_0 and μ_0 are the vacuum permittivity and the vacuum permeability, and χ_e and χ_m, are the electric and magnetic susceptibility, respectively. According to Ohm's law the macroscopic current density $\mathbf{J}$ produced by an electric field $\mathbf{E}$ is given by

$$\mathbf{J} = \sigma . \mathbf{E}. \tag{174}$$

Equations (172), (173), and (174) are known as the material equations. They are known such that because they characterize the *response functions of the medium to external excitations*: the dielectric constant ϵ, the magnetic permeability μ, and the electrical conductivity σ. In general the dielectric constant is a function of both spatial and time variables that relates the displacement field $\mathbf{D}(\mathbf{r}, t)$ to the total electric field $\mathbf{E}(\mathbf{r}', t')$:

$$\mathbf{D}(\mathbf{r}, t) = \int \int_{-\infty}^{t} \epsilon(\mathbf{r}, \mathbf{r}', t')\mathbf{E}(\mathbf{r}', t')dt'dr'. \tag{175}$$

In the following we neglect the spatial dependence of the dielectric constant and consider only its frequency dependence $\epsilon(\omega)$. Usually, the effect of the magnetic permeability $\mu(\omega)$ on optical phenomena is small and we assume that $\mu(\omega) = \mu_0 \mathbf{I}$ where $\mathbf{I}$ is a unit tensor. It should be stressed also that ϵ and μ may depend on the field strength. In such cases higher order terms in a Taylor expansion of the material parameters lead to appearance of the nonlinear effects [127]. Using the material equations and Maxwell equations it can be easily shown that

$$\epsilon = \frac{1}{\epsilon_0}\left(1 + i\frac{\sigma}{\omega}\right). \tag{176}$$

For simplicity let us consider a material of cubic structure with a magnetization $\mathbf{M}$ directed along z axis. Above the Curie temperature T_C the three components of the dielectric tensor are equal[21] so that

$$\epsilon(\omega) = \epsilon \boldsymbol{I}. \tag{177}$$

---

[21] This is the case when the dielectric components are presented in the cubic principal axes. The principal axes are the classical analogue of the local frame axes in quantum mechanics.

When the magnetization $\mathbf{M}$ appears below T_C the symmetry is lower and $\epsilon(\omega)$ becomes [128]

$$\epsilon(\mathbf{M}, \omega) = \begin{pmatrix} \epsilon_{xx} & \epsilon_{xy} & 0 \\ -\epsilon_{xy} & \epsilon_{xx} & 0 \\ 0 & 0 & \epsilon_{zz} \end{pmatrix}. \tag{178}$$

The remaining symmetry of the system depends on the orientation of the magnetization. The components of the dielectric tensor depend on the magnetization and satisfy the following Onsager relations

$$\epsilon_{i,j}(-\mathbf{M}, \omega) = \epsilon_{j,i}(\mathbf{M}, \omega), \tag{179}$$

where $i, j = x, y$ or z. These relations mean that the diagonal components of the dielectric tensor are even functions of $\mathbf{M}$, whereas the nondiagonal ones are odd functions of $\mathbf{M}$. In the lowest order in $\mathbf{M}$

$$\epsilon_{xy} \sim M, \epsilon_{zz} - \epsilon_{xx} \sim \mathbf{M}^2. \tag{180}$$

In the absence of an external current ($\mathbf{J} = 0$) and free charges ($\rho = 0$) Maxwell equations reduce to

$$\nabla \times \mathbf{E} = -\mu_0 \frac{\partial \mathbf{H}}{\partial t}, \tag{181}$$

$$\nabla \times \mathbf{H} = \epsilon \frac{\partial \mathbf{E}}{\partial t}. \tag{182}$$

After substitution of $\mathbf{E}$ and $\mathbf{H}$ in a form of plane waves

$$\mathbf{E} = \mathbf{E}_0 e^{[-i(\omega t - \mathbf{q} \cdot \mathbf{r})]}, \tag{183}$$

$$\mathbf{H} = \mathbf{H}_0 e^{[-i(\omega t - \mathbf{q} \cdot \mathbf{r})]}, \tag{184}$$

one arrives to the a secular equation

$$\begin{pmatrix} N^2 - \epsilon_{xx} & -\epsilon_{xy} & 0 \\ \epsilon_{xy} & N^2 - \epsilon_{xx} & 0 \\ 0 & 0 & N^2 - \epsilon_{zz} \end{pmatrix} \begin{pmatrix} E_x \\ E_y \\ E_z \end{pmatrix} = 0, \tag{185}$$

where ω is the frequency, $\mathbf{q}$ is the wave vector of light, and $\mathbf{N}$ is a unit vector directed along $\mathbf{q}$ ($N = \frac{q}{\omega}c$). When the light propagates along z direction, i.e., along M, $E_z = 0$, and one finds the eigenvalues

$$n_{\pm}^2 = \epsilon_{xx} \pm i\epsilon_{xy}. \tag{186}$$

This means that the normal modes of the light accounting for the response (the displacement field **D**) to the plane wave field (**E**) are

$$D_+ = n_-^2 (E_x + i E_y), \; D_- = n_+^2 (E_x - i E_y), \tag{187}$$

i.e., a left and right polarized light wave with complex refractive indices of n_- and n_+, respectively.

8.2.1 Faraday Effect

In 1845, Faraday discovered [123] that the polarization vector of linearly polarized light is rotated upon transmission through a sample of thickness l that is exposed to a magnetic field parallel to the propagation direction of light. Indeed, in a ferromagnet, the left-hand and right-hand circularly polarized lights propagate generally with different refractive indices or different velocities c/n_- and c/n_+. When the two transmitted light waves are combined at the exit surface of the sample, they yield again a linearly polarized light, but its plane of polarization is rotated by the so-called Faraday angle θ_F given by [129]

$$\theta_F = \frac{\omega l}{2c} \mathrm{Re}(n_+ - n_-). \tag{188}$$

The direction of the rotation depends on the relative orientation of the magnetization and the light propagation. If two circularly polarized waves attenuate at different rates, then after traveling through the sample, their relative amplitude change. Therefore the transmitted light becomes elliptically polarized, with an ellipticity

$$\eta_F = -\frac{\omega l}{2c} Im(n_+ - n_-). \tag{189}$$

The ellipticity η_F corresponds to the ratio of the minor to the major axes of the polarization ellipsoid and is related to the magnetic circular dichroism, which is defined by the difference of the absorption coefficient μ between the right and the left circularly polarized light

$$\Delta\mu(\omega) = \mu_+(\omega) - \mu_-(\omega) = -\frac{4\eta_F(\omega)}{l}. \tag{190}$$

8.2.2 Kerr Effect

About 30 years later, Kerr [124] observed that when linearly polarized light is reflected from a magnetic solid, its polarization plane (the major axis of the ellipse) also rotates over a small angle with respect to that of the incident light.

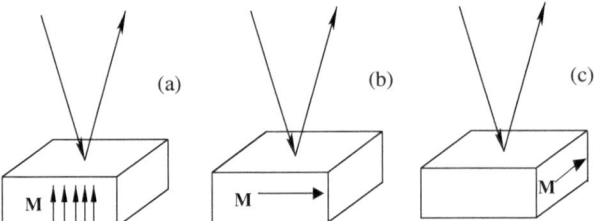

Fig. 5 The different geometries for the MO Kerr effect: (**a**) the polar Kerr effect, (**b**) the longitudinal Kerr effect, (**c**) the transversal Kerr effect

Depending on the orientation of the magnetization vector relative to the reflective surface and the plane of incidence of the light beam, three types of the magneto-optical effects in reflection are distinguished: polar, longitudinal, and transverse (equatorial) effects (Fig. 5). For linearly polarized incident light the reflected light will in general be elliptically polarized in the polar Kerr geometry (Fig. 5a). The relation between the complex polar Kerr angle and the complex refraction indices can be derived from the Fresnel relations and is given by [130]

$$\frac{1 + \tan(\eta_K)}{1 - \tan(\eta_K)} e^{2i\theta_K} = \frac{(1 + n_+)(1 - n_-)}{(1 - n_+)(1 + n_-)}. \tag{191}$$

For most materials the Kerr rotation and ellipticity are less than $1°$. For more detailed explanations for these effects and related results the reader is advised to see the chapter of reference [131].

8.2.3 The XMCD Formalism

In the previous section the response of the medium to electromagnetic waves was described in a phenomenological manner in terms of the frequency-dependent complex dielectric constant and conductivity. Within the linear response theory and using band structure methods, Callaway and Wang [22] have proposed a microscopic model for the calculations of the optical conductivity tensor

$$\sigma_{\alpha\beta}(\omega) = \frac{ie^2}{m^2\hbar V} \sum_{\mathbf{k}} \sum_{\nu\nu'} \frac{(f(\epsilon_{\nu\mathbf{k}}) - f(\epsilon_{\nu'\mathbf{k}}))}{(\omega - \omega_{\nu\nu'(\mathbf{k})+i\gamma})} \frac{M^{\alpha}_{\nu\nu'}(\mathbf{k})M^{\beta}_{\nu\nu'}(\mathbf{k})}{\omega_{\nu\nu'}(\mathbf{k})}. \tag{192}$$

It relates the macroscopic optical conductivity to the sum of interband transitions between Bloch states $\psi_{\nu\mathbf{k}}$ and $\psi_{\nu'\mathbf{k}}$ with energies $\epsilon_{\nu\mathbf{k}}$ and $\epsilon_{\nu'\mathbf{k}}$, where ν and ν' being the band indices, V the unit cell volume, $f(\epsilon_{\nu\mathbf{k}})$ the Fermi function, $\hbar\omega_{\nu\nu'}(\mathbf{k}) = \epsilon_{\nu\mathbf{k}} - \epsilon_{\nu'\mathbf{k}}$, and $\gamma = \frac{1}{\tau}$ is a phenomenological relaxation time parameter that takes into account the finite lifetime of the excited electronic states. $M^i_{\nu\nu'}(\mathbf{k})$ are the interband electronic transitions matrix elements which account for the probability of transition after an electron–photon interaction takes places. This matrix will be considered in more details later.

The real and imaginary parts $\sigma^{(1)}(\omega)$ and $\sigma^{(2)}(\omega)$ are related by the Kramers–Kronig relations [132] and can be determined separately. It is important to note that the relation (192) was derived for interband transitions, i.e., $\mathbf{q} = \mathbf{k} - \mathbf{k}'=0$. Usually the missing intraband contributions depend on lattice imperfections of the system as well as on the temperature. These contributions lie beyond the scope of our manuscript and are not considered.

In recent years the study of magneto-optical effects in the x-ray range has gained a great importance as a tool for the investigation of magnetic materials [4, 133]. The attenuation of the x-ray intensity when passing through a sample of thickness d is given by Beer's law:

$$I(d) = I_0 e^{-\mu_{\mathbf{q}\lambda}(\omega)d}, \tag{193}$$

where $\mu_{\mathbf{q}\lambda}(\omega)$ is the absorption coefficient which in general depends on the wave vector $\mathbf{q}$, the energy $\hbar\omega$, and the polarization λ of the radiation. In the x-ray regime the absorption coefficient $\mu_{\mathbf{q}\lambda}$ is related to the absorptive part of the dielectric function $\epsilon_{\mathbf{q}\lambda}$ or the optical conductivity $\sigma_{\mathbf{q}\lambda}$ via [4]

$$\mu_{\mathbf{q}\lambda}(\omega) = \frac{\omega}{c}\epsilon_{\mathbf{q}\lambda}^{(2)}(\omega) = \frac{4\pi}{c}\sigma_{\mathbf{q}\lambda}^{(1)}(\omega). \tag{194}$$

This means that $\mu_{\mathbf{q}\lambda}(\omega)$ can be evaluated[22] using Eq. (192):

$$\mu_{\mathbf{q}\lambda}(\omega) = \frac{\pi c^2}{\hbar\omega m V} \sum_{i}^{\text{occ}} \sum_{f}^{\text{unocc}} |M_{if}^{\mathbf{q}\lambda}|^2 \delta(\hbar\omega - E_f + E_i). \tag{195}$$

In contrast to Eq. (192) in which the matrix elements of the electron–photon interaction are evaluated between two Bloch states, the matrix elements $M_{if}^{\mathbf{q}\lambda}$ are calculated between a well localized initial core state i and an extended final state f. The sum over initial states i is usually restricted to one core shell which could be achieved by an experimental fine-tuning of a particular absorption edge. This important property makes x-ray absorption an element specific probe.

---

[22] Equation (195) can be considered as the limit of the real part of the matrix elements (Eq. 192) when the frequency (ω) becomes too high (x-ray regime). In this case the frequency ω can be rewritten as $\omega = \omega_0 + \delta\omega$ because of the sharp energy of the involved core levels and therefore

$$\frac{1}{\omega} = \frac{1}{\omega_0 + \delta\omega} \sim \frac{1}{\omega_0}.$$

This is why the factor $\frac{1}{\omega}$ is again present in Eq. (195).

The $M_{if}^{q\lambda}$ matrix transitions accounts for the electron–photon interaction operator

$$\hat{H}_{el-ph} = -\frac{1}{c}\mathbf{J}\mathbf{A}_{q\lambda}(\mathbf{r}) = -\frac{1}{c}\mathbf{J}\mathbf{e}_{\lambda}A e^{i\mathbf{q}\mathbf{r}}, \tag{196}$$

where $\mathbf{A}_{q\lambda}(\mathbf{r})$ is the vector potential with the wave vector $\mathbf{q}$ and polarization λ, $\mathbf{J}$ is the electronic current density operator

$$\mathbf{J} = -ec\boldsymbol{\alpha}, \tag{197}$$

and $\boldsymbol{\alpha}$ accounts for the electronic momentum operator[23]: $(\hbar/i)\boldsymbol{\nabla}$. The components of the polarization vector for linearly polarized light are given by

$$\mathbf{e}_x = \begin{pmatrix} 1 \\ 0 \\ 0 \end{pmatrix}, \mathbf{e}_y = \begin{pmatrix} 0 \\ 1 \\ 0 \end{pmatrix}, \mathbf{e}_z = \begin{pmatrix} 0 \\ 0 \\ 1 \end{pmatrix}. \tag{199}$$

For $\mathbf{q}$ pointing along the z axis, left (+) and right (-) circularly polarized lights are presented by the polarization vector

$$\mathbf{e}_{\pm} = \frac{1}{\sqrt{2}}\begin{pmatrix} 1 \\ \pm i \\ 0 \end{pmatrix}. \tag{200}$$

In order to get insight into the corresponding absorption phenomena one needs to calculate matrix elements of the form

$$M_{if}^{q\lambda} = \langle \psi_i | \hat{H}_{el-ph} | \psi_f \rangle. \tag{201}$$

It is generally argued that in the frequency range of conventional optics the amplitude of the vector potential varies only on a microscopic scale. This implies that it is sufficient to expand the exponential factor in Eq. (196)

$$e^{i\mathbf{q}\mathbf{r}} = 1 + i\mathbf{q}\mathbf{r} - \frac{1}{2}(\mathbf{q}\mathbf{r})^2 \dots, \tag{202}$$

and retain just the first constant term, in which case only the electric–dipole interaction is accounted for. For x-ray regime (XMCD) the next term in the expansion

---

[23] Within the scalar relativistic approximation (see Sect. 7.2) the total momentum operator is expressed as

$$\boldsymbol{\alpha} = \mathbf{p} + \frac{\hbar}{4mc^2}\boldsymbol{\sigma} \times \boldsymbol{\nabla}V = \frac{\hbar}{i}\boldsymbol{\nabla} + \frac{\hbar}{4mc^2}\boldsymbol{\sigma} \times \boldsymbol{\nabla}V, \tag{198}$$

while in the non-relativistic case ($c \to \infty$) this operator reduces to the electronic momentum operator.

that represents the quadripolar interaction may also be important. However, Arola et al. [134] showed that the contribution from the quadripolar interaction to the K-edge cross-sections of iron is two orders of magnitude smaller than that of the electric–dipole contribution. We have also shown that for bulk gadolinium [57] as well as for gadolinium compounds [58] (GdN) our dipolar XMCD calculations led to a good agreement with experiment without need for including the quadripolar contribution. Within the dipolar approximation the absorption coefficient reduces to

$$M_{if}^{\mathbf{q}\lambda} = \langle \psi_i | \boldsymbol{\alpha} \mathbf{e}_\lambda | \psi_f \rangle. \tag{203}$$

The ec constant is deliberately omitted. It is worth mentioning that the symmetry reduction due to the presence of spontaneous magnetization that leads to the appearance of nonzero off-diagonal components of the dielectric tensor, e.g., ϵ_{xy} in (178), occurs only if both the spin polarization and the spin–orbit coupling are simultaneously taken into account in the calculations. Technically speaking, our FLAPW-XMCD calculations are performed in two steps. First a good convergence is achieved (in term of total energy and charge density) within a scalar relativistic calculation where the SOC is included in a second variational way, after that one iteration is carried out in order to calculate the absorption coefficients using the electronic wave functions accounting for the supposed ground state. The initial core wave functions ψ_i are given by

$$
\begin{aligned}
\psi_i = \psi_{j\mu} &= \sum_{m_{sc}} C^{j\mu}_{l_c\mu-m_{sc},\frac{1}{2}m_{sc}} u_{l_c}(r) Y_{l_c\mu-m_{sc}}(\hat{\mathbf{r}}) \chi_{m_{sc}} \\
&= \sum_{m_{sc}} C^{j\mu}_{l_c m_c,\frac{1}{2}m_{sc}} u_{l_c}(r) Y_{l_c m_c}(\hat{\mathbf{r}}) \chi_{m_{sc}},
\end{aligned}
\tag{204}
$$

and the final wave functions ψ_f states are the dispersive (**k**-dependent) FLAPW valence wave functions

$$\psi_f = \psi_v^\sigma(\mathbf{k}, r) = \sum_{lm} (A_{lm}(\mathbf{k}) u_l(r) Y_{lm}(\hat{\mathbf{r}}) + B_{lm}(\mathbf{k}) \dot{u}_l Y_{lm}(\hat{\mathbf{r}})) \chi_{ms}, \tag{205}$$

where χ_{sc}, χ_s, m_{sc}, and m_s are the core spin functions, the valence spin functions, the corresponding magnetic quantum numbers, respectively. $C^{j\mu}_{l_c\mu-m_s,\frac{1}{2}}$ are the Clebsh–Gordan coefficients, j is the total momentum of the electron, l_c and l are the core and valence angular momentum quantum numbers, μ (or m_j) and m are the corresponding magnetic quantum numbers. The core and valence states are calculated separately and in a different way, that is to say that the core wave functions corresponding to deep energy levels are determined within a fully relativistic calculation while valence eigenfunctions are evaluated within a scalar relativistic calculation including the SOC as a perturbation (second variation approximation).

Let us consider one edge transitions involving the initial j states and the final l states. The $M_{if}^{q\lambda}$ matrix can be rewritten as

$$M_{if}^{q\lambda} = M_{j\mu}^{q\lambda}(\mathbf{k}) = \sum_{m,m_{sc}} C_{l_c m_c, \frac{1}{2} m_{sc}}^{j\mu}$$
$$\langle u_{l_c}(r) Y_{l_c m_c}(\hat{\mathbf{r}}) | \alpha \mathbf{e}_\lambda | (A_{lm}(\mathbf{k}) u_l(r) Y_{lm}(\hat{\mathbf{r}}) + B_{lm}(\mathbf{k}) \dot{u}_l Y_{lm}(\hat{\mathbf{r}})) \rangle \delta_{m_{sc} m_s}. \quad (206)$$

Using the relation

$$\alpha \mathbf{e}_\lambda = \frac{\mathbf{e}_r . \mathbf{e}_\lambda}{i} \frac{\partial}{\partial r} - \frac{1}{r}(\mathbf{e}_r \times \mathbf{L}).\mathbf{e}_\lambda, \quad (207)$$

where $\mathbf{L}$ is the orbital angular momentum operator, Eq. (206) becomes

$$M_{j\mu}^{q\lambda}(\mathbf{k}) = \sum_{m,m_s} C_{l_c m_c, \frac{1}{2} m_s}^{j\mu} \left(\left\langle u_{l_c} Y_{l_c m_c} \left| \frac{\mathbf{e}_r . \mathbf{e}_\lambda}{i} \frac{\partial}{\partial r} \right| (A_{lm}(\mathbf{k}) u_l Y_{lm} + B_{lm}(\mathbf{k}) \dot{u}_l Y_{lm} \right\rangle \right.$$
$$\left. - \left\langle u_{l_c} Y_{l_c m_c} \left| \frac{1}{r}(\mathbf{e}_r \times \mathbf{L}).\mathbf{e}_\lambda \right| (A_{lm}(\mathbf{k}) u_l Y_{lm} + B_{lm}(\mathbf{k}) \dot{u}_l Y_{lm}) \right\rangle \right). \quad (208)$$

Both of the terms inside the parenthesis can be separated into radial and angular part as

$$\left\langle u_{l_c} Y_{l_c m_c} \left| \frac{\mathbf{e}_r . \mathbf{e}_\lambda}{i} \frac{\partial}{\partial r} \right| (A_{lm}(\mathbf{k}) u_l Y_{lm} + B_{lm}(\mathbf{k}) \dot{u}_l Y_{lm} \right\rangle =$$
$$\left(A_{lm}(\mathbf{k}) \left\langle u_{l_c} \left| \frac{1}{i} \frac{\partial u_l}{\partial r} \right\rangle + B_{lm}(\mathbf{k}) \left\langle u_{l_c} \left| \frac{1}{i} \frac{\partial \dot{u}_l}{\partial r} \right\rangle \right) \langle Y_{l_c m_c} | \mathbf{e}_r . \mathbf{e}_\lambda | Y_{lm} \rangle, \quad (209)$$

and

$$\left\langle u_{l_c} Y_{l_c m_c} \left| \frac{1}{r}(\mathbf{e}_r \times \mathbf{L}).\mathbf{e}_\lambda \right| (A_{lm}(\mathbf{k}) u_l Y_{lm} + B_{lm}(\mathbf{k}) \dot{u}_l Y_{lm} \right\rangle =$$
$$\left(A_{lm}(\mathbf{k}) \left\langle u_{l_c} \left| \frac{1}{r} u_l \right\rangle + B_{lm}(\mathbf{k}) \left\langle u_{l_c} \left| \frac{1}{r} \dot{u}_l \right\rangle \right) \langle Y_{l_c m_c} | (\mathbf{e}_r \times \mathbf{L}).\mathbf{e}_\lambda | Y_{lm} \rangle. \quad (210)$$

It can be easily seen that the angular multiplicative factor of Eq. (209) involves the Gaunt coefficients $G_{l_c 1l}^{m_c \lambda m}$. Using the spherical harmonic relations [135]:

$$\cos(\theta) Y_{lm} = \sqrt{\frac{(l+m+1)(l-m+1)}{(2l+1)(2l+3)}} Y_{l+1m} + \sqrt{\frac{(l+m)(l-m)}{(2l+1)(2l-1)}} Y_{l-1m}, \quad (211)$$

and

$$[L_-, \cos(\theta)] = \hbar e^{-i\phi} \sin(\theta), \quad (212)$$

the angular multiplicative factor of Eq. (210) can be expressed as a function of spherical harmonics. Using the mentioned relations, after some algebraic manipulations the matrix transitions for different polarizations can be formulated as

$$
M_{j\mu}^{q+}(\mathbf{k}) = \sum_{m,m_s} C_{l_c m_c, \frac{1}{2} m_s}^{j\mu} \left(-\left(A_{lm}(\mathbf{k})\langle u_{l_c}|\frac{1}{i}\frac{\partial u_l}{\partial r}\rangle + B_{lm}(\mathbf{k})\langle u_{l_c}|\frac{1}{i}\frac{\partial \dot{u}_l}{\partial r}\rangle\right)\sqrt{\frac{4\pi}{3}}\,G_{l_c 1 l}^{m_c+1 m}\right.
$$
$$
+ \left(A_{lm}(\mathbf{k})\langle u_{l_c}|\frac{1}{r}u_l\rangle + B_{lm}(\mathbf{k})\langle u_{l_c}|\frac{1}{r}\dot{u}_l\rangle\right)\frac{1}{\sqrt{2}}\delta_{mc,m+1}
$$
$$
\left. \cdot \left((l+1)\sqrt{\frac{(l-m)(l-m-1)}{(2l-1)(2l+1)}}\,\delta_{l_c,l-1} + l\sqrt{\frac{(l+m+2)(l+m+1)}{(2l+1)(2l+3)}}\delta_{l_c,l+1}\right)\right). \tag{213}
$$

$$
M_{j\mu}^{q-}(\mathbf{k}) = \sum_{m,m_s} C_{l_c m_c, \frac{1}{2} m_s}^{j\mu} \left(\left(A_{lm}(\mathbf{k})\langle u_{l_c}|\frac{1}{i}\frac{\partial u_l}{\partial r}\rangle + B_{lm}(\mathbf{k})\langle u_{l_c}|\frac{1}{i}\frac{\partial \dot{u}_l}{\partial r}\rangle\right)\sqrt{\frac{4\pi}{3}}\,G_{l_c 1 l}^{m_c-1 m}\right.
$$
$$
- \left(A_{lm}(\mathbf{k})\langle u_{l_c}|\frac{1}{r}u_l\rangle + B_{lm}(\mathbf{k})\langle u_{l_c}|\frac{1}{r}\dot{u}_l\rangle\right)\frac{1}{\sqrt{2}}\delta_{mc,m-1}
$$
$$
\left. \cdot \left((l+1)\sqrt{\frac{(l+m)(l+m-1)}{(2l-1)(2l+1)}}\,\delta_{l_c,l-1} + l\sqrt{\frac{(l-m+2)(l-m+1)}{(2l+1)(2l+3)}}\delta_{l_c,l+1}\right)\right). \tag{214}
$$

$$
M_{j\mu}^{qz}(\mathbf{k}) = \sum_{m,m_s} C_{l_c m_c, \frac{1}{2} m_s}^{j\mu} \left(\left(A_{lm}(\mathbf{k})\langle u_{l_c}|\frac{1}{i}\frac{\partial u_l}{\partial r}\rangle + B_{lm}(\mathbf{k})\langle u_{l_c}|\frac{1}{i}\frac{\partial \dot{u}_l}{\partial r}\rangle\right)\sqrt{\frac{4\pi}{3}}\,G_{l_c 1 l}^{m_c 0 m}\right.
$$
$$
+ \left(A_{lm}(\mathbf{k})\langle u_{l_c}|\frac{1}{r}u_l\rangle + B_{lm}(\mathbf{k})\langle u_{l_c}|\frac{1}{r}\dot{u}_l\rangle\right)\delta_{mc,m}
$$
$$
\left. \cdot \left((l+1)\sqrt{\frac{(l-m)(l+m)}{(2l-1)(2l+1)}}\,\delta_{l_c,l-1} - l\sqrt{\frac{(l-m+1)(l+m+1)}{(2l+1)(2l+3)}}\delta_{l_c,l+1}\right)\right). \tag{215}
$$

The brackets in Eq. (212) denote the commutator, and θ and ϕ are the spherical angles.

Inserting Eqs. (213),(214), and (215) in Eq. (195) and performing **k**-integration (according to the Brillouin zone integration methods) one can finally calculate the corresponding absorption coefficients $\mu^{q+}(\omega)$, $\mu^{q-}(\omega)$, and $\mu^{q0}(\omega)$ for left, right, and z polarized light and therefore calculate the *key* physical quantity

$$
\Delta\mu(\omega) = \mu^{q+}(\omega) - \mu^{q-}(\omega) \neq 0. \tag{216}
$$

If x-rays are absorbed by a magnetic solid the absorption coefficients for left and right circularly polarized photons are in general different so that $\Delta\mu \neq 0$. This quantity can be measured experimentally [113] and is called x-ray magnetic circular dichroism (XMCD).

As it can easily be seen from Eqs. (213), (214), and (215) the corresponding matrix transitions elements $M_{j\mu}^{q\lambda}(\mathbf{k})$ do not vanish only if

$$
\begin{cases}
\Delta l = l - l_c = \pm 1 \\
\Delta m = m - m_c = \lambda \\
\Delta m_s = m_s - m_{sc} = 0
\end{cases} . \tag{217}
$$

These conditions are used to select the allowed transitions within the dipolar approximation and they are known as the dipole selection rules. We will give in the second appendix some details about the solution of Fresnel equation in matter.

8.3 The XMCD Sum Rules

Magnetic compounds and alloys characterization represent one of the outstanding problem in condensed matter physics. Recently, a considerable evolution of the spectroscopic techniques has been achieved and was helped by theoretical efforts. With the derivation of the sum rules by Thole and coworkers [15, 16] XMCD spectroscopy became the most used technique for studying magnetic materials. These sum rules supply a firm basis to estimate directly from XMCD spectra the orbital moment ($M_L = -\frac{\mu_B}{\hbar} \langle L_z \rangle$) and the magnetic moment ($M_S = -2\frac{\mu_B}{\hbar} \langle S_z \rangle$) contributions to the total magnetic moment associated with a specific state of given symmetry. Thus the magnetic spin and orbital moments of the absorber atom are related to the integrated absorption spectra for a specific core shell and polarization of the radiation as

$$\int_{j_+} \Delta\mu dE - \left[\frac{l_c + 1}{l_c}\right] \int_{j_-} \Delta\mu dE = \frac{N}{n_h} \left[\frac{l(l + 1) - 2 - l_c(l_c + 1)}{3l_c} \langle S_z \rangle \right. \tag{218}$$

$$\left. + \frac{l(l + 1)[l(l + 1) + 2l_c(l_c + 1) + 4] - 3(l_c - 1)^2(l_c + 2)^2}{6ll_c(l + 1)} \langle T_z \rangle \right], \tag{219}$$

and

$$\int_{j_+ + j_-} \Delta\mu dE = \frac{N}{2n_h} \left[\frac{l(l + 1) + 2 - l_c(l_c + 1)}{l(l + 1)}\right] \langle L_z \rangle, \tag{220}$$

where N is the total integrated spectrum corresponding to the unpolarized radiation (known also as the isotropic absorption contribution)

$$N = \int_{j_+ + j_-} \left(\sum_{\lambda=+,-,0} \mu^\lambda\right) dE, \tag{221}$$

$$\Delta\mu = \mu^+ - \mu^-,$$

and T_z is the magnetic dipole operator

$$T_z = \frac{1}{2}\left[\sigma - 3\hat{\mathbf{r}}(\hat{\mathbf{r}}.\sigma)\right]_z. \tag{222}$$

that will be derived in the appendix.

$\int_{j_+ + j_-}$ means that the integral is performed over both of the $j_+ = l + 1/2$ and of the $j_- = l - 1/2$ edge spectra, e.g., $j_+ = 3/2$ and $j_- = 1/2$ for the $L_{2,3}$ edges of transition metals, n_h denotes the number of holes or the number of unoccupied final states, and $\langle S_z \rangle$, $\langle L_z \rangle$, and $\langle T_z \rangle$ are, respectively, the expectation values of the magnetic moment, the orbital moment, and the magnetic dipole operator.

The expectation value of the magnetic dipole operator accounts for the asphericity of the spin magnetization. This asphericity can be considered as a magnetic anisotropy resulting from the spin–orbit coupling or crystal-field effects.

The application of these sum rules provides as with the magnetic spin and orbital moments since the expectation value of the T_z operator is determined. In order to extract these moments from the absorption spectra we have used [57] the sum rules for the different edges:

K-edge

$$\int_{E_F}^{E_{cut}} \Delta \mu \, dE = \frac{N}{n_h} \langle L_z \rangle , \tag{223}$$

where

$$N = \sum_{\lambda = +,-,0} \int_{E_F}^{E_{cut}} \Delta \mu^\lambda , \tag{224}$$

$L_{2,3}$ edges

$$\int_{E_F}^{E_{cut}} \left[(\mu_{L_3}^+ - \mu_{L_3}^-) - 2(\mu_{L_2}^+ - \mu_{L_2}^-) \right] dE = \frac{N}{3n_h} \left[\langle S_z \rangle + 7 \langle T_z \rangle \right], \tag{225}$$

$$\int_{E_F}^{E_{cut}} \left[(\mu_{L_3}^+ - \mu_{L_3}^-) + (\mu_{L_2}^+ - \mu_{L_2}^-) \right] dE = \frac{N}{2n_h} \langle L_z \rangle , \tag{226}$$

where

$$N = \sum_{\lambda = +,-,0} \int_{E_F}^{E_{cut}} (\Delta \mu_{L_3}^\lambda + \Delta \mu_{L_2}^\lambda), \tag{227}$$

and $M_{4,5}$ edges

$$\int_{E_{\mathrm{F}}}^{E_{\mathrm{cut}}} \left[(\mu_{M_5}^+ - \mu_{M_5}^-) - \frac{3}{2}(\mu_{M_4}^+ - \mu_{M_4}^-) \right] \mathrm{d}E = \frac{N}{3n_h} \left[\langle S_z \rangle + 6 \langle T_z \rangle \right], \quad (228)$$

$$\int_{E_{\mathrm{F}}}^{E_{\mathrm{cut}}} \left[(\mu_{M_5}^+ - \mu_{M_5}^-) + (\mu_{M_4}^+ - \mu_{M_4}^-) \right] \mathrm{d}E = \frac{N}{3n_h} \langle L_z \rangle, \quad (229)$$

where

$$N = \sum_{\lambda=+,-,0} \int_{E_{\mathrm{F}}}^{E_{\mathrm{cut}}} (\Delta\mu_{M_5}^\lambda + \Delta\mu_{M_4}^\lambda). \quad (230)$$

The integrations are carried out from the Fermi energy E_{F} up to an energy cut-off E_{cut}. This energy represents the energy of the top of the final magnetic states. The number of holes n_h are also calculated from the density of states, and they are determined from the integration of the unoccupied part of the partial density of final states.

In order to make a useful and relevant applications of these sum rules one should know their limitations due to the assumptions made in order to derive them. In fact, to derive the XMCD sum rules, Thole and coworkers have adopted a single ion model combined with a scalar relativistic approach. The principle assumption of these sum rules derivation is that of the two-step model [113]. Depending on the photon polarization, the XMCD transitions will be achieved in two steps. First, the core electron will choose one of the spin directions according to the core spin–orbit splitting, that is to say, depending on the encountered spin–orbit interaction and because of the conservation of the angular momentum during the absorption process the angular momentum carried out by the photon is completely or partially transferred to the photoelectron, in a second step the exchange spin splitting of the final state is different whether the spin of the incoming electron is up or down. This could simulate the eventual change of the exchange splitting resulting from the spin dependence of the incoming photoelectron. The others' assumptions of the underlying physics of the XMCD sum rules are to ignore the following [136]:

1. the exchange splitting of the core states,
2. the asphericity of the core states,
3. the difference between the radial relativistic part of the final wave functions, i.e., the radial parts $u_l(r)$ of $p_{1/2}$ and $p_{3/2}$ or $d_{3/2}$ and $d_{5/2}$ are the same, and
4. the energy dependence of the wave function.

Despite such limiting approximations, the validity of the sum rules appears to be now rather well established, at least in the cases of the $L_{2,3}$ absorption edges of $3d$ [137, 138, 10], $4d$ [139] and $5d$ [140] transition metals. However, one should keep in mind that there are some problems when applying the sum rules to XMCD spectra. The most severe one is the separation of the L_2- and L_3-spectra, e.g., because of the strong hybridization between the $2p$ N orbitals and the $4d$-Gd orbitals in GdN compound ([141] and references therein). The $5d$-Gd magnetic moment extracted from the application of the sum rules to $L_{2,3}$ edges of Gd could not account for the realistic $5d$ magnetic moment since a part of that moment is supposed to be transferred or transformed to $2p$ magnetic moment.

Apart from this weak point of the XMCD sum rules, the successful use we have made of the XMCD sum rules to calculate the magnetic moment of Gd atoms in gadolinium bulk has shown the validity and the usefulness of these sum rules for strongly localized $4f$ materials [57]. This is not surprising since $4f$ rare-earth orbitals are so localized that the hybridization with others orbital will be marginal and f states will carry the whole magnetic moment of $4f$ electrons. Therefore we expect that $4f$ magnetic materials such as rare-earth metals are well studied by XMCD investigations.

Acknowledgments We acknowledge financial support from an ANR grant ANR-06-NANO-053.

Appendix

Spin–Orbit Coupling

As it was shown in Sect. 4, the Dirac Hamiltonian can be transformed into a Schrödinger-like Hamiltonian. In this appendix, we will develop the spin–orbit contribution to the Hamiltonian. Let H be

$$H = H_0 + H_{SOC} = H_0 + \frac{e}{4m^2}\sigma.(E \wedge p), \qquad (231)$$

where H_0 is the Kohn–Sham semi-relativistic Hamiltonian, σ the Pauli matrices, p the momentum, $E = -\nabla V$ the electric field, and m the electron mass. The second variational method consists first in solving the Hamiltonian H_0, i.e., obtaining its eigenvalues and eigenvectors by diagonalization. For a system with translational symmetry, the eigenvalues of H_0 can be written as

$$H_0|n, k, \sigma\rangle = \varepsilon_{n,k}^{\sigma}|n, k, \sigma\rangle. \qquad (232)$$

In the FLAPW method, the wave functions can be given by

$$\langle r|n, k, \sigma\rangle = \Psi_{n,k}^{\sigma}(r) = \begin{cases} \frac{1}{\sqrt{\Omega}} \sum_{G} C_{nk}^{\sigma}(G)e^{i(k+G).r} & r \in I \\ \sum_{lm} \left[A_{n,k}^{lm\sigma} U_l^{\sigma}(r) + B_{n,k}^{lm\sigma} \dot{U}_l^{\sigma}(r) \right] Y_l^m(\hat{r}) & r \in \text{MT}. \end{cases} \qquad (233)$$

The index I represents the interstitial region, whereas MT represents the muffin-tin region. Now we will express the matrix elements of the SOC in both regions.

SOC in the Muffin-Tin Region

In the muffin-tin region the potential of the crystal is developed on spherical harmonics. In general, the spherical part of the potential is dominant. We will treat in a first approximation only the spherical part and in the second one the non-spherical parts of the potential.

Spherical Approximation

In the case where the potential has a spherical symmetry, the electric field is written as

$$E = -\frac{dV_0}{dr}u_r,$$

where u_r is the radial unitary vector. The Hamiltonian H_{SOC} is then given by

$$H_{SOC} = \frac{\alpha^2}{m^2}\frac{dV_0}{dr}\sigma.u_r \wedge p = \xi(r)\sigma.L, \tag{234}$$

where L is the orbital moment operator and $\xi(r) = \frac{\alpha^2}{m^2 r}\frac{dV_0}{dr}$. In order to calculate the matrix element of $\langle n, k, \sigma | H_{SOC} | n', k, \sigma' \rangle$, using the Kohn–Sham wave functions $|n, k, \sigma\rangle$, we determine the following radial integrals:

$$\zeta_{1,l}^{\sigma\sigma'} = \alpha^2 \int_{MT} r dr U_l^\sigma(r)\frac{1}{m^2}\frac{dV_0^{\sigma'}}{dr}U_l^{\sigma'},$$

$$\zeta_{2,l}^{\sigma\sigma'} = \alpha^2 \int_{MT} r dr U_l^\sigma(r)\frac{1}{m^2}\frac{dV_0^{\sigma'}}{dr}\dot{U}_l^{\sigma'}, \tag{235}$$

$$\zeta_{3,l}^{\sigma\sigma'} = \alpha^2 \int_{MT} r dr \dot{U}_l^\sigma(r)\frac{1}{m^2}\frac{dV_0^{\sigma'}}{dr}\dot{U}_l^{\sigma'}.$$

We assume that the quantization axis is along the u axis determined by the angles ϑ et φ with respect to z axis. Let $|\pm\rangle$ be the spin basis vectors and $|\pm\rangle_u$ the eigenvectors of $\sigma.u$. We can write [135, 142]

$$\begin{pmatrix} |+\rangle_u \\ |-\rangle_u \end{pmatrix} = R(\vartheta, \varphi)\begin{pmatrix} |+\rangle \\ |-\rangle \end{pmatrix}, \tag{236}$$

where

$$R(\vartheta, \varphi) = \begin{pmatrix} \cos\frac{\vartheta}{2}e^{-i\frac{\varphi}{2}} & \sin\frac{\vartheta}{2}e^{i\frac{\varphi}{2}} \\ -\sin\frac{\vartheta}{2}e^{-i\frac{\varphi}{2}} & \cos\frac{\vartheta}{2}e^{i\frac{\varphi}{2}} \end{pmatrix}. \tag{237}$$

We deduce that

$$
\begin{aligned}
(\boldsymbol{\sigma}.\boldsymbol{L})_u &= \begin{pmatrix} L_z & L_- \\ L_+ & -L_z \end{pmatrix}_u \\
&= \left[R(\vartheta, \varphi) \begin{pmatrix} L_z & L_- \\ L_+ & -L_z \end{pmatrix} R^\dagger(\vartheta, \varphi) \right]_z
\end{aligned}
\tag{238}
$$

and

$$
\langle lm|\boldsymbol{\sigma}.\boldsymbol{L}|lm'\rangle = \begin{pmatrix} \chi^1_{lm,lm'} & \chi^2_{lm,lm'} \\ \chi^{2*}_{lm,lm'} & -\chi^1_{lm,lm'} \end{pmatrix},
\tag{239}
$$

where

$$
\begin{aligned}
\chi^1_{lm,lm'} &= m\cos\vartheta\,\delta_{m,m'} + \tfrac{1}{2}\sin\vartheta\left[\sqrt{(l+m)(l-m+1)}e^{i\varphi}\delta_{m,m'+1}\right] + \\
&\quad \tfrac{1}{2}\sin\vartheta\left[\sqrt{(l-m)(l+m+1)}e^{-i\varphi}\delta_{m,m'-1}\right], \\
\chi^2_{lm,lm'} &= -m\sin\vartheta\,\delta_{m,m'} + \left[-\sin^2\tfrac{\vartheta}{2}\sqrt{(l+m)(l-m+1)}e^{i\varphi}\delta_{m,m'+1}\right] \\
&\quad + \left[\cos^2\tfrac{\vartheta}{2}\sqrt{(l-m)(l+m+1)}e^{-i\varphi}\delta_{m,m'-1}\right].
\end{aligned}
\tag{240}
$$

The matrix element of H_{SOC} can then be written as

$$
\langle n\boldsymbol{k}|H_{\mathrm{SOC}}|n'\boldsymbol{k}\rangle = \begin{pmatrix} H^{\uparrow\uparrow}_{n,n',\boldsymbol{k}} & H^{\uparrow\downarrow}_{n,n',\boldsymbol{k}} \\ H^{\downarrow\uparrow}_{n,n',\boldsymbol{k}} & H^{\downarrow\downarrow}_{n,n',\boldsymbol{k}} \end{pmatrix},
\tag{241}
$$

where

$$
\begin{aligned}
H^{\uparrow\uparrow}_{n,n'} &= \sum_{l,m,m'}\left[A^\uparrow A'^\uparrow \zeta^{\uparrow\uparrow}_{1,l} + A^\uparrow B'^\uparrow \zeta^{\uparrow\uparrow}_{2,l} + B^\uparrow A'^\uparrow \zeta^{\uparrow\uparrow}_{2,l} + B^\uparrow B'^\uparrow \zeta^{\uparrow\uparrow}_{3,l}\right]\chi^1_{lm,lm'}, \\
H^{\uparrow\downarrow}_{n,n'} &= \sum_{l,m,m'}\left[A^\uparrow A'^\downarrow \zeta^{\uparrow\downarrow}_{1,l} + A^\uparrow B'^\downarrow \zeta^{\uparrow\downarrow}_{2,l} + B^\uparrow A'^\downarrow \zeta^{\uparrow\downarrow}_{2,l} + B^\uparrow B'^\downarrow \zeta^{\uparrow\downarrow}_{3,l}\right]\chi^2_{lm,lm'}, \\
H^{\downarrow\uparrow}_{n,n'} &= \sum_{l,m,m'}\left[A^\downarrow A'^\uparrow \zeta^{\downarrow\uparrow}_{1,l} + A^\downarrow B'^\uparrow \zeta^{\downarrow\uparrow}_{2,l} + B^\downarrow A'^\uparrow \zeta^{\downarrow\uparrow}_{2,l} + B^\downarrow B'^\uparrow \zeta^{\downarrow\uparrow}_{3,l}\right]\chi^2_{lm,lm'}, \\
H^{\downarrow\downarrow}_{n,n'} &= -\sum_{l,m,m'}\left[A^\downarrow A'^\downarrow \zeta^{\downarrow\downarrow}_{1,l} + A^\downarrow B'^\downarrow \zeta^{\downarrow\downarrow}_{2,l} + B^\downarrow A'^\downarrow \zeta^{\downarrow\downarrow}_{2,l} + B^\downarrow B'^\downarrow \zeta^{\downarrow\downarrow}_{3,l}\right]\chi^1_{lm,lm'}.
\end{aligned}
\tag{242}
$$

To simplify the writing, we suppress the indices $\boldsymbol{k}$ and lm and replace $A(B)^{lm,\sigma}_{n,k}$ by $A(B)^\sigma$. We can then write the matrix of H of size $2N \times 2N$, where N is the number of band energies, and using the matrix elements of H_0

$$
\langle n, \boldsymbol{k}, \sigma|H_0|n', \boldsymbol{k}, \sigma'\rangle = \varepsilon_{n,\boldsymbol{k}}\delta_{n,n'}\delta_{\sigma,\sigma'}.
\tag{243}
$$

The diagonalization of the full Hamiltonian leads to the eigenvalues and eigen-vectors of the Hamiltonian written in terms of those of H_0.

Discussion

The non-diagonal matrix elements $H_{n,n'}^{\downarrow\uparrow}$ and $H_{n,n'}^{\uparrow\downarrow}$ couple the spin components, and consequently the spin sub-bands of majority and minority spins are not independent. In addition to the splitting of the energy levels $\varepsilon_{n\mathbf{k}}$ due to the SOC diagonal elements, there is a spin flips due to the non-diagonal elements. These spin flips are of great importance, particularly to the physics of half-metals. The splitting of the bands toward lower and higher energies is also important (see Fig. 6).

It is also important to determine the order of magnitude of the SOC. As it was shown, the SOC is of the order of α^2 multiplied by a radial integral (see Eq. (235)), and this last integral is proportional to the radial derivative of the potential, which is itself proportional to the atomic Z number. Therefore, the SOC is more impor-tant for heavy atoms. Table 1 shows the calculated radial integrals for gold and nickel. We can see that those of gold are an order of magnitude greater than those of nickel.

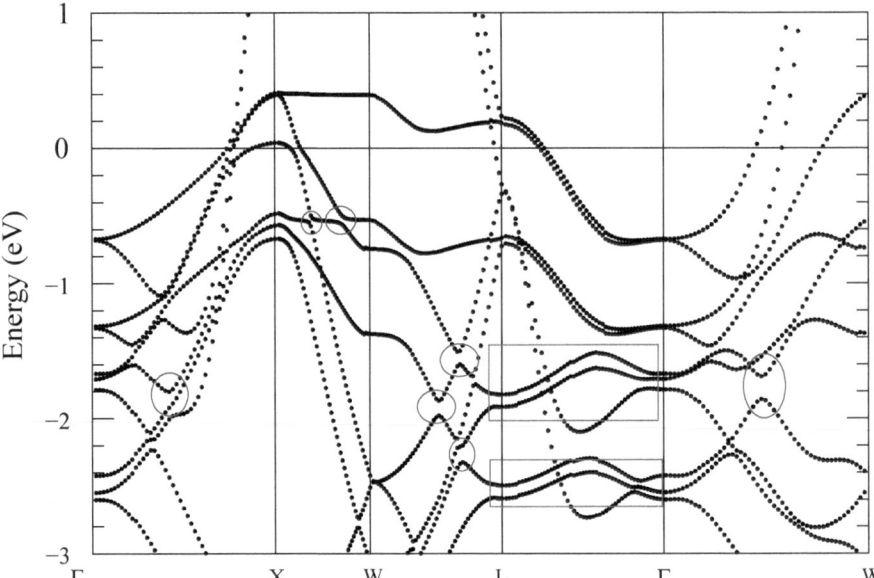

Fig. 6 Band structure of nickel. The *circles* show the splitting of the bands at the band crossing, whereas the squares show the splitting of the double degenerate bands due to the SOC

Table 1 Expectation values of the radial integrals $\zeta^{\sigma\sigma'}_{i,l}$ (in Hartrees) calculated for nickel and gold

		$\zeta^{\uparrow\uparrow}_{1,l}$	$\zeta^{\downarrow\uparrow}_{1,l}$	$\zeta^{\downarrow\downarrow}_{1,l}$	$\zeta^{\uparrow\uparrow}_{2,l}$	$\zeta^{\downarrow\uparrow}_{2,l}$	$\zeta^{\downarrow\downarrow}_{2,l}$	$\zeta^{\uparrow\uparrow}_{3,l}$	$\zeta^{\downarrow\uparrow}_{3,l}$	$\zeta^{\downarrow\downarrow}_{3,l}$
Ni	p	0.24	0.24	0.24	0.03	0.03	0.03	0.005	0.005	0.005
	d	0.095	0.090	0.093	0.106	0.106	0.106	0.122	0.127	0.127
Au	p	2.16	2.16	2.16	0.25	0.25	0.25	0.03	0.03	0.03
	d	0.66	0.66	0.66	0.66	0.66	0.66	0.65	0.65	0.65

SOC of Non-magnetic or Paramagnetic Materials

In general, to include the SOC one has to choose a quantification axis, parallel to the magnetization axis. However, in the case of non-magnetic or paramagnetic materials there is no magnetization. The choice of a quantification axis leads to an asymmetry of the physical quantities because of the symmetry reduction. To overcome this restriction we have developed a formalism of SOC where this problem is solved, i.e., the full symmetry is restored.

Let $p(\vartheta, \varphi)$ be the probability for the magnetization pointing in the direction **u**, defined by the angles ϑ and φ. In a ferromagnetic material where the magnetization points toward the direction given by ϑ_0 and φ_0, the probability $p(\vartheta, \varphi)$ is given by

$$p(\vartheta, \varphi) = \delta(\vartheta - \vartheta_0)\delta(\varphi - \varphi_0), \tag{244}$$

where δ is the Dirac distribution. In paramagnetic materials, all directions of the magnetization (spin axis) are equally probable:

$$p(\vartheta, \varphi) = \frac{1}{4\pi}. \tag{245}$$

We have determined the Hamiltonian for ferromagnetic materials (234), but we can generalize to any paramagnetic system by rewriting the SOC part of the Hamiltonian as

$$H_{\mathrm{SOC}} = \overline{H^{\vartheta,\varphi}} = \int\limits_0^\pi d\vartheta \int\limits_0^{2\pi} p(\vartheta, \varphi) R(\vartheta, \varphi) H_{\mathrm{SOC}} R^\dagger(\vartheta, \varphi). \tag{246}$$

This leads to

$$\langle lm|\boldsymbol{\sigma}.\boldsymbol{L}|lm'\rangle = \delta_{l,l'}\delta_{m,m'}\begin{pmatrix} 0 & -\frac{m}{2\pi} \\ -\frac{m}{2\pi} & 0 \end{pmatrix} = -\frac{m}{2\pi}\delta_{l,l'}\delta_{m,m'}\sigma_x. \tag{247}$$

Magnetic Dipole

The dipolar magnetic operator can be written as:

$$T_z = \frac{1}{2} [\boldsymbol{\sigma} - 3\boldsymbol{u}_r(\boldsymbol{u}_r.\boldsymbol{\sigma})]_z \tag{248}$$

or explicitly as

$$T_z = \frac{1}{2} \begin{pmatrix} 1 - 3\cos^2\theta & -3\cos\theta\sin\theta e^{-i\varphi} \\ -3\cos\theta\sin\theta e^{i\varphi} & -1 + 3\cos^2\theta \end{pmatrix} = \sqrt{\frac{2\pi}{5}} \begin{pmatrix} \sqrt{2}Y_2^0 & -\sqrt{3}Y_2^{-1} \\ \sqrt{3}Y_2^1 & -\sqrt{2}Y_2^0 \end{pmatrix}. \tag{249}$$

When the magnetization points along the quantification axis $\mathbf{z}$, T_z becomes

$$T_z = \sqrt{\frac{4\pi}{5}} \begin{pmatrix} Y_2^0 & 0 \\ 0 & -Y_2^0 \end{pmatrix}. \tag{250}$$

In the FLAPW method, we determine the mean value of the dipolar magnetic operator inside the muffin-tin region MT α, where the wave function is given by

$$\Psi_{n\boldsymbol{k}}^{\alpha,\sigma}(\boldsymbol{r}) = \sum_{lm} \left[A_{\alpha,lm}^{\sigma,n,\boldsymbol{k}} U_l^{\alpha,\sigma}(|\boldsymbol{r} - \boldsymbol{R}_\alpha|) + B_{\alpha,lm}^{\sigma,n,\boldsymbol{k}} \dot{U}_l^{\alpha,\sigma}(|\boldsymbol{r} - \boldsymbol{R}_\alpha|) \right] Y_l^m(\boldsymbol{r} \hat{-} \boldsymbol{R}_\alpha). \tag{251}$$

To simplify the notation, we restrict ourselves to the case of one atom per unit cell. The mean value of the T_z operator is given by:

$$\langle T_z \rangle = \sum_{\sigma,n,\boldsymbol{k}} \sum_{lm,l'm'} A_{\alpha,lm}^{*\sigma,n,\boldsymbol{k}} A_{\alpha,l'm'}^{\sigma,n,\boldsymbol{k}} C_{lm,l'm'}^{1,\sigma} + \\ B_{\alpha,lm}^{*\sigma,n,\boldsymbol{k}} A_{l'm'}^{\sigma,n,\boldsymbol{k}} C_{lm,l'm'}^{2,\sigma} + \\ A_{\alpha,lm}^{*\sigma,n,\boldsymbol{k}} B_{\alpha,l'm'}^{\sigma,n,\boldsymbol{k}} C_{lm,l'm'}^{3,\sigma} + \\ B_{\alpha,lm}^{*\sigma,n,\boldsymbol{k}} B_{l'm'}^{\sigma,n,\boldsymbol{k}} C_{lm,l'm'}^{4,\sigma}, \tag{252}$$

where

$$\begin{aligned}
C_{lm,l'm'}^{1,\sigma} &= \sigma\sqrt{\tfrac{4\pi}{5}} C_{lm,l'm'}^{2,0} \int r^2 dr \, U_l^\sigma(r) U_{l'}^\sigma(r), \\
C_{lm,l'm'}^{2,\sigma} &= \sigma\sqrt{\tfrac{4\pi}{5}} C_{lm,l'm'}^{2,0} \int r^2 dr \, U_l^\sigma(r) \dot{U}_{l'}^\sigma(r), \\
C_{lm,l'm'}^{3,\sigma} &= \sigma\sqrt{\tfrac{4\pi}{5}} C_{lm,l'm'}^{2,0} \int r^2 dr \, \dot{U}_l^\sigma(r) U_{l'}^\sigma(r), \\
C_{lm,l'm'}^{4,\sigma} &= \sigma\sqrt{\tfrac{4\pi}{5}} C_{lm,l'm'}^{2,0} \int r^2 dr \, \dot{U}_l^\sigma(r) \dot{U}_{l'}^\sigma(r),
\end{aligned} \tag{253}$$

and where $C_{lm,l'm}^{2,0}$ are the Gaunt coefficients.

Fresnel Equations

We use Maxwell equations for the propagation of light in an isotropic media, characterized by a magnetic or crystalline asymmetry. Maxwell equations can be written as

$$\nabla \times \boldsymbol{E} = -\frac{\partial \boldsymbol{B}}{\partial t},$$

$$\nabla \times \boldsymbol{H} = \frac{1}{c}\frac{\partial \boldsymbol{D}}{\partial t} + \frac{4\pi}{c}\boldsymbol{J},$$

$$\nabla.\boldsymbol{D} = 4\pi\rho, \tag{254}$$

$$\nabla.\boldsymbol{B} = 0.$$

Combining these equations with the equation of the current

$$\boldsymbol{J} = \hat{\sigma}\,\boldsymbol{E} = \sum_i \sigma_{ij}\,\boldsymbol{E}, \tag{255}$$

we obtain the Fresnel equations:

$$\left\{ n^2\mathbf{I} - n^2\mathbf{ss}^T - \varepsilon \right\} \boldsymbol{E} = 0 \Longleftrightarrow det\left[n^2 - n^2 s_i s_j - \varepsilon_{ij} \right] = 0, \tag{256}$$

where s is the wave vector of the transmitted wave, and n is the unit vector perpendicular to the surface of the media form which the light is reflected. The incidence of the light is described by the vector s_0. We will use the orthonormal reference frame described by the three vectors e_1, e_2 and n such that

$$e_1 = (n \times s_0) \times n,$$

$$e_2 = n \times s_0.$$

The vectors n and e_1 determine the incidence plan, and e_2 is perpendicular to this plan. In this reference frame, the wave vector of the transmitted wave s and that of the incident wave are in the incidence plan, so that

$$s = \begin{pmatrix} \sin\vartheta \\ 0 \\ \cos\vartheta \end{pmatrix},$$

and

$$s_0 = \begin{pmatrix} \sin\varphi \\ 0 \\ \cos\varphi \end{pmatrix}.$$

We can rewrite the Fresnel equation in a matrix form:

$$
\begin{pmatrix}
n^2 \cos^2 \vartheta - \varepsilon_{11} & -\varepsilon_{12} & -\frac{n^2}{2} \sin 2\vartheta - \varepsilon_{13} \\
-\varepsilon_{21} & n^2 - \varepsilon_{22} & -\varepsilon_{23} \\
-\frac{n^2}{2} \sin 2\vartheta - \varepsilon_{31} & -\varepsilon_{32} & n^2 \sin^2 \vartheta - \varepsilon_{33}
\end{pmatrix}
\begin{pmatrix}
E_1 \\
E_2 \\
E_3
\end{pmatrix}
= 0.
\tag{257}
$$

To solve this equation, one has to find the zeros of its determinant

$$
\mathbf{det}
\begin{bmatrix}
n^2 \cos^2 \vartheta - \varepsilon_{11} & -\varepsilon_{12} & -\frac{n^2}{2} \sin 2\vartheta - \varepsilon_{13} \\
-\varepsilon_{21} & n^2 - \varepsilon_{22} & -\varepsilon_{23} \\
-\frac{n^2}{2} \sin 2\vartheta - \varepsilon_{31} & -\varepsilon_{32} & n^2 \sin^2 \vartheta - \varepsilon_{33}
\end{bmatrix}
= 0,
\tag{258}
$$

$$
n^4 \left[-\varepsilon_{33} \cos^2 \vartheta - \varepsilon_{11} \sin^2 \vartheta - \frac{(\varepsilon_{31} + \varepsilon_{13})}{2} \sin 2\vartheta \right]
$$

$$
+ n^2 \left[\varepsilon_{22}\varepsilon_{33} \cos^2 \vartheta + \varepsilon_{11}\varepsilon_{33} + \varepsilon_{11}\varepsilon_{22} \sin^2 \vartheta - \varepsilon_{32}\varepsilon_{23} \cos^2 \vartheta - \varepsilon_{12}\varepsilon_{21} \sin^2 \vartheta \right]
$$

$$
+ n^2 \left[\frac{\sin 2\vartheta}{2} (\varepsilon_{12}\varepsilon_{23} + \varepsilon_{13}\varepsilon_{22} + \varepsilon_{31}\varepsilon_{22} + \varepsilon_{21}\varepsilon_{32}) - \varepsilon_{13}\varepsilon_{31} \right]
$$

$$
+ \varepsilon_{11}\varepsilon_{23}\varepsilon_{32} + \varepsilon_{22}\varepsilon_{13}\varepsilon_{31} + \varepsilon_{33}\varepsilon_{12}\varepsilon_{21} - \varepsilon_{11}\varepsilon_{22}\varepsilon_{33} - \varepsilon_{12}\varepsilon_{23}\varepsilon_{31} - \varepsilon_{21}\varepsilon_{13}\varepsilon_{32} = 0.
\tag{259}
$$

Using Snell law

$$
n_0 \mathbf{s}_0 \times \mathbf{n} = n\mathbf{s} \times \mathbf{n} \iff n_0 \sin \varphi = n \sin \vartheta,
\tag{260}
$$

we obtain:

$$
\begin{aligned}
n^2 \sin^2 \vartheta &= n_0^2 \sin^2 \varphi \\
n^2 \cos^2 \vartheta &= n^2 - n_0^2 \sin^2 \varphi.
\end{aligned}
\tag{261}
$$

Taking into account the expressions (261) in (259) equation, we obtain:

$$
n^4 \left[-\varepsilon_{33} - \frac{(\varepsilon_{31} + \varepsilon_{13})}{2} \sin 2\vartheta \right]
$$

$$
+ n^2 \left[-\varepsilon_{11}n_0^2 \sin^2 \varphi + \varepsilon_{33}n_0^2 \sin^2 \varphi + \varepsilon_{22}\varepsilon_{33} + \varepsilon_{11}\varepsilon_{33} - \varepsilon_{32}\varepsilon_{23} - \varepsilon_{13}\varepsilon_{31} \right]
$$

$$
+ n^2 \left[\frac{\sin 2\vartheta}{2} (\varepsilon_{12}\varepsilon_{23} + \varepsilon_{13}\varepsilon_{22} + \varepsilon_{31}\varepsilon_{22} + \varepsilon_{21}\varepsilon_{32}) \right]
\tag{262}
$$

$$
+ \varepsilon_{11}\varepsilon_{23}\varepsilon_{32} + \varepsilon_{22}\varepsilon_{13}\varepsilon_{31} + \varepsilon_{33}\varepsilon_{12}\varepsilon_{21} - \varepsilon_{11}\varepsilon_{22}\varepsilon_{33} - \varepsilon_{12}\varepsilon_{23}\varepsilon_{31} - \varepsilon_{21}\varepsilon_{13}\varepsilon_{32}
$$

$$
+ n_0^2 \sin^2 \varphi \left[\varepsilon_{11}\varepsilon_{22} + \varepsilon_{23}\varepsilon_{32} - \varepsilon_{12}\varepsilon_{21} - \varepsilon_{33}\varepsilon_{22} \right] = 0.
$$

If we define that

$$a_1 = -\varepsilon_{33},$$

$$a_2 = (\varepsilon_{33} - \varepsilon_{11})n_0^2 \sin^2 \varphi + \varepsilon_{22}\varepsilon_{33} + \varepsilon_{11}\varepsilon_{33} - \varepsilon_{32}\varepsilon_{23} - \varepsilon_{13}\varepsilon_{31},$$

$$a_3 = \varepsilon_{11}\varepsilon_{23}\varepsilon_{32} + \varepsilon_{22}\varepsilon_{13}\varepsilon_{31} + \varepsilon_{33}\varepsilon_{12}\varepsilon_{21} - \varepsilon_{11}\varepsilon_{22}\varepsilon_{33} - \varepsilon_{12}\varepsilon_{23}\varepsilon_{31} - \varepsilon_{21}\varepsilon_{13}\varepsilon_{32} + \\ n_0^2 \sin^2 \varphi \, [\varepsilon_{11}\varepsilon_{22} + \varepsilon_{23}\varepsilon_{32} - \varepsilon_{12}\varepsilon_{21} - \varepsilon_{33}\varepsilon_{22}],$$

$$b_1 = \varepsilon_{31} + \varepsilon_{13},$$

$$b_2 = \varepsilon_{12}\varepsilon_{23} + \varepsilon_{13}\varepsilon_{22} + \varepsilon_{31}\varepsilon_{22} + \varepsilon_{21}\varepsilon_{32}.$$

(263)

Equation (262) becomes

$$(a_1 n^4 + a_2 n^2 + a_3)^2 = (n^2 - n_0^2 \sin^2 \varphi)n_0^2 \sin^2 \varphi(b_1 n^2 + b_2)^2. \qquad (264)$$

If we define that

$$x = n^2,$$
$$c_4 = a_1^2,$$
$$c_3 = 2a_1 a_2 - b_1^2 n_0^2 \sin^2 \varphi,$$
$$c_2 = a_2^2 + 2a_1 a_3 + b_1^2 n_0^4 \sin^4 \varphi - 2b_1 b_2 n_0^2 \sin^2 \varphi,$$
$$c_1 = 2a_3 a_2 + 2b_1 b_2 n_0^4 \sin^4 \varphi - b_2^2 n_0^2 \sin^2 \varphi,$$
$$c_0 = a_3^2 + b_2^2 n_0^4 \sin^4 \varphi,$$

we must then solve the following equation:

$$c_4 x^4 + c_3 x^3 + c_2 x^2 + c_1 x + c_0 = 0. \qquad (265)$$

This equation has, in principle, four different solutions. We choose $n_\pm$ and the two corresponding vectors of propagations $s_\pm$. Using the two values of n and s in Eq. (257), we obtain the corresponding modes, i.e., for each value of n we obtain:

$$\boldsymbol{E} = E_t A e^{i\omega(\frac{n}{c}sr - t)}, \qquad (266)$$

where

$$A = \begin{pmatrix} \varepsilon_{12}\left(\left(\frac{n^2}{2}\sin 2\vartheta + \varepsilon_{13}\right)\varepsilon_{21} + (n^2\cos^2\vartheta - \varepsilon_{11})\varepsilon_{23}\right) + \\ \left(\frac{n^2}{2}\sin 2\vartheta + \varepsilon_{13}\right)\left((n^2 - \varepsilon_{22})(n^2\cos^2\vartheta - \varepsilon_{11}) - \varepsilon_{21}\varepsilon_{12}\right) \\ \\ (n^2\cos^2\vartheta - \varepsilon_{11})\left(\left(\frac{n^2}{2}\sin 2\vartheta + \varepsilon_{13}\right)\varepsilon_{21} + (n^2\cos^2\vartheta + \varepsilon_{11})\varepsilon_{23}\right) \\ \\ (n^2\cos^2\vartheta - \varepsilon_{11})\left((n^2 - \varepsilon_{22})(n^2\cos^2\vartheta - \varepsilon_{11}) - \varepsilon_{21}\varepsilon_{12}\right) \end{pmatrix},$$

(267)

and

$$E = E_t B e^{i\omega(\frac{n}{c} s r - t)}, \tag{268}$$

where

$$B = \begin{pmatrix} \varepsilon_{12}(n^2 \cos^2 \vartheta - \varepsilon_{11})\varepsilon_{23} + \left(\frac{n^2}{2} \sin 2\vartheta + \varepsilon_{13}\right)(n^2 - \varepsilon_{22})(n^2 \cos^2 \vartheta - \varepsilon_{11}) \\ (n^2 \cos^2 \vartheta - \varepsilon_{11})\left(\left(\frac{n^2}{2} \sin 2\vartheta + \varepsilon_{13}\right)\varepsilon_{21} + (n^2 \cos^2 \vartheta - \varepsilon_{11})\varepsilon_{23}\right) \\ (n^2 \cos^2 \vartheta - \varepsilon_{11})\left((n^2 - \varepsilon_{22})(n^2 \cos^2 \vartheta - \varepsilon_{11}) - \varepsilon_{21}\varepsilon_{12}\right) \end{pmatrix}. \tag{269}$$

The continuity of the parallel component of the electric field to the surface results in

$$(\mathbf{E}_i + \mathbf{E}_r) \times \mathbf{n} = \mathbf{E}_t \times \mathbf{n},$$

if

$$\mathbf{E}_i = \begin{pmatrix} E_1^{i(r)} \\ E_2^{i(r)} \\ E_3^{i(r)} \end{pmatrix} e^{i\omega(\frac{n_0}{c} s_{i(r)} \cdot \mathbf{r} - t)}.$$

At the surface we get:

$$E_1^i + E_1^r = E_t \varepsilon_{12}(n^2 \cos^2 \vartheta - \varepsilon_{11})\varepsilon_{23} + \left(\frac{n^2}{2} \sin 2\vartheta + \varepsilon_{13}\right)(n^2 - \varepsilon_{22})(n^2 \cos^2 \vartheta - \varepsilon_{11})$$

$$E_2^i + E_2^r = E_t(n^2 \cos^2 \vartheta - \varepsilon_{11})\left(\left(\frac{n^2}{2} \sin 2\vartheta + \varepsilon_{13}\right)\varepsilon_{21} + (n^2 \cos^2 \vartheta - \varepsilon_{11})\varepsilon_{23}\right). \tag{270}$$

For the normal component to the surface, we will use the discontinuity of the electric displacement

$$\varepsilon_0(E_3^i + E_3^r) = \varepsilon_{3i} E_t^i + 4\pi \Sigma. \tag{271}$$

Case of Non-magnetic Materials

We will write Eq. (257) for a non magnetic system as

$$\begin{pmatrix} n^2 \cos^2 \vartheta - \varepsilon_{11} & 0 & -\frac{n^2}{2} \sin 2\vartheta \\ 0 & n^2 - \varepsilon_{22} & 0 \\ -\frac{n^2}{2} \sin 2\vartheta & 0 & n^2 \sin^2 \vartheta - \varepsilon_{33} \end{pmatrix} \begin{pmatrix} E_1 \\ E_2 \\ E_3 \end{pmatrix} = 0. \tag{272}$$

This equation has two solutions:

$$n_1^2 = \varepsilon_{22},$$

$$n_2^2 = \frac{\varepsilon_{11}\varepsilon_{33}}{\varepsilon_{11}\sin^2\vartheta + \varepsilon_{33}\cos^2\vartheta} \qquad (273)$$

The corresponding eigenmodes are given by:

$$\boldsymbol{E}_1 = \begin{pmatrix} 0 \\ E \\ 0 \end{pmatrix} \quad \boldsymbol{E}_2 = \begin{pmatrix} -\varepsilon_{33}\cos\vartheta \\ 0 \\ \varepsilon_{11}\sin\vartheta \end{pmatrix}. \qquad (274)$$

In the case of a cubic system, where $\varepsilon_{11} = \varepsilon_{22} = \varepsilon_{33}$, the two eigenvalues are identical, and in this case there is no difference in absorption between the two polarizations of light.

In the case of tetragonal lattice, we distinguish two situations: the first, where the surface is perpendicular to the c axis, and the second situation, where the c axis is perpendicular to the incidence plan.

- *First case*

  $\varepsilon_{11} = \varepsilon_{22} = \varepsilon \neq \varepsilon_{33} = \varepsilon'$: the difference $\Delta = n_2^2 - n_1^2$ depends on the angle of incidence. If $\vartheta = 0$ $(\boldsymbol{k}//\boldsymbol{c})$, Δ is zero. There is no difference between the two modes, so for $\vartheta \neq 0$, Δ is not zero and is given by:

$$\Delta = \frac{\varepsilon' - \varepsilon}{\varepsilon\sin^2\vartheta + \varepsilon'\cos^2\vartheta}\varepsilon\sin^2\vartheta.$$

We have therefore two waves, the ordinary one corresponding to the first mode, where the propagation speed $v = \frac{c}{n_1}$ is independent of the incidence angle. The extraordinary one, where propagation speed is depend on the angle $v = \frac{c}{n_2}$. Even in the case where the system is non-magnetic, there is some difference between the two modes, which will lead to a dichroic signal (Fig. 7).

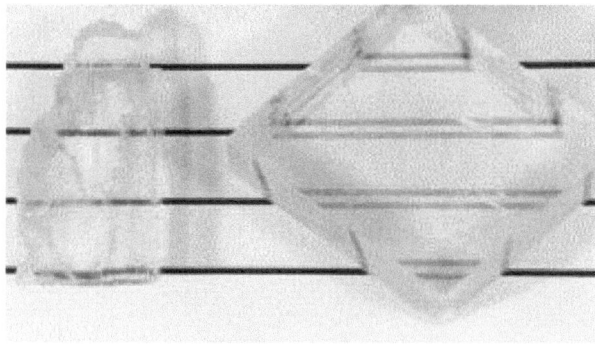

Fig. 7 Two possible modes and transmission vectors in the case of a biaxial material like calcite

- *Second case*
  $\varepsilon_{11} = \varepsilon_{33} = \varepsilon$ and $\varepsilon_{22} = \varepsilon'$: there is always a signal for any incidence angle. The difference Δ is also independent of the angle:

$$\Delta = \varepsilon - \varepsilon'.$$

We can also see that for a non-magnetic system, there is also a dichroic signal which depends on the crystal asymmetry.

Case of a Magnetic Material

Let's consider a magnetic material, limited by it surface, with a normal vector n, and a magnetization along the c direction. Let us consider (1) the polar geometry, where the c axis is parallel to the vector n, (2) the incidence plan is determined by the vector n and the vector c, such that $n \perp c$, and (3) the transverse geometry where the vector c is perpendicular to the incidence plan. The vector of incidence and transmission, in the reference frame (e_1, e_2, n), can be written as:

$$s_0 = \begin{pmatrix} \sin \varphi \\ 0 \\ \cos \varphi \end{pmatrix}, \tag{275}$$

$$s = \begin{pmatrix} \sin \vartheta \\ 0 \\ \cos \vartheta \end{pmatrix}. \tag{276}$$

Polar Geometry

In this geometry, we assume that the incidence plan is the xz plan. The dielectric tensor can be written as[24]

$$\epsilon = \begin{pmatrix} \varepsilon & \varepsilon_1 & 0 \\ -\varepsilon_1 & \varepsilon & 0 \\ 0 & 0 & \varepsilon' \end{pmatrix}. \tag{277}$$

Fresnel equation becomes

$$\begin{pmatrix} n^2 \cos^2 \vartheta - \varepsilon & -\varepsilon_1 & -n^2 \sin \vartheta \cos \vartheta \\ \varepsilon_1 & n^2 - \varepsilon & 0 \\ -n^2 \sin \vartheta \cos \vartheta & 0 & n^2 \sin^2 \vartheta - \varepsilon' \end{pmatrix} \begin{pmatrix} E_1 \\ E_2 \\ E_3 \end{pmatrix} = 0. \tag{278}$$

---

[24] such that $\varepsilon' = \varepsilon_{cc}$, $\varepsilon = \varepsilon_{aa}$ and $\varepsilon_1 = \varepsilon_{ac}$

The secular equation is then given by

$$n^4 \left[\varepsilon' \cos^2 \vartheta + \varepsilon \sin^2 \vartheta\right] - n^2 \left[\varepsilon^2 \sin^2 \vartheta + \varepsilon\varepsilon' \cos^2 \vartheta + \varepsilon\varepsilon' + \varepsilon_1^2 \sin^2 \vartheta\right] + \varepsilon^2 \varepsilon' + \varepsilon_1^2 \varepsilon' = 0, \tag{279}$$

and accepts two solutions:

$$n_{1(2)}^2 = \frac{\varepsilon^2 \sin^2 \vartheta + \varepsilon\varepsilon' \cos^2 \vartheta + \varepsilon\varepsilon' + \varepsilon_1^2 \sin^2 \vartheta \pm \sqrt{\Delta}}{2\left[\varepsilon \sin^2 \vartheta + \varepsilon' \cos^2 \vartheta\right]}, \tag{280}$$

where

$$\Delta = \left[\varepsilon^2 \sin^2 \vartheta + \varepsilon\varepsilon' \cos^2 \vartheta + \varepsilon\varepsilon' + \varepsilon_1^2 \sin^2 \vartheta\right]^2 - 4\left[\varepsilon^2 \varepsilon' + \varepsilon_1^2 \varepsilon'\right]\left[\varepsilon \sin^2 \vartheta + \varepsilon' \cos^2 \vartheta\right], \tag{281}$$

and the two modes are the eigenvectors associated to the eigenvalues n_1 and n_2.

In the particular case where $\vartheta = 0$,

$$\Delta = -4\varepsilon_1^2 \varepsilon'^2 \implies \sqrt{\Delta} = 2i\varepsilon_1\varepsilon', \tag{282}$$

and

$$n_{\pm}^2 = \varepsilon \pm i\varepsilon_1. \tag{283}$$

The associated modes are given by

$$\mathbf{E}_1 = \begin{pmatrix} 1 \\ i \\ 0 \end{pmatrix} \quad \mathbf{E}_2 = \begin{pmatrix} 1 \\ -i \\ 0 \end{pmatrix}. \tag{284}$$

Longitudinal Geometry

In this particular geometry the tensor matrix elements in the previous reference frame are given by:

$$\epsilon = \begin{pmatrix} \varepsilon' & 0 & 0 \\ 0 & \varepsilon & \varepsilon_1 \\ 0 & -\varepsilon_1 & \varepsilon \end{pmatrix}, \tag{285}$$

and Fresnel equation becomes

$$\begin{pmatrix} n^2 \cos^2 \vartheta - \varepsilon' & 0 & -n^2 \sin \vartheta \cos \vartheta \\ 0 & n^2 - \varepsilon & -\varepsilon_1 \\ -n^2 \sin \vartheta \cos \vartheta & \varepsilon_1 & n^2 \sin^2 \vartheta - \varepsilon \end{pmatrix} \begin{pmatrix} E_1 \\ E_2 \\ E_3 \end{pmatrix} = 0. \tag{286}$$

Its secular equation is then

$$n^4 \left[\varepsilon \cos^2 \vartheta + \varepsilon' \sin^2 \vartheta \right] - n^2 \left[\varepsilon^2 \cos^2 \vartheta + \varepsilon \varepsilon' \sin^2 \vartheta + \varepsilon \varepsilon' - \varepsilon_1^2 \cos^2 \vartheta \right] + \varepsilon^2 \varepsilon' + \varepsilon_1^2 \varepsilon' = 0. \tag{287}$$

The discriminant of this equation is given by

$$\Delta = \left[\varepsilon^2 \cos^2 \vartheta + \varepsilon \varepsilon' \sin^2 \vartheta + \varepsilon \varepsilon' + \varepsilon_1^2 \cos^2 \vartheta \right]^2 - 4 \left[\varepsilon^2 \varepsilon' + \varepsilon_1^2 \varepsilon' \right] \left[\varepsilon \cos^2 \vartheta + \varepsilon' \sin^2 \vartheta \right].$$

This equation has the two following eigenvalues:

$$n_{1(2)}^2 = \frac{\varepsilon^2 \cos^2 \vartheta + \varepsilon \varepsilon' \sin^2 \vartheta + \varepsilon \varepsilon' + \varepsilon_1^2 \cos^2 \vartheta \pm \sqrt{\Delta}}{2 \left[\varepsilon \cos^2 \vartheta + \varepsilon' \sin^2 \vartheta \right]}. \tag{288}$$

Transverse Geometry

In this geometry the dielectric tensor matrix elements are the same as those given by Eq. (285). However, the incident and transmitted wave vectors are, respectively, given by

$$s = \begin{pmatrix} \sin \vartheta \\ 0 \\ \cos \vartheta \end{pmatrix}, \tag{289}$$

$$s_0 = \begin{pmatrix} \sin \varphi \\ 0 \\ \cos \varphi \end{pmatrix}, \tag{290}$$

and Fresnel equation becomes:

$$\begin{pmatrix} n^2 \cos^2 \vartheta - \varepsilon & 0 & -n^2 \sin \vartheta \cos \vartheta - \varepsilon_1 \\ 0 & n^2 - \varepsilon' & 0 \\ -n^2 \sin \vartheta \cos \vartheta + \varepsilon_1 & 0 & n^2 \sin^2 \vartheta - \varepsilon \end{pmatrix} \begin{pmatrix} E_1 \\ E_2 \\ E_3 \end{pmatrix} = 0. \tag{291}$$

Its secular equation is then

$$\left[n^2 - \varepsilon' \right] \left[n^2 \varepsilon - \varepsilon^2 - \varepsilon_1^2 \right] = 0. \tag{292}$$

Its eigenvalues are given by

$$n_1^2 = \varepsilon',$$

$$n_2^2 = \varepsilon + \frac{\varepsilon_1^2}{\varepsilon}, \tag{293}$$

and the associated modes by

$$E_1 = \begin{pmatrix} 0 \\ E \\ 0 \end{pmatrix} \quad E_2 \sim \begin{pmatrix} n^2 \sin \vartheta \cos \vartheta + \varepsilon_1 \\ 0 \\ n^2 \cos^2 \vartheta - \varepsilon \end{pmatrix}. \tag{294}$$

General Case

In the case where the magnetization is in the plan, the incidence plan is determine by the vector n and another surface vector u, making an angle $\frac{\pi}{2} - \chi$ with the magnetization (c vector in this case). To find the characteristic matrix elements of the dielectric tensor, we apply a rotation of angle χ around the direction parallel to the vector n :

$$\epsilon' = R^{-1} \epsilon R,$$

where

$$R = \begin{pmatrix} \cos \chi & \sin \chi & 0 \\ -\sin \chi & \cos \chi & 0 \\ 0 & 0 & 1. \end{pmatrix}. \tag{295}$$

We then obtain

$$\epsilon' = \begin{pmatrix} \varepsilon \cos^2 \chi + \varepsilon' \sin^2 \chi & (\varepsilon - \varepsilon') \cos \chi \sin \chi & \varepsilon_1 \cos \chi \\ (\varepsilon' - \varepsilon) \cos \chi \sin \chi & \varepsilon \sin^2 \chi + \varepsilon' \cos^2 \chi & \varepsilon_1 \sin \chi \\ -\varepsilon_1 \cos \chi & -\varepsilon_1 \sin \chi & \varepsilon \end{pmatrix}, \tag{296}$$

and

$$s = \begin{pmatrix} \sin \vartheta \\ 0 \\ \cos \vartheta \end{pmatrix}, \tag{297}$$

$$s_0 = \begin{pmatrix} \sin \varphi \\ 0 \\ \cos \varphi \end{pmatrix}. \tag{298}$$

This transformation leads to the following Fresnel equation:

$$A \begin{pmatrix} E_1 \\ E_2 \\ E_3 \end{pmatrix} = 0, \tag{299}$$

where

$$
A = \begin{pmatrix}
n^2 \cos^2 \theta - \varepsilon \cos^2 \chi - \varepsilon' \sin^2 \chi & -(\varepsilon - \varepsilon') \cos \chi \sin \chi & -n^2 \cos \vartheta \sin \vartheta - \varepsilon_1 \cos \chi \\
-(\varepsilon' - \varepsilon) \cos \chi \sin \chi & n^2 - \varepsilon \sin^2 \chi - \varepsilon' \cos^2 \chi & -\varepsilon_1 \sin \chi \\
-n^2 \cos \vartheta \sin \vartheta + \varepsilon_1 \cos \chi & \varepsilon_1 \sin \chi & n^2 \sin^2 \vartheta - \varepsilon
\end{pmatrix}.
$$

$$(300)$$

In calculating the eigenvalues and eigenvectors of Fresnel equation, we obtain the indices of refraction and the corresponding modes.

Continuity Conditions

The Maxwell equations, combined with the constitutive equation of matter, lead to the continuity conditions of field, at the interface separating the two medias defined by their indices n_0 and n. Those conditions are

1. The perpendicular to the surface component of the magnetic induction must be continuous through the surface.
2. In presence of a charge surface distribution Σ, the perpendicular to the surface component of the vector electric displacement must be discontinuous, and the discontinuity is $4\pi \Sigma$.
3. The parallel component to the interface of the electric field must be continuous.
4. In presence of surface current J_s, the parallel to the surface component of the magnetic field must be discontinuous, and this discontinuity is $\frac{4\pi}{c} J_s \times n$.

Those conditions of continuity impose the equality of the phase of the propagations, as produced by Snell's equation. Starting from the preceding continuity equations and assuming that that there is no induced surface charge and electric current, we can relate the component of the incident, transmitted, and reflected electric field by the following formulas:

$$
\begin{pmatrix} E_\parallel^t \\ E_\perp^t \end{pmatrix} = \begin{pmatrix} T_{pp} & T_{ps} \\ T_{sp} & T_{ss} \end{pmatrix} \begin{pmatrix} E_\parallel^i \\ E_\perp^i \end{pmatrix},
\tag{301}
$$

and

$$
\begin{pmatrix} E_\parallel^r \\ E_\perp^r \end{pmatrix} = \begin{pmatrix} R_{pp} & R_{ps} \\ R_{sp} & R_{ss} \end{pmatrix} \begin{pmatrix} E_\parallel^i \\ E_\perp^i \end{pmatrix},
\tag{302}
$$

where the indices i, r, and t correspond, respectively, to the incident, reflected, and transmitted waves. R is the reflexion matrix and T is the transmission one, verifying:

$$
R + T = 1.
$$

The diagonal matrix elements of R and T are given by [143–145] :

$$R_{pp} = \frac{n.\cos\varphi - n_0\cos\vartheta}{n.\cos\varphi + n_0.\cos\vartheta},$$

$$R_{ss} = \frac{n_0.\cos\varphi - n.\cos\vartheta}{n_0.\cos\varphi + n.\cos\vartheta},$$

$$T_{pp} = \frac{2n_0\cos\vartheta}{n.\cos\varphi + n_0.\cos\vartheta}, \tag{303}$$

$$T_{ss} = \frac{2n_0\cos\varphi}{n_0\cos\varphi + n\cos\vartheta},$$

and the non-diagonal elements by

$$R_{ps} = \frac{-in_0(n_2 - n_1).\cos\varphi}{(n.\cos\varphi + n_0.\cos\vartheta)(n_0.\cos\varphi + n.\cos\vartheta)\cos\vartheta},$$

$$R_{sp} = \pm R_{ps},$$

$$T_{ps} = -R_{ps}, \tag{304}$$

$$T_{sp} = -R_{sp}.$$

Thus, the knowledge of the dielectric tensor matrix elements together with the geometry of the incident wave leads to the determination of the transmitted and reflected electric fields.

Circular Dichroism of X-Rays

In this section, we will discuss the x-ray absorption. The diagonal elements of the electric tensor are close to the permeability of free space ($\varepsilon_0 = n_0^2$). Since the dielectric tensor is given by

$$\epsilon = 1 + \frac{4i\pi\sigma}{\omega} = 1 - \frac{4\pi\sigma^2}{\omega} + i\frac{4\pi\sigma^1}{\omega},$$

where σ is the optical conductivity tensor (σ^1 its real part, and σ^2 its imaginary part, and ω is the frequency of the incident photons).

We would like to calculate the absorption coefficient of light for a sample of width d. To proceed, we write the solution of the propagation equation of the two modes (E_1 and E_2):

$$E^t(r, t) = E_0 e^{i\omega(\frac{n}{c}s.r - t)}. \tag{305}$$

If we use the solutions of Fresnel equation (for example, Eq. (293)), where n_1 is written as

$$n_1^2 = \varepsilon' = 1 - \frac{4\pi\sigma_{cc}^2}{\omega} + i\frac{4\pi\sigma_{cc}^1}{\omega}. \tag{306}$$

Since ϵ is very close to unity,

$$\frac{4\pi\sigma^1}{\omega} \ll 1.$$

Consequently, the index of refraction n is given by

$$n = \sqrt{\varepsilon'} \sim 1 - \frac{4\pi\sigma_{cc}^2}{\omega} + i\frac{2\pi\sigma_{cc}^1}{\omega} = \alpha + i\frac{2\pi\sigma_{cc}^1}{\omega}. \tag{307}$$

Substituting the expression of n in Eq. (305), we obtain

$$E^t(r,t) = E_0 e^{-\frac{2\pi\sigma_{cc}^1}{c}.s.r} e^{i\omega\left(\frac{\alpha}{c}s.r-t\right)}. \tag{308}$$

The intensity of light absorbed by the sample is proportional to square of vector E and is given by

$$I = |E_0|^2 e^{-\frac{4\pi\sigma_{cc}^1}{c}.s.r}. \tag{309}$$

We find Beer's law, where the absorption coefficient is given by

$$\mu = \frac{4\pi\sigma^1}{c}. \tag{310}$$

References

1. W. Kohn and L. J. Sham, Phys. Rev. **140**, 1133 (1965).
2. H. Ebert, Phys. Rev. B **38**, 9390 (1988).
3. D. S. Wang, R. Wu, and A. J. Freeman, Phys. Rev. Lett. **73**, 1994 (1994).
4. H. Ebert, Rep. Prog. Phys. **59**, 1665 (1996).
5. M. Alouani, J. M. Wills, and J. W. Wilkins, Phys. Rev. B **57**, 9502 (1998).
6. M. Asato, A. Settels, T. Hoshino, T. Asada, S. Bluegel, R. Zeller, and P. H. Dederichs, Phys. Rev. B **60**, 5202 (1999).
7. J. C. Slater and G. F. Koster, Phys. Rev. **94**, 1498 (1954).
8. O. Eriksson and J. M. Wills, *Electronic Structure and Physical Properties of Solids*, ed. H. Dreyssé (Springer-Verlag, Heidelberg, 1998), p. 247.
9. I. Galanakis, S. Ostanin, M. Alouani, H. Dreyssé, and J. M. Wills, Phys. Rev. B **61**, 599 (2000).
10. I. Galanakis, S. Ostanin, M. Alouani, H. Dreyssé, and J. M. Wills, Phys. Rev. B **61**, 4093 (2000).
11. L. Sandratskii, Adv. Phys. **47**, 91 (1998).
12. P. Ravindran, A. Delin, P. James, B. Johansson, J. Wills, R. Ahuja, and O. Eriksson, Phys. Rev. B **59**, 15680 (1999).

13. F. P. P. S. Bagus, G. Pacchioni, *Core Level Spectroscopies for Magnetic Phenomena: Theory and Experiment* (Plenum, New York, 1995).
14. H. Ebert and G. Schütz, *Spin-Orbit Influenced Spectroscopies of Magnetic Solids* (Springer-Verlag, Heidelberg, 1996).
15. B. Thole, P. Carra, F. Sette, and G. van der Laan, Phys. Rev. Lett. **68**, 1943 (1992).
16. P. Carra, B. Thole, M. Altarelli, and X. Wang, Phys. Rev. Lett. **70**, 694 (1993).
17. P. Bruno, Phys. Rev. B **39**, 865 (1989).
18. G. van der Laan, Phys. Rev. Lett. **82**, 640 (1999).
19. M. Alouani and J. M. Wills, *Electronic Structure and Physical Properties of Solids*, ed. H. Dreyssé (Springer-Verlag, Heidelberg, 1998), p. 168.
20. A. Ankudinov and J. Rehr, Phys. Rev. B **52**, 10214 (1995).
21. C. Brouder, M. Alouani, and K. Bennemann, Phys. Rev. B **54**, 7334 (1996).
22. C. S. Wang and J. Callaway, Phys. Rev. B **9**, 4897 (1974).
23. S. Halilov and Yu. A. Uspenskii, J. Phys.: Condens Matter **2**, 6137 (1990).
24. S. Halilov, J. Phys.: Condens Matter **4**, 1299 (1992).
25. T. Kraft, P. Oppeneer, V. N. Antonov, and H. Eschrig, Phys. Rev. B **52**, 3561 (1995).
26. P. Oppeneer and V. Antonov, *Spin-Orbit Influenced Spectroscopies of Magnetic Solids*, eds. H. Ebert and G. Schütz (Springer, Berlin, 1996), p. 29.
27. M. Born and J. R. Oppenheimer, Ann. Phys. **87**, 457 (1927).
28. V. Fock, Z. Phys. **61**, 126 (1930).
29. V. Fock, Z. Phys. **62**, 795 (1930).
30. L. H. Thomas, Proc. Cambridge Phyl. Soc. **23**, 542 (1927).
31. E. Fermi, Rend. Naz.Linzei **6**, 602 (1927).
32. J. C. Slater, *The Self-Consistent Field of Molecules and solids*, vol. 4 (McGraw-Hill, New York, 1974).
33. P. Hohenberg and W. Kohn, Phys. Rev. **136**, 864 (1964).
34. D. M. Ceperly and B. J. Alder, Phys. Rev. Lett. **45**, 566 (1980).
35. U. von Barth and L. Hedin, J. Phys. C **5**, 1629 (1972).
36. V. L. Moruzzi, J. F. Janak, and A. R. Williams, *Calculated Electronic Properties of Metals* (Pergamon, New York, 1978).
37. Y. Wang and J. P. Perdew, Phys. Rev. B **44**, 13298 (1991).
38. J. P. Perdew, K. Burke, and M. Ernzerhof, Phys. Rev. B **77**, 3865 (1996).
39. J. P. Perdew, K. Burke, and Y. Wang, Phys. Rev. B **54**, 16533 (1996).
40. O. Gunnarsson and B. I. Lundqvist, Phys. Rev. B **13**, 4274 (1976).
41. J. P. Perdew and A. Zunger, Phys. Rev. B **23**, 5048 (1981).
42. V. Anisomov, F. Aryasetiawan, and A. I. Lichtenstein, J. Phys.: Condens Matter **9**, 767 (1997).
43. J. Hubbard, Proc. R. Soc. London **276**, 238 (1963).
44. J. Hubbard, Proc. R. Soc. London **277**, 237 (1964).
45. J. Hubbard, Proc. R. Soc. London **281**, 401 (1964).
46. J. Hubbard, Proc. R. Soc. London **285**, 442 (1965).
47. A. I. Liechtenstein, V. I. Anisomov, and J. Zaanen, Phys. Rev. B **52**, 5467 (1995).
48. A. B. Shick, A. I. Liechtenstein, and W. E. Pickett, Phys. Rev. B **60**, 10763 (1999).
49. B. N. Harmon, V. P. Antropov, A. I. Liechtenstein, I. V. Solovyev, and V. I. Anisomov, J. Phys. Chem. Solids **56**, 1521 (1995).
50. V. Anisomov and O. Gunnarsson, Phys. Rev. B **43**, 5770 (1991).
51. W. E. Pickett, S. C. Erwin, and E. C. Ethridge, Phys. Rev. B **58**, 1201 (2005).
52. M. Cococcioni and S. de Gironcoli, Phys. Rev. B **71**, 035105 (2005).
53. Z. Szotek, W. M. Temmerman, and H. Winter, Phys. Rev. B **47**, 4029 (1993).
54. V. I. Anisomov, I. V. Solovyev, M. A. Korotin, M. T. Czyzyk, and G. A. Sawatzky, Phys. Rev. B **48**, 16929 (1993).
55. O. Bengone, M. Alouani, P. Blöchl, and J. Hugel, Phys. Rev. B **62**, 16392 (2000).

56. A. G. Petukhov, I. I. Mazin, L. Chioncel, and A. I. Lichtenstein, Phys. Rev. B **67**, 153106 (2003).
57. S. Abdelouahed, N. Baadji, and M. Alouani, Phys. Rev. B **75**, 094428 (2007).
58. S. Abdelouahed and M. Alouani, Phys. Rev. B **76**, 214409 (2007).
59. S. Abdelouahed and M. Alouani, Phys. Rev. B **79**, 054406 (2009).
60. H. Ebert, *Electronic Structure and Physical Properties of Solids*, ed. H. Dreyssé (Springer-Verlag, Heidelberg, 1998), p. 191.
61. J. D. Djorken and S. D. Drell, *Relativitic Quantum Mechanics* (Mc Graw Hill, New York, 1964).
62. E. M. Rose, *Relativistic Electron Theory* (Wiley, New York, 1961).
63. H. Ebert, Phys. Rev. B **38**, 9390 (1988).
64. J. C. Slater, Phys. Rev. **51**, 846 (1937).
65. J. C. Slater, Phys. Rev. **92**, 603 (1953).
66. M. M. Saffren and J. C. Slater, Phys. Rev. **92**, 1126 (1953).
67. J. Korringa, Physica **13**, 392 (1947).
68. W. Kohn and N. Rostoker, Phys. Rev. **94**, 1111 (1954).
69. J. Callaway, *Energy Band Theory* (Academic Press, New York, 1964).
70. J. H. Terrell, Phys. Rev. Lett. **8**, 149 (1964).
71. O. K. Andersen, Phys. Rev. B **12**, 3060 (1975).
72. T. L. Loucks, *Augmented Plane Wave Method* (Benjamin, New York, 1967).
73. S. Bluegel and G. Bihlmayer, *Full-Potential Linearized Augmented Plane Wave Method*, vol. 31 (John von Neumann Institute for Computing, Juelich, 2006).
74. H. Skriver, *The LMTO Method* (Springer-Verlag, Berlin, 1984).
75. D. Singh, Phys. Rev. B **43**, 6388 (1991).
76. D. J. Singh, *Planewaves, Pseudopotentials and the LAPW Method* (Kluwer Academic Publishers, Boston/Dordrecht/London, 1994).
77. D. R. Hamann, Phys. Rev. Lett. **42**, 662 (1979).
78. E. Wimmer, H. Krakauer, M. Weinert, and A. J. Freeman, Phys. Rev. B **24**, 864 (1981).
79. http://www.fleur.de
80. M. Weinert, J. Math. Phys. **22**, 2433 (1981).
81. J. D. Jackson, *Classical Electrodynamics* (Wiley, New York, 1975).
82. M. Brooks and D. Goodings, J. Phys. C **5**, 1279 (1968).
83. J. Jenson and A. Mackintosh, *Rare Earth Magnetism* (Oxford University Press, Oxford, 1991).
84. J. H. van Vleck, Phys. Rev. **52**, 1178 (1937).
85. A. B. Shick, V. Drchal, J. Kudrnovský, and P. Weinberger, Phys. Rev. B **54**, 1610 (1996).
86. A. R. Makintosh and O. K. Andersen, *Electron at the Fermi Surface*, ed. M. Springford (Cambridge University Press, Cambridge, 1980).
87. D. Koelling and B. Harmon, J. Phys. C **10**, 3107 (1977).
88. C. Cohen-Tannoudji, B. Diu, and F. Laloë, *Mécanique Quantique*, vol. I, chap. IV, complément B (Hermann, Paris, 1996), pp. 418–422.
89. R. Wu and A. J. Freeman, J. App. Phys. **79**, 6209 (1996).
90. S. Ostanin, J. B. Staunton, S. S. A. Razee, C. Demangeat, B. Ginatempo, and E. Bruno, Phys. Rev. B **69**, O64425 (2004).
91. A. B. Shick, D. L. Novikov, and A. J. Freeman, Phys. Rev. B **56**, 14259 (1997).
92. G. van der Laan, J. Phys.: Condens Matter **10**, 3239 (1998).
93. G. H. O. Daalderop, P. J. Kelly, and M. F. H. Schuurmans, Phys. Rev. B **41**, 11919 (1990).
94. D.-S. Wang, R. Wu, and A. J. Freeman, Phys. Rev. Lett. **70**, 869 (1993).
95. X. Wang, R. Wu, D.-S. Wang, and A. J. Freeman, Phys. Rev. B **54**, 61 (1996).
96. A. Hubert, W. Unger, and J. Kranz, Z. Phys. **148**, 224 (1969).
97. P. Escudier, Ann. Phys. **9**, 125 (1975).
98. H. L. Skriver, Phys. Rev. B **31**, 1909 (1985).

99. A. I. Lichtenstein, M. I. Katsnelson, V. P. Antropov, and V. A. Gubanov, J. Magn. Magn. Mater. **67**, 65 (1987).
100. O. L. Bacq, O. Eriksson, B. Johansson, P. James, and A. Delin, Phys. Rev. B **65**, 134430 (2002).
101. J. J. M. Franse and R. Gersdorf, Phys. Rev. Lett. **45**, 50 (1980).
102. W. Röntgen, Physik. Med. Ges. **137**, 132 (1895).
103. E. Basseler, nachgewiesen mittels Sekundärstrahlung: Annalen der Phydik **28**, 808 (1909).
104. J. Chapman, Phil. Mag. **25**, 792 (1913).
105. J. Becker, Phys. Rev. **20**, 134 (1922).
106. W. Kartschagin and E. Tschetwerikova, für Physik **39**, 886 (1926).
107. D. Froman, Phys. Rev. **41**, 693 (1932).
108. C. Kurylenko, J. Phys. Rad. **1**, 133 (1940).
109. D. Coster, Book Rev. **3**, 160 (1948).
110. J. Hrdý, E. Krouský, and O. Renner, Phy. Stat. Sol. (a) **53**, 143 (1979).
111. G. Schütz, E. Zech, E. Hagn, and P. Kienle, Hyp. Int. **16**, 1039 (1983).
112. E. Keller and E. Stern, *Proceedings of the EXAFS and Near Edge Structure III Conference*, eds. K. Hodgson, B. Hedman, and J. Penner-Hahn (Springer-Verlag, Berlin, 1984), p. 507.
113. G. Shütz, W. Wagner, W. Wilhelm, P. Kienle, R. Zeller, R. Frahm, and G. Materlik, Phys. Rev. Lett. **58**, 737 (1987).
114. G. Shütz, R. Frahm, P. Mautner, R. Wienke, W. Wagner, W. Wilhelm, and P. Kienle, Phys. Rev. Lett. **62**, 2620 (1989).
115. A. Rogalev, F. Wilhelm, N. Jaouen, J. Goulon, and J. P. Kappler, *Magnetism: A Synchrotron Radiation Approach*, chap. 4, eds. E. Beaurepaire, H. Bulou, F. Scheurer, and J.-P. Kappler (Springer, Berlin, 2006), pp. 71–89.
116. J. Erskine and E. Stern, Phys. Rev. B **12**, 5016 (1975).
117. B. T. Thole, G. van der Laan, and G. A. Sawatzky, Phys. Rev. Lett. **55**, 2086 (1985).
118. R. D. Cowan, J. Opt. Soc. Am. **58**, 808 (1968).
119. B. Thole, G. van der Laan, J. C. Fuggle, G. Sawatzky, R. Karnatak, and J.-M. Esteva, Phys. Rev. B **32**, 5107 (1985).
120. C. T. Chen, F. Sette, Y. Ma, and S. Modesti, Phys. Rev. B **42**, 7262 (1990).
121. C. T. Chen, N. V. Smith, and F. Sette, Phys. Rev. B **43**, 6785 (1991).
122. N. V. Smith, T. C. Chen, F. Sette, and L. F. Mattheiss, Phys. Rev. B **46**, 1023 (1992).
123. M. Faraday, Phil. Trans. R. Soc. **136**, 1 (1846).
124. J. Kerr, Philos. Mag. **3**, 321 (1877).
125. H. R. Hulme, Proc R. Soc. (London) Ser. A **135**, 237 (1932).
126. J. D. Jackson, *Classical Electrodynamics* (John Willey, New York, 1975).
127. N. Bloembergen, *Nonlinear Optics* (Benjamin, New York, 1965).
128. W. H. Kleiner, Phys. Rev. **142**, 318 (1966).
129. A. K. Zvezdin and V. A. Kotov, *Modern Magnetooptics and Magnetooptical Materials* (Institute of Physics Publishing, Bristol and Philadelphia, 1997).
130. W. Reim and J. Schoenes, *Ferromagnetic Materials*, vol. 5, eds. E. P. Wohlfarth and K. H. J. Buschow (Springer, Berlin, 1990), p. 133.
131. A. Yaresko, A. Perlov, V. Antonov, and B. Harmon, *Magnetism: A Synchrotron Radiation Approach*, chap. 6, eds. E. Beaurepaire, H. Bulou, F. Scheurer, and J.-P. Kappler (Springer, Berlin Heidelberg, 2006), pp. 121–141.
132. H. S. Bennett and E. A. Stern, Phys. Rev. **137**, A448 (1965).
133. V. N. Antonov, B. N. Harmon, A. N. Yaresko, and A. P. Shpak, Phys. Rev. B **75**, 184422 (2007).
134. E. Arola, P. Strange, and B. Gyorffy, Phys. Rev. B **55**, 472 (1997).
135. C. Cohen-Tannoudji, B. Diu, and F. Laloë, *Mécanique Quantique*, vol. I, chap. VI, (Hermann, Paris, 1996), pp. 689, 690.
136. H. Ebert, *Spin-Orbit-Influenced Spectroscopies of Magnetic Solids*, chap. 9, eds. H. Ebert and G. Schütz (Springer, Herrsching, Germany, 1995), pp. 159–177.

137. D. W. R. Wu and A. J. Freeman, Phys. Rev. Lett. **71**, 3581 (1993).
138. C. T. Chen, Y. U. Idzerda, H.-J. Lin, N. V. Smith, G. Meigs, E. Chaban, G. H. Ho, E. Pellegrin, and F. Sette, Phys. Rev. Lett. **75**, 152 (1995).
139. J. Vogel, A. Fontaine, V. Cros, F. Petroff, J.-P. Kappler, G. Krill, A. Rogalev, and J. Goulon, Phys. Rev. B **55**, 3663 (1997).
140. W. Grange, M. Maret, J.-P. Kappler, J. Vogel, A. Fontaine, F. Petroff, G. Krill, A. Rogalev, J. Goulon, M. Finazzi et al., Phys. Rev. B **58**, 6289 (1998).
141. F. Leuenberger, A. Parge, W. Felsch, F. Baudelet, C. Giorgetti, E. Dartyge, and F. Wilhelm, Phys. Rev. B **73**, 214430 (2006).
142. C. Cohen-Tannoudji, B. Diu, and F. Laloë, *Mécanique Quantique*, vol. II (Hermann, Paris, 1996).
143. R. P. Hunt, J. App. Phys. **38**, 1652 (1967).
144. Z. Q. Qiu and S. D. Badera, Rev. Sci. Instrum. **71**, 1243 (2000).
145. M. Born and E. Wolf, *Principles of Optics* (Pergamon Press, Oxford, 1980).

Effect of Spin–Orbit Coupling on the Magnetic Properties of Materials: Results

M. Alouani, N. Baadji, S. Abdelouahed, O. Bengone, and H. Dreyssé

Abstract This contribution concerning the effect of spin–orbit coupling on the magnetic properties of materials is divided into two sections. In the first section we review the method based on the density functional theory (DFT) within the local density approximation (LDA) used to compute the electronic structure, the magnetic anisotropy, the x-ray absorption spectra, and the x-ray magnetic circular dichroism. We give the major approximations used to derive the Kohn–Sham equations with or without the Hubbard interaction for correlated orbitals. We give also a brief introduction to the generalized gradient approximation (GGA). We then provide a solution of the latter equations using the full-potential linear augmented plane wave (FLAPW) basis set and discuss the so-called LDA+U method, where the Hubbard U is included for localized orbitals. We show how the relativistic effects, such as the spin–orbit coupling, can be introduced into band structure calculations and show their effect on magnetism, i.e., magnetic anisotropy energy (MAE),magnetooptical properties, and x-ray magnetic circular dichroism (XMCD).

M. Alouani (✉)
Institut de Physique et Chimie des Matériaux de Strasbourg, 23 Rue du Loess, BP 43, F-67034 Strasbourg Cedex 2, France, mebarek.alouani@ipcms.u-strasbg.fr

N. Baadji
Institut de Physique et Chimie des Matériaux de Strasbourg, 23 Rue du Loess, BP 43, F-67034 Strasbourg Cedex 2, France, baadji@ipcms.u-strasbg.fr

S. Abdelouahed
Institut de Physique et Chimie des Matériaux de Strasbourg, 23 Rue du Loess, BP 43, F-67034 Strasbourg Cedex 2, France, abdelouahed@ipcms.u-strasbg.fr

O. Bengone
Institut de Physique et Chimie des Matériaux de Strasbourg, 23 Rue du Loess, BP 43, F-67034 Strasbourg Cedex 2, France, olivier.bengone@ipcms.u-strasbg.fr

H. Dreyssé
Institut de Physique et Chimie des Matériaux de Strasbourg, 23 Rue du Loess, BP 43, F-67034 Strasbourg Cedex 2, France, hugues.dreysse@ipcms.u-strasbg.fr

Alouani, M.: *Effect of Spin–Orbit Coupling on the Magnetic Properties of Materials: Results*. Lect. Notes Phys. **795**, 309–341 (2010)
DOI 10.1007/978-3-642-04650-6_9

Then we show a brief derivation of the force theorem for the calculation of the magnetic anisotropy as well as a description of its application to the MAE calculations and show the details of the calculation of the XMCD matrix elements in the electric dipole approximation. The second section of this contribution includes some applications of the method to the computation of the electronic, magnetic, and spectroscopic properties of spintronics materials. In particular, we investigate the electronic structure and x-ray magnetic circular dichroism (XMCD) of Sr_2FeMoO_6 (SFMO for short) and other useful ferromagnetic half-metals with 100% spin polarization, materials useful for spin injection. In particular, we show that the spin–orbit coupling reduces the spin polarization, while the intra-site electronic correlations tend to increase it. For example, SFMO is found to be a half-metallic ferrimagnet with a gap in the spin-up channel. The calculated spin magnetic moments on iron and Mo sites confirm the ferromagnetic ordering and settle the controversy existing between the earlier experimental works. The orbital magnetism at the Fe and Mo sites agrees quite well with the recent experimental XMCD measurements. The computed $L_{2,3}$ XMCD at the Fe and the Mo sites compares fairly well with the experiment. The XMCD sum rule computed spin and orbital magnetic moments are in good agreement with the values obtained from the direct self-consistent calculations. In the last application, we focus on the GGA+U treatment of the electronic and magnetic structure of Gd and Gd-related compounds, such as GdN and $GdFe_2$. We compare the calculated density of states to the experimental photoemission and inverse photoemission spectra (XPS and BIS) and determine the Fermi surface with and without the Hubbard U and spin–orbit coupling. The GGA+U is found to be the most appropriate for treating the $4f$ Gd electrons. We have investigated the bulk properties and calculated the XMCD spectra at the $L_{2,3}$ edges at the Gd site of GdN. The agreement of the calculated spectra with experiment is the indication of the relevance of the XMCD formalism within the one-electron picture. The results also show that the ground-state electronic structure of GdN is that of a half-metal. Finally our computational method is used to determine the magnetic anisotropy aspect of Gd and its compounds GdN and $GdFe_2$. Using force theorem, we have calculated the MAE of Gd, GdN, and $GdFe_2$ for different directions of the magnetization. Indeed, owing to the nil spin–orbit interaction of the $4f$ half-filled shell, the force theorem is expected to be efficient for Gd and Gd compounds' MAE calculations. This theorem allows a considerable computational effort gain since the spin–orbit coupling could be calculated only for one self-consistent iteration. Once again, the GGA+U method is found to be the most adequate approach for the force theorem calculations of the Gd MAE. The GGA and GGA-core model treatments of the $4f$ states have led to a wrong MAE. It turns out that the electronic properties and the magnetic properties of $4f$ systems are tightly related, and the $4f$ electrons play a crucial role in the computed magnetic anisotropy. Although the Gd MAE is found to be similar to that of a typical $3d$ transition metal like hcp Co, the GdN and $GdFe_2$ cubic crystal MAEs are found to be different from that of a pure $3d$ cubic material like fcc Ni.

1 Magnetic Anisotropy of Transition Metal Compounds

Spectacular results have been obtained recently in the ground-state determination of the magnetic anisotropy energy (MAE) of materials. Thin films of Ni on Cu(001) have been found experimentally to display an unexpected set of magnetic phases. In contrast to a simple description, the magnetization has been found to be *in-plane*, i.e., lying in the (001) plane for the first few ad-layers of Ni. For increasing Ni thickness it rotates perpendicular to the interface plane over, and finally, for much higher Ni coverage, the magnetization rotates back in-plane due to a stronger shape anisotropy. This behavior is linked to a tetragonal distortion of the Ni films. The lattice mismatch between Ni and Cu is small enough to accommodate a perfect 2D growth, but large enough to lead to a contraction of the Ni(001) films.

Using the spin-polarized Korringa–Kohn–Rostoker (KKR) method, Uiberacker et al. [1] have described Ni films to a thickness up to 15 layers. Considering an uniform relaxation, the spin reorientation occurs at about seven layers of Ni, in good agreement with the experimental results (for the value of the lattice relaxation and for the critical thickness of Ni). These band structure frameworks compute the magnetic anisotropy energy as the sum of individual terms. In the precited work [1], the different contributions of the MAE are clearly identified for the Ni/Cu(001) system. The internal contribution (which acts as a bulk term) is counterbalanced by the surface and interface term which favor in-plane orientation. When the dipolar term is added, a subtle balance occurs. The description of the Ni films proposed by Uiberacker *et al.* has been confirmed by Spisak and Hafner [2] where all Ni planes are relaxed. In this latter work, the Vienna ab initio simulation package has been used, and the equilibrium distances found are in agreement with [1], with an additional surface inward relaxation of 3% which should not affect the trends obtained in [1]. In addition, Spisak and Hafner have demonstrated that the film's structure is probably more complex: the formation of a surfactant overlayer of Cu on top of the Ni films is energetically favored, in agreement with LEED data.

When determining the magnetic anisotropy properties, atomic relaxations play a key role. For that reason all-electron full-potential approaches are necessary; the interatomic distances have not to be assigned a priori. The FP-LMTO method has led to very interesting results. A presentation of such a framework can be found in the contribution of Eriksson and Wills in [3]. For instance, Galanakis et al. [4] have performed a systematical study of magnetic anisotropy energy of FCT $Fe_{0.5}Pd_{0.5}$ alloy versus the lattice parameters a and c; their results compare nicely with experiments in thin films. This method has been used to determine the MAE of many transition metal alloys within the local spin density approximation and the generalized gradient approximation [5]. Figure 1 shows the MAE of many transition metals calculated within the LSDA and the GGA compared to the available experimental results [5]. We can deduce directly from this figure that both LSDA and GGA produce the same tendencies as we pass from one system to another. But there are systems like $MnPt_3$, $CoPt_3$, and $MnAu_4$ where the two functionals present strong deviations. For all the other binary alloys the MAE values calculated within the two

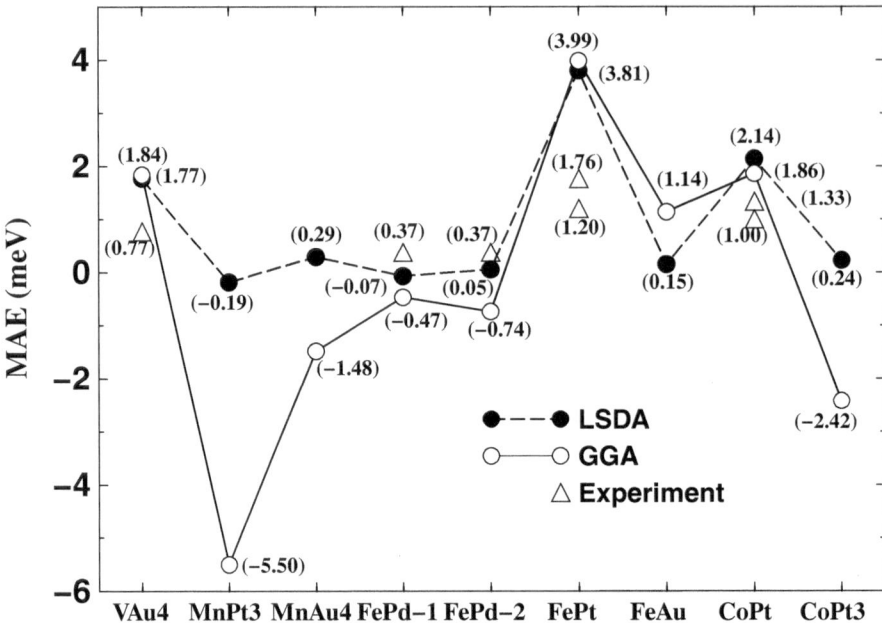

Fig. 1 The local spin density approximation (LSDA) (*filled circles*) and the generalized gradient approximation (GGA) (*empty circles*) calculated magnetic-anisotropy energy (MAE) of XY, XPt_3, and $Mn(V)Au_4$ (X= Fe, Co, Mn; Y= Pd, Pt, Au) ordered alloys compared to experimental results [5] (*open triangles*). The easy axis for the $L1_0$ structure alloys and the VAu_4 is the [001] axis and for XPt_3 the [111]. In the case of the FePt, CoPt, FeAu, and VAu_4 alloys, the theory always favors the perpendicular axis. The other binary alloys show different behavior depending on the type of approximation to the exchange-correlation potential

approximations differ by less than 1 meV, but when the values are close to zero, as is the case for FePd, it is possible that the LSDA and GGA predict a different magnetization axis. The MCA results obtained using the LSDA and GGA are, in most cases, different which led us to the conclusion that there is no general rule favoring either LSDA or GGA as the better description of the MAE of magnetic alloys. The calculated orbital moment anisotropy is similar for both LSDA and GGA and cannot explain the differences in the calculation of the MAE. Nevertheless, from this study it seems that the LSDA results are slightly in better agreement with the available experimental results. To confirm this claim further experimental data are needed.

These results indicate the present limitations of ab initio band calculations. Physical systems of interest usually include a large number of non-equivalent atoms (few tens at least). The need of an ab initio method with a better CPU timescaling with the number of atoms is necessary. The recent development of Beiden et al. [6] allows new interesting possibilities. In this new approach [6, 7] a local interaction zone (LIZ), embedded in a large supercell, for solving the quantum mechanical problem is considered, while the Poisson equation is solved in the whole space, using a screened reference medium [8]. The KKR matrices become sparse and thus a LIZ of more than thousand atoms can be considered. This real-space scheme has only been tested for simple crystallographic arrangements but a priori it could be used for much larger systems. The implementation of a full-potential version which exists already for the standard KKR method [9] will make this approach an attractive one.

2 Thin Films and Alloys

Ab initio band structure methods provide a nice tool to elucidate the behavior of adsorbed atoms and are now widely used due to the availability of efficient numerical codes [10]. It is largely recognized that interdiffusion occurs during the growth of transition metal on a substrate. A recent paper [11] on the initial growth of Co on Cu(001) combines an experimental study and a FPLAPW calculation to show that Co atoms occupying substitutional sites in the Cu substrate act as pinning centers for subsequent island nucleation. The description of magnetic nanostructures on a noble metal Ag(001) has been also investigated by means of KKR, illustrating the large number of possible magnetic arrangements [12]. For $4d$ and $5d$ elements,

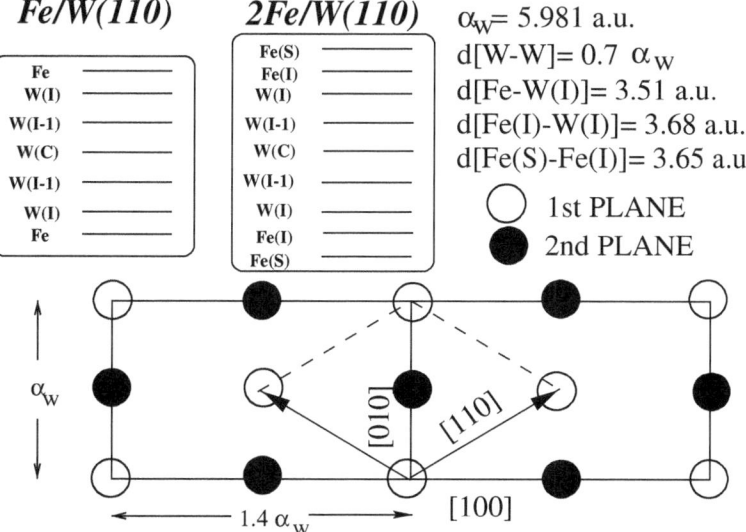

Fig. 2 One and two Fe layer slabs on W substrate (*upper drawings*) and surface view of one and two layers of Fe on W(110) substrate. The α_W is the lattice constant of bulk fcc W. The calculated interlayer distances between the W and the Fe d[Fe–W(I)] is 3.51 a.u. and that for the two-layer system d[Fe–W(I)] is 3.68 a.u., and the surface Fe and Fe(I) distance d[F(S)–Fe(i)] is 3.65 a.u. The vacuum spacing between the two slabs for the mono-layer and for the bi-layer systems is 3.5 a_W and 5.5 a_W, respectively. The surface Fe atom is represented by a *filled circle* and the Fe(I) by an *empty circle*

unexpected enhancement of the local magnetic moments is obtained. The adsorption on the (001) surface has been widely investigated (see for instance [13–15]). Experimental results involving middle series transition metals, particularly Mn, are still not clearly understood and a dense activity is developed [16]. The (111) cubic surface also attracts interest since it displays subtler behavior due to frustrations of antiferromagnetic coupling [16]. Figure 2 shows the Fe/W(110) surface and relaxed Fe–Fe and Fe–W layers [17]. The FP-LMTO is capable of obtaining the correct spin alignment of the Fe layers by calculating the MAE (see Table 1).

Table 1 Calculated Fe/W(110) magnetic anisotropy energy (MAE) for the one, two, and three Fe layer systems. The MAE is decomposed into magnetic surface anisotropy (MSA), magneto-elastic anisotropy (MEA), and shape anisotropy (SA) due to the interactions of the spins. In the case of the two-layer system, the shape anisotropy rotates the magnetization in-plane

	1ML	2ML	3ML
MSA+MEA	3.35	−0.05	1.45
SA	0.08	0.26	0.43
MAE	3.43	0.21	1.88

The determination of the electronic structure of semi-infinite ordered alloy requires the same techniques used in the previously reported films' studies. The only change is a larger number of inequivalent atoms and, in some cases, the multiplicity of numerical solutions [18]. For semi-infinite disordered alloys the coherent potential approximation [19] is definitely the right approach. Two contributions deserve special notice [20, 21]. In [20], Turek et al. predicted a new class of magnetic materials. Taking two bulk non-magnetic transition metals such as V, Ru, Rh, and Pd, they show that the (001) surfaces of the RuV, RhV, and PdV binary alloys in the

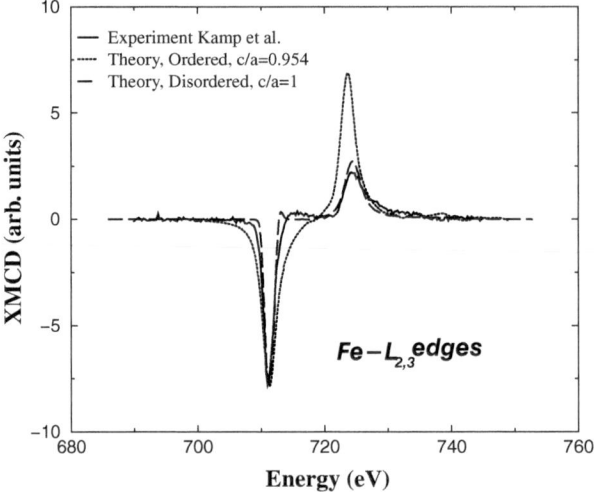

Fig. 3 Theoretical XMCD spectrum for the Fe $L_{2,3}$ edge of the ordered compressed fct FePd alloy and of the disordered fcc alloy compared with the experimental result for the FePd thick film deposited on MgO(001) at 623 K, see [22] for more details

bcc structure are magnetic over a broad concentration range; the magnetic moments are mainly located at V surface sites and are as large as 1 μ_B for alloy with 75% V concentration. For FePd the disorder is shown to play a major role for the determination of the magnetic properties. Figure 3 presents the calculated Fe $L_{2,3}$ XMCD spectrum of the ordered compressed face-centered tetragonal (fct) alloy (c/a=0.954) and of the disordered face-centered cubic (fcc) alloy together with the experiment of the fct alloy. We notice that the spectrum of the disordered alloy is in a better agreement with experiment than that of the ordered alloy. This is a clear evidence that the disorder in the FePd alloy is important and should be taken into account for the determination of its physical properties. The theory for the ordered alloy underestimates the L_3/L_2 branching ratio, L_3/L_2=1.12 (theory) and 1.32 (experiment), while theory for the disordered alloy gives a branching ratio of 1.25 in much better agreement with experiment [22].

3 Electronic Structure, XMCD of Sr$_2$FeMoO$_6$

Sr$_2$FeMoO$_6$ (in the following abbreviated to SFMO) is a case of special interest due to its technological potential as a spintronics material and due to the many diverging reports, both theoretical and experimental, on its electronic and magnetic structure. SFMO is a magnetic metal with a gap in one spin channel. It is therefore a half-metal with a Curie temperature T_C of 418 K, exceeding room temperature. The half-metallic electronic structure causes very high, in principle total, spin polarization of the charge carriers. This, in turn, may give rise to a low-field magneto-resistive effect based on inter-grain tunneling, which can be explained in the following way. In a simple model of how the magneto-resistive effect comes about, imagine a system with a microstructure consisting of half-metallic mono-domain grains dispersed in an insulating matrix. When the magnetic field is zero, the magnetic moments of the magnetic grains are randomly ordered and the tunneling from grain to grain becomes low since the moments of two adjacent grains in general are not aligned. As the grain magnetic moments align due to an increasing external magnetic field, the resistance goes down. More important is a low coercivity of the material (otherwise one needs a very large magnetic field in order to turn the grain magnetic moments) and a high enough Curie temperature.

SFMO crystallizes in a body-centered tetragonal structure consisting of slightly distorted oxygen octahedra with alternating Fe and Mo ions in the center. The voids in-between the octahedra are occupied by Sr atoms.

The magnetic structure of SFMO single crystal was investigated by two groups [23, 24] using x-ray magnetic circular dichroism at the $L_{2,3}$ edges of Fe and Mo, in which they obtain contradicting results regarding the moment induced in the Mo site. Photoemission and x-ray absorption spectroscopic studies on the Mo-based double perovskites concluded mixed valence state for Fe [25], whereas recent neutron diffraction measurements report Fe^{3+} and Mo^{5+} valence states [26, 27]. On the other hand, many other spectroscopic studies also demand Fe to be in the 3+ state [28–31]. The experimentally observed magnetic structure is ferrimagnetic (FiM), with large Fe spin moments antiparallel to small Mo spin moments.

In the ionic model, the Fe atoms are in the 3+ valence state and the $3d$ shell is exactly half filled, giving a spin moment of 5 μ_B and zero orbital moment per Fe atom. The Mo atoms are monovalent with one d electron of t_{2g} symmetry in the $4d$ shell. If this electron were completely localized, it would result in a spin moment of 1 μ_B on the Mo site. In the real material the Mo moments are quite small, and one may therefore conclude that the Mo d electron is partly delocalized, leading to quenching of both the spin and orbital moments. The half-filling of the Fe $3d$ shell leads to an interesting effect. Since it is half-filled, only electrons of the opposite spin can hop into the Fe d shell due to the Pauli principle. This means that we must have an antiferromagnetic coupling between the Fe and Mo d states. In energy terms, the kinetic energy in SFMO is thus minimized when the local Fe moments are parallel to each other and antiparallel to the itinerant Mo spins, resulting in the observed ferrimagnetic structure [32].

On a more detailed level, in order to explain the anomalously high Curie temperature or the general stability of the magnetic structure, many models have been suggested for the electronic structure, for example, a tight-binding dynamical mean-field model [33], the presence of strong enhancement of the intra-atomic exchange strength on the Mo site [34], or strong Coulomb correlation effects in both the Fe $3d$ and O $2p$ states [35].

It remains a fact that a simple DFT calculation using the experimental structure and the minimal unit cell correctly reproduces the half-metallic gap and reasonable values for the spin moments.

We have calculated the electronic properties, the x-ray absorption (XAS), and XMCD spectra at the Fe $L_{2,3}$ and Mo $L_{2,3}$ edges of SFMO [36]. Using the theoretical XMCD spectra and the sum rules, we calculate the spin and orbital magnetic moments and compare them to the moments resulting from the direct self-consistent calculation. The comparison with experiment allows us to understand the magnetic coupling between the Fe and Mo sites in order to understand the ferrimagnetic ground state of SFMO. In particular, we have found that the Fe and Mo spin moments are antiferromagnetically aligned in excellent agreement with the experimental results. For the details of the calculations see [36].

The calculated partial density of states, shown in Fig. 4, agree well with earlier calculations [34, 37–39]. The main features of the calculated DOS are summarized below.

In the spin-up channel, the O $2p$ states are positioned between −8 and −2 eV relative to the Fermi level. The (nearly) cubic symmetry of the octahedral co-ordination of the oxygen atoms around the transition metals splits, in a simplified picture, the d levels into one peak of t_{2g} states, and another peak of e_g states, with the t_{2g} states having the lower energy. The hybridization between the crystal field split Fe t_{2g} and e_g states with the O $2p$ states is clearly evident just below the Fermi level. The narrow bands lying above the Fermi level are the Mo t_{2g} states. Thus we have a gap between the Fe e_g states and Mo t_{2g} states. In the spin-down channel, the O $2p$ states are fully occupied and the states lying between −1 and 1.5 eV are mainly Fe t_{2g} and Mo t_{2g} states, followed by the Fe e_g and Mo e_g states up to 4 eV.

Fig. 4 Calculated spin-resolved partial density of states (DOS) of Sr_2FeMoO_6 at the experimental lattice constants in units of states/eV/fu (formula unit). The partial DOS of 3d-Fe is the *continuous line*, that of 4d-Mo is the *dashed line*, and that of 2p-O is the *dotted line*. The majority spins are represented in the positive scale part of the plot and that of minority spins in the negative part

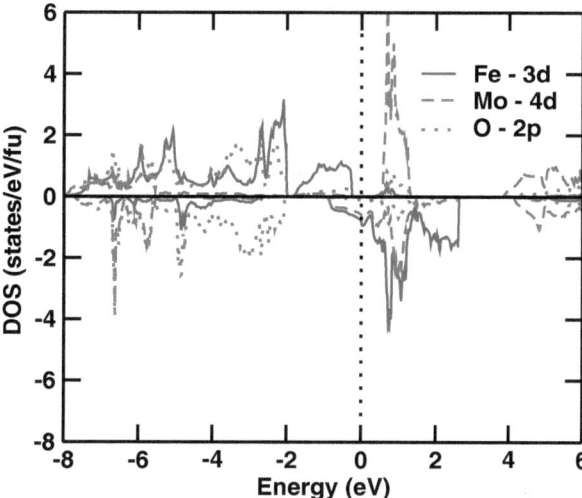

By comparing the spin-resolved partial DOS in Fig. 4 with a calculation where the SOC is excluded [38], we find that the SO coupling induces a small splitting of the Fe t_{2g} and Mo t_{2g} states. SFMO remains half-metallic with the SOC included, which is in contrast to the Re-based double perovskites, where the inclusion of SOC eventually destroys the band gap resulting in a pseudo half-metallic ground state [40, 41]. A possible reason could be that the SOC parameter of Mo is slightly smaller when compared to the 5d transition metal Re. In addition to that, the half-metallic gap in SFMO is significantly larger than that of the Re compounds which helps in preserving the half-metallic ground state. We now turn to a discussion of the magnetic and orbital moments. The calculations give a total spin moment per unit cell of 4 μ_B, in agreement with previous calculations. The Fe spin and orbital moments are parallel, whereas the spin and orbital Mo moments are antiparallel, in accordance with Hund's third rule. In Table 2, we have listed the calculated magnetic moments for each atomic site. The calculated spin magnetic moment of 3.72 μ_B for Fe is in good agreement with the earlier values of 3.8 μ_B [37], 3.79 μ_B [38], and 3.8 μ_B [39]. Also the orbital Fe magnetic moment of 0.042 μ_B is in good agreement with the recent FP-LMTO results of Jeng and Guo (0.043 μ_B [39]). The calculated spin moment of Mo atom, -0.29 μ_B, is also consistent with the earlier results [38]. The orbital magnetic moment of the Mo site is found to be 0.020 μ_B which is slightly lower than that obtained by Jeng and Guo [39]. Finally, we also mention that the calculations give a minute induced spin moment on the oxygen atom of around 0.09 μ_B.

The main results of the present work are the calculated x-ray absorption (XAS) and XMCD spectra of Fe $L_{2,3}$ and Mo $L_{2,3}$ edges, shown in Figs. 5 and 6, respectively. We convoluted the spectra using a Lorentzian followed by a Gaussian, both of full-width at half maximum (FWHM) of 0.25 eV for Fe and 0.5 eV for Mo. The Gaussian and Lorentzian broadenings represent, respectively, the experimental

Table 2 Spin and orbital d magnetic moments in μ_B/atom for Sr_2FeMoO_6 obtained from the self-consistent (SC) calculation and from the sum rules (SR) along with the experimental values taken from [24]

Spin				Orbital		
	SC	SR	Expt.	SC	SR	Expt.
Fe	3.72	3.67	3.05 ± 0.2	0.042	0.052	0.02
Mo	−0.29	−0.23	-0.32 ± 0.05	0.020	0.042	-0.05 ± 0.05

Fig. 5 Calculated XAS and XMCD spectra of Fe $L_{2,3}$ edge (*full lines*) as compared to the experimental data of [24] (*dashed lines*). We have used a FWHM of 0.25 eV to broaden the spectra

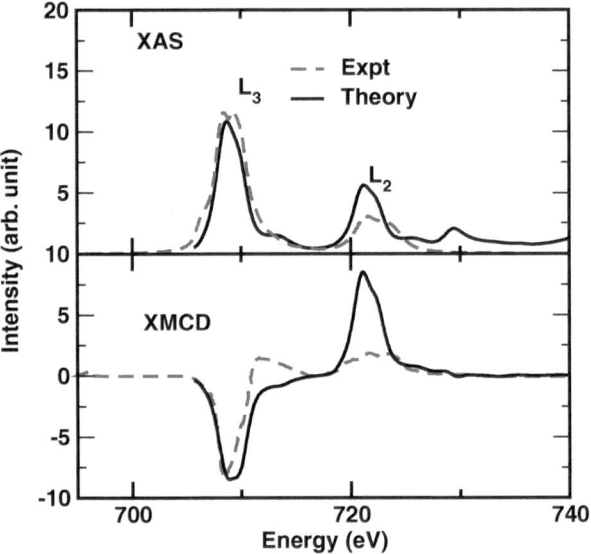

Fig. 6 Calculated XAS and XMCD spectra of Mo $L_2\ L_3$ edges (*full lines*) as compared to the experimental results of [24] (*dashed lines*). We have used a FWHM of 0.5 eV to broaden the spectra

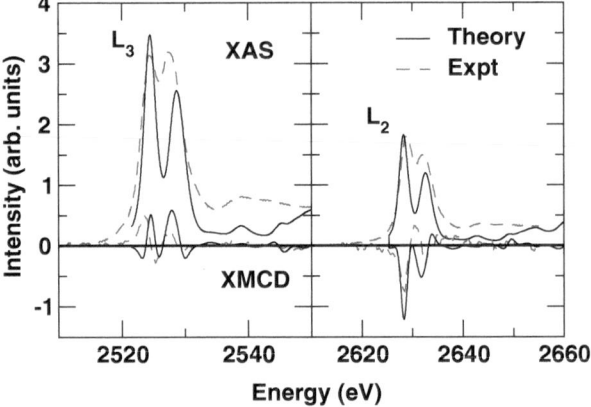

resolution and the width of the core hole. The calculated spin–orbit splitting of the Fe $2p$ core states is 12.52 eV, in good agreement with the experimental separation between the L_2 and L_3 edges of 12.5 eV [24]. The corresponding splitting for Mo was found to be 106.5 eV. The number of d holes used to compute the Fe magnetic moment using the sum rule is 4.

The upper panel of Fig. 5 shows the XAS spectra of Fe $L_{2,3}$ edge and the lower panel the XMCD spectra together with the experimental spectra of Besse et al. [24]. The calculations reproduce most features of the experimental spectra, but at the qualitative level. We find that the L_2 intensity is underestimated in the absorption spectra. The same situation prevails in many other compounds [42] and can be improved by taking into account the core–hole interaction.

Interestingly, at both Fe absorption edges in the experimental spectra of Besse et al. [24] there is a slight doublet structure present, interpreted in the paper by Besse et al. as signaling the presence of both Fe^{2+} (d^6) and Fe^{3+} (d^5) in SFMO. In the XMCD spectra by Ray et al. [23], however, a corresponding doublet structure is not visible. We speculate that the doublet structure is sensitive to the exact composition of the sample, e.g., the amount of anti-site disorder and/or oxygen and other vacancies and not primarily connected to the intrinsic electronic structure of ideal SFMO. This conclusion is supported by the self-interaction corrected calculations by Szotek et al. [43], which basically rule out the possibility of *any* Fe^{2+} valence in SFMO. They find that the Fe^{3+} valence is the most energetically favorable one, with Fe^{4+} 0.83 eV more unstable and Fe^{2+} 1.66 eV more unstable than the Fe^{3+} valence.

The Fe XMCD spectrum shown in the lower panel of Fig. 5 reveals a sharp signal indicating the large value of the Fe moment. Using the XMCD sum rules on the calculated spectra, we find a Fe spin magnetic moment of 3.67 μ_B which is slightly lower than what we get from the direct calculation, but still significantly higher than experiment. The lower experimental spin moment is most likely due to the Fe–Mo anti-site disorder [44], because in the ionic picture SFMO is expected to have a total spin moment of 4 μ_B per formula unit due to a ferrimagnetic coupling between the Fe^{3+} $3d^5$ and Mo^{1+} $4d^1$ electronic configurations. As for the very small (0.052 μ_B) Fe orbital moment re-calculated from the spectra in the same way, we find that it compares well with the direct calculation, but is much larger than the experimental one. The XAS and the XMCD spectra of the Mo $L_{2,3}$ edge are shown in Fig. 6. The calculated spectra are in surprisingly good agreement with that of Besse et al. [24], especially considering that the Mo moments are nearly quenched in this system. Though the single Mo d electron is delocalized, an appreciable electron density is still present at the Mo site which results in a pronounced XMCD signal, evident in Fig. 6. The spin moment obtained from the XMCD sum rule is somewhat smaller than the direct calculation and the experimental value. As for the orbital moment, the sum rule produced a much higher value compared to the direct calculation. The experimental value has a different sign, but because the absolute error is as large as the value itself, it is very difficult to draw any conclusion.

4 Electronic Structure, XMCD, and Magnetic Anisotropy of Rare Earth Compounds

4.1 Electronic Structure of Gd and GdN

4.1.1 Gadolinium

In the last few decades there have been considerable improvements in designing and manufacturing electronic devices. Especially those based on the spin degrees of freedom, labeled nowadays spintronic(s) devises. It is the functionalization efforts of the electronic spin degrees of freedom together with the charge degrees of freedom which led to such interesting electronic devices. In particular, mastering the spin degrees of freedom might be beneficial at the nanoscale, increased data processing speed, decreased electric power consumption, and increased integration densities [45].

Nowadays mostly $3d$ magnetic materials are studied for such interesting applications while only few attention is paid to rare earth magnetic materials. Due to their $4f$ localized orbitals rare earth materials exhibit a strong magnetism. These materials might be, therefore, promising candidates for the above-mentioned applications. Because of its half-filled $4f$ shell, gadolinium (Gd) is certainly the most important among these kinds of materials. With the evolution of computational resources, modern electronic structure methods are going to be more and more used for studying magnetic materials.

Since the pioneering work of Dimmock and Freeman [46] where the Gd electronic structure has been calculated using the core model for the treatment of the $4f$ electrons, there has been a few more band structure calculations for Gd. In this simple model, while the $4f$ bands have been successfully *removed* from the conduction band at the vicinity of the Fermi level, the hybridization of the $4f$ states with the other states was not accounted for. Some years later, the self-consistent calculations of Sticht and Kubler [47] have shown that the standard LDA potential leads to a smaller lattice parameter because of the spurious presence of the $4f$ minority states close to the Fermi level. Later, Temmerman and Sterne [48] have found a very large sensitivity to the treatment of the extended $5p$ core states as semicore states. Afterward, Singh [49] has shown that the LDA does not provide a fully satisfactory description of Gd. This reflects particularly the complexity of the Gd electronic structure due to the presence of $4f$ electrons. The conduction electrons of Gd consist of three kinds of electrons: the $4f$ strongly localized electrons, the $5p$ and $5s$ semicore electrons, and the itinerant $3d$ and $6s$ electrons.

In addition, it is unclear whether the Gd magnetism is that of a typical Stoner-like magnetism [50, 51] or that of a Heisenberg-like magnetism [52]. It turns out that the electronic and therefore the magnetic properties are far from clearly being understood, and further theoretical investigations are therefore called for.

The failure of the LDA for the description of localized electron systems was already proved, i.e., the so-called Mott insulators were found to be metallic within the LDA calculations. Indeed, unlike in pure $3d$ transition metals, the Mott insulators

3d electrons, such as NiO, are localized because of the presence of neighboring oxygen. This means that an *extra*-Coulomb interaction between these electrons should take place. It is this interaction which is missing in the LDA scheme and one should therefore come up with a method which allows an appropriate representation of those localized electrons.

During the last decade, first-principles calculations within the LDA(GGA)+U methods have provided a good description and allowed a better understanding of the electronic properties of strongly correlated 4f electron materials and Mott–Hubbard insulators. In this section we present results of LDA(GGA)+U calculations. The choice of the J value of 0.7 eV is justified by the early electronic properties study of Harmon et al. [53], within the APW method. They evaluated the strength of the 4f-conduction electron exchange interaction inside the muffin-tin spheres for the 4f–6s and 4f–5d, and they obtained J_{4f-5d} of 0.5 eV and a J_{4f-6s} of 0.2 eV and hence a total J of 0.7 eV. As stated in the previous section, the value of the Hubbard interaction U is much more difficult to estimate because constrained LSDA calculation does not necessarily provide the ultimate value to be used in an LDA+U or GGA+U study. It is interesting to notice that a value of U larger by 1 eV than the constrained one produced the experimental splitting between the spin up and spin down of the 4f energy levels. However, to compare with the XPS and BIS data, we had to rigidly shift the occupied and empty density of states by 1.7 eV toward higher energies.

Figures 7 and 8 present our LDA+U and GGA+U total DOS calculations, which are in good agreement with the XPS and BIS experimental results [54], after a rigid shift of the occupied and empty density of states by 1.7 eV toward higher energies.

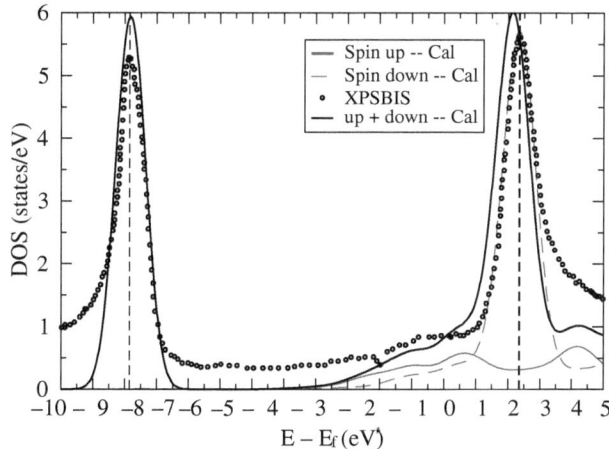

Fig. 7 LDA+U total DOS; the *thin grey (orange) curve* is the spin-up part, the *dashed (red) curve* is the spin-down part, the *thick black curve* is the sum of the up and down parts as compared to the XPS and BIS experimental data *(dotted black curve)* of [54]. The calculated spectra are rigidly shifted toward higher energies by 1.7 eV to facilitate the comparison with experiment. The DOS broadened using a full-width at half maximum Gaussian smearing of 0.25 eV (For color figure, see online version)

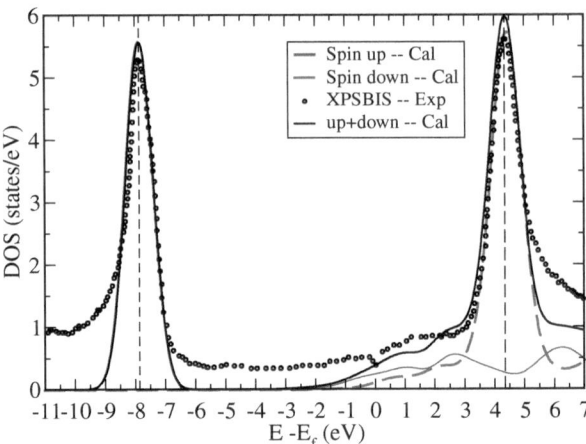

Fig. 8 The same as the previous figure but the calculations are done within the GGA+U method

The large calculated exchange spin splitting of 11.97 eV obtained using the LDA+U potential and that of 12.2 eV using the GGA+U potential is a direct consequence of the U effect of 7.7 eV. The angular ℓ resolved DOS's (not presented here) showed that the pronounced peaks of the total DOS's presented above are almost of f character. In comparing our LDA(GGA)+U DOS's with those of the LSDA(GGA), calculated and reported in many recent papers [55–57] but not reported here, we noticed that the minority (spin-down) $4f$ states are shifted away from the Fermi energy to higher energies, and the majority (spin-up) $4f$ states are shifted to lower energies giving rise to the experimental exchange spin splitting of $\Delta = 12.2$ eV. Although the two methods LDA+U and GGA+U produced a good description of the energy distribution of the electronic states, particularly the energy positions of the minority and majority $4f$ states, the GGA+U result is about 0.2 eV larger, in good agreement with the experimental data. We assign the slight improvement of the GGA+U to the fact that the spin-dependent exchange-correlation GGA potential describes a bit better the electron–electron interactions involving the strongly localized and correlated $4f$ electrons. With the *help* of the U interaction, the relative position of the $4f$ majority states with respect to the $4f$ minority states is in a better agreement with experiment leading to a good exchange $4f$ spin-splitting value. However, we cannot state for certain that the GGA+U is significantly better than the LDA+U since the small relative accuracy of these two methods can be debated. This is because the DFT error can be much larger than this energy difference.

Through the band structure plots reported hereafter, we would like to convey the adequacy of the GGA+U method, compared to the GGA $4f$-core model for the description of the gadolinium electronic structure.

The two upper panels of Fig. 9 represent the GGA+U band structure without including the SOC for the majority and minority spin states along high symmetry directions for the positions of the high symmetry points in the BZ. The distinguishable dispersionless atomic-like character of the states located at about 2.7 eV above

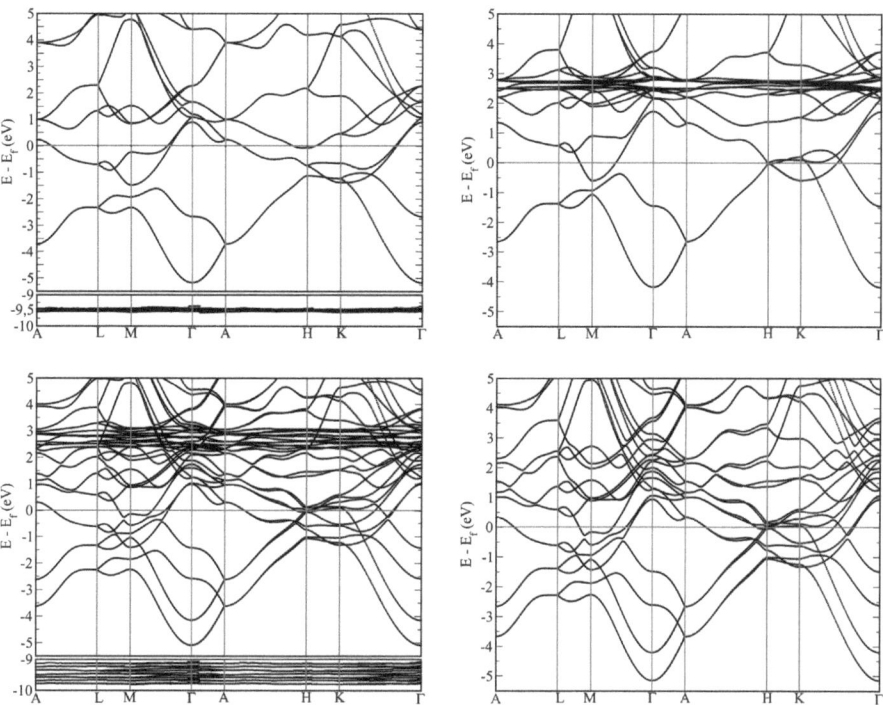

Fig. 9 The upper left and right panels represent, respectively, the GGA+U band structure plot of the majority and minority spin states without SOC along some high symmetry directions. The lower-left panel represents the total band structure (majority and minority spin states) including the SOC, whereas the lower right panel represents the GGA total band structure (majority and minority spin states) within the $4f$-core model including the SOC

the Fermi level for the minority spin states (top right and bottom left panels of Fig. 9) and 9.5 eV below the Fermi level for the majority spin states (top left or bottom left panels of Fig. 9) are that of the $4f$ states as is the case experimentally. Despite the crystallographic environment these states behave as in the free atom case due to the fact that the $4f$ electrons are tightly bound to the atom and hence do not overlap appreciably with the neighboring atoms. It is worth mentioning here that the states at the vicinity of the Fermi level are mostly of hybridized $6s$–$5d$ character. The $6s$ band width is larger than that of $5d$ states which is similar to the situation in transition metals, in agreement with the early results reported in [58]. The lower-left panel shows the effect of the SOC on the GGA+U calculation, in addition to the lifting of the degeneracy for some bands (because the spin is no longer a good quantum number), it is easily seen that the splitting of the $4f$ bands broadened the occupied (majority) bands from 0.2 eV to around 0.8 eV and the unoccupied (minority) bands from 0.3 eV to about almost 0.6 eV. This difference in the splitting of the majority and minority parts can be explained by the large relativistic effects of deep states, and the transformation of some $4f$ band character to $6s$ and $5d$ band character via hybridization effects, making the splitting mechanism more difficult for the minority

electrons than for the majority ones. Apart from the effect of splitting, there is no
further large effect of the SOC because the $4f$ spin-majority band is fully occupied
and the $4f$ spin-minority band is almost completely empty.

To analyze further the hybridization mechanism between the $4f$ states and the
$6s$ and $5d$ ones, we reported in the lower-right panel of Fig. 9 the SOC $4f$-core
model bands to be compared to the previous SOC GGA+U calculation within the
$4f$ band model. From the upper and lower-right panels we can see that the majority
$4f$ state removal (in the $4f$-core model) does not affect the filled states and those
lying just above the Fermi level. Hence, the two models provide a similar description for all states lying up to 1 eV above the Fermi energy. However, the minority
$4f$ states' (located at 2.7 eV) removal affects considerably the surrounding $6s$ and
$5d$ bands. In particular, in the $4f$-core panel of Fig. 9 (bottom right panel), we
observe that along the high symmetry direction A–L, the band that starts from 2.4
eV at the A point and ends at 3.5 eV at the L point is almost of $5d$ character. It
starts much higher, in the GGA+U $4f$ model, from 3.1 eV from the A point and
ends at 3.7 eV at the L point. This contraction and small shift of the s–d bands is
ascribed to the hybridization with $4f$ bands. Although the $4f$-core model removes
the unphysical minority $4f$ states contribution to the valence states, it neglects the
effects of hybridization of the f states with the other states. Therefore, the GGA+U
band model produces the experimental energy positions of the $4f$ minority states
and reduces their hybridization with the other states and is more physical than the
$4f$-core model.

It is also worth noting that while the GGA+U or the LDA+U methods improved
considerably the spin splitting between the spin-up and spin-down $4f$ electrons with
respect to the GGA results it did not affect the spin splitting of the $5d6s$ bands.
This splitting of about 1 eV is found in good agreement with the recent spin- and
angle-resolved photoemission result [59] of 0.9 eV. Thus our calculations partially
support their conclusion concerning the band structure nature of $5d6s$ states and that
the $4f$ correlation does not change the dispersion of these bands below the Fermi
level.

It is of great interest to study the Fermi surface of gadolinium using different
types of approximations to describe the electron–electron interaction. In Fig. 10
we compared the GGA and GGA+U band structure at the vicinity of the Fermi
level along some high symmetry directions, and in Fig. 11 we plotted the three-
dimensional representation of the Fermi surface per spin of each band cutting the
Fermi level. The calculation is done using the GGA+SOC, the GGA, and GGA+U
without spin–orbit coupling. The first three rows of the figures (a), (b), and (c) are
the majority spin Fermi surfaces of the first three bands cutting the Fermi level, and
(d) and (e) are those of minority spins. The total Fermi surface for all bands cutting the Fermi level is represented in the last row (f). To determine the quantitative
change of the Fermi surface computed using different ab initio methods, we used
the linear tetrahedron method to calculate the Fermi surface area of each band. One
has to add up over the full BZ the surface areas cutting each tetrahedron for each
band crossing the Fermi level. To obtain Fermi surface areas converged to within a
relative error of 10^{-4} we used 1372 **k** points in the irreducible BZ. The results of the

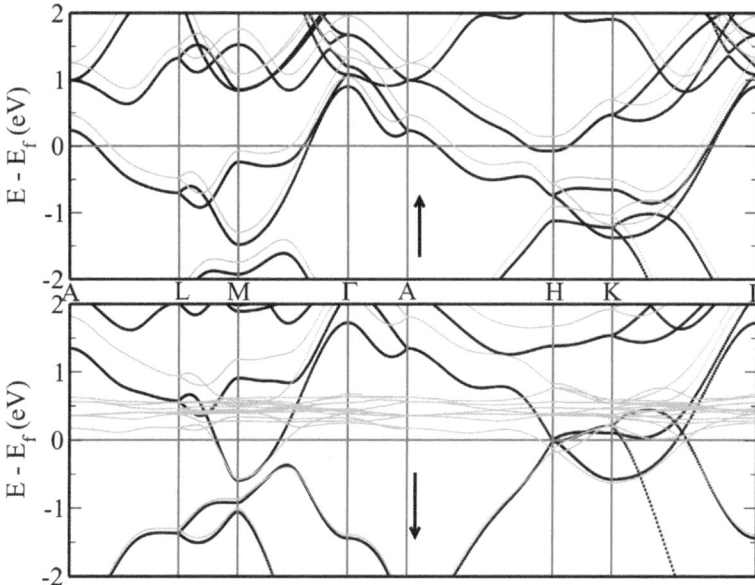

Fig. 10 The *upper and lower panels* represent the comparison of the Gd band structure within the GGA in grey (*orange*) and GGA+U (*black*) along some high symmetry directions. The spin up is represented in the *upper panel* and spin down in the *lower panel*. The Fermi level is at the zero of energy scale (For color figure, see online version)

calculation are displayed in Table 3. Notice that for the calculations including the spin–orbit coupling, the spin is not a good quantum number. Because of the small amount of spin mixing in each Fermi surface we can still use the spin-up and spin-down notation. It is interesting to notice that the spin–orbit coupling reduces slightly the areas of all Fermi surfaces, whereas the Hubbard U has a much important effect. First, it reduces also the Fermi surface areas of both spin-up and spin-down bands crossing the Fermi level. Second, because of the shifting of the energy bands toward low energies (see Fig. 10) a new electron pocket with a sizeable Fermi surface area appeared at the H high symmetry point. We can therefore conclude that both the SOC and the U parameter have an effect on physical properties involving the Fermi surface, like electronic and thermal transport or crystalline magnetic anisotropy. It should be of great interest to study the effect of electron–electron interaction on the dHvA frequencies and masses. We can already say that these frequencies which are proportional to the Fermi surface cross-sectional areas will be reduced when either the SOC or the Hubbard U is included in the calculations.

4.1.2 Magnetic Order of Gd

According to the experimental investigations of Jensen [60], it is known that the localized spin moments of the gadolinium couple through a Ruderman–Kittel–Yosida (RKKY)-type exchange interaction to form a ferromagnetic (FM) Heisenberg system with a bulk Curie temperature (T_c) of 293 K. In order to discuss the

Fig. 11 Three-dimensional Fermi surface per spin of gadolinium calculated, respectively, in GGA+SOC (*left column*), GGA (*middle column*), and GGA+U (*right column*) methods. The (**a**), (**b**), and (**c**) represent the majority spin Fermi surfaces, the (**d**) and (**e**) the spin minority, and (**f**) the total Fermi surface

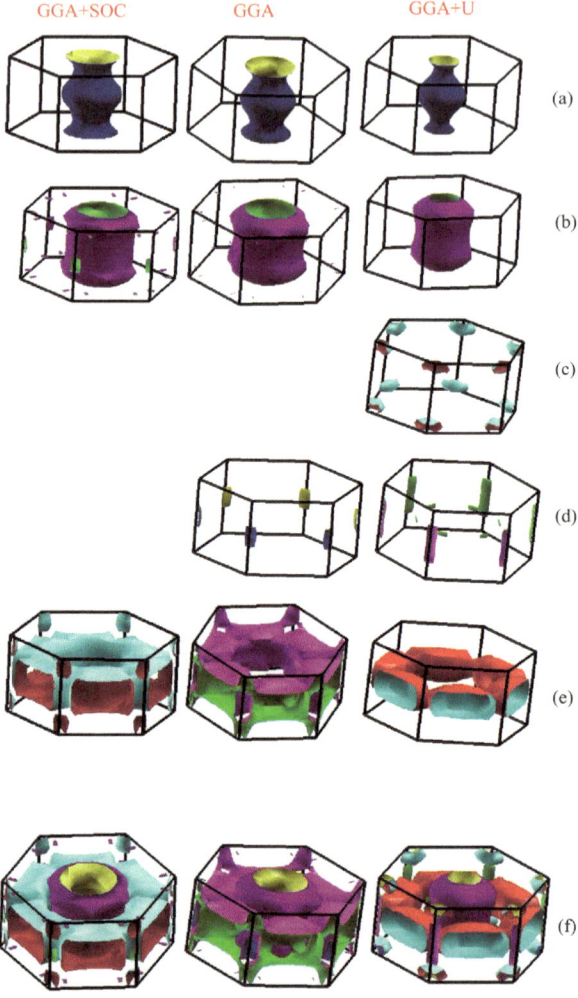

Table 3 Calculated Fermi surface areas in a.u.$^{-2}$ within the GGA, GGA+SOC, GGA+U, and GGA+U+SOC. Notice that for the calculations with spin–orbit coupling, the spin is not a good quantum number and therefore there is a small spin mixing in all the bands

	10 down	11 down	12 down	25 up	26 up	27 up
GGA	0.109	3.157	0.018	0.710	1.258	0.000
GGA+U	0.242	1.895	0.000	0.522	0.908	0.590
GGA+SOC	0.000	2.903	0.000	0.792	1.393	0.000
GGA+U+SOC	0.000	2.290	0.000	0.618	1.319	0.000

ground-state magnetic configuration of FM gadolinium, we have carried out total energy first-principles calculations for the FM and AFM configurations. The AFM configuration calculations were done by reversing the magnetization sign of every second close-packed plane of atoms along the c direction.

Table 4 Total energy difference between the AFM ($\uparrow\downarrow$) and the FM ($\uparrow\uparrow$) configurations using different approximations to the exchange and correlation all-electron potential

	LSDA	LDA+U	GGA	GGA+U
$\mathcal{E}_{(\uparrow\downarrow-\uparrow\uparrow)}$ (meV)	−24.48	78.62	−20.00	70.93

The total energy differences $\mathcal{E}_{(\uparrow\downarrow-\uparrow\uparrow)}$ between the AFM configuration $\uparrow\downarrow$ and the FM $\uparrow\uparrow$ one, recapitulated in Table 4, show that the LDA(GGA)+U favor the experimental FM configuration over the AFM one. These results are in agreement with the calculations of [57], using the force theorem. Those authors have shown that within the LSDA(GGA), the f states strongly prefer the AFM order whereas the p and d states prefer the FM order and concluded that this is due to the unphysical partial occupation of the minority $4f$ states. Using the LDA(GGA)+U and thus removing the emphasized unphysical partial occupation of the $4f$ states, our calculations provide the correct experimental FM order. Our results are also in qualitative agreement with the results of [56]. Thus, our calculation supports the fact that the force theorem, developed by Andersen and coworkers [61], which was initially used to provide an analytical expression for the hydrostatic pressure of materials, works quite well for producing the total energy differences between different magnetic configurations.

4.1.3 Gadolinium Nitride

The emergence of spintronics and the great interest which aroused the scientific community toward magnetic diluted semiconductors (DMS), particularly half-metals because of their applications for spin injections, have motivated us to study the GdN compound. Unlike the classical DMS, where the magnetism is due to $3d$ electrons, GdN compound proposes a semi-itinerant magnetism due to the f electrons. Furthermore, GdN, following the nature of the substrate on which it is grown, covers a big range of electronic properties, half-metal, semimetal, or semiconductor. The understanding of such systems properties is thus of interest to spintronics.

Within the same FLAPW computational framework we have carried out first-principles calculations of the GdN electronic, magnetic, and structural properties. The corresponding results show that the ground-state electronic structure of GdN is that of a half-metal.

Figure 12 presents our GGA and GGA+U total DOS calculations compared to the XPS experimental results [62]. The high peak of the strongly localized $4f$ projected DOS is easily distinguishable from the rest of the DOS in both theory and experiment. It is clear that the GGA energy position of the $4f$ states is in total disagreement with experiment. Therefore a GGA+U calculation using the values of $U = 9.9$ eV and $J = 1.2$ eV produces the best agreement between the calculated $4f$ states, and the corresponding experimental XPS data in either the energy position (as an effect of the Hubbard U interaction) or the width of the $4f$ DOS (as a consequence of the exchange parameter J and somehow the SOC effect) are in good agreement with the recently used values in the LDA+U FP-LMTO study of Larson

Fig. 12 GGA+U total DOS
(*full curve*) for the optimal
value of U compared to the
XPS experimental data
(*dotted curve*) of [62] and to
the GGA calculation (*dashed
curve*) and for a smaller value
of $U = 6$ eV. The calculated
spectra are convoluted using
a 1.5 eV full-width at half
maximum Gaussian
broadening

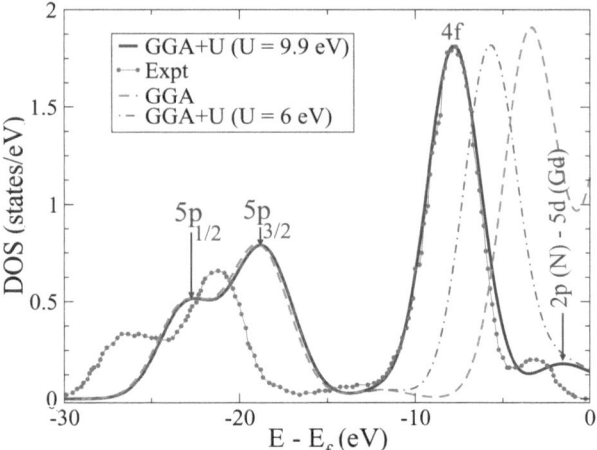

et al. [63]. Notice that one has to optimize the value of U to describe the electronic properties and that a value of somehow reasonable value of $U = 6$ eV, shown by the dot-dashed curve, does not lead to the best agreement with experiment. This is the reason for optimizing the value of U to produce as many electronic properties as possible as in the work of Bengone et al. [64].

The contribution of the hybridized Gd $5d$ states and N $2p$ states lying just below the Fermi level is somewhat broadened and shifted toward higher energies with respect to experiment. The Gd $5s$ and $5p$ shallow states were treated as valence semicore states using local orbitals [65]. The main disagreement between our results and the experimental data is that of the energy positions of the $5p$ semicore states. Nevertheless, the two peaks of the split $5p$ states are easily recognizable in our calculated DOS and are in qualitative agreement with experiment.

The splitting of the $5p$ states into $5p_{3/2}$ and $5p_{1/2}$ with a 1.6 branching ratio is due to the SOC effect. This branching ratio is smaller than the 2:1 experimental one. The difference between these ratios could be the result of the self-interaction contribution to the calculated exchange-correlation potential. The energy positions of the $5p$ semicore states can be improved using the so-called Slater transition state [66], as was used by Aryasetiawan and Gunnarsson [67] to study $3d$ semicore states in ZnSe, GaAs, and Ge. Their Slater transition state calculated energies compare favorably with those of the GW self-consistent LMTO method. It was shown that both methods provide good agreement with experiment, with a slightly better agreement for the GW-LMTO method. The main advantage of the Slater transition state is that it physically simulates and accounts, in a transparent way, for a realistic description of *intermediate* states and charge distributions induced during the experimental probe. We believe therefore that such a method is useful for providing the correct energy positions of the $5p$ semicore states of gadolinium in GdN. While a detailed calculation lies outside the scope of our study, a qualitative estimation of the change in the binding energy of the $5p$ states in the presence of one or one-half hole was carried out within an atomic calculation. We found that 1/2 hole produces an energy shift of 1.0 eV of the $5p$ states toward lower energies, with a SOC splitting of 4.48 eV between the $5p_{1/2}$ and the $5p_{3/2}$ states whereas the presence of one hole produces a shift of 1.9 eV and a splitting of 4.54 eV. These estimations fall somehow short of the required 2.5 eV and follow the same trend as the ZnSe $3d$ state shifts of 3 eV as obtained in [67]. Therefore, a full supercell calculation of Slater transition state, which allows for the valence electron screening of the core hole, is desirable.

4.2 Magnetic Anisotropy of Gd, GdN, and GdFe₂

The last part of our computational investigations is devoted to the magnetic anisotropy aspect of Gd and its compounds GdN and GdFe₂. The rotation or the deviation of the magnetization in a large variety of materials, e.g., permanent magnetic materials, ultrathin films, low-dimensional magnetic nanostructures or atomic chain, influences the magnetic and therefore the electronic properties of these materials. The energy required to rotate the magnetization of a magnetic crystal is defined as the magnetocrystalline anisotropy energy (MAE).

Using the force theorem, we have calculated the MAE of Gd, GdN, and GdFe₂ for different directions of the magnetization. Indeed, owing to the nil spin–orbit interaction of the $4f$ half-filled shell, the force theorem is expected to be efficient for Gd and Gd compounds' MAE calculations. This theorem allows a considerable computational effort gain since the spin–orbit coupling could be calculated only for one iteration.

Once again, the GGA+U method is found to be the most adequate approach for the force theorem calculations of the Gd MAE. The GGA and GGA-core model treatments of the $4f$ states have led to a wrong Gd MAE. It turns out that the electronic properties and the magnetic properties of $4f$ systems are tightly related, and the $4f$ electrons have a crucial role in the rare earth magnetic anisotropy.

In the last two decades, the force theorem [68] has been an important and efficient tool for computing the MAE [69–71]. As proposed by Van Vleck [72] so early, the magnetocrystalline anisotropy originates mainly from the SOC. Its variation might lead to interesting tuning of the orbital magnetic moments and MAE of complex materials and may lead to the violation of Hund's third rule [73]. Indeed, the force theorem-based calculations save an appreciable computational effort and computer CPU time. This is because the simulation of the magnetization direction changes via the SOC, which requires only one single iteration of the Kohn–Sham equations. The basic idea of the force theorem is to introduce the spin–orbit interaction as a perturbation to the scalar relativistic Hamiltonian. It is shown that the rotation of the spins is such a tiny perturbation that the electron–electron interaction hardly changes. We expect therefore that most of the contributions to the total energy remain unchanged, and subsequently the total energy difference between two spin configurations with the magnetization along two different polarization directions is given approximately by the difference between the sums of the eigenvalues up to the Fermi energy. Because this change of the total energy using a frozen potential approximation is given by the sum of one-electron energy difference [68], one can calculate this difference, with less computational effort, by *switching on* the SOC to diagonalize the relativistic Hamiltonian. This is the way in which we proceeded during our evaluations of the MAE, i.e., we first make a self-consistent calculation with a scalar relativistic potential without spin–orbit interaction, then we calculated the eigenvalues including the spin–orbit interaction for a given spin axis without allowing the self-consistent potential to change. Notice that one has to make sure that the scalar relativistic calculations are converged with the same number of **k** points as those used to determine the MAE.

4.2.1 The Gd (0001) Magnetization Easy Axis

In this section, we discuss the Gd MAE within the GGA+U method. Figure 13a shows the MAE calculations for different angles θ, i.e., the difference of the eigenvalue sums as a function of the angle θ between the c axis and the magnetization axis. The reference energy is at $\theta = 0°$. The GGA+U MAE calculations are in black circles and those calculated according to the Bruno's model are in violet squares (Eq. 1). As can be easily seen from this figure, the minimum of the difference of the eigenvalue sums is obtained for $0°$ and the maximum for $90°$. These results show clearly that the easy axis of magnetization is lying at $\theta = 0°$ and the hard one at $\theta = 90°$. These calculations were carried out using a sampling of around 16,000 **k** points in the whole Brillouin zone. In order to justify the convergence of this Gaussian broadening sampling [75], we have performed MAE calculations up to 18,000 **k** points in the BZ. Figure 13b represents the MAE convergence according to the set of **k** points. This MAE is defined as the difference energy between the hard and the easy axis of magnetization. The overall shape of the MAE presented in Fig. 13b

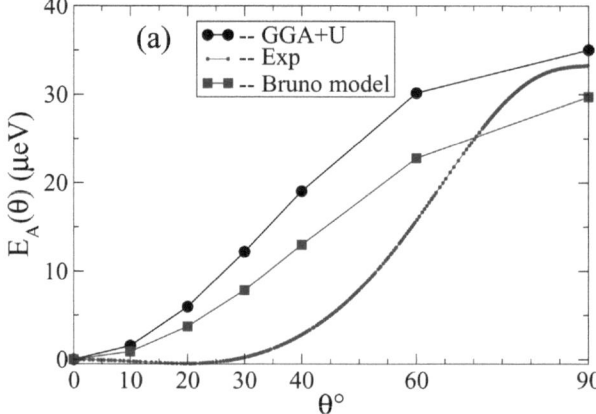

Fig. 13 The GGA+U calculated MAE of Gd as a function of the angle θ from the c axis (*circles*) and the Bruno model MAE (Eq. 1) (*squares*) compared to the experimental one (*solid curve*) [74]. The continuous curves are guides for the eye

shows that this latter is sensitive to the **k** points number up to the set of 16,224 **k** points. The largest number considered is 18,928 **k** points and it yields a MAE that deviates by less than 2% from the MAE using 16,224 **k** points. We have checked the force theorem MAE by directly calculating the total energy including the spin–orbit coupling in a self-consistent manner. The results of the calculations showed that the MAE is about 32.14 μeV using 16,224 **k** points, in good agreement with the converged force theorem calculation (see Fig. 13b). We note here that though the force theorem allows a saving of considerable computational effort, it still requires a considerable computational time because of the fine grid of **k** points one should use to assess the tiny MAE.

Figure 14 summarizes the MAE calculations for the different ways in which the 4f electrons are treated. In order to compare the GGA+U (Fig. 13a) MAE to the other methods this latter is represented with the GGA, and the GGA-core. It is worth mentioning here the controversial debate concerning whether the Gd 4f states should be considered as localized core states or whether it should be allowed to hybridize as band states ([76] and references therein). As it can be easily seen from Fig. 14, the Gd MAE calculated within the GGA+U scheme is in much better agreement with experiment (Fig. 13a). The value of 520 μeV of our MAE, calculated using the standard GGA potential, is in good agreement with the FP-LMTO calculation of 571 μeV by Colarieti-Tosti et al. [77]. However, within our FLAPW framework, the core treatment of the 4f electrons leads to a MAE of 87 μeV, while within the FP-LMTO one [77] it is of about only 24 μeV in disagreement with our calculation. In order to understand the SOC magnetic anisotropy in more detail, we have applied Bruno's model [78] to calculate the Gd MAE. According to this model

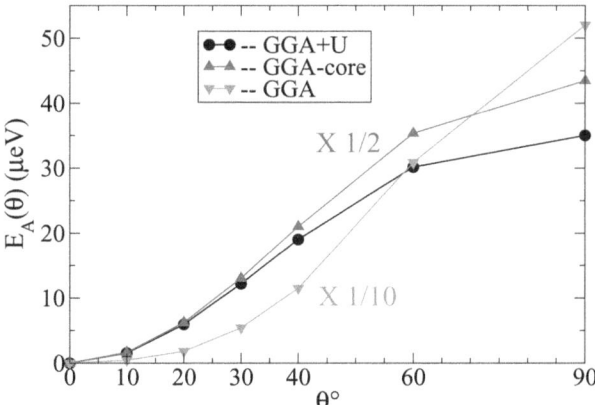

Fig. 14 Calculated Gd MAE for the different treatments of the 4f states. The calculation within the GGA+U method is shown in *black circles* and is the same as that of Fig. 13. The GGA core, where the 4f states are considered as core electrons, is shown in red up triangles, while the standard GGA, where the 4f electrons are allowed to relax as valence bands, is shown in *green down triangles*. Notice that the GGA and GGA-core curves are scaled, respectively, by a factor of 1/10 and 1/2 to fit into the graph. The *continuous curves* are guides for the eye (For color figure, see online version)

the MAE stems completely from the spin–orbit contribution and the anisotropy of orbital magnetic moments and is given by

$$E_A(\theta) = \Delta E(\theta, 0^\circ) = -\frac{\xi}{4\mu_B} \left([\mu^\uparrow_{\text{orb}}(\theta) - \mu^\downarrow_{\text{orb}}(\theta)] - [\mu^\uparrow_{\text{orb}}(0^\circ) - \mu^\downarrow_{\text{orb}}(0^\circ)] \right), \quad (1)$$

where ξ is the spin–orbit parameter for $5d$-Gd orbitals and μ^σ_{orb} the orbital moment of the spin σ. We have presented in Fig. 13a (violet curve) the corresponding calculations. The spin–orbit coupling parameter we have used to calculate the MAE according to the Bruno's model is that of the $5d$ orbitals and is found to be of $\xi = \xi_d = 71.15$ meV. As can be seen from this figure the overall behavior of the estimated MAE of the model is too similar to that of the GGA+U. The MAE calculated according to Bruno's model is somewhat situated between our GGA+U calculations and the experimental one. Bruno's model predicts a MAE maximum of 30 μeV. Given the fact that the spin–orbit parameter ξ and the orbital moment μ^σ_{orb} used in Eq. (1) are those of the GGA+U calculations, the agreement of the full calculation with the model is not surprising. However, this implies that the MAE is essentially due to the orbital anisotropy. Gd is such a complex metal, and we have seen that the energy position of the $4f$ states is crucial for the strength of the MAE. We can only conclude here that once the $4f$ levels are well positioned, the MAE is mainly due to the orbital moment anisotropy as suggested by Bruno's model.

Bruno's model validity for describing the spin–orbit magnetic anisotropy of Gd should reflect the fact that the magnetic anisotropy of Gd is too similar to that of a typical $3d$ transition metal such as hcp Co. However, there are additional terms which are related to the magnetic dipole operator due to the anisotropy of the field of the spin. This additional contribution was derived by van dar Laan [79]. The strong magnetic moment of the $4f$ electrons might give rise to this latter contribution. The resulting exchange field of that $4f$ spin is large enough to be sufficient to polarize significantly the remaining conduction electrons. In others words, the $4f$ magnetic field makes, in particular, the Gd $5d$ magnetic moment parallel to that of the $4f$. Despite this high magnetic field, the van dar Laan contribution for Gd is found to be negligible compared to that expected from Bruno's model. In fact, this contribution is only considerable for non-half-filled systems where spin flips among the $4f$ electrons occur.

However, though the GGA+U calculations using the force theorem have reproduced the experimental magnitude of MAE of 34 μeV, they did not show that the easy axis of the magnetization makes 20° away from the c axis as experimentally observed, instead they show that it is along the c direction. If we believe our calculation, which is in agreement with Bruno's model and in disagreement with the FP-LMTO calculation using $4f$ states as core states [77], then the deviation of the magnetization from the c axis could be only explained if one invokes symmetry breaking lattice imperfections of the hcp structure of Gd, like presence of intrinsic defects, impurities, or dislocations. We suspect the erroneous GGA energy positions of the $4f$ minority states [76] to be at the origin of the corresponding predicted large MAE. The presence of these states near the Fermi level leads to the erroneous MAE.

The integration of the one-electron energies includes an *extra* contribution coming from a strong mixing of the $4f$ states with the other states at the Fermi level. Using the GGA+U method these $4f$ states are moved away from the Fermi level (U effect), resulting in a more realistic assessment of the MAE. The MAE is therefore sensitive to the electronic structure around the Fermi level and a better representation of the electronic structure could lead to a precise evaluation of the MAE. Compared to the GGA and GGA-core, the GGA+U method is once more the best one for the MAE calculations. Given the adequacy of the GGA+U, we have proceeded in the same way to calculate the MAEs of GdN and GdFe$_2$.

4.2.2 GdN and GdFe$_2$ Magnetic Anisotropy

In order to get insight into the magnetic anisotropy of Gd compounds, we have applied the force theorem to calculate the MAE of the GdN pnictide and the metallic compound GdFe$_2$. Using the GGA+U method, we have recently shown that the GdN compound is a half-metal for the experimental lattice constant [80]. A better understanding of the magnetic anisotropy of this compound would be useful for future spin-injection applications.

In this section the MAE, $E_A(\theta)$, is defined as in the previous section: $E_A(\theta) = E_{\theta,\phi} - E_{0°,0°}$. Unlike Gd, the GdN compound crystallizes in the cubic rock-salt structure and its magnetic anisotropy will depend not only on θ but also on ϕ. In order to determine the easy and the hard axes of magnetization, we have calculated the MAE as a function of spherical coordinate angles θ or ϕ by keeping one of them fixed and varying the other one.

Figure 15 shows the MAEs of GdN and GdFe$_2$ as a function of the spherical coordinate angles θ or ϕ. The black circles curve in Fig. 15a represents the GdN MAE versus θ for $\phi = 0°$, the red up triangles curve in the same figure represents the GdN MAE versus ϕ for $\theta = 55°$. According to the black circle curve, the easy axis of magnetization is along the direction (001) defined by ($\theta = 0°, \phi = 0°$), and according to the red up triangles curve the hard axis of magnetization is along the direction (111) defined by ($\theta = 55°, \phi = 45°$).

The GdFe$_2$ MAE (see Fig. 15c,d) is found to exhibit a similar behavior to that of GdN MAE with the same axis of easy and hard magnetization, but with a higher MAE. The GdN MAE is only of 0.38 μeV, while that of the GdFe$_2$ is of about 9 μeV. It is worth mentioning here that although Gd monocrystal MAE is very similar to that of a $3d$ transition metal, the MAEs of its GdN and GdFe$_2$ compounds seem to be different from that of a cubic transition metal, like Ni. It is well known that in a fcc transition metal like Ni, the (111) direction is that of the easy axis of magnetization and the hard axis is found to lie along one of the symmetry equivalent (001), (010), or the (100) directions. Our results suggest the opposite for GdN and GdFe$_2$ compounds. This peculiar behavior of the magnetic anisotropy of the Gd compounds shows that even in the presence of another non-magnetic (N) or magnetic (Fe) atoms is the Gd strong magnetism which manages indirectly the magnetic anisotropy in these compounds. Indeed, because of the zero spin–orbit coupling of the $4f$ half-filled shell, the $4f$ magnetic moment should not be involved directly

Fig. 15 The GdN and
GdFe$_2$; (**a**) the GGA+U
calculated MAE of GdN, (**b**)
the Bruno model estimation
of the GdN MAE, (**c**) the
GGA+U calculated MAE of
GdFe$_2$, (**d**) the Bruno model
estimation of the GdFe2
MAE. These MAE's are
calculated as a function of the
angles θ, ϕ for varying ϕ
while keeping $\theta = 55°$ in *red
triangles* and varying θ while
keeping $\phi = 0°$ in *black
circles*. The *continuous
curves* are guides for the eye
(For color figure, see online
version)

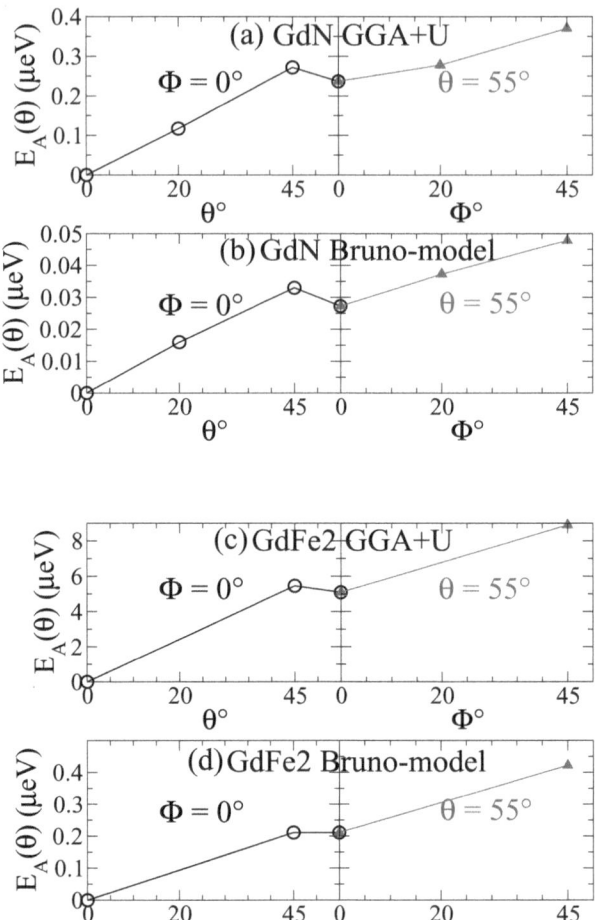

in the MAE but only through hybridization and polarization of the other valence orbitals. One could therefore easily notice that the $4f$ strong magnetic moment is to some extent *decoupled* from the crystal structure. However, due to the strongly localized character of these orbitals, the $4f$ states carry a strong magnetic moment that polarizes strongly the remaining valence electron bands. Therefore, despite their strong localized character and zero orbital moment, their energy positions in the band structure are directly related to the strength of MAE. As discussed in the previous section, there is a big difference between the GGA+U MAE and the GGA or the GGA-core MAEs, i.e., one is left with a wrong magnetic anisotropy of three times that of the GGA+U if the $4f$ orbitals are *prevented* to hybridize correctly with the other orbitals, and one order of magnitude if they hybridize too much, like in the GGA calculation. In the case of the GdN compound not only the $5d$-Gd orbital but also the $3p$-N orbitals would be affected by the $4f$ exchange magnetic

Fig. 16 (**a**) The GGA+U calculation of the MAE and (**b**) the corresponding Bruno-model determination of the MAE for a Gd crystal with an fcc structure. The *continuous curves* are guides for the eye

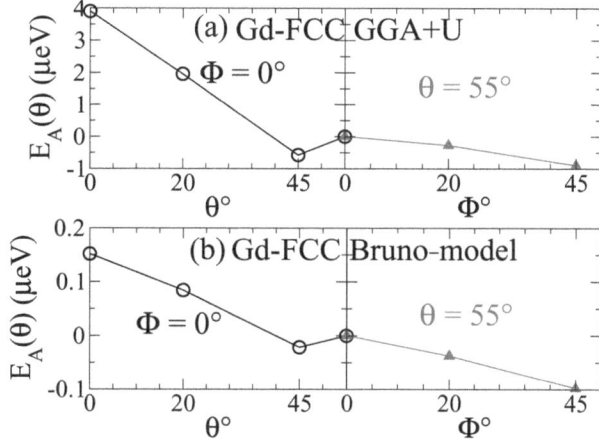

field. This happens because of the hybridization effect of the $5d$-Gd orbitals with $2p$-N orbitals [80]. For the $GdFe_2$ compound the same scenario happens to the $3d$-Fe orbitals. This interesting property would make of Gd a good candidate for high-performance ferromagnets. Indeed, if we could make materials with different $4f$ energy positions in order to change the hybridization and induce large spin polarization in other orbitals, we will be able to tune the MAE of Gd magnetic materials. In order to simulate the effect of the crystal symmetry and the presence of nitrogen on the magnetic anisotropy of Gd we have carried out a GGA+U calculation for the MAE of the Gd fcc crystal. The calculations are performed using the same lattice parameter of the GdN and the same GGA+U parameters (U and J) of the Gd hcp crystal. Figure 16a,b represents the GGA+U MAE together with that of the Bruno model. From this figure one can easily recognize a magnetic anisotropy with the same characteristics of a typical $3d$ material such as Ni. Both Fig. (16a,b show that the easy axis of magnetization for an fcc Gd crystal is that of the (111) direction. Once more the magnetic anisotropy of a pure Gd is too much similar to that of a transition metal. Therefore, the Gd fcc material would be a very good candidate for industrial application of the magnetic anisotropy with the advantage of a strong magnetic moment.

4.3 XMCD of GdN

The appreciable progress in spectroscopic techniques such as those of the x-ray magnetic circular dichroism (XMCD) and the several investigations this spectroscopy had led to have motivated our implementation of the XMCD within the ab initio Fleur code [81] using the full-potential linear augmented planewave method (FLAPW) [82, 83]. In fact, XMCD spectra have two useful properties for magnetic materials characterization. The first one is that of the chemical or atomic selectivity, i.e., each chemical element and each core orbital has its own absorption edge(s), the second one is that of the final shell or of the final states selectivity, i.e., the transitions involved during the x-ray absorption are selected according to the selection rules. Since the initial states are chosen, only the transitions for which the final states satisfy the dipolar selection rules may happen. In this respect, we should remind the powerful advantage of the XMCD sum rules. Nowadays, there are many techniques for magnetic properties measurement. Most of them are sensitive to the *total* magnetization and do not distinguish between the different atomic contributions of alloys or their magnetic spin and orbital moments. With the derivation of the sum rules by Thole and coworkers [84, 85] XMCD spectroscopy became the most efficient technique for studying magnetic materials. The sum rules allow the extraction of both the spin and orbital magnetic moments from the absorption spectra. In order to figure out the spin and orbital moments from the x-ray absorption spectra we have implemented these sum rules.

X-ray spectroscopy is one of the most useful and powerful methods used in modern experimental research of magnetic materials. In particular, x-ray magnetic circular dichroism (XMCD) characterized the magnetic properties of a large variety of rare earth materials [62, 86, 87]. A parallel theoretical effort is therefore of great interest since the dichroic spectrum is the result of a microscopic radiation–matter interaction. Due to the major presence of the $5d$-Gd states (which are strongly hybridized with the N $2p$ states) in the conduction (valence) bands of GdN, the revelation of these states by means of XMCD would be enriching. In this part, we compare our calculated $L_{2,3}$ x-ray absorption and XMCD spectra with those of the recent experimental work of Leuenberger et al. [86] for GdN bulk-like layers. Our spectra are calculated within the electric-dipole approximation using polar geometry [76]. The calculated cross-sections of the Gd L_2 and L_3 absorption coefficients involving the electronic transitions from the $2p_{1/2}$ and $2p_{3/2}$ core levels, respectively, toward the Gd $5d$ and $6s$ unoccupied conduction states are presented in Fig. 17 together with the experimental spectra obtained by Leuenberger and coworkers [86]. In order

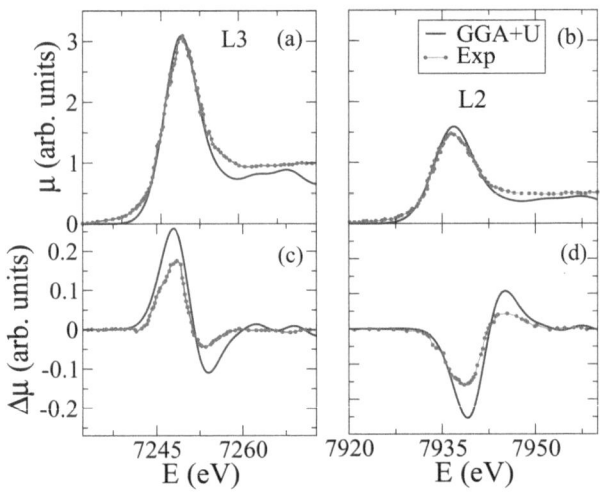

Fig. 17 The GGA+U calculated XAS absorption spectra and XMCD spectra (*full blue curve*) compared with the measured spectra (*dotted red curve*) of [86], (**a**) and (**b**) represent the XAS absorption spectra, (**c**) and (**d**) the XMCD spectra of GdN at the Gd L_3 and L_2 edges. The calculated spectra are convoluted using a 2.5 eV full-width at half maximum Gaussian broadening (For color figure, see online version)

to compare the calculated spectra with experimental results, whose intensities are in arbitrary units, we adjusted only the height and the energy position of the first peak of the L_3 absorption edge to the experimental one and scaled the L_2 absorption edge and corresponding XMCD spectra accordingly. Our XAS spectra are in surprisingly good agreement with the corresponding experimental spectra, e.g., our XAS spectra reproduce not only the shape of the main peak but also the branching ratio of the L_3 and L_2 edges of 2:1 emanating from the electronic population of the $2p_{1/2}$ and $2p_{3/2}$ core states of 2 and 4 electrons, respectively. It is worth mentioning here that this ratio has not been accessible within the recent LMTO calculations in the atomic sphere approximation (ASA) and a frozen core approximation of Antonov et al. [88]. Those authors have added a step function to the XAS spectra at the $L_{2,3}$ edges of Gd to achieve a good agreement with the same experimental data [86]. This shows that the FLAPW method provides more realistic core and conduction state wave functions used to calculate the electronic transition matrix elements from the $2p_{1/2}$ (L_2) and $2p_{3/2}$ (L_3) core states toward the $5d$ and $6s$ unoccupied conduction states. The branching ratio of the two opposite peaks of the L_3 XMCD spectrum, due to the difference of the DOS resulting from $J = 5/2$, $m_j=-5/2$ and $-3/2$ and $J = 5/2$, $m_j = 5/2$ and $3/2$, d splitted states, is much larger than that of Antonov et al. [88] but is still smaller than experiment [86]. The difference between the two calculations could be due to a more accurate evaluation of the electronic transition matrix elements using the FLAPW method, because the partial DOS alone without dipole matrix elements will result in a branching ratio of 1:−1 (see [88] for further details). Similar, but opposite, branching ratio is obtained for the L_2 MCD spectrum.

5 Conclusion

To conclude, we have shown that the new developments of ab initio methods have allowed a quantitative determination of physical properties of magnetic materials. For the magnetic anisotropy energy the MAE results obtained using the LSDA and GGA are in the most cases different which led us to the conclusion that there is no general rule favoring either LSDA or GGA as the better description of the MAE of magnetic alloys. However, for Gd and GdN the GGA+U or the LDA+U methods describe correctly both the experimental XPS and BIS. The LDA+U and GGA+U methods were able also to provide the correct magnetic configuration of gadolinium. Both LDA+U and GGA+U produced smaller Fermi surface areas than the LDA or the GGA methods. This could motivate further first-principles calculations of the other physical properties of gadolinium or related compounds using either the GGA+U or the LDA+U.

For GdN the electronic structure explorations provided the optimal U and J parameters ($U = 9.9$ eV, $J = 1.2$ eV) that allowed the GGA+U method to describe approximately the experimental XPS. We have shown that the spin-dependent x-ray absorption cross-section, implemented within the electric-dipolar approximation, describes nicely both the XAS and the XMCD at the L_2 and L_3 edges in good agreement with the experimental spectra [62] without need for quadripolar transitions, showing indirectly that in gadolinium compounds the quadripolar transitions are marginal. The GGA+U and GGA calculated relative total energies are very similar due to the absence of the $4f$ states in the vicinity of the Fermi level in both calculations.

We have carried out first-principles calculations of the MAE within the GGA, GGA-core, and GGA+U methods for the purpose of representing accurately the $4f$ electrons of Gd. It is shown that the MAE is very sensitive to the electronic structure details at the Fermi level, i.e., the failure of the GGA method to account for the correct $4f$ energy position results in an overestimation of the Gd MAE. To the contrary, the GGA+U, which produced the best position of the $4f$ states of Gd, reproduced the best MAE of Gd. Indeed, the force theorem MAE results of the GGA+U produced the best agreement with the experimental MAE magnitude. The results of GGA+U are also in good agreement with Bruno's model, where the MAE is obtained from the anisotropy of the orbital magnetic moments. The calculation did not, however, find any deviation of the easy axis from the crystal c direction as shown in experiment and in the calculation of Colarieti-Tosti and coworkers [77]. Based on our GGA+U calculations, Bruno's model, and the symmetry of the hcp lattice, we did not find any good argument for the deviation of the easy axis from the hcp crystal c direction. We can only speculate that this deviation might be the result of symmetry breaking imperfections in the hcp structure.

The comparison of the GGA-core MAE and the GGA+U MAE with experiment has indirectly demonstrated that the $4f$ hybridization with the rest of the valence orbitals, resulting in an induced polarization, is the key parameter for the tuning of the MAE of Gd- or Gd-based compounds. This parameter is tuned by the energy position of the $4f$ states in each compound. Indeed, within the GGA+U scheme we have shown that for both GdN and GdFe$_2$ compounds, the Gd $4f$ states through hybridization and induced strong polarization of, respectively, the nitrogen p and Fe $3d$ states change drastically the MAE. Unlike $3d$ transition metals fcc structure like Ni, GdN, and GdFe$_2$ magnetizations are found to lie along one of the symmetry equivalent (100), (010), or (001) direction. It will be of great interest to perform experimental measurements of MAE for GdN or GdFe$_2$ to check our theoretical predictions.

In addition, many physical properties of thin films are now becoming possible to understand by means of ab initio calculations. Namely, adsorbed atoms on surfaces [13–15], pinning centers for cluster nucleation [11], as well as the magnetic state of surfaces [12, 16].

Acknowledgments We acknowledge financial support from the ANR, grant ANR-06-NANO-053.

References

1. C. Uiberacker, J. Zabloudil, P. Weinberger, L. Szunyogh, and C. Sommers, Phys. Rev. Lett. **82**, 1289 (1999).
2. D. Spisak and J. Hafner, J. Phys.: Condens Matter **12**, L139 (2000).

3. O. Eriksson and J. M. Wills, *Electronic Structure and Physical Properties of Solids*, ed. H. Dreyssé (Springer-Verlag, Heidelberg, 1998), p. 247.
4. I. Galanakis, S. Ostanin, M. Alouani, H. Dreyssé, and J. M. Wills, Phys. Rev. B **61**, 599 (2000).
5. I. Galanakis, M. Alouani, and H. Dreyssé, Phys. Rev. B **62**, 6475 (2000).
6. S. Beiden, W. Temmerman, Z. Szotek, and G. Gehring, J. Phys.: Condens Matter **57**, 14247 (1998).
7. I. A. Abrikosov, A. M. Niklasson, S. I. Simak, B. Johansson, A. V. Ruban, and H. L. Skriver, Phys. Rev. Lett. **76**, 4203 (1996).
8. R. Zeller, P. Dederichs, B. Ujfalussy, L. Szunyogh, and P. Weinberger, Phys. Rev. B **52**, 8807 (1995).
9. M. Asato, A. Stells, T. Hoshino, T. Asada, S. Blügel, R. Z. R., and P. Dederichs, Phys. Rev. B **60**, 5202 (1999).
10. The availability and the free diffusion of the LMTO and the LAPW numerical codes to the scientific community have intensified the use of ab-initio methods. For the diffusion of the LMTO code see the web page: www.mpi-stuttgart.mpg.de, for the FLAPW code called FLEUR see www.flapw.de, and for the Wien2k code see www.wien2k.at
11. F. Nouvertné, U. May, M. Bamming, A. Rampe, U. Korte, G. Güntherodt, R. Pentcheva, and M. Scheffler, Phys. Rev. B **60**, 14382 (1999).
12. V. Stepanyuk, W. Hergert, P. Rennert, K. Wildberger, R. Zeller, and P. Dederichs, Phys. Rev. B **59**, 1681 (1999).
13. H. Dreyssé and C. Demangeat, Prog. Surf. Sci. **28**, 65 (1997).
14. S. Handschuh and S. Blügel, Solid State Commun. **105**, 633 (1998).
15. J. Izquierdo, A. Vega, O. Elmouhssine, H. Dreyssé, and C. Demangeat, Phys. Rev. B **59**, 14510 (1999).
16. P. Krüger, M. Taguchi, and S. Meza-Aguilar, Phys. Rev. B **61**, 15277 (2000).
17. I. Galanakis, M. Alouani, and H. Dreyssé, Phys. Rev. B **62**, 3923 (2000).
18. F. Amalou, M. Benakki, A. Mokrani, and C. Demangeat, Europhys. Lett. **9**, 149 (1999).
19. I. Turek, V. Drchal, J. Kudrnovsky, M. Sob, and P. Weinberger, *Electronic Structure of Disordered Alloys, Surfaces and Interfaces* (Kluwer, Borton, 1997).
20. I. Turek, S. Blügel, and J. Kudrnovsky, Phys. Rev. B **57**, R11065 (1998).
21. I. Turek, P. Weinberger, M. Freyss, D. Stoeffler, and H. Dreyssé, Phil. Mag. **78**, 637 (1998).
22. I. Galanakis, S. Ostanin, M. Alouani, H. Dreyssé, and H. Ebert, Comp. Mat. Sci. **17**, 455 (2000).
23. S. Ray et al., Phys. Rev. Lett. **87**, 09720 (2001).
24. M. Besse et al., Europhys. Lett. **60**, 608 (2002).
25. J. S. Kang et al., Phys. Rev. B **66**, 113105 (2002).
26. Y. Moritoma et al., J. Phys. Soc. Jpn. **69**, 1723 (2000).
27. D. Sanchez et al., Phys. Rev. B **65**, 104426 (2002).
28. M. S. Moreno et al., Solid State Comm. **161**, 104426 (2001).
29. K. Kuepper et al., Phys. Stat. Sol. **15**, 3252 (2004).
30. J. Herrero-Martin et al., J. Phys.: Condens Matter **16**, 6877 (2004).
31. J. Herrero-Martin et al., Physica Scripta **T115**, 471 (2005).
32. G. Jackeli, Phys. Rev. B **68**, 092401 (2003).
33. A. Chattopadhyay and A. J. Millis, Phys. Rev. B **64**, 024424 (2001).
34. D. D. Sarma et al., Phys. Rev. Lett. **85**, 2549 (2000).
35. S. Ray et al., Phys. Rev. B **67**, 085109 (2003).
36. V. Kanchana, G. Vaitheeswaran, M. Alouani, and A. Delin, Phys. Rev. B **75**, 220404 (2007).
37. T. Saha-Dasgupta and D. D. Sarma, Phys. Rev. B **64**, 064408 (2001).
38. K. I. Kobayashi et al., Nature **395**, 677 (1998).
39. H. T. Jeng and G. Y. Guo, Phys. Rev. B **67**, 094438 (2003).
40. G. Vaitheeswaran, V. Kanchana, and A. Delin, Appl. Phys. Lett. **86**, 032513 (2005).
41. G. Vaitheeswaran, V. Kanchana, and A. Delin, J. Phys. Conf. Ser. **29**, 50 (2006).
42. V. Kanchana, G. Vaitheeswaran, and M. Alouani, J. Phys.: Condens Matter **18**, 5155 (2003).

43. Z. Szotek et al., Phys. Rev. B **68**, 104411 (2003).
44. A. Ogale, S. B. Ogale, R. Ramesh, and T. Venkatesan, Appl. Phys. Lett. **75**, 537 (1999).
45. S. A. Wolf, D. D. Awschalom, R. A. Buhrman, J. M. Daughton, S. von Molnar, M. L. Roukes, A. Y. Chtchelkanova, and D. M. Treger, Science **294**, 1488 (2001).
46. J. O. de Dimmock and A. J. Freeman, Phys. Rev. Lett. **13**, 750 (1964).
47. J. Sticht and J. Kübler, Solid State Commun. **53**, 529 (1985).
48. W. Temmerman and P. A. Sterne, J. Phys.: Condens Matter **2**, 5529 (1990).
49. D. J. Singh, Phys. Rev. B **44**, 7451 (1991).
50. B. Kim, A. B. Andrews, J. L. Erskine, K. J. Kim, and B. N. Harmon, Phys. Rev. Lett. **68**, 1931 (1992).
51. D. Li, J. Pearson, S. D. Bader, D. N. Mcllory, C. Waldfried, and P. A. Dowben, Phys. Rev. B **51**, 13895 (1995).
52. K. Maiti, M. C. Malagoli, A. Dallmeyer, and C. Carbone, Phys. Rev. Lett. **88**, 167205 (2002).
53. B. N. Harmon and A. J. Freeman, Phys. Rev. B **10**, 1979 (1974).
54. J. K. Lang, Y. Baer, and P. A. Cox, J. Phys. F: Metal Phys. **11**, 121 (1981).
55. A. B. Shick, A. I. Liechtenstein, and W. E. Pickett, Phys. Rev. B **60**, 10763 (1999).
56. A. B. Shick, W. E. Pickett, and C. S. Fadley, Phys. Rev. B **61**, 9213 (2000).
57. G. B. Ph. Kurz and S. Blügel, J. Phys.: Condens Matter **14**, 6353 (2002).
58. A. J. Freeman, Magnetic Properties of Rare Earth Metals, chap. 6, ed. R. J. Elliott (Plenum, London, 1972).
59. K. Maiti, M. C. Malagoli, E. Magnano, A. Dallmeyer, and C. Carbone, Phys. Rev. Lett. **86**, 2846 (2001).
60. J. Jensen and A. R. Mackintosh, *Rare Earth Magnetism* (Clarendon Press, Oxford, 1991).
61. H. Skriver, *The LMTO Method* (Springer-Verlag, Berlin, 1984).
62. F. Leuenberger, A. Parge, W. Felsch, K. Fauth, and M. Hessler, Phys. Rev. B **72**, 14427 (2005).
63. P. Larson and W. R. L. Lambrecht, Phys. Rev. B **75**, 45114 (2007).
64. O. Bengone, M. Alouani, P. Blöchl, and J. Hugel, Phys. Rev. B **62**, 16392 (2000).
65. D. Singh, Phys. Rev. B **43**, 6388 (1991).
66. J. C. Slater, *The Self-Consistent Field of Molecules and solids*, vol. 4 (McGraw-Hill, New York, 1974).
67. F. Aryasetiawan and O. Gunnarsson, Phys. Rev. B **54**, 17564 (1996).
68. A. R. Makintosh and O. K. Andersen, *Electron at the Fermi Surface*, ed. M. Springford (Cambridge University Press, Cambridge, 1980).
69. G. H. O. Daalderop, P. J. Kelly, and M. F. H. Schuurmans, Phys. Rev. B **41**, 11919 (1990).
70. D.-S. Wang, R. Wu, and A. J. Freeman, Phys. Rev. Lett. **70**, 869 (1993).
71. X. Wang, R. Wu, D.-S. Wang, and A. J. Freeman, Phys. Rev. B **54**, 61 (1996).
72. J. H. van Vleck, Phys. Rev. **52**, 1178 (1937).
73. I. G. P. M. Oppneer, P. Ravindran, L. Nordström, P. James, M. Alouani, H. Dreyssé, and O. Eriksson, Phys. Rev. B **63**, 172405 (2001).
74. J. J. M. Franse and R. Gersdorf, Phys. Rev. Lett. **45**, 50 (1980).
75. M. J. Gillan, J. Phys.: Condens Matter **1**, 689 (1981).
76. S. Abdelouahed, N. Baadji, and M. Alouani, Phys. Rev. B **75**, 094428 (2007).
77. M. Colarieti-Tosti, S. I. Simak, R. Ahuja, L. Nordström, O. Eriksson, D. Aberg, S. Edvardsson, and M. S. S. Brooks, Phys. Rev. Lett. **91**, 157201 (2003).
78. P. Bruno, Phys. Rev. B **39**, 865 (1989).
79. G. van der Laan, J. Phys.: Condens Matter **10**, 3239 (1998).
80. S. Abdelouahed and M. Alouani, Phys. Rev. B **76**, 214409 (2007).
81. http://www.fleur.de
82. D. R. Hamann, Phys. Rev. Lett. **42**, 662 (1979).
83. E. Wimmer, H. Krakauer, M. Weinert, and A. J. Freeman, Phys. Rev. B **24**, 864 (1981).
84. B. Thole, P. Carra, F. Sette, and G. van der Laan, Phys. Rev. Lett. **68**, 1943 (1992).
85. P. Carra, B. Thole, M. Altarelli, and X. Wang, Phys. Rev. Lett. **70**, 694 (1993).

86. F. Leuenberger, A. Parge, W. Felsch, F. Baudelet, C. Giorgetti, E. Dartyge, and F. Wilhelm, Phys. Rev. B **73**, 214430 (2006).
87. J. Jenson and A. Mackintosh, *Rare Earth Magnetism* (Oxford University Press, Oxford, 1991).
88. V. N. Antonov, B. N. Harmon, A. N. Yaresko, and A. P. Shpak, Phys. Rev. B **75**, 184422 (2007).

Nanostructural Units in Disordered Network-Forming Materials and the Origin of Intermediate Range Order

C. Massobrio

Abstract Disordered network-forming materials are characterized by structural order extending well beyond the first shell of neighbors. For these reasons, reliable atomic-scale modeling is ideally suited to complement experiments in the search of the microscopic origins of this behavior. A key to understand why these systems have specific structural properties is to focus on the nanostructural units by which they are composed. By analyzing the role played by these units, one is able to put forth a valuable rationale accounting for the occurrence of intermediate range order. In this review, we present recent results obtained via first-principles molecular dynamics on a set of disordered network-forming materials, with special emphasis on the prototypical system $GeSe_2$. In a short introduction we begin with explicit examples of differences, at the structure factor and pair correlation level, between networks exhibiting intermediate range order and those purely disordered at any length scale. Concerning our theoretical approach, we rely on density functional theory and first-principles molecular dynamics to follow the time trajectories at finite temperature of these networks and obtain statistical averages to be compared with the experimental quantities. Specific methodological issues pertaining to the simulation of disordered materials are analyzed in detail (size of the computational cell, role of exchange–correlation functional, and production of an amorphous phase). Then, three specific points are addressed by considering both experimental and simulation results: first, the atomic-scale signature of intermediate range order as it manifests itself via the appearance of the first sharp diffraction peak in the total neutron structure factor; second, the correlation existing between fluctuations of concentration on the intermediate distances scale and the shape taken by the partial structure factors; and third, the establishment of the nanostructural units responsible for the occurrence of the first sharp diffraction peak in the concentration–concentration structure factor. All these examples are substantiated by extensive reference made to existing and ongoing first-principles molecular dynamics simulations.

C. Massobrio (✉)
Institut de Physique et Chimie des Matériaux de Strasbourg, 23 rue du Loess, BP 43, F-67034 Strasbourg Cedex 2, France, `Carlo.Massobrio@ipcms.u-strasbg.fr`

Massobrio C.: *Nanostructural Units in Disordered Network-Forming Materials and the Origin of Intermediate Range Order*. Lect. Notes Phys. **795**, 343–374 (2010)
DOI 10.1007/978-3-642-04650-6_10   © Springer-Verlag Berlin Heidelberg 2010

1 Introduction

By its very definition, a disordered system is characterized by the loss of structural order on long-range distances, while a reminiscent order subsist at short distances, typically involving nearest neighbors. Disordered network-forming materials, such as those belonging to the AX_2 family (A=Ge, Si: X=O, Se, S), are notable exceptions to this rule, since they feature intermediate range order (IRO) extending on distances in the range 5–15 Å [1–4]. Is there a specific signature indicating the establishment of the IRO and readily accessible to both experiments and computer modeling..? We find instructive to open this review by addressing this question directly through the observation of the total neutron structure factor $S_T(k)$ for liquid $GeSe_2$ at $T = 1,050$ K and $T = 1,373$ K. These quantities have been measured by the team of P.S. Salmon at different temperatures and calculated by the present authors by using first-principles molecular dynamics (FPMD), as detailed later in this same review. Neutron scattering data at $T = 1,050$ K can be found in [5] and the corresponding FPMD calculations in [6]. Analogously, results at $T = 1373$ K are available in [7] (neutron scattering experiments) and in [8] (FPMD calculations).

In both experiments and theory, the aim is to correlate changes in the profile of the total neutron structure factor to specific structural features, to be identified by observing the corresponding behavior of the pair correlation functions $G_T(r)$. Both $S_T(k)$ at $T = 1,050$ K and $T = 1,373$ K are characterized by a double peak in the region $2\,\text{Å}^{-1} < k < 4\,\text{Å}^{-1}$, followed by a damped oscillating pattern for larger values of k (Fig. 1). A striking difference is noticeable for $k < 2\,\text{Å}^{-1}$ and, in particular,

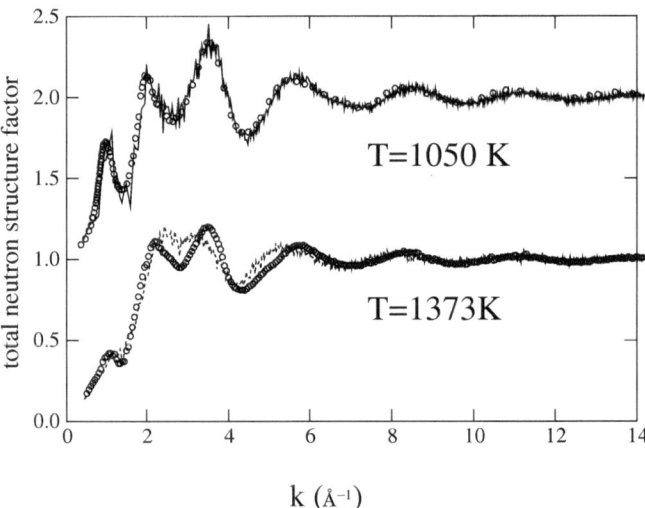

Fig. 1 Total neutron structure factors for liquid $GeSe_2$ at $T = 1,050$ K and $T = 1,373$ K. *Open circles*: experimental results at $T = 1,050$ K [5] and at $T = 1,373$ K [7] *Full line* (*upper part*) and *dotted line* (*lower part*): FPMD calculation, $T = 1,050$ K [6], $T = 1,373$ K [8]

at $k \sim 1$ Å^{-1}, where a prominent peak appears at $T = 1,050$ K and becomes negligible at $T = 1,373$ K. By terming this feature FSDP (first sharp diffraction peak), one underlines its peculiar location in k space, associated to positions in real space typical of next-nearest neighbors interactions, i.e., representative of intermediate (as opposed to short) range order. This point can be better understood by taking advantage of the relationship $k \cdot r \approx 7.7$ that allows to relate the position r of a peak in real space to the position k of a corresponding peak in Fourier space [the above relationship being based on the location of the first maximum of the spherical Bessel function $j_0(kr)$]. By inserting the value $k \sim 1$ Å^{-1} in $k \cdot r \approx 7.7$, it appears clear that a structural order involving distances as large as ~ 8 Å exists at $T = 1,050$ K but it becomes much smaller at $T = 1,373$ K where the intensity of the FSDP vanishes.

It is worth pointing out that the relationship between the FSDP and specific structural arrangements has long been the object of several interpretation schemes [1–3, 9–22]. Two of them are frequently invoked [1, 18]. The first considers the FSDP as a distinct signature of crystalline-like layers, its position being related to the interlayer separation [11–14, 18]. In particular, it was emphasized that quasi-lattice planes do occur in amorphous silica [18]. The second approach highlights the occurrence of characteristic low-density regions in covalent glasses, by successfully accounting for the position of the FSDP in a variety of AX$_2$ disordered systems [1, 10, 3]. In this picture, basic structural units ("clusters") are decorated by interstitial "voids," leading to correlation distances typical of IRO. The general validity of such models for the appearance of the FSDP in the total neutron structure factor has been recently tested within the framework of accurate first-principles molecular dynamics calculations [23].

Observation of the corresponding $G_T(r)$ in real space shows a more structured profile of the total pair correlation function at lower temperatures, consistent with the presence of higher correlation among neighbors. This is somewhat expected, but it does not explain the reasons underlying the appearance of the FSDP at $T = 1,050$ K and its absence at $T = 1,373$ K, since the patterns observed in Fig. 2 are also encountered in systems that do not exhibit any signature of IRO. Given this open issue, a more refined analysis based on structural units is very much needed to understand the microscopic origins of the IRO in connection with the atomic structure of disordered network-forming materials (DNFM hereafter). DNFM materials can be considered as collections of well-defined units, linked to each other so as to obtain the best compromise between chemical order (absence of homopolar bonds and miscoordinations), thermodynamic conditions, and optimal bonding. What makes DNFM so peculiar with respect to other disordered materials is the fact that, in principle, the connectivity among structural units can result in extended spatial correlations, as those visualized in Fig. 3. The present work shows how modern atomic-scale models (first-principles molecular dynamics) contribute to a precise understanding of the structure of these materials that, in turn, become benchmark systems to assess the predictive power of the approaches employed.

Fig. 2 Total pair correlation
function $G_T(r)$ of liquid
GeSe$_2$ obtained via FPMD at
$T = 1,050$ K (*dotted line*)
and at $T = 1,373$ K (*full
line*).

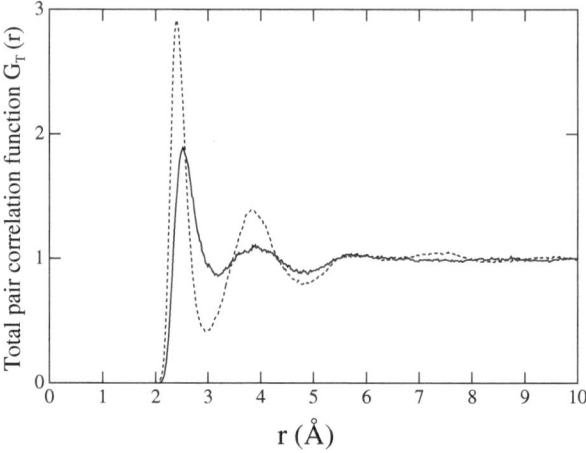

Fig. 3 Snapshot of a subset
of atoms forming the liquid
GeSe$_2$ network. Ge atoms are
dark and Se atoms are *grey*.
An edge-sharing connections
is visible, characterized by
two adjacent fourfold
Ge-Se-Ge-Se rings

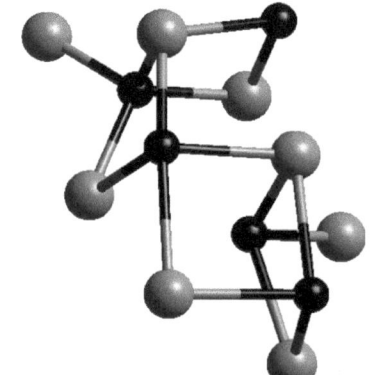

2 Generalities on the Methodology

Molecular dynamics (MD) consists in the solution of the equations of motion for
a collection of particles and can be taken as the ideal tool to describe systems at
atomic scale in two respects. First, it allows a direct access to the mechanisms of
motion. Within this class of phenomena, one can think (in a non-exhaustive list) of
diffusion processes, nucleation and melting processes, and the elementary steps of
a chemical reaction on surfaces. Second, it provides the macroscopic quantities that
correspond to the measurable property accessible to experiments via the equilibrium
time averages of the microscopic ones. In this context, the term "microscopic" indicates a suitable function of the atomic positions and velocities. Realistic modeling of
materials is based on the assumption that the interatomic forces derived to describe
the motion reflect the true nature of chemical bonding and of its variations as a
function of the temperature and/or of external conditions. We have summarized
in this sentence the challenge one has to face to confer MD methods a predictive

power, being able to complement experiments and even to go beyond them when experimental findings are lacking or difficult to interpret. In this section devoted to methodology, we recall in a first step the basic principles of the first-principles molecular dynamics technique (FPMD). Then, we shall describe specific features of its application to disordered network-forming materials in the liquid and the glassy states.

3 First-Principles Molecular Dynamics

3.1 An Overview of the Kohn–Sham DFT Approach

It is instructive to review the theorems that are at the basis of the Density Functional Theory (DFT) which allows us to find ground-state properties of a system without dealing directly with the many-electrons state $|\psi\rangle$ [24]. We deal with a system of N electrons moving in a static potential and adopt a conventional normalization in which $\langle\psi|\psi\rangle = N$.

As a result of the Born–Oppenheimer approximation [25], the Coulomb potential arising from N_{at} nuclei is treated as a static external potential $V_{ext}(\mathbf{r})$:

$$V_{ext}(\mathbf{r}) = -\sum_{\alpha=1}^{N_{at}} \frac{Z_\alpha}{|\mathbf{r} - \mathbf{R}_\alpha|}.$$

We define the remainder of the electronic Hamiltonian

$$F = T_e + V_{ee}$$

such that $H = F + V_{ext}$ where

$$V_{ext} = \sum_i V_{ext}(\mathbf{r}_i).$$

F is the same for all N electron systems, so that the Hamiltonian, and hence the ground state $|\psi_o\rangle$, is completely determined by N and $V_{ext}(\mathbf{r})$. The ground state $|\psi_o\rangle$ for this Hamiltonian gives rise to a ground-state electronic density $n_o(\mathbf{r})$:

$$n_o(\mathbf{r}) = \langle\psi_o|\hat{n}|\psi_o\rangle = \int \prod_{i=2}^{N} d\mathbf{r}_i |\psi_o(\mathbf{r}, \mathbf{r}_2, \mathbf{r}_3, \ldots, \mathbf{r}_N)|^2.$$

Thus, the ground state $|\psi_o\rangle$ and density $n_o(\mathbf{r})$ are both functionals of the number of electrons N and of the external potential $V_{ext}(\mathbf{r})$. Density Functional Theory makes two remarkable statements which are expressed in the two following theorems. The first states that the external potential $V_{ext}(\mathbf{r})$ is uniquely determined by the corresponding ground-state electronic density, to within an additive constant. In the second, given a density $n(\mathbf{r})$, $E_V[n] \geq E_o$ where E_o is now the ground-state energy for N electrons in the external potential $V(\mathbf{r})$.

The remarkable result of DFT is the existence of the universal functional $F[N]$, which is independent of the external potential. This allows to deal with a function of only three variables (the density) replacing a function of $3N$ variables (the many-electron wave function). The complexity of the problem has been much reduced. The exact form of the universal functional $F[N]$ is unknown. The failure to find accurate expressions for F is a result of the complexity of the many-body problem. For an electron gas the effects of the exchange and correlation (XC) are crucial for an accurate description of its behavior. The antisymmetry of the wave function requires that particles with the same spin occupy distinct orthogonal orbitals, and this results in the particles becoming spatially separated. Moreover, the interactions cause the motion of the particles to become correlated to further reduce the energy of the interaction.

The main contribution of Kohn–Sham (KS) was to recast a many-body problem in the solution of a non-interacting particle problem [26]. In fact, it is possible to assume that for any non-interacting system of electrons, there exists an auxiliary system of non-interacting particles such that the exact ground-state density of the interacting system equals the ground-state density of the auxiliary non-interacting system $\rho_0 = \rho_s$. If the ground state of the Hamiltonian of the non-interacting particles is non-degenerate, then ρ_s and ρ_0, by assumption, possess an unique representation

$$\rho_0(\mathbf{r}) = \sum_i^{\text{occ}} f_i |\psi_i(\mathbf{r})|^2, \tag{1}$$

where the $\psi_i(r)$ are the N single-particle orthonormal orbitals obtained from the Schrödinger equation

$$\mathcal{H}_{\text{KS}} \psi_i = \left(-\frac{1}{2}\nabla^2 + v_s\right)\psi_i = \epsilon_i \psi_i. \tag{2}$$

For simplicity the occupation numbers f_i will be considered all equal and will be omitted in the following.

Thus, the total energy functional E_v of the interacting system in the external potential v can be written as

$$E_v[\rho_0] = T_s[\rho_0] + \int d\mathbf{r}\, v(\mathbf{r})\rho_0(\mathbf{r}) + \frac{1}{2}\int d\mathbf{r}\, d\mathbf{r}'\, \frac{\rho_0(\mathbf{r})\rho_0(\mathbf{r}')}{|\mathbf{r} - \mathbf{r}'|} + E_{\text{xc}}[\rho_0], \tag{3}$$

where on the right-hand side the four terms are, respectively, the kinetic energy of the corresponding non-interacting system, the interaction energy with the external potential, the classical electrostatic energy, and the exchange and correlation term. The last term is introduced and defined in Eq. (3) and takes into account all the remaining quantum interactions between the electrons. In this case the auxiliary potential which generates ρ_0 is given by

$$v_s(\mathbf{r}) = v(\mathbf{r}) + \int d\mathbf{r}' \frac{\rho_0(\mathbf{r}')}{|\mathbf{r} - \mathbf{r}'|} + \frac{\delta E_{xc}}{\delta \rho(\mathbf{r})}. \tag{4}$$

Since v_s depends on the density, the KS equations (1), (2), and (4) must be solved in a self-consistent fashion.

The only term that can be a source of problems is the term E_{xc} that is not known. An approximation to E_{xc} that has been widely used in the literature is the local density approximation (LDA). The XC energy density is approximated by the corresponding expression of the homogeneous electron gas ($e_{xc}(n)$) with the replacement of the constant density n by the local density $\rho(\mathbf{r})$ of the actual inhomogeneous system

$$E_{xc} = \int d\mathbf{r}\, \rho(\mathbf{r}) e_{xc}[\rho(\mathbf{r})].$$

Application of the local density scheme gives better results when applied to a system with a slowly varying density. Otherwise, it is better to resort to more sophisticated approximations, like the generalized gradient approximation (GGA), in which E_{xc} depends not only on the electron density but also on the magnitude of the gradient of this density

$$E_{xc} = \int d\mathbf{r}\, f_{xc}(\rho(\mathbf{r}), |\nabla \rho(\mathbf{r})|)$$

with f_{xc} depending on the particular GGA used. In the section devoted to methodology, we provide more detail on the implications of the choice of the XC functional for the specific cases considered in this chapter. In particular, we shall consider two XC functionals, the one due to Perdew and Wang (PW) and the other due to Becke (B) for the exchange energy and Lee, Yang, and Parr (LYP) for the correlation energy [27–29].

Thus, since the set of Eqs. (1), (2), and (4) describe the single-particle Schrödinger equations for the Kohn–Sham eigenstates $\psi_i(\mathbf{r})$, these functions can be interpreted as equations for an electron moving in an effective potential due to atom cores and other electrons. The solution of these equations give the ground-state electron density and the total energy of the system. It is important to stress that the $\psi_i(\mathbf{r})$ functions have no physical meaning and they can be used only to build the electron density. However, the eigenvalues associated with the $\psi_i(\mathbf{r})$ functions reproduce quite well the experimental electron energy spectra, apart from a systematic underestimation of the bandgap in semiconductors, which can be attributed to the XC term. Therefore, in practice the $\psi_i(\mathbf{r})$ functions are often considered as the true electron wave functions.

To solve the KS equation numerically, a basis set must be chosen to expand the electron wave functions. One of the basis set most used in literature is the plane wave (PlW) set that represents the natural choice when the periodic boundary conditions (PBC) are used. In fact, in this case, the effective potential is periodic in three dimensions and the KS eigenstates become Bloch functions.

Furthermore, other advantages of the PlW expansion are the following:

- it permits the use of FFT techniques which are computationally very efficient;
- since PlWs do not depend on atomic positions, the forces acting on the atoms can be easily computed via the application of the Hellman–Feynman theorem [30];
- the convergence of the PlW calculations can be controlled by just the number of Fourier components included in the expansion of the ψ_i.

Some disadvantages of the PlW expansion rely on the large number M of basis functions needed to represent the electronic orbitals, compared for example to basis set of gaussian or atomic-like orbitals.

The use of PlW expansion calls for the use of pseudopotentials to describe the atomic core, since the number of PlWs needed to describe localized core states is too large and cannot be afforded from a computational point of view.

As a consequence of the periodicity introduced by PBC, the single-particle orbitals satisfy the Bloch theorem and can be expanded in PlWs:

$$\psi_i^k = e^{i\mathbf{k}\cdot\mathbf{r}} \sum_{\mathbf{g}} c_i^{\mathbf{k}}(\mathbf{g}) e^{i\mathbf{g}\cdot\mathbf{r}}, \tag{5}$$

where $\mathbf{g}$ is the reciprocal lattice vector of the MD cell and the wavevector $\mathbf{k}$ lies in the Brillouin Zone (BZ) of the reciprocal lattice. It will be shown in the section devoted to the Car-Parrinello method that the Fourier components $c_i^{\mathbf{k}}$ are time-dependent degrees of freedom during the molecular dynamics calculations. The basis set specified in Eq. (5) is truncated including only those PlWs with a kinetic energy $E_k = \frac{1}{2}(k + g)^2$ less than a given cutoff energy E_{cut}. The value of E_{cut} depends on the specific system and in particular upon the choice of the pseudopotential for the description of the core–valence interaction. For a given pseudopotential the choice of E_{cut} determines the accuracy of the calculation.

The computation of the electronic density $(\rho_e(\mathbf{r}))$ involves an integral over the BZ:

$$\rho_e(\mathbf{r}) = \sum_{\mathbf{k}} \omega_{\mathbf{k}} \sum_i |\psi_i^{\mathbf{k}}(\mathbf{r})|^2,$$

where $\omega_{\mathbf{k}}$ is the $\mathbf{k}$-point weight. The summation over $\mathbf{k}$ can be reduced taking into account only the Γ point ($\mathbf{k}=(0,0,0)$) because, in general, disordered systems do not preserve the same symmetry during all the molecular dynamics simulation. Therefore, the computation of the finite-temperature properties of disordered systems is most appropriately performed with large cells and the use of Γ point for the BZ sampling, reducing in this way the overall computational effort.

Another advantage in the use of the Γ point is that it is possible to choose the single-particle orbitals $\psi_i(\mathbf{r})$ to be real, since the phase factor of the wave function is arbitrary. Thus, for each reciprocal lattice vector $\mathbf{g}$, the Fourier components $c_i(\mathbf{g})$ of the orbitals satisfy the symmetry relation $c_i(-\mathbf{g}) = c_i^*(\mathbf{g})$. Taking advantage of this property it is possible to reduce the independent Fourier components by a factor 2, reducing then also the computational workload.

The numerical effort needed to solve the KS equations increases with the number of electrons involved, and moreover since the core electronic states are very localized they need a very high number of PlWs in the Fourier expansion. These problems can be overcome observing that in most molecular and solid-state systems, the electronic core states are only slightly involved in the interactions among the atoms and that they can be considered unchanged with respect to the free atom. Thus, it is possible to separate core electrons from valence electrons and replacing their atomic Coulomb potential, designed in such a way that the effect (essentially the orthogonality condition) of core states on the valence states is effectively reproduced.

The construction of the ionic pseudopotential starts with the resolution of the radial KS equation for the single atom. This gives the radial part $R_{nl}(r)$ of the atomic orbital with principle quantum number n, angular quantum number l. Then, a pseudo wave function $R_l^{PP}(r)$ is constructed from the (valence) wave function $R_{nl}(r)$ satisfying the following constraints:

- $R_l^{PP}(\mathbf{r})$ and $R_{nl}(\mathbf{r})$ are equal outside a core radius r_c;
- the amount of charge inside the core radius is equal for the two wave functions;
- the eigenvalues of $R_l^{PP}(\mathbf{r})$ and $R_{nl}(\mathbf{r})$ are equal;
- $R_l^{PP}(\mathbf{r})$ contains no nodes.

The second constraint is the norm-conserving condition: it ensures that the two wave functions outside r_c are identical not only in shape but also in magnitude, i.e., the pseudo wave function produces the correct amount of charge outside the core region.

The formed pseudopotential $v_l(\mathbf{r})$ depends on the angular quantum number and is a nonlocal operator given by

$$v_{ps}(\mathbf{r}) = \sum_{lm} v_{lm}(\mathbf{r}) \, \hat{P}_l,$$

where $\hat{P}_l$ is a projector operator onto the lth angular momentum.

A further requirement for pseudopotential is to be as smooth as possible, so that it would require the lowest possible E_{cut}.

3.2 FPMD: Basic Concepts

In this section we will introduce the key concepts to perform FPMD simulations using the electronic structure calculations discussed in the previous sections. In fact, once the electronic ground state for a fixed ion has been obtained, the quantum mechanical forces on ions can be calculated by means of the Hellman–Feynman theorem. It states that in the electronic ground states the forces on the ions are given by minus the partial derivative of the total energy with respect to the ion coordinates. In principle a first-principle molecular dynamics can be performed following four steps:

- calculate Hellman–Feynman forces;
- move ions according to Newton's law;
- solve Kohn–Sham equations;
- repeat until satisfied.

An alternative approach was proposed by Car and Parrinello (CP) [31]. The key point of the CP technique is the definition of a fictitious dynamical system associated with the physical one. The fictitious system is devised in such a way that the trajectory generated by its dynamics reproduce very closely those of the physical system. The physical system has a Lagrangian given by the sum of the ionic kinetic energy and the ionic potential energy

$$\mathcal{L}_{cl} = \frac{1}{2} \sum_I M_I \dot{\mathbf{R}}_I^2 - \mathcal{V}[\{R_I\}], \tag{6}$$

where M_I are the physical masses of the ions. The generalized classical Lagrangian of the fictitious system is

$$\mathcal{L} = \sum_i^{occ} \int d\mathbf{r}\, \mu_i |\dot{\psi}_i(\mathbf{r})|^2 + \frac{1}{2} \sum_I M_I \dot{R}_I^2 - E[\{\psi_i\}, \mathbf{R}_I] + \sum_{ij} \Lambda_{ij} \left(\int d\mathbf{r}\, \psi_i^* \psi_j - \delta_{ij} \right),$$

where μ_i are arbitrary parameters of units (mass) × (length)^2 which play the role of generalized masses for the electronic degrees of freedom. For simplicity we will consider a unique μ for the ψ_i, independent from the electronic state, even though this is not necessary. The first and second terms in the Lagrangian are the electronic K_e and ionic K_I kinetic energies, respectively. E is the potential energy of the coupling among electronic and ionic degrees of freedom. The Lagrangian multipliers Λ_{ij} are used to impose the orthogonality condition on the ψ_i (they are simply holonomic constraints).

The Euler equations associated with the Lagrangian of the fictitious system are

$$\begin{aligned}
\mu \ddot{\psi}_i &= -\frac{\delta E}{\delta \psi_i^*} + \sum_j \Lambda_{ij} \psi_j, \\
M_I \ddot{\mathbf{R}}_I &= -\frac{\partial E}{\partial \mathbf{R}_I}.
\end{aligned} \tag{7}$$

On the contrary the Euler equations from the Lagrangian of the physical system (Eq. (6)) are

$$M_I \ddot{\mathbf{R}}_I = -\frac{\partial \mathcal{V}[\{R_I\}]}{\partial \mathbf{R}_I}. \tag{8}$$

In general the ionic trajectory generated by Eqs. (8) and (7) do not coincide, unless $E[\{\psi_i\}, R_I]$ is at the instantaneous minimum. If the parameter μ and the initial conditions $\{\psi_i\}_0$ and $\{\dot{\psi}_i\}_0$ are properly chosen, it is possible to have the two sets of classical degrees of freedom, ions and electrons, weakly coupled. In this way the transfer of energy between them is small enough to allow the electrons to follow adiabatically the ionic motion, remaining close to the Born–Oppenheimer surface. This kind of dynamics, that is usually called classical adiabatic dynamics, reproduce in a computationally effective way what happens in real life: the electrons follow in adiabatic way the movement of the ions.

The CP method can reproduce the properties of the quantum eigenvalue spectrum of the electrons. In particular, it is possible to show that for small deviations from the ground state, the dynamics of the KS orbitals can be well described as a superposition of oscillations whose frequency is given by

$$\omega_{ij} = \left[\frac{2(\epsilon_j - \epsilon_i)}{\mu}\right]^{1/2},$$

where $\epsilon_j(\epsilon_i)$ is the eigenvalue of an empty (occupied) state. The lowest frequency that in a regime of small deviations from the ground state can appear in the system is $\omega_{\min} = (2E_g/\mu)^{1/2}$, where E_g is the energy gap. If the energy gap is different from zero and μ is sufficiently small, it is possible to have characteristic electronic frequencies much higher than those characteristic of the ionic motion. This assures that the fast electronic degrees of freedom can follow adiabatically the ionic motion. An important point to underscore is that, if the conditions for the adiabatic motion of the electrons are satisfied, the instantaneous value of the forces does not coincide with the Hellman–Feynman forces but their average value does to a very high degree of accuracy.

4 Practical Implementation of FPMD to Disordered Network-Forming Materials

The practical implementation of the ideas and concepts inherent in the FPMD to the case of disordered systems is far from being a straightforward application of a mere simulation recipe. Depending on the extent of residual structural order, the description of structural correlations may require system sizes capable of accounting for range of distances beyond nearest neighbors. Also, the length of the temporal trajectories should allow to capture relevant diffusion mechanisms and becomes highly critical when systems configurationally arrested are created by rapid quench. Finally, different DFT schemes differing by the form taken by the XC functionals might lead to well-distinct structural descriptions, suggesting that the comparison between theory and experiments can provide clues on the reliability of these various DFT schemes. In what follows we shall list and detail a number of critical issues to be dealt with to ensure realistic modeling of structural properties in DNFM. In particular, we shall focus on Ge_xSe_{1-x} liquid and glasses by referring mostly to the prototypical case of $GeSe_2$.

4.1 Size of the Periodic System

Hereafter, we refer to FPMD simulations performed on Ge_xSe_{1-x} liquid and glasses for which not more than $N = 144$ atoms have been employed, $N = 120$ being the standard value in most cases. The sizes of the periodically repeated cubic cells [15.7 Å, for liquid $GeSe_2$ (l-$GeSe_2$ hereafter)] are taken to match experimental densities and are sufficiently large to cover the region of wavevectors in which the FSDP occurs. In the case of l-$GeSe_2$, the smallest wavevector compatible with the supercell, k_{min}=0.4 Å^{-1}, is smaller than the FSDP wavevector k_{FSDP}=1.0 Å^{-1}. In addition, the region of wavevectors in which the FSDP appears is described by as much as eight discrete wavevectors compatible with the periodicity of our supercell. Regardless of these considerations, it appears desirable to carry out calculations on larger system sizes, since the existence of spurious boundary effects can only be avoided by a careful study of the dependence of FPMD results on the total number of atoms. The computational cost of FPMD calculations on systems encompassing several hundredths of atoms (say $N = 400$–500) was prohibitive when the first FPMD simulations on disordered network-forming materials appeared in the early nineties. Increase of computational power and the advent of massively parallel computers have made these calculations within the reach of realistic projects and are currently in progress. To provide an example, one can compare the cost of 1 ps of FPMD for $N = 120$ and $N = 480$ on the IBM SP6 parallel computer exploited by the two French national computer centers IDRIS and CINES. In both calculations, a time step of $\delta t = 0.25$ fs was adopted. In the first case ($N = 120$), 19 h/processor were necessary, while in the second ($N = 480$) the above hours/processor ratio rose to 2,100, leading to a computational effort larger by a factor close to 100. To understand this increase one can consider that the FPMD computational effort follows a scaling law based on the $N_g N_s^2$ product, N_g being the number of Fourier components for each orbital (i.e., plane waves) and N_s^2 the number of orbitals (i.e., Kohn–Sham eigenstates). By taking into account the linear relationship between N_g and the volume, together with the number of the eigenstates (four times more numerous when $N = 480$), the computational effort is expected to increase by a factor 64 when going from $N = 120$ to $N = 480$. However, for a parallel computer, these estimates hold only in the case of an ideal communication among processors with no residual time besides that devoted to the calculations.

By coming back to the rationale used to validate our first choice of the system size ($N = 120$), we mention the indications of an analysis of the range of real-space correlations which are responsible for the appearance of the FSDP in the total and in the partial structure factors [6]. For a given partial pair correlation function $g_{\alpha\beta}(r)$, this range can be determined by truncating $g_{\alpha\beta}(r)$ at decreasing distances r_c, monitoring the behavior of the corresponding Fourier-transformed structure factor $S_{\alpha\beta}^{FT}(k)$ and comparing $S_{\alpha\beta}^{FT}(k)$ to the $S_{\alpha\beta}(k)$ directly calculated in reciprocal space. In [6], we showed that the FSDP is present in $S_{\alpha\beta}^{FT}(k)$ when r_c extends up to $\sim \sqrt{2}L/2$, L being the size of our cubic simulations cells. For these values of r_c, reliable statistics can be collected for distances between independent atoms in the supercell [32]. These conditions are largely met in all calculations presented in recent years on Ge_xSe_{1-x} and closely related systems.

4.2 Temporal Trajectories and the Production of an Amorphous System

Statistical mechanics require accurate sampling of extended temporal trajectories to ensure reliable evaluation of macroscopic properties. In the case of liquid systems, this is readily obtained, provided the average spatial length of the trajectories spanned by each individual atom is larger than a few interatomic distances. As a customary practice, it is worthwhile to produce N_{st}-independent liquid configurations by selecting uncorrelated starting points, such us disordered configurations created ad hoc. Access to statistical errors can be obtained as follows. First, one takes separate averages over N_{st} well-equilibrated portions lasting at least 10 ps at the desired temperature. Then, from this set of N_{st} partial averages, one extracts global averages (mean values). For each subtrajectory, a specific standard deviation can be obtained. In order to reflect the variations found among the whole population of the given N_{st} partial averages, we express the statistical errors of the mean values σ_{mean} as $\sigma/\sqrt{N_{st}-1}$, where σ is half the largest difference among the N_{st} partial averages.

The production of glassy phases by quench from the liquid state and its level of reliability when comparing with experimental samples is the subject of a long-standing controversy. Widespread criticism as regards to a procedure that consists in producing configurations kinetically arrested by a reduction of temperature have stimulated two basic directions of debate. First, the very rapid interval of cooling is suspected to make the structure of the system dependent on the procedure followed to reduce the temperature, the "true" glassy structure being the one that corresponds to the comparatively much slower experimental quench rates. Second, the supposed similarities shared between the structure of the parent liquid and the amorphous phase are claimed to render any structural analysis performed on this latter devoid of real significance. A few considerations are in order to fully capture the advantages and the drawbacks encountered when producing a glassy phases via computer simulations. The rapid quench strategy has to be understood as a compromise between the search of an accurate description of the chemical bonding (intrinsic to FPMD) and the penalizing temporal span of FPMD trajectories. We found that the reliability of this approach is larger in systems characterized by a high degree of chemical order, as shown by a comparison with measured structural properties such as the total and partial pair correlation functions and the total and partial structure factors [6, 33–38]. For glassy systems known to exhibit an extent of chemical order different than in the liquid phase, it is crucial to rely on the relaxation effects accessible on the available time scale. In the case of amorphous $GeSe_2$, the impact of relaxation on the structural properties has been investigated in detail [36]. First, we selected several uncorrelated liquid configurations. Then, we annealed the system at the target temperature for intervals not shorter than $\simeq 10$ ps. Significant changes in the structure were observed, proving that it is possible to minimize memory effects. In the case of $GeSe_2$, the main effect of a quench from the liquid is to restore chemical order by reducing the number of miscoordinations and homopolar bonds [36]. Taken altogether, these ideas suggest that different quench rates may alter the relative proportions of the various structural units in the glass. However, at least in the case of systems not too chemically disordered in the liquid state, the identity and the number of these units is not expected to be drastically different when moving from the liquid to the glass.

4.3 The Role of the Exchange–Correlation Functional

The most delicate issue to be addressed when planifying FPMD calculations on disordered network-forming materials is the choice of the exchange–correlation functional to be used within DFT. In this section, we provide an historical overview of the reasons underlying this choice. Even though this section is not intended to present results in their detail, we found convenient to make explicit reference to some of them in this context. A suitable electronic structure-based scheme has to account for the fact that some of these networks are composed by atoms that are prone to charge transfer processes when bound to interact. For instance, this is the case of $Ge_x Se_{1-x}$ systems, meaning that A (Ge) atoms have a strong tendency to transfer electrons to the X(Se) atoms, the amount and the spatial localization of this transfer depending on the relative concentration considered. By leaving aside for a while the complexity of an electronic structure-based model, let us assume one prefers to select an alternative (less expensive) option, focusing on interactions among localized charges of opposite sign. This choice has the obvious advantage of making more affordable large-scale calculation and is currently being pursued by some groups, in an attempt to improve upon early interatomic potential models [19–21, 39, 40]. When the first classical molecular dynamics models appeared for these systems, collections of tetrahedra in which a highly "cationic" atom was connected to four highly "anionic" atoms were considered, in principle, ideally suited to describe the structural network [19–21].

However, an analysis of the early literature for DNFM liquid and glasses, reveals that the main hypotheses legitimating a predominant ionic interaction are bound to fail when the difference in electronegativity among the systems components is not too large. The key observation unraveling the weaknesses of effective (otherwise called empirical, or semiquantitative) interatomic potential is the presence of homopolar bonds, implying that atoms of the same kind are nearest neighbors. This feature cannot be reproduced by a potential in which formal charges are assigned to each atomic site, since the energetic cost required to surmount the repulsive interaction is prohibitive, unless the temperature is increased up to unrealistic values. Homopolar bonds are found in most disordered network materials, at various percentages. An example is given in Fig. 4, showing a snapshot of the structure of l-GeSe$_2$ obtained via the FPMD approach, found to be much more consistent with experimental evidence than the interatomic potentials.

Having established that the account of the electronic structure is of crucial importance for most DNFM, we stress that in its DFT–LDA version, FPMD was found unable to correctly describe the structure of a prototypical DNFM, l-GeSe$_2$ [34]. In Fig. 5, a comparison is given among the experimental total neutron structure factor and its theoretical counterparts calculated by using LDA and the PW generalized gradient approximation, obtained at the same temperature. It appears

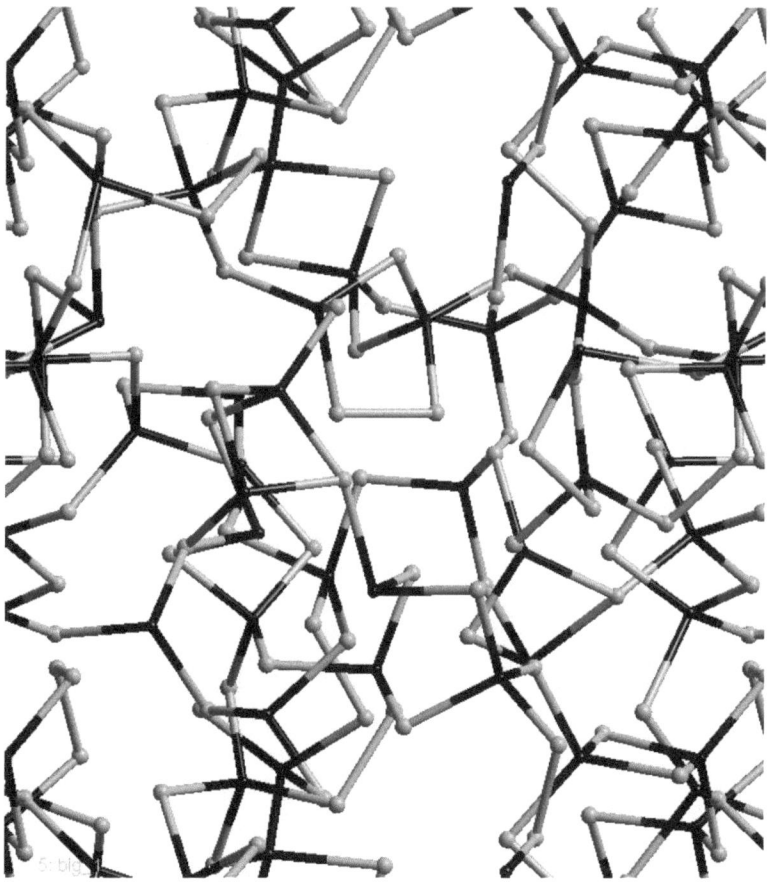

Fig. 4 Snapshot of a configuration of liquid GeSe$_2$ at $T = 1,050$ K, as obtained by using FPMD and the PW exchange–correlation functional. *Dark sticks* depart from Ge atoms and *grey sticks* depart from Se atoms. Several Ge–Ge and Se–Se homopolar bonds can be observed

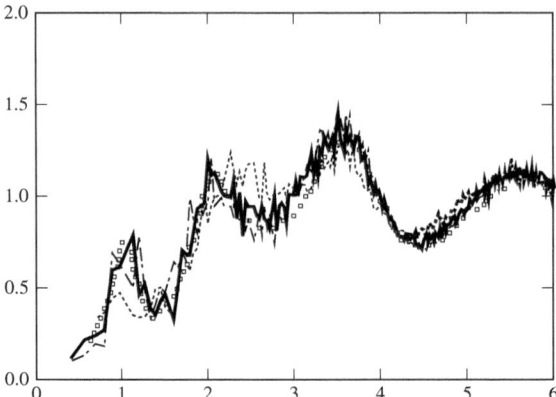

Fig. 5 Total neutron structure factor of liquid GeSe$_2$ calculated at $T = 1,050$ K by using three different exchange–correlation functionals, namely LDA (*dotted line*), PW (*full line*), and BLYP (*dash dotted line*). *Open squares* are the experimental data from [5]

that LDA is unable to describe intermediate range correlation (FSDP absent in the LDA results), while the situation is drastically improved by introducing the GGA approximation. Quite recently, we have analyzed the impact of the GGA functionals on the structure of l-GeSe$_2$ by considering a third GGA functional, the one due to Lee, Yang, and Parr for the correlation energy. The reasons underlying this study can be expressed as follows.

The adoption of the PW scheme due to Perdew and Wang as a generalized gradient approximation [33, 34, 6] allows to go beyond the local density approximation (LDA) that uses an analytic representation of the correlation energy $\varepsilon_c(\rho)$ for a uniform electron gas. The PW representation introduces variations of $\varepsilon_c(\rho)$ as a function of ρ and the spin polarization [27]. Turning our attention to the FPMD approach and to the case of l-GeSe$_2$, it is worth recalling the indications collected through the use of the local density approximation (LDA) within DFT. The absence of the FSDP (first sharp diffraction peak) in the total neutron structure factor could be correlated to the lack of a predominant structural unit (the GeSe$_4$ tetrahedron), with comparable percentages of Ge atoms twofold, threefold, fourfold, and fivefold coordinated [34]. We shall come back to this point in the section devoted to the results, by showing that the predominant presence of the tetrahedra is an unambiguous fingerprint of the appearance of the FSDP and the concomitant establishment of intermediate range order. The improvements brought about by the GGA in the PW form was due to the ionic character of bonding, as shown in [34] through an analysis of the contour plots for the valence charge densities. The larger ionicity of bonding characteristic of the PW approach manifests itself through a larger depletion of the valence charge at the Ge sites and a larger accumulation around the Se atoms, the covalency being very close in the LDA and in the PW scheme [34].

This success was partially undermined by the detailed comparison of the partial correlations and the observation that residual differences persisted between theory and experiment. These shortcomings were ascribed to an inadequate description of Ge−Ge correlations, as shown by the shape of the calculated Ge−Ge correlation function, much less structured than its experimental counterpart and by the long (15% more than the experimental value) first-neighbors Ge−Ge distances. Longer interatomic Ge−Ge distances and less structured Ge−Ge pair correlation functions were correlated to an overestimate of the metallic character in l-GeSe$_2$, the calculated Ge−Ge distances being close to those of Ge liquid.

In an attempt to further improve upon LDA, the generalized gradient approximation after Becke (B) for the exchange energy and Lee, Yang, and Parr (LYP) for the correlation energy [28, 29] was selected for the following reasons. In this GGA scheme, no reference to the uniform electron gas is made in the derivation of the correlation energy, that is expressed according to a formula due to Colle and Salvetti, recast in terms of the electron density and of a suitable Hartree–Fock density matrix [41, 29]. This scheme enhances the localized behavior of the electron density at the expenses of the electronic delocalization effects that favor the metallic character. Which systems and bonding situations are likely to be better described by BLYP than by PW ..? By focusing on multicomponent systems $A_n B_{(1-n)}$ of concentrations n, the BLYP scheme has to be preferred to treat bonding situations characterized by a moderate difference of electronegativity (termed Δ_{el} hereafter) among the system components. This is especially true for compositions at which optimal coordination between the species A and B occurs (for instance GeSe$_2$ within the Ge$_n$Se$_{(1-n)}$ family). In the case of a large Δ_{el}, the ionic contribution to bonding is sufficiently large to ensure effective charge transfer in a way essentially independent on the details of the exchange–correlation functional. This is the case of disordered SiO$_2$ ($\Delta_{el} = 1.54$), well described as a corner-sharing network within LDA [42, 43]. For lower values of Δ_{el}, the amount of the valence charge density in the bond directions becomes non-negligible and the relative weight of the ionic and covalent character is less straightforward to quantify. The case of l-GeSe$_2$ is a prototype of this situation, being characterized by $\Delta_{el} = 0.54$. The choice of BLYP to improve upon PW (and LDA) is the practical realization of a strategy aimed at minimizing the excess electronic delocalization effects preventing a correct description of competing ionic and covalent contributions.

5 Structural Properties and the Intermediate Range Order

Comparing the structural properties of different models of a prototypical DNFM can help to elucidate the origin of the intermediate range order, its occurrence being associated with the presence of the FSDP in the total neutron structure factor. In what follows we apply this strategy to l-GeSe$_2$ for which three DFT descriptions are available, differing in the form taken by the exchange–correlations functionals (LDA, PW, and BLYP). As shown in Fig. 5, LDA is unable to provide a significant FSDP in the total neutron structure factor, clearly distinguishable with both PW and BLYP. As a by-product, considering three XC functional will also allow to determine whether the structure of l-GeSe$_2$ is sensitive to the quality of the functional and in which directions some improvements can be gained. We point out that DFT results on l-GeSe$_2$ have also been obtained by employing a framework based on a nonself-consistent electronic structure scheme, the local density approximation of DFT, and a minimal basis set [44, 45]. A comparison of the structural properties obtained within these two approaches is provided in a recent paper for the case of amorphous GeSe$_2$ [46].

Partial pair correlation functions $g_{\alpha\beta}(r)$ and their experimental counterparts [5] are shown in Fig. 6. Peak positions and number of neighbors within given integration ranges are displayed in Table 1 for the three XC models. From the experimental point of view, a first maximum indicative of homopolar bonds can be seen in

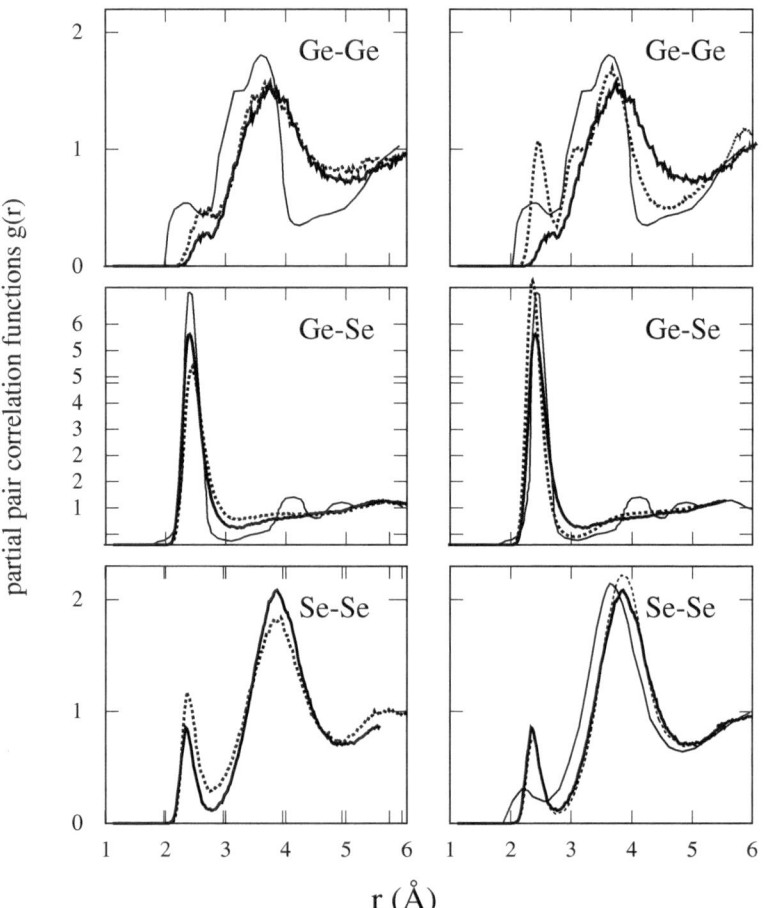

Fig. 6 On the *left*: partial pair correlation functions obtained by FPMD calculations by using the LDA (*dotted line*) and the PW (*thick line*) exchange–correlation functionals. The *thin line* is the experimental results [5]. On the *right*: partial pair correlation functions obtained by FPMD calculations by using the BLYP (*dotted line*) and the PW (*thick line*) exchange–correlation functionals. The *thin line* is the experimental results [5]

Table 1 First (FPP) and second (SPP) peak positions in experimental [5, 7] and theoretical $g_{\alpha\beta}(r)$. BLYP, PW, and LDA denote results obtained with different exchange–correlation functionals. The integration ranges corresponding to the coordination numbers $n_{\alpha\beta}$ and $n'_{\alpha\beta}$ are 0–2.6 Å, 2.6–4.2 Å for $g_{GeGe}(r)$, 0–3.1 Å, 3.1–4.5 Å for $g_{GeSe}(r)$, and 0–2.7 Å, 2.7–4.8 Å for $g_{SeSe}(r)$

$g_{\alpha\beta}(r)$	FPP (Å)	$n_{\alpha\beta}$	SPP (Å)	$n'_{\alpha\beta}$
$g_{GeGe}^{BLYP}(r)$	2.45±0.10	0.22±0.01	3.67±0.10	2.70±0.06
$g_{GeGe}^{PW}(r)$	2.70±0.10	0.04±0.01	3.74±0.05	2.74±0.06
$g_{GeGe}^{LDA}(r)$	2.7±0.1	0.08±0.01	3.65±0.10	2.89±0.06
$g_{GeGe}^{exp}(r)$	2.33±0.03	0.25±0.10	3.59±0.02	2.9±0.3
$g_{GeSe}^{BLYP}(r)$	2.36±0.10	3.55±0.01	5.67±0.02	3.85±0.06
$g_{GeSe}^{PW}(r)$	2.41±0.10	3.76±0.01	5.60±0.01	3.72±0.03
$g_{GeSe}^{LDA}(r)$	2.45±0.10	3.68±0.01	5.70±0.02	4.32±0.06
$g_{GeSe}^{exp}(r)$	2.42±0.02	3.5±0.2	4.15±0.10	4.0±0.3
$g_{SeSe}^{BLYP}(r)$	2.38±0.02	0.33±0.01	3.83±0.02	8.9±0.06
$g_{SeSe}^{PW}(r)$	2.34±0.02	0.37±0.01	3.84±0.02	9.28±0.04
$g_{SeSe}^{LDA}(r)$	2.36±0.07	0.56±0.02	3.82±0.05	8.9±0.06
$g_{SeSe}^{exp}(r)$	2.30±0.02	0.23±0.05	3.75±0.02	9.6±0.3

$g_{GeGe}^{exp}(r)$, followed by a main peak and a deep minimum, showing that the Ge sub-network is organized in distinct shells of neighbors. This trend is not accurately reproduced either by $g_{GeGe}^{LDA}(r)$ or by $g_{GeGe}^{PW}(r)$, which are both characterized by the absence of a clear first maximum, a larger distance for homopolar bonds, a broader main peak, and a much less pronounced first minimum. All these features are indicative of a less structured shape for $g_{GeGe}^{PW}(r)$ and $g_{GeGe}^{LDA}(r)$ when compared to $g_{GeGe}^{exp}(r)$. Among the three-pair correlation functions, $g_{GeGe}(r)$ is the one most affected by the choice of the exchange–correlation functional. The BLYP scheme improves upon the PW one by yielding a clear first maximum, due to homopolar Ge–Ge bonds, and a very pronounced first minimum, closely reproducing the trends observed in $g_{GeGe}^{exp}(r)$. In $g_{GeGe}^{BLYP}(r)$ the position of the first peak approaches the experimental value ($r = 2.45$ Å, BLYP; $r = 2.70$ Å, PW, $r = 2.70$ Å, LDA; $r = 2.33$ Å, [7]). The shape of $g_{GeGe}^{BLYP}(r)$ reproduces the shoulder in the main peak occurring at $r \sim 3.1$Å, indicative of edge-sharing (ES) connections among tetrahedra [5]. Equally favorable is the BLYP prediction of the number of Ge in the first-neighbor shell (0.22, see Table 1, to be compared with 0.25 [7]). This number is clearly underestimated within LDA and PW (0.08 and 0.04, respectively).

The Ge–Se pair correlation function $g_{GeSe}^{exp}(r)$ is characterized by a prominent main peak and a deep minimum. This behavior is well reproduced in $g_{GeSe}^{PW}(r)$, though the maximum and the minimum are found to be less pronounced. At higher distances, $g_{GeSe}^{PW}(r)$ shows less structure than the experimental curve, with a flat second maximum at $r = 5.5$ Å. In $g_{GeSe}^{BLYP}(r)$, the position of the main peak is slightly displaced toward shorter distance, by 0.05 Å. The BLYP scheme is able to reproduce accurately the height of the first maximum and the abrupt decay from the first, sharp maximum down to vanishing values. The first shell of coordination extracted from $g_{GeSe}^{BLYP}(r)$ has a number of neighbors (3.55) in very good agreement with experiments (3.50, see Table 1). Little improvement is found for larger distances, both $g_{GeSe}^{BLYP}(r)$ and $g_{GeSe}^{PW}(r)$ lacking of the second, small maximum visible in $g_{GeSe}^{exp}(r)$.

In the case of Se$-$Se correlations, $g_{SeSe}^{PW}(r)$ follows closely the experimental $g_{SeSe}^{exp}(r)$ for $r > 3$ Å. Although the first peak is sharper than in experiment, $g_{SeSe}^{PW}(r)$ yields an accurate value for the first-neighbor coordination number n_{SeSe}. This holds true also for $g_{SeSe}^{BLYP}(r)$. Se$-$Se correlations are very similar in the PW and BLYP schemes, as shown by the close numbers for the first coordination shell neighbors (see Table 1). Homopolar Se$-$Se bonds are found at a distance only 2% larger than in the PW case.

Considering the results obtained in the LDA, we observe the following. On one hand, the overall shape of the three LDA partial correlation functions is remarkably similar to the PW ones. On the other hand, the LDA curves show a higher number of homopolar neighbors and are generally less structured than in the PW scheme.

Focusing on the results in real space obtained within the PW scheme, one can conclude that they are intrinsically better than those obtained via the LDA. However, PW persists in giving a broader distribution of Ge$-$Ge bond lengths, which on average are longer by as much as 15% with respect to experimental values (see Table 1). Longer interatomic Ge$--$Ge distances and less structured Ge$-$Ge pair correlation functions were recently also obtained for liquid GeSe within the same first-principles framework used here [47]. We note that such longer Ge$-$Ge bond lengths are characteristic of the metallic liquid Ge. For this system, first-principles calculations correctly reproduce the experimental bond lengths (theory:[48, 49] $2.63-2.75$ Å; expt.:[50, 51] $2.66-2.75$ Å). This suggests that the PW still overestimates the metallic character in l-GeSe$_2$, thereby legitimating the use of a XC functional better suited to localize the electron charge density on the atomic sites. This is exactly the case of the BLYP recipe. We obtain partial (n_{Ge}, n_{Se}) and average (n) coordination numbers from the first-neighbor coordination numbers, n_{GeGe}, n_{GeSe}, and n_{SeSe}, given in Table 2. The resulting theoretical values are compared to experimental data in Table 2. As pointed out in [6], a compensation occurred in the PW case between the underestimated value of n_{GeGe} and the overestimated value of n_{GeSe}, leading to a good agreement for n_{Ge}. In the BLYP case, the small difference

Table 2 Experimental and theoretical values for the partial coordination numbers n_{Ge} and n_{Se} and the average coordination number n of liquid GeSe$_2$ at $T = 1,050$ K. BLYP, PW, and LDA denote results obtained with different exchange–correlation functionals. The coordination numbers n_{Ge} and n_{Se} are given by $n_{GeGe} + n_{GeSe}$ and $n_{SeSe} + n_{SeGe}$, respectively (see the values reported in Table 1 for n_{GeGe}, n_{GeSe} and n_{SeSe}, where $n_{GeSe} = 2n_{SeGe}$). The average coordination number n is equal to $c_{Ge}(n_{GeGe} + n_{GeSe}) + c_{Se}(n_{SeSe} + n_{SeGe})$.

	n_{Ge}	n_{Se}	n
BLYP	3.77±0.02	2.11±0.02	2.66±0.02
PW	3.80±0.02	2.25±0.02	2.77±0.02
LDA	3.76±0.02	2.40±0.02	2.85±0.02
[7]	3.75±0.3	1.98±0.15	2.57±0.20

between n_{Ge}(BLYP) and its experimental counterpart is the result of close values for each single contributions, i.e., n_{GeGe}, n_{GeSe}, and n_{SeSe}. As a result, the calculated and experimental average coordination numbers n differ by only 3.5%. As expected, LDA features the worst agreement with experiments.

Overall, BLYP calculations improve the short-range structure of l-GeSe$_2$, featuring more structured $g_{GeGe}^{exp}(r)$ and $g_{GeSe}^{exp}(r)$. In particular, the BLYP approach provides much shorter Ge$-$Ge distances and a well-defined first shell of Ge neighbors, bringing pair correlation functions in better agreement with experiments.

In order to link the appearance of the FSDP to a real-space feature, one can find a specific structural motif (or a combination of them) that are predominant when the FSDP is observed. In the present case, two set of results (PW and BLYP) do feature the FSDP that is largely understimated within LDA. A comparative analysis of the three corresponding structures is expected to provide clues into the atomic-scale origins of intermediate range order when the chemical nature of the components is not taken into account (i.e., at the level of the total structure factor). To this end we resort to the quantity $n_\alpha(l)$, defined as the average number of atoms of species α l–fold coordinated (see Table 3), where α are Ge or Se atoms.

We use a cutoff distance of 3 Å which corresponds to the first minimum in the Ge$-$Se pair correlation function and well describes the first shell of neighbors also for Ge$-$Ge and Se$-$Se correlations. The results for $l = 4$ relative to Ge atoms are those that can be most easily interpreted in terms of deviations from a perfect tetrahedral network. The lowest percentage of Ge atoms fourfold coordinated (about 55%) is found in LDA, this number increasing beyond 60% in the case of PW and BLYP. The same holds for Se atoms with percentages of Se atoms twofold coordi-

Table 3 Average number $n_\alpha(l)$ (expressed as a percentage) of Ge and Se atoms l–fold coordinated at a distance of 3.0 Å. For each value of $n_\alpha(l)$, we give the identity and the number of the Ge and Se neighbors. For instance, GeSe$_3$ with $l = 4$ means a fourfold coordinated Ge with one Ge and three Se nearest neighbors. Values smaller than 1 are reported only for sake of comparison with corresponding values equal or larger than 1. BLYP, PW, and LDA denote results obtained with different exchange–correlation functionals

Ge	$l = 2$	BLYP	PW	LDA	$l = 3$	BLYP	PW	LDA
	Se$_2$	4.0	5.2	5.4	GeSe$_2$	0.8	2.6	3.1
					Se$_3$	13.5	19.8	23.6
	$l = 4$	BLYP	PW	LDA	$l = 5$	BLYP	PW	LDA
	GeSe$_3$	23.3	7.0	13.9	Ge$_2$Se$_3$	2.4	0.4	0.8
	Se$_4$	41.8	53.8	42.1	GeSe$_4$	11.7	5.9	5.0
					Se$_5$	0.6	(4.6)	4.2
Se	$l = 1$	BLYP	PW	LDA	$l = 2$	BLYP	PW	LDA
	Ge	0.9	1.7	1.3	Se$_2$	3.6	2.8	4.2
					SeGe	20.7	21.9	22.9
					Ge$_2$	59.2	45.6	27.6
	$l = 3$	BLYP	PW	LDA	$l = 4$	BLYP	PW	LDA
	Se$_2$Ge	2.4	3.2	8.0	SeGe$_3$	0.3	1.0	1.8
	SeGe$_2$	5.8	8.6	13.2				
	Ge$_3$	6.4	13.1	14.0				

nated that are larger than 60% only in the PW and BLYP cases. Deviations form a perfect tetrahedral order correspond to the presence on Table 3 of a large variety of units. We observe that the BLYP scheme is characterized by a higher proportion of Ge—GeSe$_3$ connections (as much as 23%), contributing to a percentage of Ge four-fold coordinated atoms moderately larger than in the PW case (66% against 61%). The increase of Ge–Ge homopolar bonds and Ge—GeSe$_3$ connections takes place at the expenses of the undefective GeSe$_4$ tetrahedra, lowering from 53.8 (PW) to 41.8% (BLYP). The distribution of miscoordinated Ge atoms is different in the three situations. LDA favors chemical disorder with a large number of Ge atoms and Se threefold coordinated. This is indicative of a network in which the tetrahedra coexist with other geometrical units, preventing intermediate range correlations from estab-lishing. PW and BLYP favor threefold ($l = 3$) and fivefold ($l = 5$) connections, respectively. In particular, in Ge—GeSe$_4$ connections a homopolar Ge–Ge bond is found to coexist with an adjacent tetrahedral arrangement. Coordination of Se atoms reflects the increase of chemical order occurring within the BLYP scheme. The number of twofold coordinated Se atoms becomes more than 10% larger, mostly due to the predominant Se—Ge$_2$ configuration. A corresponding decrease in the number of miscoordinations (in particular Se—SeGe$_2$), lowering from 13 to 6.4%, is noticeable in Table 3.

Overall, the three models stand for three distinct network configurations. Within LDA, the Ge and Se atoms cannot organize themselves on a full extent in a tetrahe-dral network since the ionic contribution to bonding is insufficient. This leads to a multitude of structural units and to a distribution of coordinations, the coordination four (two) for Ge (Se) being barely the most likely. Some of these shortcomings are partially corrected within PW, for which the tetrahedral coordination becomes predominant. As a result, the intermediate range order can settle due to a restored regular connectivity among tetrahedra. BLYP corresponds to a further improvement since it allows to achieve the best compromise between the existence of regular tetra-hedra and the presence of Ge–Ge homopolar bonds. This results from the higher charge localization properties of this description, favoring not only ionicity but also the formation of covalent bonds among atoms of the same kind. In summary, inter-mediate range order appears to be related to the existence of a predominant structural units accounting for the majority of the atomic coordinations.

6 Chemical Sensitivity to Intermediate Range Order

As a further step toward precise understanding of the origins of the intermediate range order, we can consider the case of the presence of the FSDP in a partial struc-ture factor [52]. This means that the order extending beyond the nearest neighbor can be associated to an atomic component or, alternatively, to a specific variable resulting from a linear combination of the atomic components. In what follows we shall focus on the case of the Bhatia–Thornton concentration–concentration partial structure factor $S_{CC}(k)$, a quantity representative of chemical ordering. For a binary system made of A and X atoms in concentrations c_A and c_X, respectively, the Bhatia–Thornton concentration–concentration partial structure factor $S_{CC}(k)$ is defined as

$$S_{CC}(k) = c_A c_X + c_A^2 c_X^2 \{ [S_{AA}(k) - S_{AX}(k)] \\ + [S_{XX}(k) - S_{AX}(k)] \}, \qquad (9)$$

where $S_{AA}(k)$, $S_{AX}(k)$, and $S_{XX}(k)$ are the Faber–Ziman partial structure factors. Fluctuations of concentration are reflected by the presence of positive or negative peaks in $S_{CC}(k)$. These correspond to preferred correlations among atoms of the same kind or of different kind on length scales associated with the value of k.

The significance of the appearance of an FSDP in the concentration–concentration partial structure factor $S_{CC}(k)$ is a long-standing matter of debate. Neutron-diffraction measurements showed an FSDP in the structure factor $S_{CC}(k)$ of both l-GeSe$_2$ and amorphous (a-GeSe$_2$) [5, 53]. However, first-principles molecular dynamics did not show any FSDP in the $S_{CC}(k)$ of l-GeSe$_2$ yet featuring excellent agreement for the total neutron structure factor over the entire k range [33]. According to recent study, it is possible to rationalize the occurrence of an FSDP in the $S_{CC}(k)$ as a signal for the departure from chemical order and to identify systems of three different classes [54]. Class I features networks showing perfect chemical order and the absence of any FSDP in the $S_{CC}(k)$. To this class belong SiO$_2$ and GeO$_2$ [42, 43, 55]. In class II one finds networks with a distinct FSDP in the $S_{CC}(k)$. An extended set of such networks have been found experimentally [56]. In this class the disordered systems have a very moderate departures from chemical order, as confirmed by the structure of l-GeSe$_4$ and l-SiSe$_2$ obtained by simulation [54, 37, 57]. A network structure of class III has so far only been encountered in first-principles molecular dynamics simulations of l-GeSe$_2$. This is exactly the system considered in the previous section at the level of the total neutron structure factor with three different schemes for the XC functional. Like for class I, no feature appears at the FSDP location in the $S_{CC}(k)$. However, contrarily to class I, the associated network shows a rich variety of structural motifs in the first-neighbor coordination shells [33]. This latter case precedes the disappearance of the FSDP from the total neutron structure factor as occurring for systems with a high degree of structural disorder, such as for l-GeSe$_2$ at high temperatures [8].

The question arises on the correlation existing between the fluctuations of concentration and the fluctuations of charge in a DNFM. The first are described by the concentration–concentration structure factor, while the second are expressed in terms of the charge–charge structure factor. Is it possible to gain insight into the presence of the FSDP in $S_{CC}(k)$ by seeking an analogy with a charge distribution..? To elucidate this issue, the present section is devoted to the calculation of the charge–charge structure factor. We calculate the charge–charge structure factor $S_{zz}(k)$ for three AX_2 networks: l-SiO$_2$, a-SiSe$_2$, and l-GeSe$_2$. These systems are representative of the three classes (l-SiO$_2$: class I, a-SiSe$_2$: class II, l-GeSe$_2$: class III) which were introduced to relate the appearance of an FSDP in the $S_{CC}(k)$ to the different degrees of departure from chemical order [54]. Our calculations reveal that no FSDP appears in any of the calculated charge–charge structure factors $S_{zz}(k)$. These results provide evidence in support of the postulate advanced in [54] that no charge ordering is observed at IRO length scales irrespective of fluctuations of concentration occurring at the same length scales.

In order to calculate $S_{zz}(k)$, one has to consider the self-consistent valence electron density in the definition of the total charge density composed of ionic and electronic parts, $\rho_t(\mathbf{r}) = \sum_i z_i \delta(\mathbf{r} - \mathbf{r}_i) + \rho_e(\mathbf{r})$:

$$S_{zz}(k) = N^{-1} \langle z_v^2 \rangle^{-1} \int d\mathbf{r} d\mathbf{r}' \rho_t(\mathbf{r}) \rho_t(\mathbf{r}') e^{i\mathbf{k}\cdot(\mathbf{r}-\mathbf{r}')}. \tag{10}$$

In a pseudopotential formulation, only the valence electrons are accounted for in $\rho_e(\mathbf{r})$, the ionic charges z_i being $z_A = +4$ and $z_X = +6$. In Eq. (10) the spherical average over the orientations of $\mathbf{k}$ is assumed implicitly, N is the number of atoms and $\langle z_v^2 \rangle$ is an appropriate normalization factor, $\langle z_v^2 \rangle = z_{vA}^2 c_A + z_{vX}^2 c_X$. In the expression for $\langle z_v^2 \rangle$, $z_{vA} = +4$ and $z_{vX} = -2$ are the charges attributed to A and X atoms within a pointlike charge model (PLC). In the limit $k \to \infty$, $S_{zz}(\infty) = \langle z^2 \rangle / \langle z_v^2 \rangle$, with $\langle z^2 \rangle = z_A^2 c_A + z_X^2 c_X$. For AX_2 systems this leads to $S_{zz}(\infty) = 3.66$.

When the point-like charge (PLC) approximation is adopted, the total charge density becomes $\rho_t(\mathbf{r}) = \sum_i z_{vi} \delta(\mathbf{r} - \mathbf{r}_i)$. The charge–charge structure factor $S_{zz}(k)$ is now proportional to $S_{CC}(k)$ and reads

$$S_{zz}^{PLC}(k) = N^{-1} \langle z_v^2 \rangle^{-1} \sum_{ij} z_{vi} z_{vj} e^{i\mathbf{k}\cdot(\mathbf{r}_i - \mathbf{r}_j)}$$

$$= (c_A c_X)^{-1} S_{CC}(k). \tag{11}$$

6.1 Liquid SiO₂

In liquid SiO_2, the Si atoms are at the center of tetrahedra linked by corner-sharing O atoms, at least at not too high temperatures [42, 43]. The total neutron structure factor of l-SiO_2 exhibits an FSDP at $k = 1.6$ Å^{-1} [42, 43]. In Fig. 7, we compare the structure factors $S_{zz}(k)$ and $S_{CC}(k)$ for l-SiO_2, the latter being normalized to 1 for $k \to \infty$ [see Eq. (11)]. No feature appears in the $S_{CC}(k)$ at the FSDP location, showing that fluctuations of concentration on the intermediate range scale do not arise in a system characterized by perfect short-range order. According to the classification introduced above, this network belongs to class I. The charge–charge structure factor $S_{zz}(k)$ differs from the concentration–concentration structure factor $S_{zz}^{PLC}(k)$. The main peak the $S_{zz}(k)$ is located at $k_M = 5.1$ Å^{-1}, followed by one deep minimum and shallow oscillations. Within statistical accuracy, the FSDP is absent. A departure from charge neutrality is therefore not expected for distances beyond $r \sim 1.5$ Å, of the order of the nearest-neighbor distances.

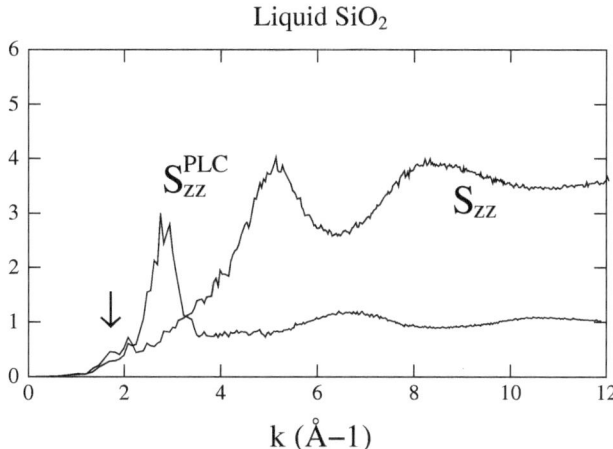

Fig. 7 Charge–charge structure factor $S_{zz}(k)$ and concentration–concentration structure factor $S_{CC}(k)$ of liquid SiO_2 at $T = 3,500$ K. $S_{CC}(k)$ is normalized as in Eq. (11), i.e., $S_{zz}^{PLC}(k) = (c_A c_X)^{-1} S_{CC}(k)$. The *arrows* indicate the location of the FSDP in the total neutron structure factor $(k_{FSDP} = 1.6$ Å$^{-1})$

The differences found between $S_{CC}(k)$ and $S_{zz}(k)$ is a clear demonstration that structural order and charge order are two distinct physical properties as a consequence of the spatially distributed nature of the electron charge.

6.2 Amorphous SiSe₂

Amorphous $SiSe_2$ forms a network representative of class II. In this network there are a small number of structural defects. Its total neutron structure factor exhibits an FSDP, while no experimental partial structure factors are currently available. In the atomic structure of a-$SiSe_2$, edge-sharing connections are predominant, with a majority of Si atoms that belong to one or two fourfold rings [38]. Experimental evidence and first-principles molecular dynamics indicate that at least 1% of the bonds involving Si or Se atoms are homopolar [38, 58, 59]. Figure 8 shows that a prominent peak is clearly visible at the FSDP location in the $S_{zz}^{PLC}(k)$ structure factor of a-$SiSe_2$.

The charge–charge structure factor $S_{zz}(k)$ does not show any peak at the FSDP location (Fig. 9). However, a feature can be seen in the $S_{zz}(k)$ at a k value corresponding to the main peak in the $S_{zz}^{PLC}(k)$, $k_M \sim 2$ Å^{-1}. It appears that fluctuations of concentrations and fluctuations of charge occur at short-range length scales. However, while fluctuations of charge are suppressed for distances beyond $r \sim 3.8$ Å, fluctuations of concentration persist for larger distances, as proved by the peak at $k_{FSDP} = 1$ Å^{-1}. The behavior found for a-$SiSe_2$ shows that the fluctuations of concentration over IRO length scales are sensitive to the presence of a small

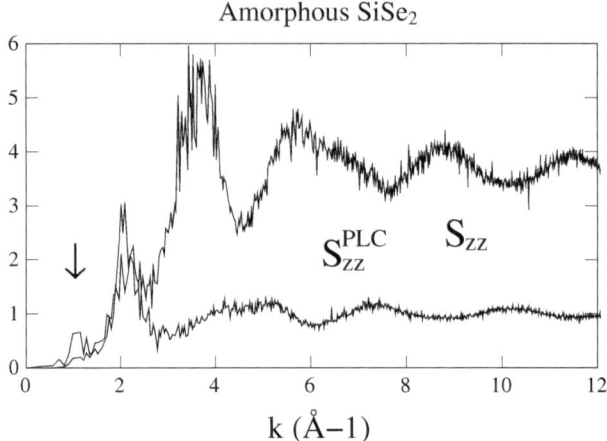

Fig. 8 Charge–charge structure factor $S_{zz}(k)$ and concentration–concentration structure factor $S_{CC}(k)$ of amorphous SiSe$_2$ at $T = 300$ K. $S_{CC}(k)$ is normalized as in Eq. (11), i.e., $S_{zz}^{PLC}(k) = (c_A c_X)^{-1} S_{CC}(k)$. The *arrows* indicate the location of the FSDP in the total neutron structure factor ($k_{FSDP} = 1.0$ Å^{-1})

number of structural defects. These defects are responsible for the appearance of an FSDP in the $S_{CC}(k)$. Interestingly, the condition of charge neutrality over IRO length scales is not affected by the small deviation from chemical order found in a-SiSe$_2$.

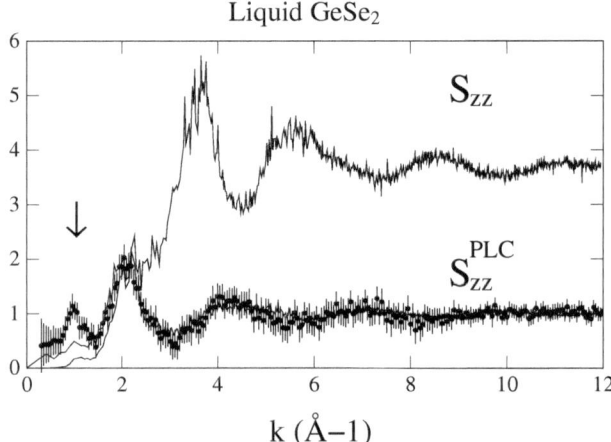

Fig. 9 Charge–charge structure factor $S_{zz}(k)$. and concentration–concentration structure factor $S_{CC}(k)$ of liquid GeSe$_2$ at $T = 1,050$ K. $S_{CC}(k)$ is normalized as in Eq. (11), i.e., $S_{zz}^{PLC}(k) = (c_A c_X)^{-1} S_{CC}(k)$. *Dots* with error bars: experimental results [5]. The *arrows* indicate the location of the FSDP in the total neutron structure factor ($k_{FSDP} = 1.0$ Å^{-1})

6.3 Liquid GeSe₂

The partial structure factors of l-GeSe$_2$ have been measured by Penfold and Salmon using the method of isotopic substitution in neutron diffraction [5]. A prominent FSDP characterizes the $S_{CC}(k)$, due to IRO correlations involving mostly Ge—Ge interactions. Accordingly, this network belongs to class II. The situation is somewhat different from the theoretical point of view. Regardless of the XC scheme employed, the FSDP is vanishing in the calculated $S_{zz}^{PLC}(k)$ (Fig. 9). This disagreement is a further manifestation of residual inaccuracies in the description of Ge—Ge correlations, partially corrected when changing from PW to BLYP but having less effect when considering intermediate range properties. By exploiting the very definition of $S_{zz}^{PLC}(k)$ in terms of Faber–Ziman structure factors, the low intensity in the FSDP region is due to an underestimate of the height of the FSDP in the Ge—Ge partial structure factor. While this underestimation does not affect the total neutron structure factor because of compensation effects related to the other partial structure factors, this discrepancy is magnified in the $S_{CC}(k)$ [i.e., in $S_{zz}^{PLC}(k)$].

As shown in the section devoted to the structure of l-GeSe$_2$, calculations predict a predominant fourfold GeSe$_4$ coordination coexisting with a large variety of structural motifs [33]. This marked departure from perfect tetrahedral order shows that fluctuations of concentration on IRO distances are suppressed in the presence of a high concentration of defective units. Therefore, the absence of the FSDP in $S_{CC}(k)$ might result from two drastically different origins, i.e., from the establishment of perfect chemical order (as in l-SiO$_2$, class I) or from the occurrence of a high level of structural disorder (as in the simulations of l-GeSe$_2$, class III).

The charge–charge structure factor $S_{zz}(k)$ of l-GeSe$_2$ and of a-SiSe$_2$ behave similarly. No charge ordering occurs at IRO scales in disordered network-forming materials (no FSDP visible), and this holds irrespective of the presence of fluctuations of concentration on the same length scale. The absence of any feature below the shoulder at $k_s \sim 2$ Å^{-1} in the $S_{zz}(k)$ is indicative that no departure from charge neutrality is expected for distances beyond ~ 3.8 Å.

In summary, for the three DNFM liquid SiO$_2$, amorphous SiSe$_2$ and liquid GeSe$_2$ (all characterized by IRO, that is by an FSDP in the total neutron structure factor) three distinct behaviors have been identified in the concentration–concentration structure factor $S_{CC}(k)$ at low k values. When the network is made of nondefective tetrahedral units, the FSDP in the $S_{CC}(k)$ is absent (class I). A moderate departure from perfect chemical order is reflected by a distinct feature at the FSDP location in the $S_{CC}(k)$ structure factor (class II). This mark disappears again for higher levels of disorder (class III).

7 Origin of the FSDP in $S_{CC}(k)$

In the search of the atomic-scale origins of intermediate range order, two sets of results have been presented so far. In the first, we have shown that the FSDP in the total neutron structure factor is associated to the presence of a predominant structural unit, the tetrahedra in the case of AX_2 DNFM materials. This rationale has been developed by taking l-$GeSe_2$ as a prototypical example. In the second set of results, we have provided evidence for the correlation existing between a moderate departure from chemical order and the appearance of FSDP in the $S_{CC}(k)$ concentration–concentration structure factor. In view of the above findings, we are left with the open issue of finding the microscopic origin for the presence of the FSDP in $S_{CC}(k)$. Is there a specific structural unit accounting for this feature..? In what follows we address this issue by exploiting the FPMD trajectories obtained for l-$GeSe_2$ with the PW XC functional.

Our analysis is based on the observation that the FSDP height (FSDP-h hereafter) in $S_{CC}(k)$ varies during the course of the simulation between a minimum of 0.04 and a maximum of 0.36, showing similar fluctuations as the FSDP-h in the total structure factor [6]. The evolution of the FSDP-h allows a partition of the instantaneous atomic configurations occurring during the motion into two sets, corresponding to values of the FSDP-h in $S_{CC}(k)$, respectively, smaller and larger than the average 0.136. We shall refer to these two sets as $\mathcal{E}_{low}$ (FSDP-h in $S_{CC}(k) \leq 0.136$) and $\mathcal{E}_{high}$ (FSDP-h in $S_{CC}(k) > 0.136$).

In Fig. 10, $S_{GeGe}(k)$ averaged over $\mathcal{E}_{low}$ and $\mathcal{E}_{high}$ is compared to neutron scattering data [5]. $S_{SeSe}(k)$ and $S_{GeSe}(k)$ are not shown since the results obtained for $\mathcal{E}_{low}$ and $\mathcal{E}_{high}$ do not differ substantially. The improvement obtained for $S_{GeGe}(k)$ in the $\mathcal{E}_{high}$ is remarkable in the height of the FSDP, now very close to the experimental value. It should be added that the total neutron structure factor corresponding to $\mathcal{E}_{low}$ and $\mathcal{E}_{high}$ are very similar. This is consistent with the fact that the FSDP in the total neutron structure factor appears in the two subtrajectories, being both characterized by a predominant tetrahedral arrangement.

To understand which structural properties can be correlated to the different behaviors of $S_{GeGe}(k)$ and $S_{CC}(k)$ over $\mathcal{E}_{low}$ and $\mathcal{E}_{high}$, one has to identify which structural differences exist between the two subsets. The one that appears the most notable is the number of Ge atoms belonging to two fourfold rings, termed Ge(2).

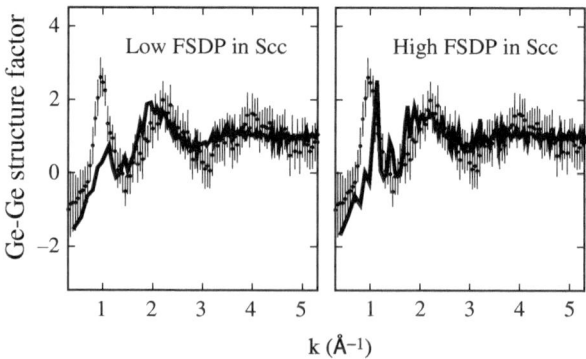

Fig. 10 Ge–Ge partial structure factor for liquid $GeSe_2$: experiment (*dots* with error bars) [5] compared to the theoretical result obtained by averaging separately over all configurations in $\mathcal{E}_{low}$ (*left panel*) and $\mathcal{E}_{high}$ (*right panel*)

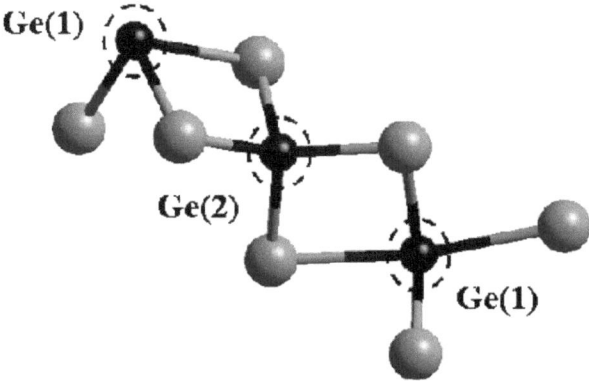

Fig. 11 Snapshots of structural subunits found in liquid GeSe$_2$. Ge atoms are *dark* and Se atoms are *light*. Bonds are drawn when two atoms are separated by less than 3 Å, the first minimum in the Ge–Se radial pair distribution function. Ge atoms forming a Ge* subunit are surrounded by a *dashed line*

The typical configuration associated with a Ge(2) atom involves two four-fold rings forming a chain and having in common *only* the Ge(2) atom, i.e., a Ge(1)—Ge(2)—Ge(1) chain (see Fig. 11). These sequences are of particular interest since, by omitting the four internal Ge—Se bonds, their total valence (four) equals that of a single Ge atom. We term these Ge(1)—Ge(2)—Ge(1) subunits Ge*. Note that in Fig. 11 Ge* has a total valence of four but it is threefold coordinated due to a miscoordination affecting one of the Ge(1) atoms. Most of the Ge(2) form Ge* subunits, the number of Ge* in $\mathcal{E}_{high}$ remaining significantly larger than in $\mathcal{E}_{low}$ (3.4 ± 0.1 vs. 3.0 ± 0.1, 8.5 vs. 7.5%, respectively).

In the search of a correlation between Ge(n) and FSDP-h in the partial structure factors, we recorded the instantaneous values of this latter quantity over the full trajectory. Then, we found the average FSDP-h and its corresponding error bar associated to instantaneous configurations with a given amount of Ge(n), for different values of n. No correlation is found between the amount of Ge(n) ($n = 0, 1$) and the FSDP-h in the partial structure factor $S_{GeGe}(k)$. The same holds for $S_{GeSe}(k)$ and $S_{SeSe}(k)$ for all n, consistent with their low sensitivity to the separation of configurations in $\mathcal{E}_{low}$ and $\mathcal{E}_{high}$. On the contrary, the FSDP-h of $S_{GeGe}(k)$ grows linearly with the number of Ge* units. As a consequence, $S_{CC}(k)$ also increases with the number of Ge* (Fig. 12). This allows us to identify Ge* as the structural feature responsible for the appearance of the FSDP in $S_{CC}(k)$.

The relationship between the FSDP-h in $S_{GeGe}(k)$ and the number of pairs of connected fourfold rings implies that Ge—Ge distances in these subunits are relevant to intermediate range order. This point is substantiated by the distribution of distances between pairs of Ge atoms at the opposite sides in each Ge* unit (Fig. 12). Indeed, a pronounced peak stands out at $\sim$6.5 Å in Fig. 12. This value is compatible with the relationship between the position r of a peak in real space and the position k of a corresponding peak in Fourier space, $k \cdot r \approx 7.7$. This identifies the location

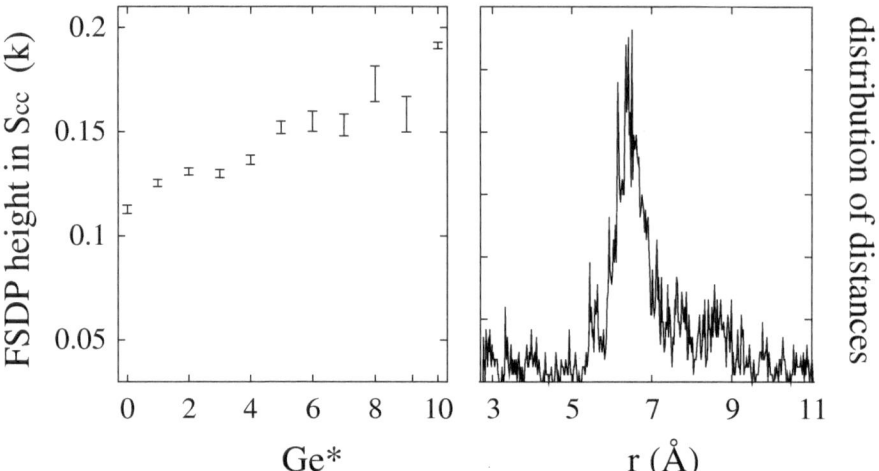

Fig. 12 On the *left*: Height of the FSDP in the Bhatia–Thornton concentration–concentration structure factor $S_{CC}(k)$ as a function of the number of Ge* subunits. On the *right*: Distribution of distances between pairs of Ge atoms at the opposite ends of Ge* subunits

of the first maximum of the spherical Bessel function. The FSDP position in our model is at 1.13 Å^{-1}. Therefore, we conclude that the structural motif consisting of a sequence of tetrahedra connected in a edge-sharing fashion is at the origin of the presence of the FSDP in the concentration–concentration structure factor.

8 Conclusions

In disordered network-forming materials the extent of structural order is larger than in ordinary liquid and glasses. An intermediate range order is observed through the appearance of a first sharp diffraction peak in the total neutron structure factor and, for some cases, in some of the partials. By employing first-principles molecular dynamics, the goal is to correlate these specific features to a predominant nanostructural unit or to a collection of them. The explicit account of the electronic structure proved necessary to describe bonding in systems where there is a delicate balance of ionic and covalent contributions. Furthermore, within a well-established DFT Kohn–Sham framework, the choice of the exchange–correlation functional was found to be crucial. In this review, the first-principles molecular dynamics tool has been proved useful in two respects. First, it has allowed to elucidate the atomic structure of the networks, by associating specific structural features to the shapes taken by the structure factors (in reciprocal space) and by the pair correlation functions (in real space). Second, the use of distinct exchange–correlation functionals has made available three atomic-scale descriptions differing by the spatial distribution of the electronic charge density, LDA being the most delocalized and BLYP being the most localized on the atomic sites. By combining these two approaches, one is able to recover the structural features and changes encountered when adopting these schemes at an increased level of accuracy, i.e., going from LDA to PW to BLYP for the exchange–correlation part of the DFT Kohn–Sham functional. Three crucial issues have been addressed and solved: (a) what is the atomic configuration corresponding to the onset of intermediate range order through the appearance of the first sharp diffraction peak? This question has to do with the organization of the network irrespective of its chemical composition, (b) what are the building blocks of a network leading to an FSDP in the concentration–concentration structure factor?, and (c) what are the structural units associated to an FSDP in the concentration–concentration structure factor? These last two points are intimately related with the chemical nature of the systems. By focusing mostly on a prototypical disordered network-forming material (liquid $GeSe_2$), we have shown that (a) the first sharp diffraction peak in the total neutron structure factor is linked to the existence of a predominant structural unit (the tetrahedra for a AX_2 system), (b) the first sharp diffraction peak in the concentration–concentration structure factor occurs for moderate levels of chemical disorder, and (c) its occurrence is due to the existence of chains of tetrahedra connected in an edge-sharing fashion.

Acknowledgments The author acknowledges the contribution of several colleagues that have played a major role within the research work presented in this chapter. In particular, the crucial impact of Alfredo Pasquarello (EPFL, Lausanne) has to be underlined. The author also acknowledges the help of Massimo Celino (ENEA, Rome) in the writing of the methodological part.

References

1. S. R. Elliott, Nature **354**, 445 (1991).
2. D. L. Price, S. C. Moss, R. Reijers, M. L. Saboungi, S. Susman, J. Phys. C **21**, L1069 (1988).
3. S. R. Elliott, Phys. Rev. Lett. **67**, 711 (1991).
4. P. Boolchand, W. J. Bresser, Phil. Mag. B **80**, 1757 (2000).
5. I. T. Penfold, P. S. Salmon, Phys. Rev. Lett. **67**, 97 (1991).
6. C. Massobrio, A. Pasquarello, R. Car, Phys. Rev. B **64**, 144205 (2001).
7. I. Petri, P. S. Salmon, W. S. Howells, J. Phys. Condens. Matter **11**, 10219 (1999).
8. C. Massobrio, F. H. M. van Roon, A. Pasquarello, S. W. De Leeuw, J. Phys. Condens. Matter **12**, L697 (2000).
9. S. C. Moss, D. L. Price, *Physics of Disordered Materials*, ed. D. Adler, H. Fritzsche, S. R. Ovshinsky (Plenum, New York, 1985), p. 77.
10. S. R. Elliott, J. Phys. Condens. Matter **4**, 7661 (1992).
11. L. E. Busse, S. R. Nagel, Phys. Rev. Lett. **47**, 1848 (1981).
12. L. E. Busse, Phys. Rev. B **29**, 3639 (1984).
13. P. M. Bridenbaugh et al., Phys. Rev. B **20**, 4140 (1979).
14. J. C. Phillips, J. Non-Cryst. Solids **43**, 37 (1981).
15. A. C. Wright et al., Diffusion Defect Data, **53–54**, 255 (1987).
16. P. S. Salmon, Proc. R. Soc. London A **445**, 351 (1994).
17. M. Wilson, P. A. Madden, Phys. Rev. Lett. **72**, 3033 (1994).
18. P. H. Gaskell, D. J. Wallis, Phys. Rev. Lett. **76**, 66 (1996).
19. P. Vashishta, R. K. Kalia, G. A. Antonio, I. Ebbsjö, Phys. Rev. Lett. **62**, 1651 (1989).
20. P. Vashishta, R. K. Kalia, J. P. Rino, I. Ebbsjö, Phys. Rev. B **41**, 12197 (1990).
21. H. Iyetomi, P. Vashishta, R. K. Kalia, Phys. Rev. B **43**, 1726 (1991).

22. M. Wilson, P. A. Madden, Phys. Rev. Lett. **80**, 532 (1998).
23. C. Massobrio, A. Pasquarello, J. Chem. Phys. **114**, 7976 (2001).
24. P. Hohenberg, W. Kohn, Phys. Rev. **136**, B864 (1964).
25. M. Born, R. Oppenheimer, Ann. Phys. (Leipzig). **84**, 457 (1927).
26. W. Kohn, L. J. Sham, Phys. Rev. **140**, A1133 (1965)
27. J. P. Perdew, Y. Wang, Phys. Rev. B **45**, 13244 (1992).
28. A. D. Becke, Phys. Rev. A **38**, 3098 (1988).
29. C. Lee, W. Yang, R. G. Parr, Phys. Rev. B **37**, 785 (1988).
30. R. Feynman, Phys. Rev. **56**, 340 (1939).
31. R. Car, M. Parrinello, Phys. Rev. Lett. **55**, 2471 (1985).
32. G. Galli, M. Parrinello, J. Chem. Phys. **95**, 7504 (1991).
33. C. Massobrio, A. Pasquarello, R. Car, Phys. Rev. Lett. **80**, 2342 (1998).
34. C. Massobrio, A. Pasquarello, R. Car, J. Am. Chem. Soc. **121**, 2943 (1999).
35. C. Massobrio, A. Pasquarello, Phys. Rev. B **75**, 014206 (2007).
36. C. Massobrio, A. Pasquarello, Phys. Rev. B **77**, 144207 (2008).
37. M. J. Haye, C. Massobrio, A. Pasquarello, A. De Vita, S. W. De Leeuw, R. Car, Phys. Rev. B **58**, R14661 (1998).
38. M. Celino, C. Massobrio, Phys. Rev. Lett. **90**, 125502 (2003).
39. J. C. Mauro, A. K. Varshneya, J. Am. Ceram. Soc. **89**, 2323 (2006).
40. J. C. Mauro, A. K. Varshneya, J. Am. Ceram. Soc. **90**, 192 (2007).
41. R. Colle, D. Salvetti, Theor. Chim. Acta **37**, 329 (1975).
42. J. Sarnthein, A. Pasquarello, R. Car, Phys. Rev. Lett. **74**, 4682 (1995).
43. J. Sarnthein, A. Pasquarello, R. Car, Phys. Rev. B **52**, 12690 (1995).
44. M. Cobb, D. A. Drabold, R. L. Cappelletti, Phys. Rev. B **54**, 12162 (1996).
45. M. Cobb, D. A. Drabold, Phys. Rev. B **56**, 3054 (1997).
46. C. Massobrio, A. Pasquarello, Phys. Rev. B **77**, 144207 (2008).
47. G. Kresse, J. Hafner, Phys. Rev. B **49**, 14251 (1994).
48. R. V. Kulkarni, W. G. Aulbur, D. Stroud, Phys. Rev. B **55**, 6896 (1997).
49. P. S. Salmon, J. Phys. F **18**, 2345 (1988).
50. J. P. Gabathuler, S. Steeb, Z. Naturforsch. Teil A **34**, 1314 (1979).
51. F. H. M. van Roon, C. Massobrio, E. de Wolff, S. W. de Leeuw, J. Chem. Phys. **113**, 5425 (2000).
52. The relationship between the three sets of partial structure factors commonly used (Faber-Ziman, Ashcroft-Langreth and Bhatia-Thornton) can be found in Y. Waseda, *The Structure of Non-Crystalline Materials*, (McGraw-Hill, New York, 1980).
53. I. Petri, P. S. Salmon, H. E. Fischer, Phys. Rev. Lett. **84**, 2413 (2000).
54. C. Massobrio, A. Pasquarello, Phys. Rev. B **68**, 020201(R) (2003).
55. D. L. Price, M. L. Saboungi, A. C. Barnes, Phys. Rev. Lett. **81**, 3207 (1998).
56. P. S. Salmon, Proc. R. Soc. London A **437**, 591 (1992).
57. C. Massobrio, M. Celino, A. Pasquarello, J. Phys. Condens. Matter **15**, S1537 (2003).
58. P. Boolchand, W. J. Bresser, Philos. Mag. B **80**, 1757 (2000).
59. R. W. Johnson, D. L. Price, S. Susman, M. Arai, T. I. Morrison, G. K. Shenoy, J. Non-Cryst. Solids **83**, 251 (1986).